I0041813

Agri-Tech Approaches for Nutrients and Irrigation Water Management

This book includes concepts, methodologies, and techniques used in soil nutrients and irrigation water management with regional and global prospects. This book accommodates up-to-date approaches to agricultural technologies along with future directions and compiles a wide range of chapters ranging from soil moisture flow, nutrient dynamics, crop water estimation techniques, approaches to improve crop water productivity and soil health, crop simulation modeling, and remote sensing/ GIS applications. The book also includes chapters on climate-resilient agriculture, advances in big data and machine-learning techniques, IoT, plasma technology, seed priming, and precision farming techniques and their environmental/economic impacts.

Features:

- Discusses applications of sustainable technologies for soil nutrients and irrigation water management at multi-scale.
- Covers the applications of remote sensing/GIS, big data and machine learning, IoT, plasma technology, seed priming, and precision farming techniques for nutrients and water management.
- Reviews concepts, methodologies, and techniques being used in soil nutrients and irrigation water management.
- Provides up-to-date information as well as future directions in the field of nutrients and agricultural water management.

This book is aimed at researchers and graduate students in agriculture, water resources, environment, and irrigation engineering.

Agri-Tech Approaches for Nutrients and Irrigation Water Management

Edited by
Shivam Gupta, Sushil Kumar Himanshu
and Pankaj Kumar Gupta

CRC Press
Taylor & Francis Group
Boca Raton London New York

CRC Press is an imprint of the
Taylor & Francis Group, an **informa** business

Designed cover image: © Shutterstock

MATLAB® and Simulink® are trademarks of The MathWorks, Inc. and are used with permission. The MathWorks does not warrant the accuracy of the text or exercises in this book. This book's use or discussion of MATLAB® or Simulink® software or related products does not constitute endorsement or sponsorship by The MathWorks of a particular pedagogical approach or particular use of the MATLAB® and Simulink® software.

First edition published 2024
by CRC Press
2385 NW Executive Center Drive, Suite 320, Boca Raton FL 33431

and by CRC Press
4 Park Square, Milton Park, Abingdon, Oxon, OX14 4RN

CRC Press is an imprint of Taylor & Francis Group, LLC

© 2024 selection and editorial matter, Shivam Gupta, Sushil Kumar Himanshu and Pankaj Kumar Gupta; individual chapters, the contributors

Reasonable efforts have been made to publish reliable data and information, but the author and publisher cannot assume responsibility for the validity of all materials or the consequences of their use. The authors and publishers have attempted to trace the copyright holders of all material reproduced in this publication and apologize to copyright holders if permission to publish in this form has not been obtained. If any copyright material has not been acknowledged please write and let us know so we may rectify in any future reprint.

Except as permitted under U.S. Copyright Law, no part of this book may be reprinted, reproduced, transmitted, or utilized in any form by any electronic, mechanical, or other means, now known or hereafter invented, including photocopying, microfilming, and recording, or in any information storage or retrieval system, without written permission from the publishers.

For permission to photocopy or use material electronically from this work, access www.copyright.com or contact the Copyright Clearance Center, Inc. (CCC), 222 Rosewood Drive, Danvers, MA 01923, 978-750-8400. For works that are not available on CCC please contact mpkbookspermissions@tandf.co.uk

Trademark notice: Product or corporate names may be trademarks or registered trademarks and are used only for identification and explanation without intent to infringe.

ISBN: 9781032450230 (hbk)
ISBN: 9781032578224 (pbk)
ISBN: 9781003441175 (ebk)

DOI: 10.1201/9781003441175

Typeset in Times
by Deanta Global Publishing Services, Chennai, India

Contents

Preface

Irrigation is a major sector of water consumption and requires special attention for the implementation of policies and practices to use water judiciously to optimize its utilization. Similarly, nutrient management is another aspect of agricultural practice, and effective nutrient management is crucial for agricultural sustainability. This book compiles various aspects of nutrients and irrigation water management with regional and global prospects. The state-of-the-art technologies and methods being used for nutrients and irrigation water management are compiled, along with case studies from across the globe. The advanced agro-technologies for nutrients and irrigation water management, such as artificial intelligence, big data analytics, remote sensing, crop modeling, seed priming, and the Internet of Things (IoT) are also key features of this book.

Climate change and its repercussions are severely affecting agriculture, and this is likely to have major implications for global food security, drinking water, and sanitation. Precision agriculture can be a solution to this challenge, which ensures optimized use of water and nutrients for maximizing yield and farm income. This book compiles information regarding the recent key findings on climate change implications on agriculture and related sectors, and strategies to mitigate its impact, along with economic assessments. Advanced remote sensing and GIS-based approaches used for monitoring crop health, managing nutrients, and estimating irrigation water requirements for optimum utilization of water resources are the need of the hour, and these all aspects are well addressed in this book. Simulation-based case studies on crop responses under different irrigation and field practices are also included.

In conclusion, this AI-focused agri-tech book takes the concept of traditional agri-technologies to the next level. By incorporating the IoT, UAVs, artificial intelligence, and machine-learning algorithms, this book can provide more personalized and targeted insights for farmers and others involved in agriculture. This can help farmers make more informed decisions about when to plant, water, nutrients, and harvest their crops, as well as identify potential problems before they become too severe. Overall, this book has the potential to revolutionize the way that farmers approach agriculture, making it more efficient, sustainable, and profitable.

MATLAB® is a registered trademark of The MathWorks, Inc. For product information, please contact:

The MathWorks, Inc.
3 Apple Hill Drive
Natick, MA 01760-2098 USA
Tel: 508 647 7000
Fax: 508-647-7001
E-mail: info@mathworks.com
Web: www.mathworks.com

Editors

Shivam Gupta is Assistant Professor in the Department of Irrigation and Drainage Engineering, Mahamaya College of Agricultural Engineering and Technology, Acharya Narendra Dev University of Agriculture and Technology, Ayodhya, Uttar Pradesh. Previously he worked as Assistant Professor at Central Agricultural University Imphal, Manipur, India. He completed his Ph.D. at the Indian Institute of Technology Guwahati, India, with a specialization in Water Resources Engineering and Management. During his Ph.D., he worked on the impact of climate change on extreme climate events and water resources for the Himalayan River Basin. He has worked as Research Scientist at the National Remote Sensing Center, Indian Space Research Organization, Hyderabad, where he worked on the development of an operational snowmelt model for the Himalayan region. His area of research includes climate change, downscaling of GCM datasets, bias corrections, climatic projections for various climate change scenarios, water resources management, extreme events analysis, snowmelt modeling, and currently he is working on irrigation water management, evapotranspiration estimation, and remote sensing applications in agriculture. He has published seven research articles and four book chapters in prestigious internationally reputed journals. He has reviewed research articles for international journals such as *SERRA, TAAC, AJGS, Acta-Geophysica,* and *CATENA.*

Sushil Kumar Himanshu is Assistant Professor in the Department of Food, Agriculture and Bioresources, School of Environment, Resources and Development, Asian Institute of Technology (AIT), Pathum Thani, Thailand. Dr. Himanshu has more than eight years of international experience in research, consultancy, and capacity building in the areas of precision farming, climate-resilient agriculture systems, on-farm irrigation water management, remote sensing and GIS applications in agriculture, applications of unmanned aerial systems (UAS) and wireless sensors in agricultural applications, big data analysis and applications, machine-learning applications in agriculture, and hydrologic/cropping system modeling. He obtained his M.Tech. degree in Hydrology and Ph.D. in Water Resources Development and Management from the Indian Institute of Technology (IIT) Roorkee, India. His Ph.D. research was focused on evaluating satellite-based precipitation estimates for hydrological modeling. Before joining AIT, he was working as Postdoctoral Research Associate at the Texas A&M AgriLife Research Center (Texas A&M University System), US, where his research was focused on developing and evaluating strategies that conserve soil and water, promote water use efficiency, and protect soil and water quality in diverse agro-ecosystems. He also worked as a research scientist at the National Remote Sensing Center of the Indian Space Research Organization (ISRO) in Hyderabad, India, where he worked on the operationalization of a national-level hydrological modeling framework for in-season hydrological water balance components at daily/weekly/fortnightly time steps. Dr. Himanshu has received several awards and honors, including the Distinguished Alumni Award (2022, Indian Institute of Technology Roorkee – India), Outstanding Reviewer Award (2020, Transactions of the ASABE), Best Water Resources Student Award (2018, Indian Water Resources Society), International Travel Award (2017, Science and Engineering Research Board – Government of India), Bergen Summer Research School Award (2016, University of Bergen – Norway), MHRD Scholarship (2010–2012 and 2013–2017, Indian Institute of technology Roorkee-India), and Indian Council of Agricultural Research Fellowship (2006–2010). He has authored/co-authored more than 55 research publications in high-impact journals. He also edited one book, published eight popular press articles, and presented his research to several international conferences/meetings in various countries. He reviewed more than 80 research articles from high-repute journals.

Pankaj Kumar Gupta is a Ramanujan fellow, upholding one of the honored scientific positions at the Indian Institute of Technology (IIT) in Delhi, India. As a Ramanujan fellow, he has undertaken an exceptional project "Engineered Microbiome (E-Biome Project)", a unique combination of microbiology, hydrogeology, and chemical sciences for restoration of polluted sites, funded by the SERB, Govt. of India. He is also an adjunct assistant professor at the University of Waterloo, Canada.

Dr. Gupta is a contaminant hydrogeologist, after having completed his Ph.D. at IIT Roorkee, he has been a postdoctoral fellow at the University of Waterloo, Canada. He diversified his expertise in hydrogeology and soil-water quality during his Ph.D./PDF. He is the co-founder of the Society of Young Agriculture and Hydrology Scholars of India (SYAHI) and is also an editorial member of the *Frontiers in Water Journal, Biochar,* and *Carbon Research.* etc. He has also edited three books entitled *Soil-Water, Agriculture, and Climate Change: Exploring Linkages, Advances in Remediation Techniques for Polluted Soils and Groundwater* (Elsevier), *Fate and Transport of Subsurface Pollutants.*

Dr. Gupta holds his in-depth experience in incorporating novel technologies (viz. geophysical investigations, groundwater modeling, aquifer mapping, water quality assessment, microbiome, etc.) to map soil-water systems of 30+ sites in India. Dr. Gupta has also worked as a field officer at the Centre for Development Communication, Jaipur, and Salasar temple (Churu, Rajasthan) to improve water and wastewater management practices in the desert villages. Dr. Gupta is passionate about interdisciplinary research and teaching to understand multi-scale interactions between different components of the subsurface environment, especially the soil-groundwater-pollutant-microbes system.

He also teaches and guides Master's and Ph.D. students with diversified knowledge. He has received an AGU Travel Grant (2017), JPGU Travel Grant (2018), and EXCEEDSWINDON and DAAD Germany Grant (2018). He serves as an editorial board member for SN Applied Sciences, Biochar, Carbon Research and Frontiers in Water journals. He has published more than 20 research papers, approximately 30 book chapters, edited two books and two popular science articles. He has reviewed several research articles for reputed journals including *RSC Advances*, *Groundwater for Sustainable Development*, *ASCE JEE*, *ASCE* Irrigation and Drainage, *Env. Sc.*, *Pollution Research*, *Eco-Hydrology*, *Journal of Contaminant Hydrology*, etc. Further, he has led many site restoration and remediation consultancy projects at polluted industrial sites in India. He is a core member of the Society of Young Agriculture and Hydrology Scholars of India (www.syahindia.org), a network of early-career researchers from around the globe.

Contributors

Shubham Awasthi
Centre of Excellence in Disaster
    Management and Mitigation
Indian Institute of Technology Roorkee
Uttarakhand, India

Mani Bhushan
Department of Civil Engineering GEC
    Khagaria
Bihar, India

Lodsna Borkotoky
CSIR – North East Institute of Science
    and Technology
Jorhat, Assam, India

Kashyapi Chakravarty
Department of Languages
Literature and Cultural Studies
Manipal University
Jaipur, Rajasthan, India

Poulomi Chakravarty
Department of Environmental
    Sciences
Central University of Jharkhand
Jharkhand, India

Urjani Chakravarty
Department of Business
    Communication
Indian Institute of Management
    Bodh Gaya
Bihar, India

Gulab Chand
Department of Languages, Literature
    and Cultural Studies
Manipal University
Jaipur, Rajasthan, India

Indra Jeet Chaudhary
Department of Environmental
    Science
Savitribai Phule Pune University
Pune, Maharashtra, India

Debesh Das
Agrotechnology Discipline
Khulna University
Khulna, Bangladesh

Bhabani Shankar Dash
Department of Agricultural
    Engineering
Centurion University of Technology
    and Management
Odisha, India

Avishek Datta
Department of Food, Agriculture
    and Bioresources
School of Environment, Resources
    and Development
Asian Institute of Technology
Pathum Thani, Thailand

Thiyam Tamphasana Devi
Department of Civil Engineering
National Institute of Technology
    Manipur
Manipur, India

Anshu Gangwar
Krishi Vigyan Kendra
Parsauni,
East Champaran-II (DRPCAU, Pusa)
Bihar, India

Sai Gattupalli
College of Education
University of Massachusetts
Amherst, United States

Uttam Puri Goswami
Department of Geography, Earth, and
  Environmental Sciences
University of Northern British
  Columbia
Canada

Ankit Gupta
Central Academy for State Forest Service
Dehradun, Uttarakhand, India

Sushindra Kumar Gupta
Department of Civil Engineering
Chandigarh University
Mohali, Punjab, India

Hemanta Hazarika
Nabajyoti College
Kalgachia, Barpeta
Assam, India

Sushil Kumar Himanshu
Department of Food, Agriculture and
  Bioresources
School of Environment, Resources and
  Development,
Asian Institute of Technology
Pathum Thani, Thailand

Kamal Jain
Centre of Excellence in Disaster
  Management and Mitigation
Indian Institute of Technology Roorkee
Uttarakhand, India

Shubham Jain
Texas A&M University
College Station
Texas, United States

Mohammad Abdul Kader
Centre for Irrigation and Water
  Management (CIWM)
Rural Development Academy (RDA)
Bogura – 5842
Bangladesh

Bareerah Khalid
Department of Food, Agriculture and
  Bioresources
School of Environment, Resources and
  Development
Asian Institute of Technology
Pathum Thani, Thailand

Sandhip Khundrakpam
Department of Civil Engineering
National Institute of Technology
Manipur – 795004
India

Mridu Kulwant
Department of Environmental
  Studies
The Maharaja Sayajirao University of
  Baroda
Vadodara, Gujarat, India

Anil Kumar
Disaster Preparedness Mitigation and
  Management
School of Engineering and
  Technology
School of Environment, Resources
  and Development
Asian Institute of Technology
Pathum Thani, Thailand

Ashok Kumar
Department of Agricultural
  Engineering
Bihar Agricultural University
Sabour, Bhagalpur
Bihar, India

Brijesh Kumar
Department of Agricultural
  Engineering
Bihar Agricultural University
Sabour, Bhagalpur
Bihar, India

Hemendra Kumar
College of Agriculture and Natural
  Resources
University of Maryland
College Park, MD
United States

Ranjan Kumar
School of Management
Asian Institute of Technology
Pathum Thani, Thailand

Sanoj Kumar
Department of Agricultural Engineering
Bihar Agricultural University
Sabour, Bhagalpur
Bihar, India

Sunny Kumar
Department of Civil Engineering
Chandigarh University
Mohali, Punjab, India

Tarun Kumar
Krishi Vigyan Kendra, Saraiya
Muzaffarpur (DRPCAU, Pusa)
Bihar, India

Jasmeet Lamba
Department of Biosystems Engineering
Auburn University
Auburn, AL
United States

Moniruzzaman
Agricultural Economics Division
Bangladesh Agricultural Research
  Institute
Gazipur, Bangladesh

Divya Patel
Department of Environmental Studies
The Maharaja Sayajirao University of
  Baroda
Vadodara, Gujarat, India

Bhavna Nigam
School of Environment and Sustainable
  Development
Central University of Gujarat
Gandhinagar, Gujarat, India

Denish Okram
Department of Civil Engineering
National Institute of Technology
Manipur, India

Brenda V. Ortiz
Department of Crop, Soil and
  Environmental Sciences
Auburn University
Auburn, AL
United States

Akshay Pandey
Centre of Excellence in Disaster
  Management and Mitigation
Indian Institute of Technology Roorkee
Uttarakhand, India

Sharad Patel
Department of Environmental and
  Water Resources Engineering
University Teaching Department
Chhattisgarh Swami Vivekanand
  Technical University
Chhattisgarh, India

Md. Sadique Rahman
Department of Management and
  Finance
Sher-e-Bangla Agricultural
  University
Dhaka, Bangladesh

Ashish Rai
Krishi Vigyan Kendra
Parsauni
East Champaran-II
(DRPCAU, Pusa)
Bihar, India

Jitendra Rajput
Division of Agricultural Engineering
ICAR-Indian Agricultural Research
  Institute
Pusa, New Delhi, India

Dipankar Roy
Department of Civil Engineering
Madanapalle Institute of Technology &
  Science
Madanapalle, Andhra Pradesh, India

Bibhuti Bhusan Sahoo
Department of Agricultural
  Engineering
Centurion University of Technology
  and Management
Odisha, India

Sayantan Samanta
Texas A&M University
College Station
Texas, United States

Md. Shariot-Ullah
Department of Irrigation and Water
  Management
Bangladesh Agricultural University
Mymensingh, Bangladesh

Arvind Kumar Singh
Krishi Vigyan Kendra
Parsauni
East Champaran-II (DRPCAU, Pusa)
Bihar, India

Bhaskar Pratap Singh
Krishi Vigyan Kendra
Amethi (ANDAUT, Ayodhyaya)
Uttar Pradesh, India

Kanwarpreet Singh
Department of Civil Engineering
Chandigarh University
Mohali, Punjab

Ronald Singh
National Remote Sensing Centre
Indian Space Research Organization
Hyderabad, India

Vijay Kumar Singh
Department of Minor Irrigation
Government of Uttar Pradesh
India

Vikas Kumar Singh
Acharya Narendra Dev University of
  Agriculture and Technology
Ayodhya, Uttar Pradesh, India

Puneet Srivastava
College of Agriculture and Natural
  Resources
University of Maryland
College Park, MD

Prakash Subedi
Department of Food, Agriculture and
  Bioresources
School of Environment, Resources and
  Development,
Asian Institute of Technology
Pathum Thani, Thailand

K. N. Tiwari
Agricultural & Food Engineering
  Department
Indian Institute of Technology
  Kharagpur
Kharagpur, West Bengal

Hayat Ullah
Department of Food, Agriculture and
  Bioresources
School of Environment, Resources and
  Development
Asian Institute of Technology
Pathum Thani, Thailand

Mangesh M. Vedpathak
Shriram Institute of Information
  Technology
Solapur, Maharashtra, India

Sakron Vilavan
Department of Food, Agriculture and
    Bioresources
School of Environment, Resources and
    Development,
Asian Institute of Technology
Pathum Thani, Thailand

Pema Wangmo
Department of Food, Agriculture and
    Bioresources
School of Environment, Resources and
    Development
Asian Institute of Technology
Pathum Thani, Thailand

1 AI for Sustainable Agriculture in the Face of Climate Change

*Poulomi Chakravarty, Sai Gattupalli,
Kashyapi Chakravarty, Gulab Chand,
and Urjani Chakravarty*

INTRODUCTION

The population of Earth surpassed 8 billion on November 15, 2022, according to the United Nations report by the Department of Economic and Social Affairs (2022). It took just over a decade for the human population to grow from 7 billion in 2011 to 8 billion (Chakravarty et al., 2017). This overwhelming population rise calls for action in the sector of agriculture to ensure global food security. According to Ben Ayed and Hanana (2021), the elevated population by the year 2050 will require a colossal increase in global food production by 60% to 110%, which is just sufficient to cater to the human population of over 9 billion. Therefore, it is crucial to have sustainable agriculture in modern times to solve food security issues and address hunger as well as poverty worldwide (Rockström et al., 2017). The challenges faced by humanity are not only limited to population and food security but extend to more complex issues such as climate change, water scarcity, lack of land resources, pollution, and infectious diseases. Due to the excruciating burden of anthropogenic forcings on climate, natural resources, ecosystems, and the environment, there is a dire need for swift and strategic action to fortify global agricultural communities.

Now the question arises of how one can achieve this gigantic feat of providing nourishment to billions of people in times of disaster such as the raging COVID-19 pandemic. How to deal with the perpetual human struggles of climate change, water scarcity, resource and energy shortages? The answer to this would be to think and act "smart". The current discussion contends that to act "smart" there is also a need to understand the transdisciplinary aspect of the problem. Even though in the 21st century we are able to effectively use technology in all sectors and we can use smart technology in the field of agriculture as well, the question still remains of understanding the ethical underpinnings related to smart technology. Agriculture is not

DOI: 10.1201/9781003441175-1

only the provider of nourishment for the global population but also engages a large population in agrarian employment. Now is the time to be proactive and welcome innovative measures to increase yield in agriculture, produce crops that are resistant to diseases and climate change, and produce harvests that are sustainable in nature. Sustainable agricultural practices can be adopted by farmers aided by digital and engineering technologies such as AI, the Internet of Things (IoT), cloud computing, and machine learning (Ben Ayed & Hanana, 2021).

SMART AGRICULTURE FOR A CHANGING CLIMATE

Agriculture is the practical and scientific method used to cultivate the topsoil layer to grow crops along with practices to raise livestock. Agricultural practices are dependent on various deterministic factors. Climate and topography are among the most important aspects that govern agricultural practices of any region. Climate elements like solar insolation, temperature, rainfall, humidity, and climate extremes (such as droughts, floods, storms, cyclones, tornadoes, and snow) are determinants of the success or failure of agriculture. Water availability for irrigation, water quality and the evapotranspiration process are other major aspects of agriculture. Agriculture requires a medium for the growth of plants and livestock. As we know the topsoil is the key layer for our survival and soil fertility, soil moisture, overall soil health, and the organisms present in the soil determine the yield and productivity of agricultural produce (Mauget et al., 2021). Biotic components such as symbiotic flora and fauna, disease-causing insects, pests, and parasites also have a critical role in the health and yield of agricultural produce. Figure 1.1 portrays the factors that determine agriculture and clearly depicts the role played by the above-mentioned aspects on holistic agricultural practices. Smart agricultural approaches need to maintain these factors for agricultural production in a sustainable fashion.

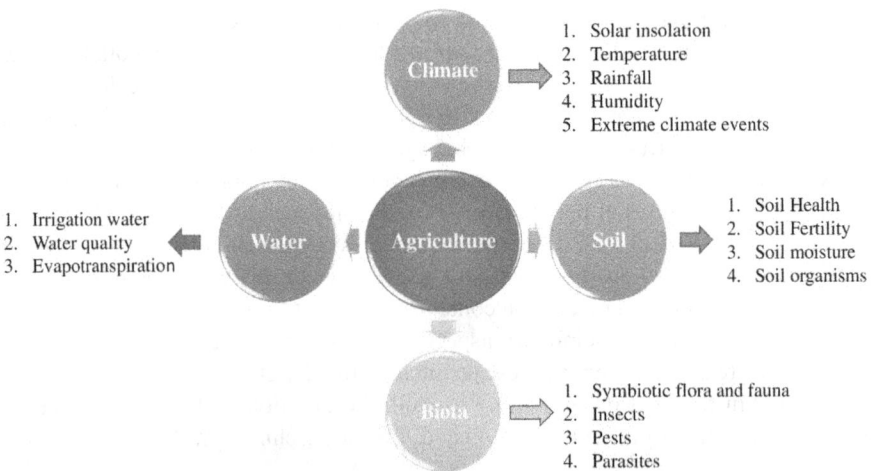

FIGURE 1.1   Factors critical to sustainable agriculture

GLOBAL ISSUES ON CLIMATE CHANGE

Climate change is one of the most pressing threats in the 21st century and its negative impacts are observed in all sectors worldwide and the most devastating impact is observed in the agricultural, which directly affects the socio-economic sectors of nations (Jug et al., 2018; Aune, 2012). As the climate change effects are now being felt worldwide there is an urgent need to address issues related to drought, floods, low groundwater table, saltwater intrusion, desertification, and biodiversity loss, among others. In the face of global climate change, worrisome issues have been cropping up all over the planet. One such example is the recent, severe drought in Europe which peaked in the summer of 2022 and has been a multi-year drought initiated from the year 2018. Rakovec et al. (2022) in their paper identify this drought to be a high-intensity, unprecedented benchmark that has remained active for more than two years and covered a shocking 35.6% of area over a period of 12.2 months with the air temperature at the near surface being +2.8° Kelvin. This current persistent drought is worse than any historical droughts in Europe over the past 250 years. As drought periods are identified by below-normal rainfall and high temperatures, it is safe to say that agriculture, which is completely dependent on water for irrigation and optimum temperature and other climate elements, is severely affected in European nations. More intense or equally intense droughts are projected to occur on the continent of Europe by the assessment of soil moisture simulations of the years 2021 to 2099 (Rakovec et al., 2022). African nations like Kenya, which are considered to be within the "Horn of Africa," are also suffering from long drought periods peaking in 2022 that is the worst recorded drought for the nation in 40 years, resulting in drying of 90% of water sources, mass-scale death of livestock, and huge agricultural losses due to parched land bereft of rains as reported by World Meteorological Organization (WMO, 2022). Floods in Pakistan in the year 2022, and wildfires in the United States of America are other examples of climate-related extreme events that have afflicted the socio-economic structures of the affected nations (Buchholz et al., 2022; Devi, 2022; Sharma et al., 2022). The rise in infectious diseases in plants and pests is another major adverse impact of climate change as projected by IPCC reports on food security (2022). Recently, swarms of locusts from African nations have been migrating to Asian nations and destroying agricultural crops, which is just one such example of pest infestation affecting agricultural lands across continents (IPPC Secretariat and FAO, 2021). All these climate-related issues have had a negative influence on the agricultural community around the world therefore it is necessary to understand the link between climate and agriculture and find solutions to end the negative impacts of climate change for the development of the agricultural sector.

CONNECTING CLIMATE AND AGRICULTURE

Globally agriculture is responsible for 6.4% of the world economy and in nine countries, agriculture is the backbone of the economy (Pathan et al., 2020). In the case of India, agriculture is the contributor to 70% of the country's economy (Chakravarty & Kumar, 2020). India is also the second highest populated country in the world, closely following China, therefore the requirement for resources is extremely high. Larger population density requires food security which should be independent of climatic extremes and resistant to droughts, floods, infectious diseases, and pestilence. Climate change is a global issue that affects us all, but it has an especially profound impact on agriculture. With rising temperatures, low rainfall, and weather patterns becoming more unpredictable, farmers must adapt to new conditions in order to ensure a successful harvest. Connecting climate and agriculture is essential for understanding the effects of climate change and developing strategies to mitigate its impacts on food production. By understanding how climate affects agricultural production, we can develop sustainable farming practices that will help ensure food security in the face of a changing climate. Jug et al. (2018) suggest that "conservation agriculture" has the potential to change the way farmers produce their goods, but it has yet to be adopted on a global scale. The main reasons behind this include lack of knowledge, traditional values and prejudices, and inadequate policies that don't recognize the need for subsidies. Poorly selected machinery, land size not optimally matched to machine cost, minimal profit returns, ineffective training, and guidance programs lead to a need for better planning and execution of any farming operation (Farooq & Siddique, 2014; Jat et al., 2014). Mendelsohn (2009) painted an alarming picture of the higher negative impacts of climate change on developing nations including India in comparison to developed nations like the United States. The study further stated that agriculture in tropical and sub-tropical nations is more sensitive to the changing climate than their temperate counterparts. Even a nominal rise in temperature over Latin American and African regions would result in damaged crop production as the temperature in these latitudes is already on the higher side. Similarly, rain-fed agriculture systems are more vulnerable to climate change impacts than agriculture with developed irrigated systems. The study concluded that all regions across the globe will not face similar impacts on agriculture due to climate change and the scenarios will vary from region to region depending on local climate. Therefore it is highly recommended to study an agricultural sector in each country to study regional climate patterns and devise strategies to overcome climate change impacts on the basis of specific issues. It is also important for sustainable agricultural development to have contingent plans for irrigation, soil health, and overall plant health to succeed in scenarios with warmer temperatures, uncertain weather and rains.

EXPLORING CHALLENGES OF THE AGRARIAN COMMUNITY

Challenges in the agricultural sector are attributed to climate change, land degradation, water scarcity, and socio-economic constraints. It is important to understand that just as climate change impacts are region-specific, challenges faced by the agricultural sector in each country or region are also case-specific. Mendelsohn (2009) explains that in case of climate change, temperate regions might benefit from a warming climate due to a higher number of days for agriculture in colder countries when global surface temperature rises. Similarly some regions are said to receive more rainfall as an impact of climate change which might be beneficial to agriculture but too much rainfall can cause floods resulting in crop damage. This is where science and technology come to the rescue of the farming community.

Smart agricultural practices have to be adapted by harnessing the available scientific resources to overcome challenges faced by the agricultural community in the face of climate change. Practices such as conservation agriculture, phytoremediation, agroforestry, and irrigation management practices have to be judiciously used with the assimilation of 21st-century tools like AI to have beneficial results.

HARNESSING THE POWER OF AI FOR A GREEN EARTH

The advent of industrialization has escalated the domino effect of problems starting from greenhouse gas emissions (GHGs), over population, global warming leading to climate change, pollution, biodiversity loss, and mass-scale extinction. Anthropogenic impacts on the environment and climate have affected all living beings on the planet. To maintain an equilibrium of ecosystems, it is necessary that we understand that humans are a part of the entire biosphere and try to live in harmony with other species. The United Nations Sustainable Development Goals (SDG) follow the same principle of judicious living to safeguard our planet and its resources for future generations. Scientists have joined hands with policymakers and stakeholders to achieve zero hunger for the globe, which is one of the United Nation's SDGs to be achieved by the year 2030. This can be only attained if drastic measures are employed. As discussed above, the challenges faced are numerous in terms of weather unpredictability, GHG emissions, and water scarcity amidst economic uncertainties (US EPA, 2022). This herculean task can be achieved when we think out of the box and move toward unconventional and sustainable agricultural practices which will ultimately lead to equity and viability in socio-economic structures, positive acceleration in productivity, and negate adverse impacts of changing climate, rising population, and pollution (Lakshmi & Corbett, 2020). It is important now, to understand how we can utilize the power of intelligent technology to help us achieve these goals.

ROLE OF ARTIFICIAL INTELLIGENCE IN THE 21ST CENTURY

The 21st century has ushered in an era of unprecedented technological advancement, with AI playing an increasingly important role. AI is a field of computer science dedicated to the research of methods and algorithms that have the ability to mimic human intelligence. Pannu (2015) calls AI the intelligence "exhibited by machines or software". We discuss a general understanding of subject technology and cover some of the recent AI-enabled innovations that have transformed the traditional way of practice and being. We see how these innovations have a direct impact on the "end user", in that they allow users to make sense of their agricultural business practices and related ethics, and in implementing new practices that are sustainable in the long run. We summarize with real-world examples how various AI-enabled technologies are shaping our world in new and sustainable ways, and in delaying the critical climate change tipping point. In 2022, Armstrong and colleagues found that Mother Earth has already crossed the tipping point thresholds due to the rise in global warming temperatures. This news brought panic to science communities. News such as this tends to get attention from social media climate activists such as Greta Thunberg, who use their social media powers to bring awareness to the youth populations. We discuss the implications of AI for sustainable agriculture in the face of climate change.

A 2016 White House report highlights the significance of AI and the necessity of a clear roadmap and strategic investment in this area (Intelligence, 2016). With an immense potential to revolutionize the way humans interact with machines and the environment, AI technology's presence is rapidly growing in our daily lives, in businesses we visit online or in person. Especially in the research domain, the field has come a long way since its beginnings in the 1950s, and today it is used in a variety of applications, from self-driving cars to monitoring climate change. For example, AI technology has already saved human lives by identifying hidden or complex patterns in diagnostic data to detect acute diseases earlier and improve treatments (Jamshidi et al., 2020), and helped scientists develop complex computing models and stunned the astronomical community by revealing the first-ever picture of a "black hole" (Event Horizon Telescope Collaboration, 2019). These are only two examples of the several ways in which technology has revolutionized our societies. Researchers thought achievements such as these were impossible to achieve only a few decades ago, but now we are really realizing the hidden potential of advanced technologies such as artificial intelligence. Many businesses have sprung up by harnessing the power of AI to create virtual assistants that many people use daily. Although there is a growing number of concerns relating to AI-ethics and its usage by notorious government agencies, the benefits continue to play an important role in improving our lives now, and in the future. Figure 1.2 represents the application of AI in various sectors such

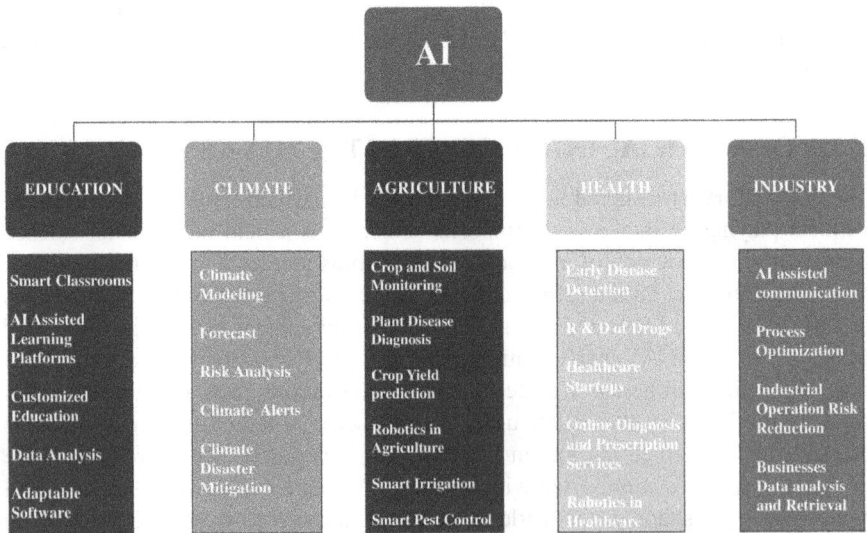

FIGURE 1.2   Applications of AI in various sectors

as education, climate, agriculture, health, and industry. In this chapter, the authors focus on and expand AI and its applications in agriculture, climate, education, and communication, and briefly summarize the healthcare and industrial aspects of AI.

In industries such as healthcare, researchers are using AI technology for early detection of glaucoma in patients (Al-Aswad et al., 2022), while others such as Liang et al. (2020) use AI to predict the tolerance of chemotherapy drugs for cancer patients and help radiologists map and target tumorous areas accurately. Again, these evolutions in science are directly saving lives; a few decades ago, we thought it would be impossible to achieve anything of this magnitude. In manufacturing, the use of AI and robotics continues to revolutionize the industry. Using AI, engineers automate product assembly processes thereby reducing costs, increasing production speed, and improving overall product quality (see Fahle et al., 2020). The evolution that AI has brought to the manufacturing sector has not only made the industry efficient and productive, but also sustainable (Ghobakhloo, 2020). This next generation of industrial systems is often regarded as "Industry 4.0" (Peres et al., 2020). In the section below, we present more AI-enabled tools in environmentally focused areas such as agriculture and climate. As AI by itself is a software at its core, new businesses have appeared promoting software-as-a-service (SaaS) (Bisong, 2019), to bring new insights into their business decisions and developments.

IMPLEMENTATION OF AI IN EDUCATION

As AI offers previously unimaginable possibilities in research and practice in many industries, we take a look at some of the most notable examples. We explore some applications of AI developed for a variety of purposes, ranging from creating awareness of climate change in educational outcomes for children, to detecting diseases early on so they can be cured. In the education pipeline, AI currently provides young math learners with a personalized learning experience tailored to their individual preferences and needs (Arroyo et al., 2014). More recently, scientists have developed learning technologies that use AI to enhance learner engagement in online math learning by observing the facial expressions of the learner (Lee et al., 2022). These examples are instances of AI-enabled technology that focus on the learner. Ironically, AI is far from practical when it comes to climate change education. Climate change education has been a hot topic in educational research for over a decade, and prominent climate change education researchers (see Monroe et al., 2019; Henderson & Drewes, 2020; Shepardson et al., 2012; Feinstein & Kirchgasler, 2015) have all argued the importance of climate change education in our school system. Pogue (2021) in their book *How to Prepare for Climate Change: A Practical Guide to Surviving the Chaos* says that about two-thirds of all US students say that they were never taught about climate change in school. Although there is a lack of proper climate change education in the US school system, there are plenty of open educational resources on the internet. The National Aeronautics and Space Administration (NASA, 2023) lists student and educator resources related to global climate change and includes interactive multimedia. The resources also show how the space research agency is trying to mitigate, and take action on the climate crisis.

Other climate education resources include the United Nations Climate Action online portal (UN Climate Action, 2023), which lists not only the causes and effects of climate change, but also thorough and well-drafted "Initiatives for Action", and broader goals for participating nations; whereas on the European continent, the centrally funded climate change education initiatives such as the European Climate Change Curriculum (2023) and the Erasmus+ youth educational and training program (2020) provide several climate change learning resources for young children. In Table 1.1, we discuss more AI-enabled tools meant to bring awareness and educate the agricultural community. In developing nations such as India, curricula related to climate education and a sustainable future are practically non-existent within the primary and secondary education systems. The curriculum is restricted to a few theoretical concepts in school textbooks on atmosphere and climate change. However, in higher education, there is scope as institutions offer graduate-level degrees in climate science and sustainable development. The Centre for Environment Education (CEE) has taken initiatives promoting climate change education by joining hands with the UN Framework Convention on Climate Change (UNFCC) after the Paris Agreement in 2016 as reported by the climate change portal of the CEE (2022). This initiative has helped with the creation of a basic framework for research and development of climate action for empowerment in India. It is our argument that in order to accomplish a sustainable future, it is necessary that change begins in our schools, and hence leads to the question of climate change education. In the big picture, we are full of hope that children, in their adulthood, are aware of the environmental and atmospheric changes, sustainable practices, monitor their carbon footprint, and make decisions appropriately and ethically, which is also the aim of the climate change education community and further aids in holistic sustainable practices.

AI–HUMAN INTERFACE: ANSWERS FROM ENVIRONMENT COMMUNICATION

Agriculture is a human activity. Humans as social animals learn everything through communication. From the time of its inception, AI too has been harnessing the structure of human communication to create the human–machine interface. So, any discussion on the roles of AI in the development of sustainable agriculture would remain incomplete without looking at the overall role of communication in environment management. Over time an exclusive branch of communication deals with issues related to the environment. AI interfacing with environment communication opens up the field to transdisciplinary research (Slovic et al., 2019). This arose in the 1980s and 1990s. As it is a relatively uncharted field, there are various opinions on how to define the term "environment communication". A change in approach from the 19th-century Romantics that instead of only appreciating nature the focus is on preservation. One prominent scholar in the area, Flor (2004) and later Senecah (2007) explores the scope of "environment communication" for conservation as well as management of the environment. Their thesis on environment communication follows the idea of "mutual understanding" and a "holistic view" of communication at the center of the discipline and not just as part of environment management or even conservation.

One philosophical basis of "environment communication" is "ecological ethics" which studies the relations between human beings and nature. Moreover, it also deals with the ecological crisis, and attempts to find out the cause that leads to the crisis. Donald Worster (1993) points out that the cause of ecological crises faced around the world is not created by ecosystems, but through a lack of proper human ethical systems. He suggests the removal of the crises in possible recognition and reformation of the said ethical system. What is this ethical system? How can we relate this to AI? The area of ecological ethics brings forth the concept of "who has the right?", which was earlier limited to the human world, and expands the same "rights" to the non-human world, i.e., "ecological holism". Going further into this discussion to follow "ecological holism", it is fundamentally about the rights of the natural world, the rights of human beings and their obligations, and in general morality interfacing with the AI systems. As a result, an AI application can be considered conducive when it preserves the human and non-human community's integrity, stability, and beauty leading to no effect on the overall ecology of the Earth. If the extension of the subject of "rights" from humans to the whole of the natural world is cultural and moral progress, the extension of the subject of "rights" from the natural world to the AI world can also be called moral progress. Whereas when it goes to the opposite, it can lead to something wrong.

The above Figure 1.3 shows the implications of ecological ethics for AI applications; further it traces how AI is bridging the gap between environmental communication and ecological practices, i.e., AI–human interface. As technological advancements are tested, Weder et al. (2021) discuss creating a space for "transformative sustainability" and revolutionizing the basic criterion of value present in

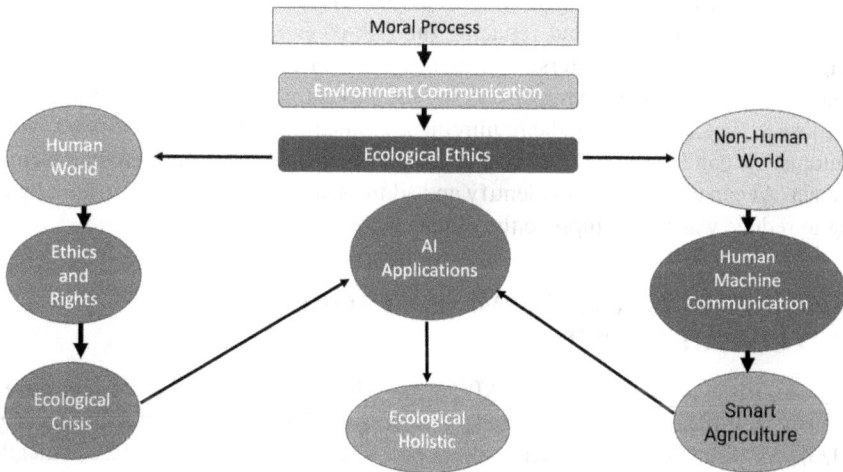

FIGURE 1.3 Ecological holism for smart agriculture

"holism". Regarding any single part of the ecosystem, be it human beings or AI as a center, does not conform to the law of "ecological holism". Similarly, any role of AI for smart agriculture is not to deny the role of human existence, but to assist humans in judging all things keeping in mind the interests of the ecosystem and keeping themselves within the bounds of ecological law. It is better to see the interface as an "environmental construct" (Chakravarty & Gaur, 2011) where the background knowledge of one can be well executed by the other. We, however, have many specific problems that need to be solved. Wherein the AI system facilitates support and structure to human endeavor. This is definitely possible with the help of what Bartz (2021) calls the "intent–action gap". Human consciousness or lack of it is the main reason for this gap. As the ecosystem is immensely complex, the question of which human conduct follows or does not follow the ecological law is still worth investigating. And how can and should the value of the natural environment be assessed in relation to human needs and goals? The true concern, finally, ought not to be with only general representations, but with humans and whatever it is in the natural world working together interfaced with AI systems.

PLANTING THE SEEDS OF A SUSTAINABLE FUTURE

Anthropogenic impacts on the environment can be reduced by a comprehensive assessment of cause and effects, then solutions can be achieved via joint efforts of nature, man, and machine. The challenges posed to the Earth and humanity are numerous thus the solutions have to be diverse in nature that address a variety of issues at once.

Hence, AI has become increasingly relevant in the 21st century, as the world is faced with a growing number of complex societal challenges. AI is being used to tackle issues in areas such as healthcare, climate change, and sustainable agriculture. It can also be used to create predictive models for climate change, helping to improve the accuracy of forecasts and reducing the risk of extreme weather events; hence helping us prepare for disasters, making mitigation strategies, and finally rehabilitation after disaster events. AI is also being used to automate various processes, such as those in factories making the requirement of manpower in many sectors obsolete. Automation can help to reduce costs, increase efficiency, and reduce environmental impact. AI can also be used to identify and address problems before they occur, helping to reduce waste and improve the efficiency of operations.

AI APPLICATIONS IN AGROMETEOROLOGICAL AND CLIMATE SERVICES

The multifaceted applications of AI can be witnessed in the research and development of climate sciences, agrometeorology, and climate education (Chakravarty, 2023). As discussed in the earlier sections, AI can prove to be a successful ally when used judiciously for sustainable agriculture by taking into consideration climate forcings on the land surface. AI has found applications in climate sciences and agrometeorology in many ways and in turn helps agriculture by predicting weather patterns (such as Huntingford et al., 2019). Agrometeorological advisories generated by AI help farmers make decisions about when to irrigate on the basis of water availability information (see Perea et al., 2019), when to harvest (such as Kim et al., 2019), and when to store crops (Kumar et al., 2022). By using AI-powered systems, farmers can gain an understanding of how different farming practices affect the environment. For example, agriculture is a contributor to GHG emissions and causes environmental pollution as agricultural runoff causes eutrophication (Chakravarty & Kumar, 2019). By leveraging data-driven insights, farmers can identify which practices are most sustainable and which are most damaging. In the same way, AI can also help farmers find ways to reduce their water usage, reduce their carbon footprint, and reduce their reliance on chemical fertilizers and pesticides. Mendelsohn (2009) lays emphasis on region-specific climate strategies for successful sustainable agriculture. AI can prove to be an excellent tool in recognizing and predicting specific climate trends a region has and can overlay this information on demographic data, agricultural data, and socio-economic statistics to create a composite analysis of the region that will help policymakers make educated decisions on their agricultural policies.

Implementation of AI as an efficient tool for agrometeorological services is depicted by an innovative mobile application called *Meghdoot* developed in India, which improves access to climate information services. It is essential for farmers to receive timely advice from agronomists in order to make appropriate decisions and reduce the risk of crop failure due to unfavorable weather which helps them maximize their yields and increase their income. The *Meghdoot* application provides essential weather details as well as crop-specific advisories, thus enabling users to make better decisions regarding their crops (Dhulipala et al., 2021). This mobile application was jointly developed by the India Meteorological Department (IMD) along with IMD's District Agrometeorological Advisory Service (DAAS), the Indian Council for Agricultural Research (ICAR), the Indian Institution for Tropical Meteorology (IITM), and the International Crops Research Institute for the Semi-Arid Tropics (ICRISAT). This application is available in English and 12 other Indian languages (Hindi, Bangla, Tamil, Telugu, Mizo, Assamese, Gujarati, Malayalam, Kannada, Odia, Punjabi, and Marathi) making it accessible to the farmers. This application covers 717 districts of India providing information on real-time weather alerts and crop advisory based on specific point location and has sent out 3,54,312 advisories since its launch in July 2020.

Kumar et al. (2022) carried out a case study on the efficacy of the *Meghdoot* mobile app for the region of Ladakh, which has low accessibility via roads and a very cold and dry climate with minimal precipitation in the form of snowfall in winter. A hundred farmers were sampled randomly, and it was reported that 50% of farmers acknowledged the user-friendly local language on the application and 70% of registered farmers followed the agrometeorological advisory alerts. The application proved superior by giving 16.71% more yield of crops for farmers who used this application over non-users since 2020. As communication is a major barrier in information dissemination, the *Meghdoot* application has bridged that gap and has helped users in a country where there are multitudes of local languages and dialects. Thus, we can safely say that AI strategically correlates climate statistics with current agricultural practices and helps with point-specific solutions to each region.

IMPLEMENTATIONS OF AI TECHNOLOGY IN AGRICULTURE

In recent years, AI has become an important tool for sustainable agriculture. AI systems are increasingly being used to monitor soil health, predict crop yields, and find ways to reduce greenhouse gas emissions (Bawa et al., 2023). By using data-driven algorithms and machine learning, AI can provide valuable insights into soil composition and crop production. For farmers, identifying the best practices for their fields is crucial, so they rely on AI to make educated decisions and save money. The software enables them to optimize their crop yields and reduce their reliance on chemical fertilizers and pesticides. AI-powered systems can monitor soil health, water usage, and even monitor crop growth and identify areas of the field that need more attention. With this data, farmers can make informed decisions about when to plant, weed, and harvest, as well as which varieties of crops to plant in each field.

AI has had a major impact on the field of sustainable agriculture, with many potential benefits to the industry, research communities and governments. AI has the potential to increase crop yields, identify pests, optimize irrigation and fertilizer use, and reduce food waste. Additionally, AI is able to minimize human error, increase efficiency, and even predict the weather (Xu et al., 2021). Researchers and farmers alike may also benefit from AI-powered autonomous vehicles and monitoring systems, which reduce the need for manual labor. As such, researchers should consider the implications of AI-driven systems and regulations must be in place to govern its use. Some examples of AI-powered tools that have a direct impact on sustainable agriculture include *Masa*, which is a mobile app for newcomers to the agricultural sector to get a first-hand walk-through guide containing all the information they need to get started and be successful and sustainable farming (Ogubuike et al., 2021). DroneDeploy is used by farmers for surveying their agricultural land area seeks to better predict the impact of weather systems (Giri et al., 2020). Table 1.1 summarizes some of the AI-driven tools and technologies in the field of agriculture.

The potential of AI to benefit the agricultural community has long been evident. With the ability to increase yields for farmers, optimize pest management, and reduce food waste, AI offers opportunities for more efficient and cost-effective agricultural practices. While some members of the agricultural community argue that AI offers more benefits than risks, others are more cautious. Despite the potential benefits of AI, there are also potential drawbacks. AI-driven decisions can have serious implications and if autonomous robotics are not properly regulated, they could lead to negative environmental consequences. A major drawback of AI robotics in agriculture would be unemployment to humans as a major number of repetitive tasks would be assigned to AI-driven machinery and human involvement would shrink to only supervision purposes (Ben Ayed & Hanana, 2021).

TABLE 1.1

Application of Artificial Intelligence and Related Tools in the Agriculture Sector

AI Tools	Applications	Methods	Benefits	Country	Source
Meghdoot Mobile Application	Weather alerts agro-advisory forecasts	Satellite imagery Automated weather stations District-wise advisory offices	Information to farmers Higher yields Beneficial alerts for storms, heatwaves, coldwaves, cyclones, hail, snow, rain, lightning	India	Dhulipala et al. (2021) Kumar et al. (2022)
Agrilyst	Precision agriculture prediction	Harvest period projection Sowing cycles regulation	Good yield by prediction and guidance system	US	Prescott (2016)
MDFC–ResNet	Crop disease recognition systems	Fine-grained Disease Recognition on three levels Residual neural networks with multi-dimensional feature compensation applied	Species recognition Disease recognition Disease level recognition	Mongolia	Hu et al. (2020)
Fast Image Processing, (FIP) Robust Crop Row Detection, (RCRD)	Weed management	Weed detection in maize fields through cameras for herbicide dosage	95% weed detection 80% crop row detection	Spain	Burgos-Artizzu et al. (2011)

(Continued)

TABLE 1.1 (CONTINUED)
Application of Artificial Intelligence and Related Tools in the Agriculture Sector

AI Tools	Applications	Methods	Benefits	Country	Source
Alesca Life	Precision farming modular food production	System driven environmental controlled crop growth containers	Production of specific edible plants in the controlled environment of the consumer	China	Kakani et al. (2020)
John Deere	AI-powered tractors, farm equipment, machinery, and tools	Utilizes computer-vision sensors on machinery	Autonomous weed identification and eradication Hands-free farm tractor driving	30 countries including the US and Canada	Chattopadhyay et al. (2022)
Trace Genomics	Analysis of agricultural data	Machine learning application to gather metadata	Soil metadata provides information on the overall health and condition of soil	US	Kakani et al. (2020)
TEAPEST	Identification of pests in tea plant	Neural networks and objective-based pest management system	Pest management in tea plantations	India	Ghosh and Samanta (2003)
Drone Deploy	Survey of agricultural land	Aerial map creation through drones	Information provided to stakeholders regarding agricultural land	Available worldwide	Giri et al. (2020)
Masa mobile App	Sustainable agriculture guidance system	Walk-through guidance for novices in agriculture	Promotes sustainable agriculture for beginners in farming	Available online	Ogubuike et. al (2021)

While research continues, it is important for agricultural decision-makers to strike a balance between these two perspectives. In many cases, existing agricultural practices can be augmented with the use of AI, while ensuring that any potential risks are minimized.

ROLE OF AI IN CONTROL OF AGRICULTURAL DISEASES AND PESTS

A major challenge faced by farmers in agricultural practices is damage to crops due to plant diseases and pests. Two significant aspects of agriculture are pest management and disease management (Sharma, 2021). There are three main components to the pest and disease management process namely prevention, detection, and control. Conventional methods have been applied in the past to deal with these aspects of agricultural practices. However, there are certain limitations to human interventions that are now being curtailed using AI in order to achieve maximum protection from pests and diseases and in turn reduce agricultural losses.

PEST MANAGEMENT

The most troubling issue in farming leading to heavy economic loss is pest infestation of crops. For the longest time, scientists have tried to devise various methods through computational processes in order to classify different pests and ways to identify and manage their infestations. Mostly these methods are confusing and do not provide a proper solution (Pasqual & Mansfield, 1988). Therefore many logic-based systems of expertise were devised to solve these issues (Saini et al., 2002; Siraj & Arbaiy, 2006). A method based on an objective-oriented technique was employed to create an expert system based on a rule by Ghosh and Samanta (2003) for pest management in tea called TEAPEST. Categorical identification and implementation of the consultation process was carried out along with the restructuring of the system by applying multifaceted back proliferation neural networks (Samkanta & Ghosh, 2012). Later on, this was modified by utilizing a functional prototype based on the radial technique in order to achieve higher rates of sorting (Banerjee et al., 2017). Characteristics ranging from the route of pest infestation to future pest attacks are now being predicted by utilizing advanced programming AI by companies and the government for pest control in agriculture. Drone footage of crops can now be used to constantly monitor the presence of diseases, pests, poor soil, and unusual degradation in plants, which will help farmers get better pest management results. Data on any particular pest-infested area can be obtained through this and this information can be used to control the further spread of any pest or disease at the earliest stage possible.

DISEASE MANAGEMENT

Plant diseases in their countless forms pose a significant threat to the agricultural economy, environment, and in turn to global consumers' health. A massive loss of crops of approximately 35% occurs in India alone due to destruction caused by plant diseases and pests. Uncontrolled application of pesticides is also hazardous to human health as they could be biomagnified and spread toxicity. All these issues can be dealt with through proper surveillance of agricultural crop plants in order to look for potential plant diseases and the application of customized treatments at the earliest. A significant level of expertise is essential to detect diseased plants and decide a course of action for recovery. Figure 1.4 broadly illustrates the steps involved in the detection of diseases in crop plants by application of AI and laboratory analysis (Sharma, 2021). For this, a computerized systematic approach is used globally for disease analysis and suitable crop recovery.

In order to detect the disease, sensing and analysis of images are carried out to categorize them into external disease-free regions, background imagery, and the diseased portion of the leaf. Samples of infected regions are then retrieved and analyzed further in the laboratory. This process greatly aids in the detection of pests and the presence of nutrient deficiency. Earlier, the rule-based systems were designed by Byod and Sun (1994) for disease management in agricultural crop plants. Francl and Panigrahi (1997) established an artificial neural network model for plant disease eradication in a variety of harvested crop plants. Along with these certain hybrid control systems were also applied. A neural network model in conjunction with an image dispensation prototype was developed to map out and detect the phalaenopsis sapling disease (Huang, 2007).

The margin for human error in disease and pest detection in agricultural crop plants is greatly reduced by the application of digital image processing techniques of AI. Therefore, AI greatly enhances the identification and categorization of diseases and pests in plants and aids in the application of early methods of treatment and in turn reduces agricultural crop damages and economic losses. Implementation of AI systems in pest and disease management in agriculture is a newer domain and still has a vast untapped territory of advanced applications. At present a combination of

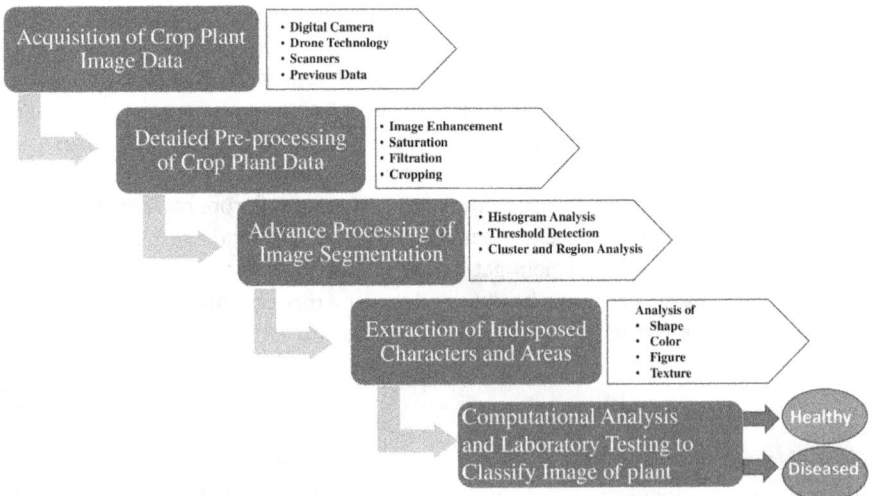

FIGURE 1.4 Crop plant disease detection process

AI and laboratory testing of samples provides suitable pest and disease monitoring and control. For future analyses completely based on AI, much more research and development are necessary to overcome the limitations of current methods of detection and management.

RECOMMENDATIONS FOR AI IN AGRICULTURE

As climate change continues to be a global concern, the agricultural sector is quickly undergoing significant changes in ways that it is making the industry sustainable and ethical. Advances in AI technologies are furthering the sector to meet its sustainability goals in the face of climate change. Using AI as a vehicle, farmers are able to make better sense of their farming data, detect patterns, and respond to changes in the environment in sustainable ways. The following are our recommendations for making the most out of the technology.

Start with education. AI has already brought many benefits to the education field, especially in computing and math literacy programs. However, there is rarely any mention of AI being used in literacy programs aligned with addressing climate change and sustainable agriculture. This can be addressed by creating learning modules that provide real-life climate change events, including the causes and effects. Such modules are worthless if they are not taught in our school curricula. Again, this recommendation is aimed at stakeholders from climate change education communities along with members of AI education to address. Next, we recommend changes in agricultural practices that have traditionally been carried out since the Industrial Revolution. We are currently living in the information age, where even farming tractors come equipped with sensors that provide real-time digital dashboards for farmers. Intelligent decisions are made based on AI-powered algorithms that allow personnel to make educated decisions that are not only cost-effective, efficient, and productive, but also address sustainability and ethical concerns. Lately, many researchers have been researching the use of modular sensory equipment that plugs into existing farming machinery (Thomasson et al., 2019) for the benefit of the equipment operator. Companies such as Toradex are developing modular attachments that enable equipment operators to make educated decisions for saving costs and for a sustainable future. Our next recommendation is aimed at our collective realization that it is our ethical responsibility to be more conscious of our impact on the environment and to do what we could to reduce it. We believe that in order to be aware of the ecological implications of smart agriculture, we need to first be aware ethically. Here, we recommend that AI-enabled technologies in environmental communication take the lead in facilitating support and structure to human endeavor. One way to achieve this is by proactively taking steps to harness the benefits of AI, such as in remote areas as they most commonly tend to be out of the applicable communication channels and may need the use of local vernacular versions to reach all corners of the society. Although the recommendations mentioned here are for climate change communities, we encourage anyone who is directly involved in farming activities to find use in them. The AI-powered tools discussed in this chapter are currently being used by many farmers across the world, and we hope you do too.

CONCLUSION

The chapter is an attempt at a transdisciplinary approach to not only AI applications but also to sustainable agriculture. The investigations deal with multifaceted fields ranging from climate change to ethics to capture the relevance and solutions being implemented. Nonetheless, the benefits and implications AI technology brings to agriculture are phenomenal, which allow us to make ethical decisions focused on sustainable agriculture and in the face of climate change. AI continues to revolutionize the way we conduct farming, and at the same time influences how we treat the environment, for the betterment of human civilization.

FUNDING

The author(s) received no financial support for the research, authorship, and/or publication of this article.

DECLARATIONS

Conflict of Interest: The authors declare no competing interests.

REFERENCES

Al-Aswad, L. A., Ramachandran, R., Schuman, J. S., Medeiros, F., Eydelman, M. B., Abramoff, M. D., Antony, B. J., Boland, M. V., Chauhan, B. C., Chiang, M., & Goldberg, J. L. (2022). Artificial intelligence for glaucoma: Creating and implementing artificial intelligence for disease detection and progression. *Ophthalmology Glaucoma*, 5(5), e16-e25.

Alesca Life. (2013). *The Future of Agriculture: Small Scale, (Almost) No Land*. https://www.alescalife.com/food_for_thought/the-future-of-agriculture-small-scale-almost-no-land/ (Accessed January 11, 2023).

Armstrong McKay, D. I., Staal, A., Abrams, J. F., Winkelmann, R., Sakschewski, B., Loriani, S., Fetzer, I., Cornell, S. E., Rockström, J., & Lenton, T. M. (2022). Exceeding 1.5 C global warming could trigger multiple climate tipping points. *Science*, 377(6611), eabn7950.

Arroyo, I., Woolf, B. P., Burelson, W., Muldner, K., Rai, D., & Tai, M. (2014). A multimedia adaptive tutoring system for mathematics that addresses cognition, metacognition and affect. *International Journal of Artificial Intelligence in Education*, 24(4), 387–426.

Artificial Intelligence. (2016). Automation, and the economy. *Executive Office of the President*, 18–19.

Aune, J. B. (2012). Conventional, organic and conservation agriculture: Production and environmental impact. In *Agroecology and Strategies for Climate Change* (pp. 149–165). Springer Netherlands. https://doi.org/10.1007/978-94-007-1905-7_7.

Banerjee, G., Sarkar, U., & Ghosh, I. (2017). A radial basis function network based classifier for detection of selected tea pests. *International Journal of Advanced Research in Computer Science and Software Engineering*, 7(5), 665–669.

Bartz, J. (2021, December). The power of communication affecting environmental impact (Video). TEDx Talks. https://www.ted.com/talks/Jeannine _bartz_the_ power_of_ communication_in_affecting_environmental_impact.

Bawa, A., Samanta, S., Himanshu, S. K., Singh, J., Kim, J., Zhang, T., … Ale, S. (2023). A support vector machine and image processing based approach for counting open cotton bolls and estimating lint yield from UAV imagery. *Smart Agricultural Technology, 3,* 100140.

Ben Ayed, R., & Hanana, M. (2021). Artificial intelligence to improve the food and agriculture sector. *Journal of Food Quality,* vol. 2021, Article ID 5584754, 7 pages, 2021. https://doi.org/10.1155/2021/5584754.

Bisong, E. (2019). *Building Machine Learning and Deep Learning Models on Google Cloud Platform: A Comprehensive Guide for Beginners.* Apress.

Boyd, D. W., & Sun, M. K. (1994). Prototyping an expert system for diagnosis of potato diseases. *Computers and Electronics in Agriculture, 10*(3), 259–267.

Buchholz, R. R., Park, M., Worden, H. M., Tang, W., Edwards, D. P., Gaubert, B., Deeter, M. N., Sullivan, T., Ru, M., Chin, M., & Levy, R. C. (2022). New seasonal pattern of pollution emerges from changing North American wildfires. *Nature Communications, 13*(1), 1–9.

Burgos-Artizzu, X. P., Ribeiro, A., Guijarro, M., & Pajares, G. (2011). Real-time image processing for crop/weed discrimination in maize fields. *Computers and Electronics in Agriculture, 75*(2), 337–346.

Chakravarty, P. (January 3, 2023). Role of AI in the future of climate education Jalvayu: Environment and climate talk. https://blog.cpoulomi.com/role-of-ai-in-the-future-of -climate-education.

Chakravarty, P., & Kumar, M. (2020). Spectral studies of ground-based observations of wind components, temperature & analysis of flux parameters during pre-monsoon thunderstorms period at Ranchi. *International Journal of Emerging Technologies, 11*(2), 763–769.

Chakravarty, P., & Kumar, M. (2019). Floral species in pollution remediation and augmentation of micrometeorological conditions and microclimate: An integrated approach. In *Phytomanagement of Polluted Sites* (pp. 203–219). Elsevier. https://doi.org/10.1016/ B978-0-12-813912-7.00006-5.

Chakravarty, P., Bauddh, K., & Kumar, M. (2017). Phytoremediation: A multidimensional and ecologically viable practice for the cleanup of environmental contaminants. In *Phytoremediation Potential of Bioenergy Plants* (pp. 1–46). Springer. https://doi.org /10.1007/978-981-10-3084-0_1.

Chakravarty, U., & Gaur, R. (2011). Environment constructs as a communication tool. *Atlantic Literary Review, 2*(2), 86–95.

Chattopadhyay, P., Patel, H. P., & Parmar, V. (2022, August). Internet of things (IoT) in smart agriculture. In *2022 3rd International Conference on Electronics and Sustainable Communication Systems (ICESC)* (pp. 536–540). IEEE.

Climate Action Initiatives. (2023). *United Nations | Climate Action.* United Nations. https:// www.un.org/en/climatechange/climate-action-coalitions (Accessed January 10, 2023).

Climate Change. (2022). *Centre for Environment Education. Centre of Excellence of the Ministry of Environment, Forest & Climate Change.* Government of India. https:// www.ceeindia.org/climate-change (Accessed January 12, 2023).

Climate Change Education. (2020). *Climate Change Education | Erasmus+.* European Union. https://ccedu.erasmus-projects.eu/ (Accessed January 9, 2023).

Devi, S. (2022). Pakistan floods: Impact on food security and health systems. *The Lancet, 400*(10355), 799–800.

Dhulipala, R. K., Gogumalla, P., Karuturi, R., Palanisamy, R., Smith, A., Nagaraji, S., Rao, S. A., Vishnoi, L., Singh, K. K., Bhan, S. C., & Whitbread, A. M. (2021). *Meghdoot—A Mobile App to Access Location-Specific Weather-Based Agro-Advisories.* PAN (CGIAR Research Program on Climate Change, Agriculture and Food Security Working Paper).

European Perspectives on Climate Education. (2023). *European Climate Change Curriculum.* European Union. https://climateperspectives.eu/ (Accessed January 9, 2023).

Event Horizon Telescope Collaboration. (2019). First M87 event horizon telescope results. IV. Imaging the central supermassive black hole. *arXiv preprint arXiv:1906.11241.*

Fahle, S., Prinz, C., & Kuhlenkötter, B. (2020). Systematic review on machine learning (ML) methods for manufacturing processes–Identifying artificial intelligence (AI) methods for field application. *Procedia CIRP, 93*, 413–418.

Farooq, M., & Siddique, K. H. M. (eds.) (2014). *Conservation Agriculture.* Springer International.

Feinstein, N. W., & Kirchgasler, K. L. (2015). Sustainability in science education? How the next generation science standards approach sustainability, and why it matters. *Science Education, 99*(1), 121–144.

Flor, A. G. (2004). *Environmental Communication: Principles, Approaches, Strategies of Communication Applied to Environmental Management.* Office of Academic Support and Instructional Services, UP Open University.

For Educators. (2023). *Global Climate Change | Vital Signs of the Planet.* NASA. https://www.jpl.nasa.gov/edu/teach/tag/search/Climate+Change (Accessed January 14, 2023).

Francl, L. J., & Panigrahi, S. (1997). Artificial neural network models of wheat leaf wetness. *Agricultural and Forest Meteorology, 88*(1–4), 57–65.

Ghobakhloo, M. (2020). Industry 4.0, digitization, and opportunities for sustainability. *Journal of Cleaner Production, 252*, 119869.

Ghosh, I., & Samanta, R. K. (2003). TEAPEST: An expert system for insect pest management in tea. *Applied Engineering in Agriculture, 19*(5), 619.

Giri, A., Saxena, D. R. R., Saini, P., & Rawte, D. S. (2020). Role of artificial intelligence in advancement of agriculture. *International Journal of Chemical Studies, 8*(2), 375–380.

Henderson, J., & Drewes, A. (2020). Teaching climate change in the United States. In *Teaching Climate Change in the United States* (pp. 1–10). Henderson, J., & Drewes, A. (Eds.) Routledge.

Hu, W. J., Fan, J., Du, Y. X., Li, B. S., Xiong, N., & Bekkering, E. (2020). MDFC–ResNet: An agricultural IoT system to accurately recognize crop diseases. *IEEE Access, 8*, 115287–115298.

Huang, K. Y. (2007). Application of artificial neural network for detecting Phalaenopsis seedling diseases using color and texture features. *Computers and Electronics in Agriculture, 57*(1), 3–11.

Huntingford, C., Jeffers, E. S., Bonsall, M. B., Christensen, H. M., Lees, T., & Yang, H. (2019). Machine learning and artificial intelligence to aid climate change research and preparedness. *Environmental Research Letters, 14*(12), 124007.

Intergovernmental Panel on Climate Change. (2022). Food security. In *Climate Change and Land: IPCC Special Report on Climate Change, Desertification, Land Degradation, Sustainable Land Management, Food Security, and Greenhouse Gas Fluxes in Terrestrial Ecosystems* (pp. 437–550). Cambridge University Press. https://www.ipcc.ch/srccl/chapter/chapter-5/.

IPPC Secretariat. (2021). Scientific review of the impact of climate change on plant pests – A global challenge to prevent and mitigate plant pest risks in agriculture, forestry and ecosystems. FAO on behalf of the International Plant Protection Convention Secretariat. https://doi.org/10.4060/cb4769en.

Jamshidi, M., Lalbakhsh, A., Talla, J., Peroutka, Z., Hadjilooei, F., Lalbakhsh, P., Jamshidi, M., La Spada, L., Mirmozafari, M., Dehghani, M., & Sabet, A. (2020). Artificial intelligence and COVID-19: Deep learning approaches for diagnosis and treatment. *IEEE Access, 8*, 109581–109595.

Jat, R. A., Sahrawat, K. L., & Kassam, A. H. (eds) (2014). *Conservation Agriculture: Global Prospects and Challenges.* CABI.

Jug, D., Jug, I., Brozović, B., Vukadinović, V., Stipešević, B., & Đurđević, B. (2018). The role of conservation agriculture in mitigation and adaptation to climate change. *Poljoprivreda, 24*(1), 35–44.

Kakani, V., Nguyen, V. H., Kumar, B. P., Kim, H., & Pasupuleti, V. R. (2020). A critical review on computer vision and artificial intelligence in food industry. *Journal of Agriculture and Food Research, 2,* 100033.

Kim, N., Ha, K. J., Park, N. W., Cho, J., Hong, S., & Lee, Y. W. (2019). A comparison between major artificial intelligence models for crop yield prediction: Case study of the mid-western United States, 2006–2015. *ISPRS International Journal of Geo-Information, 8*(5), 240.

Kumar, Y., Fatima, K., Raghuvanshi, M. S., Nain, M. S.,& Sofi, M. (2022). Impact of Meghdoot mobile app - A weather-based agro-advisory service in cold arid Ladakh. *Indian Journal of Extension Education, 58*(3), 142–146. https://doi.org/10.48165/IJEE .2022.58329.

Lakshmi, V., & Corbett, J. (2020). How artificial intelligence improves agricultural productivity and sustainability: A global thematic analysis. https://doi.org/10.24251/hicss .2020.639

Lee, W., Allessio, D., Rebelsky, W., Satish Gattupalli, S., Yu, H., Arroyo, I., … Woolf, B. P. (2022). Measurements and interventions to improve student engagement through facial expression recognition. In *International Conference on Human-Computer Interaction* (pp. 286–301). Springer.

Liang, G., Fan, W., Luo, H., & Zhu, X. (2020). The emerging roles of artificial intelligence in cancer drug development and precision therapy. *Biomedicine and Pharmacotherapy, 128,* 110255.

Mauget, S. A., Himanshu, S. K., Goebel, T. S., Ale, S., Lascano, R. J., & Gitz III, D. C. (2021). Soil and soil organic carbon effects on simulated southern high plains dryland cotton production. *Soil and Tillage Research, 212,* 105040.

Mendelsohn, R. (2009). The impact of climate change on agriculture in developing countries. *Journal of Natural Resources Policy Research, 1*(1), 5–19.

Monroe, M. C., Plate, R. R., Oxarart, A., Bowers, A., & Chaves, W. A. (2019). Identifying effective climate change education strategies: A systematic review of the research. *Environmental Education Research, 25*(6), 791–812.

Ogubuike, R., Adib, A., & Orji, R. (2021, October). Masa: AI-adaptive mobile app for sustainable agriculture. In *2021 IEEE 12th Annual Information Technology, Electronics and Mobile Communication Conference (IEMCON)* (pp. 1064–1069). IEEE.

Pannu, A. (2015). Artificial intelligence and its application in different areas. *Artificial Intelligence, 4*(10), 79–84.

Pasqual, G. M., & Mansfield, J. (1988). Development of a prototype expert system for identification and control of insect pests. *Computers and Electronics in Agriculture, 2*(4), 263–276.

Pathan, M., Patel, N., Yagnik, H., & Shah, M. (2020). Artificial cognition for applications in smart agriculture: A comprehensive review. *Artificial Intelligence in Agriculture, 4,* 81–95.

Perea, R. G., Poyato, E. C., Montesinos, P., & Díaz, J. A. R. (2019). Optimisation of water demand forecasting by artificial intelligence with short data sets. *Biosystems Engineering, 177,* 59–66.

Peres, R. S., Jia, X., Lee, J., Sun, K., Colombo, A. W., & Barata, J. (2020). Industrial artificial intelligence in industry 4.0-systematic review, challenges and outlook. *IEEE Access, 8,* 220121–220139.

Pogue, D. (2021). *How to Prepare for Climate Change: A Practical Guide to Surviving the Chaos.* Simon & Schuster.

Prescott, N. (2016). Agroterrorism, resilience, and indoor farming. *Animal Law, 23*, 103.

Rakovec, O., Samaniego, L., Hari, V., Markonis, Y., Moravec, V., Thober, S., … Kumar, R. (2022). The 2018–2020 multi-year drought sets a new benchmark in Europe. *Earth's Future, 10*(3), e2021, EF002394.

Rockström, J., Williams, J., Daily, G., Noble, A., Matthews, N., Gordon, L., Wetterstrand, H., DeClerck, F., Shah, M., Steduto, P., & de Fraiture, C. (2017). Sustainable intensification of agriculture for human prosperity and global sustainability. *Ambio – A Journal of the Human Environment, 46*(1), 4–17.

Saini, H. S., Kamal, R., & Sharma, A. N. (2002). Web based fuzzy expert system for integrated pest management in soybean. *International Journal of Information Technology, 8*(1), 55–74.

Samanta, R. K., & Ghosh, I. (2012). Tea insect pests classification based on artificial neural networks. *International Journal of Computer Engineering Science (IJCES), 2*(6), 1–13.

Senecah, S. L. (2007). Impetus, mission, and future of the environmental communication commission/division: Are we still on track? Were we ever? *Environmental Communication, 1*(1), 21–33.

Singh, T., Bhadwaj, H., Verma, L., Navadia, N. R., Singh, D., Sakalle, A., & Bhardwaj, A. (2022). Applications of AI in agriculture. *Challenges and Opportunities for Deep Learning Applications in Industry 4.0*, 181.

Siraj, F., & Arbaiy, N. (2006). Integrated pest management system using fuzzy expert system. In *Proceedings of Knowledge Management International Conference & Exhibition (KMICE)*, 6–8 June 2006 Legend Hotel Kuala Lumpur, Malaysia. Universiti Utara Malaysia, Sintok, pp. 169-176. ISBN 9833282903.

Slovic, S., Rangarajan, S., & Sarveswaran, V. (Eds.) (2019). *Routledge Handbook of Ecocriticism and Environmental Communication.* Routledge.

Sharma, A. R., Jain, P., Abatzoglou, J. T., & Flannigan, M. (2022). Persistent positive anomalies in geopotential heights promote wildfires in western North America. *Journal of Climate, 35*(19), 2867–2884.

Sharma, R. (2021). Artificial intelligence in agriculture: A review. In *2021 5th International Conference on Intelligent Computing and Control. Systems (ICICCS)* (pp. 937–942). IEEE.

Shepardson, D. P., Niyogi, D., Roychoudhury, A., & Hirsch, A. (2012). Conceptualizing climate change in the context of a climate system: Implications for climate and environmental education. *Environmental Education Research, 18*(3), 323–352.

Thomasson, J. A., Baillie, C. P., Antille, D. L., Lobsey, C. R., & McCarthy, C. L. (2019). *Autonomous Technologies in Agricultural Equipment: A Review of the State of the Art* (pp. 1–17). American Society of Agricultural and Biological Engineers.

UN ESA. (2022, November 15). As the world's population hits 8 billion people, the UN calls for solidarity in advancing sustainable development for all. UN.org. https://www.un.org/en/desa/world-population-hits-8-billion-people (Accessed January 10, 2023).

US EPA. (2022, August 5). Sources of greenhouse gas emissions. https://www.epa.gov/ghgemissions/sources-greenhouse-gas-emissions (Accessed January 9, 2023).

WMO. (2022, September 8). State of climate in Africa highlights water stress and hazards .WMO.org. https://public.wmo.int/en/media/press-release/state-of-climate-africa-highlights-water-stress-and-hazards.Press. Release Number:09082022 (Accessed January 12, 2023).

Worster, D. (1993). *The Wealth of Nature: Environmental History and the Ecological Imagination.* Oxford University Press.

Weder, F., & Milstein, T. (2021). Revolutionaries needed! Environmental communication as a transformative discipline. In *The Handbook of International Trends in Environmental Communication* (pp. 407–419). Routledge.

Xu, Y., Liu, X., Cao, X., Huang, C., Liu, E., Qian, S., Liu, X., Wu, Y., Dong, F., Qiu, C. W. and Qiu, J. (2021). Artificial intelligence: A powerful paradigm for scientific research. *The Innovation*, 2(4), 100179.

2 Development of a Modeling Approach for Agriculture Crop Type Classification Aiming at Large-Scale Precision Agriculture by Synergistic Utilization of Fused Sentinel-1 and Sentinel-2 Datasets with UAV Datasets

Akshay Pandey, Shubham Awasthi, and Kamal Jain

INTRODUCTION

Most developing countries like India are agriculture-based economies. Presently, the agriculture sector plays an essential role in India by providing livelihoods to more than 60% of the population (Sharma et al. 2010). It provides food to the people and ensures food security for the country. This sector delivers raw materials for many industries and contributes to the country's food exports (Palanisami et al. 2019). India's food security depends on producing cereal crops. The agriculture sector has been showing decent growth in the past few decades, which has led to increased productivity and exports (Seelan et al. 2003). Despite this high growth rate, the agricultural sector is still facing various challenges and issues (Jain & Pandey 2020). The biggest problem in the Indian agricultural sector is its low efficiency resulting in low agricultural yield in terms of productivity, due to the traditional agricultural

DOI: 10.1201/9781003441175-2

practices and lack of precise agricultural crop information (Pandey and Jain 2022a) at different temporal stages in the crop season.

On average, India's per-hectare agriculture yield is 30–50% lower than in developed countries. India produces 106.19 million tons of rice a year from a land cover of 44 million hectares. The yield rate of 2.4 tons per-hectare places India in 27th place out of 47 countries. However, the yield rates in China and Brazil are 4.7 t/ha and 3.6 t/ha, respectively (Raghavan 2014). Hence, despite the involvement of more than 60% of the total population, the contribution of the agriculture sector to India's GDP is only 17.8% (PIB Delhi Ministry of Finance 2021). The major portion of India's agricultural land is still dependent on monsoon rains for irrigation, which are uncertain and unreliable. Many states like Rajasthan, Haryana, Telangana, Karnataka, Gujarat, and Uttar-Pradesh are facing water scarcity for irrigation due to the limited ground channel-based irrigation facilities like canals and rivers (Dhawan 2017). The insufficient availability of water results in low agricultural yield (Tiwari & Jaga 2012). Small agrarian farms, outdated farming practices, crop diseases (Pandey and Jain 2022b), and soil fertility decrease due to overuse of fertilizers and pesticides are other major factors contributing to low agricultural productivity and rising input costs (Alam Iqbal 2018).

Precision agriculture techniques can be adopted to increase agricultural yield, lower input cost, and improve overall efficiency in agriculture (Maha et al. 2019). In precision agriculture, detailed specific information regarding crop type, crop phenological stage, crop water stress, weeds, and pests in the crop at the parcel level is required (Puri et al. 2017). In India 70% of the agricultural farms are less than 1 hectare and the national average is less than 2 hectares. Therefore, manual retrieval of this precise crop information is quite difficult. Remote sensing has shown potential in various applications of earth observation (Awasthi et al. 2019; Awasthi, Thakur et al. 2020; Awasthi, Jain et al. 2020; Awasthi et al. 2021; Awasthi, Jain et al. 2022; Awasthi, Varade, Thakur et al. 2022; Awasthi, Varade, Bhattacharjee et al. 2022). Remote sensing techniques using spaceborne platforms (satellite datasets) and Unmanned Arial Vehicles (UAVs) can be used for collecting specific information on the crops (Stafford 2000; Primicerio et al. 2012; Jain & Pandey 2020; Pandey & Jain 2022; Bawa et al. 2023). Precision agriculture refers to the process of farmers managing crops to ensure the efficiency of various inputs in the crop such as water requirement and fertilizer need, and maximizing productivity, quality, and crop yield (Radoglou-Grammatikis et al. 2020). This can be implemented by providing very precise and exact crop information utilizing UAV remote sensing (Daponte et al. 2019).

Significant development has been made in the development of methodologies for agriculture monitoring and crop type classification using satellite remote sensing and UAV-based techniques (Husak et al. 2008; Ustuner et al. 2014; Waldner et al. 2015; Park & Park 2015; Lottes et al. 2017; Orynbaikyzy et al. 2019; Théau et al. 2020). Satellite remote sensing has been tremendously applied for agriculture monitoring using the multispectral (Heupel et al. 2018; Defourny et al. 2019; Piedelobo et al. 2019; Momm et al. 2020; Praseartkul et al. 2022), hyperspectral (Wang et al. 2019; Hong & Abd El-Hamid 2020), and synthetic aperture radar (SAR) (Phung et al.

2020; Chauhan et al. 2020) sensors, and also by fusion (Yadav et al. 2022) of these satellite datasets (Orynbaikyzy et al. 2020; Moumni & Lahrouni 2021). Agricultural crop type classification on a precise scale for a large area is still for specific crop area mapping and crop yield estimation. Significant development has been made in developing methodologies for agriculture monitoring and crop type classification using satellite remote sensing and UAV-based techniques. Studies have been done for crop phenological state estimation and crop yield predictions using time-series satellite datasets (Vicente-Guijalba et al. 2014; De Bernardis et al. 2015; Mascolo et al. 2016). These satellite remote sensing-based techniques focus on macro information retrieval and provide long-term agriculture monitoring of larger-swath coverage with high temporal repeatability (Delegido et al. 2011; Sonobe et al. 2018). On the other hand, UAV-acquired datasets retrieve agriculture crop information at the micro-level (Tsouros et al. 2019). Highly specific and precise agriculture crop information is captured using UAV-based platforms, used explicitly for precision agriculture applications (Colomina & Molina 2014; Wang et al. 2019). These UAV-based crop-monitoring techniques have the limitations of smaller swath coverage and low temporal repeativity. Also, considerable human resources are required during the UAV data acquisition, and the size of the data from UAV imageries is comparatively large. The large size of the data from UAVs takes relatively more processing time and has high-end hardware requirements compared to satellite images. The repetitive use of UAVs results in high user-cost consumption (Gago et al. 2014; Tsouros et al. 2019). Hence, there is a need to develop a modeling approach that utilizes UAV and satellite datasets simultaneously to provide precise agricultural crop information along with broader coverage (Mitchell et al. 2012; Puri et al. 2017).

This study focuses on developing a modeling approach for agriculture crop type classification aiming at large-scale precision agriculture by simultaneous utilization of satellite and UAV datasets. The proposed model will utilize information from UAV datasets for crop type identification and satellite datasets for the classification of crops to a large extent. The proposed methodology will be useful in avoiding the repetitive usage of drones for large-scale precise crop information retrieval. This chapter is structured as follows: the next section describes the study area and dataset details. In the following section, a detailed description of the model development and the methodology workflow is added. The next section is dedicated to results and discussion. The conclusion of the study is given in the last section.

STUDY AREA AND DATASETS DESCRIPTION

The agriculture area selected for this study is situated in the Roorkee Tehsil of Haridwar District in Uttarakhand state in India. The location center area lies at 29.85^0 N Latitude and 77.88^0 E Longitude in the Indo-Gangetic Plains. The geographical extent of the study area map is shown in Figure 2.1. The soil in this region is very fertile, and most of the area is covered with agricultural lands. Various crops like sugarcane, rice, wheat, and pulses are grown in different seasons. During the monsoon season in July–August, rice is the major crop cultivated in this region. Other crops like fodder are also grown in this season. In the winter season, during

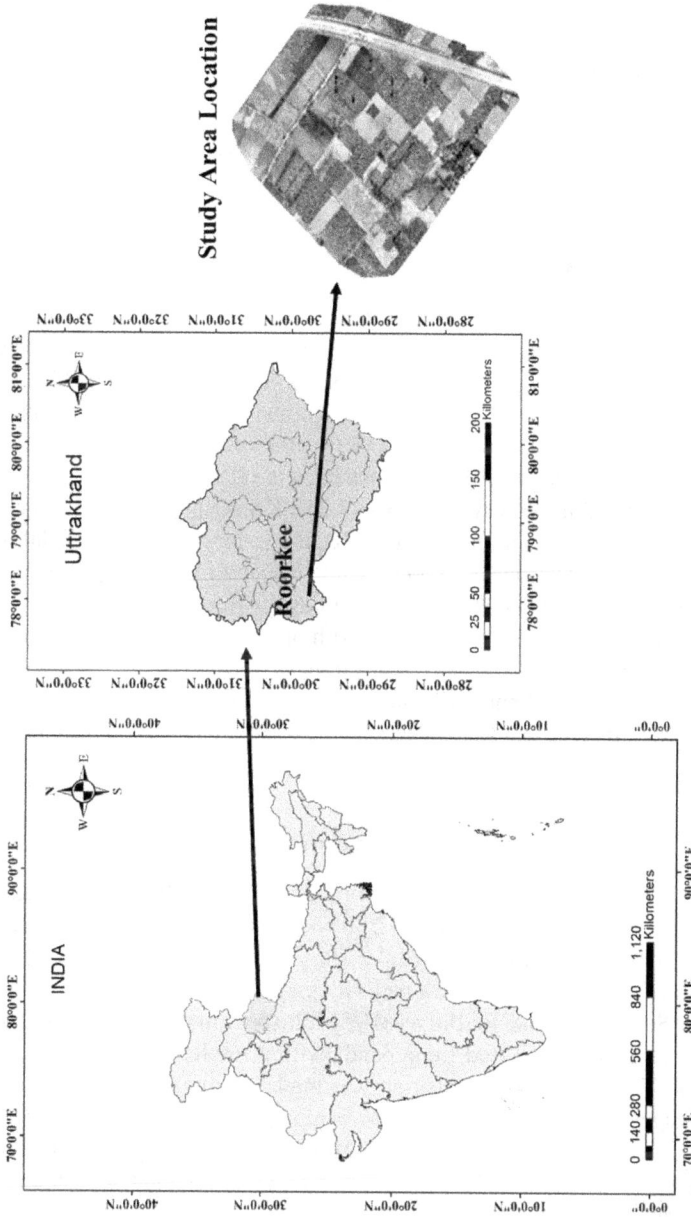

FIGURE 2.1 Study area map

December–March, wheat crop is grown. Meanwhile, sugarcane with a total growth span of ten months is cultivated from February till December. The average maximum high temperature in the study area rises to 36°C in summer, and the average minimum temperature drops to 8°C in winter.

DATASET DESCRIPTION

In the preliminary field of study, two different remote sensing platforms were used: the first satellite imagery (SAR and multispectral datasets) and the second image datasets acquired using the Unmanned Aerial Vehicle:

SATELLITE DATASET DETAILS

For this study, the freely available Sentinel-1 SAR and Sentinel-2 multispectral (Level 1C processing) datasets from August 18, 2019, are used. Both these satellites are polar-orbiting, part of the Copernicus mission of the European Space Agency. Sentinel-1 is a constellation of two satellites Sentinel-1A and -1B, operating in the C-band at an operating frequency of 5.405 GHz, providing an overall temporal resolution of six days. The Sentinel-1 dual-polarimetric GRDH SAR product's spatial resolution is 10 m and it has a swath width of 250 Km. Sentinel-2 is also a constellation of two satellites i.e., Sentinel-2A and -2B. It provides multispectral datasets of high spatial–spatial resolution (10 to 60 m) with 13 spectral bands. It provides global coverage of the earth's land surface every ten days with one satellite and five days combining both satellites. The total swath width of Sentinel-2 datasets is 290 km by applying an approximately 20° entire field-of-view (Drusch et al. 2012; Clevers & Gitelson 2013). Hence, the Sentinel-2 multispectral satellite has a high temporal and spatial resolution. It has three Red-Edge bands, which are best for agricultural crop monitoring (Delegido et al. 2011).

UNMANNED AERIAL VEHICLES DATASET DESCRIPTION

For this study, UAV data acquisition was made using a modified low-weight DJI-Inspire 2 T650A UAV shown in Figure 2.2 (a) with image MicaSense RedEdge-M™ Multispectral camera shown in Figure 2.2 (b) on August 18, 2019. The mounted ZENMUSE-X4S 20MP RGB camera had 4K imaging capability with a frame rate of 60 frames/Sec. MicaSense RedEdge-M™ Multispectral camera acquired data in Blue, Green, Red, NIR, and Red-Edge bands with a resolution of 6.8 cm from the 100 m height. This UAV data acquisition was made on August 18, 2019. A total of 560 RGB images were acquired during the campaign covering an area of 9 square km. This UAV was able to fly by remote control or autonomously equipped with the aid of its GPS (Global Position System) receiver, and the waypoint navigation system was used. Flight planning was performed on-site to observe the area to be imaged and to determine appropriate points for starting and landing using the Pix4D capture application on a handheld device.

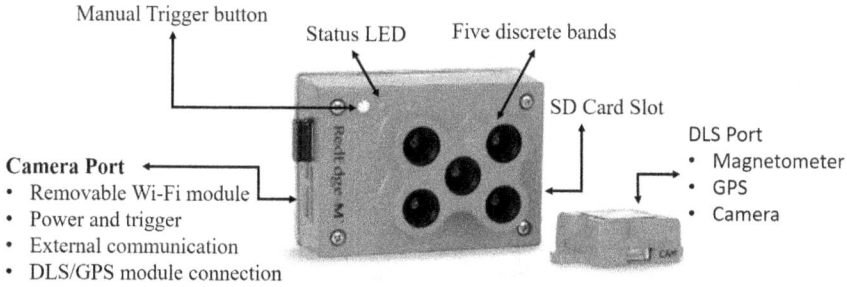

FIGURE 2.2 (a) UAV DJI-Inspire 2T 650A, (b) MicaSense Red-Edge-M™ multispectral RGB camera

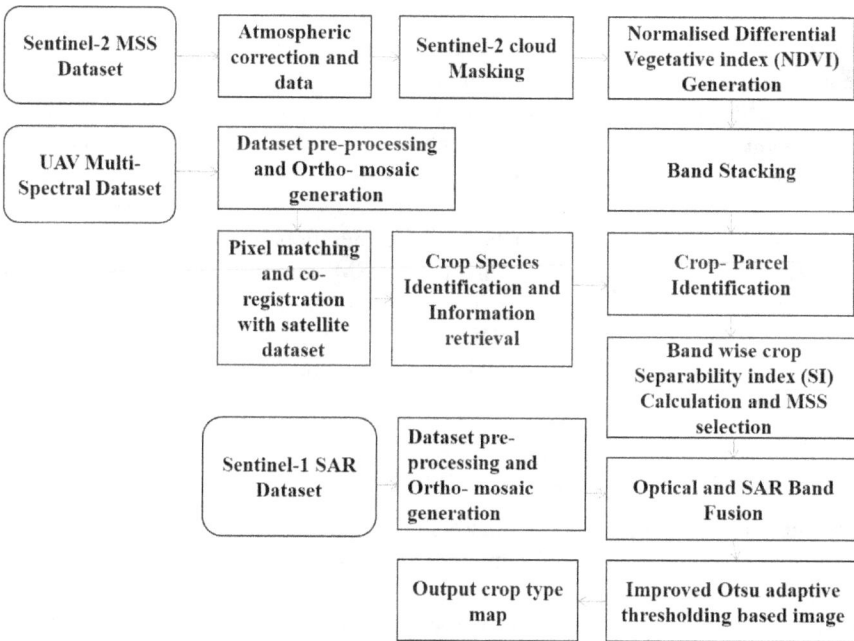

FIGURE 2.3 Methodology flow diagram

Data Processing and Methodology Implementation

Here, Figure 2.3 shows the overall methodology and the workflow steps adopted in this study.

UAV DATA PROCESSING

The UAV-acquired raw RGB datasets of the study area are preprocessed using Pix4D mapper. The UAV image processing involves three major processing steps, i.e., Step 1: Point Cloud Generation, Step 2: Point Cloud Filtering, and Step 3: Digital Surface Model (DSM) and Orthomosaic Generation. In the first step, the process of Point Cloud Generation is performed. During this process, camera calibration and the image orientation are done using key point information in the images. The optimal camera and lens calibration are performed to compensate for the effect of camera noise generated during the flight due to temperature differences, weather conditions, vibrations, and shocks. After this, the key points are generated, which are further used for relative orientation and triangulation utilizing the process of bundle block adjustment. This results in the generation of a point cloud. These matching points and the orientation information from the inertial measurement unit (IMU) are used for the bundle block adjustment for reconstructing the exact camera position and orienting the camera for every acquired image. The information from the GPS inbuilt with UAV is utilized for geocoding the generated point cloud. The positional accuracy of geocoding is improved using the ground control points (GCPs) of the target pixels in each scene. During the generation of the point cloud, noisy and erroneous points are produced, resulting in the era of outliers.

In the second step, point cloud filtering and smoothing of the generated georeferenced point clouds are performed by applying noise filtering algorithms to develop a surface from the point cloud. This generated surface contains erroneous small bumps, corrected by using a surface smoothing algorithm. The third step involves the Digital Surface Model and Orthomosaic Generation. The digital surface model is generated from the point clouds, and these DSM images are used to generate the orthomosaic images. In orthomosaic images, the image's perspective distortions are removed using the 3D model to blend the orthorectified images. This process is called orthomosaic blending. The final ortho mosaic image output is used for further processing.

Sentinel-1 Data Processing

The Sentinel-1 dual-polarimetric datasets in the VV and VH polarization channels are taken as the input. The polarimetric calibration of the polarimetric datasets is done for doing the thermal noise correction and radiometric calibration. The calibrated sigma VV (σ^0_{VV}) and sigma VH (σ^0_{VH}) backscatter products were generated after the polarimetric calibration. The Range-Doppler terrain correction was implemented using SRTM DEM for adding the geocoded information to the product. The backscatter polarization ratio ($\sigma^0_{VH} / \sigma^0_{VV}$) was generated as an additional component. The geocoded σ^0_{HH}, $\sigma^0_{VH,}$ and ($\sigma^0_{VH} / \sigma^0_{VV}$) calibrated geocoded products were used for further analysis.

Sentinel-2 Data Processing

For this study, Sentinel-2 Level-2A multispectral datasets are used. These ortho-image datasets are bottom-of-atmosphere (BOA) radiometric corrected reflectance products. This process of radiometric calibration focuses on parametrizing the atmospheric correction and outputs the atmospherically corrected dataset. The calibration parameters are estimated by incorporating various land cover targets such as vegetation, irrigated crops, forests, and manmade structures in different atmospheric conditions. During this data calibration, the DN values are scaled for a range of values, resulting in quantifying the datasets. Cloud shadow masking is another essential process that is performed to detect the cloud shadow-contaminated and cloud-free pixels in the Sentinel-2 multispectral image. In Sentinel-2 sensor datasets, the band-2 reflectance threshold-based method is used to identify cloud pixels along with the SWIR reflectance in band-11 and band-12 to avoid false detection due to snow/cloud identification (Kolecka et al. 2018).

SENTINEL-2 BAND SELECTION

In this study, Sentinel-2 multispectral band selection is done on the criteria of vegetation susceptibility as discussed in the previous studies (Clevers & Gitelson 2013; Varade et al. 2019). The multispectral bands that have specific absorption and reflection properties for agricultural crops can be utilized for mapping and classification of agricultural crops. In Sentinel-2 sensor, Blue band (B-2), Green band (B-3), Red band (B-4), NIR band (B-8), SWIR-1 band (B-11), SWIR-2 (B-12) bands are highly sensitive to the vegetation information (Clevers & Gitelson 2013). In the False Color Composite (FCC) representation, the NIR band (B-8), Red band (B-4), and Green band (B-3) are used to represent red, green, and blue color channels, respectively. This FCC representation is employed to efficiently monitor the health of the vegetation by utilizing its spectral properties. The NIR band (B-8) shows strong reflection from the vegetation (Sun et al. 2020). NIR band is used for assessing plant safety through spectral signatures, biomass content, crop marks, and subtle differences in vegetation. The Red band (B-4) has strong absorption in the vegetation. The Green band (B-3) pertains to the property of discriminating against large groups of plants and identifying the plant material. The SWIR bands (B-11, B-12) enhance the sense of agricultural plant growth in vegetation studies. Hence, based on these vegetation sensitivities and susceptibility parameters, Sentinel-2 multispectral bands are selected for further analysis (Clevers & Gitelson 2013).

VEGETATION INDICES EXTRACTION USING SENTINEL-2 BAND:

$$NDVI = \frac{(NIR - Red)}{(NIR + Red)} \qquad (2.1)$$

Spectral Indices are the combinations of two or more wavelengths of spectral reflection showing the relative abundance of characteristics of interest vegetation indices that are most common. Such spectral indices are used in multispectral imagery for automatically defining image features like vegetation, urban (Tiwari et al.), and water features (Frampton et al. 2013). The NIR band (Band-8) is strongly reflected from the vegetation, and the vegetation absorbs the Red band (Band-4). Both these bands are particularly good for highlighting and quantifying the dense vegetation that appears dark green (Kobayashi et al. 2020). The Normalized Difference Vegetation Index is for the quantitative analysis of vegetation (Dedeoğlu et al. 2020). It is represented as shown in Equation 2.1:

This Normalized Differential Vegetation Index (NDVI) value ranges between –1 and 1. For green vegetation, the typical range is 0.2 to 0.8.

BAND STACKING

The process of band stacking is performed for Blue band (B-2), Green band (B-3), Red band (B-4), NIR band (B-8), SWIR-1 band (B-11), SWIR-2 (B-12) bands, and NDVI band derived from Sentinel-2 datasets along with backscatter bands σ^0_{HH}, σ^0_{VH}, and backscatter ratio ($\sigma^0_{VH} / \sigma^0_{VV}$) retrieved from Sentinel-1 datasets. The incorporation of the Sentinel-1 SAR bands along with the Sentinel-2 multispectral bands results in the inclusion of the backscatter information of the SAR, which is highly sensitive to the shape and size structure of the target features. Hence, it can efficiently help in identifying the different agricultural crops (Park et al. 2017). The further development of the modeling approach for the classification of crops is done using this stack of bands.

MODEL DEVELOPMENT FOR CROP CLASSIFICATION

This section describes the modeling approach developed for the classification of different agricultural crops. In this study, rice, sugarcane, and fodder crops are considered for the development of the proposed modeling approach. The developed modeling approach is divided into two parts: a) band-wise estimation of the Separability Index (*SI*) for each crop, and b) the adaptive thresholding for spectral separation of each crop during the large-scale crop classification.

SEPARABILITY INDEX

$$SI_x = \frac{\left| \mu_i - \mu_j \right|}{\sigma_i + \sigma_j} \qquad (2.2)$$

After identifying different crop types using the UAV images, the calculation of the crop-wise separability index is done (Townshend et al. 1991). This is done by extracting the corresponding spectral values for each of the three crops corresponding to Blue, Green, Red, NIR, SWIR1, SWIR2, and NDVI bands (Yadav 2022). The statistical parameters of the spectral values for each class, like standard deviation (σ) and mean (μ), are calculated during the estimation of *SI*. This Separability index measure will help in identifying the best band to separate the individual crop types (Mishra & Singh 2014). This separability index is the normalized distance of the mean and standard deviation between the two classes shown in Equation 2.2 given below:

where μ and σ are the statistical mean and standard deviation of classes i and j, respectively. For a band, if the *SI* value is greater than 1.5, it is suitable for the classification of the two classes. For the classification of three classes, individual calculations of SI_1, SI_2, and SI_3 are done for estimating the band to be utilized for the variation analysis and distinguishing between the crops (Townshend et al. 1991). For individual values with the conditions SI_1, SI_2, and $SI_3 > 1.5$, the bands are selected for the classification of the individual crop classes. Further, in Table 2.1, the band-wise separability index has been calculated for each band for the three crops considered in the study.

OPTICAL AND SAR BAND FUSION

Synthetic aperture radar datasets are generated from the active microwave sensors. Polarimetric electromagnetic waves are transmitted from the SAR sensor toward the target, and the SAR sensor receives the backscattered electromagnetic waves from the targets. These polarimetric electromagnetic waves backscattered from the targets are sensitive toward surface roughness, size, shape, and orientation, shape moisture content, and dielectric properties of the targets (Hedayati & Bargiel 2018). Hence, these backscattered electromagnetic waves characterize the structural properties of various target features and identify different targets in the dataset on the basis of their surface characteristics. This results in significant spatial information. In contrast, optical sensors are passive sensors that receive a portion of solar illumination reflected from the earth's targets (Manakos et al. 2019). These targets are imaged depending on their multispectral reflectivity and spectral properties. The fusion of optical and SAR datasets provides an opportunity to utilize the advantages of both techniques, which could indeed surely be helpful in retrieving the feature information using spectral properties along with the structural information like size, shape, and orientation of the object (Manakos et al. 2019). Based on the separability index, the selected NIR, SWIR-1, and NDVI bands are fused with the SAR backscatter ratio ($\sigma^0_{VH} / \sigma^0_{VV}$) band to increase the efficiency of crop identification. SAR backscatter has sensitivity toward the agricultural crops' structural and dielectric properties (Hedayati & Bargiel 2018; Manakos et al. 2019). Although the SAR backscatters (σ^0_{VV}), (σ^0_{VH}) and ($\sigma^0_{VH} / \sigma^0_{VV}$) do not provide a high separability Index, it is observed that the fusion of backscatter ratio ($\sigma^0_{VH} / \sigma^0_{VV}$) with the optical band's NIR, SWIR-1, and NDVI increases the crop identification efficiency and separability index of the individual bands (Dimov et al. 2017; Adrian et al. 2021).

ADAPTIVE THRESHOLDING USING AN IMPROVED IMAGE SEGMENTATION ALGORITHM BASED ON THE OTSU METHOD

In the process of thresholding, the segmentation of an image is achieved by setting the value of the pixels whose intensity values are more significant than the threshold to a foreground. Various types of traditional and adaptive thresholding approaches are used in the various applications of image segmentation (Roy et al.

TABLE 2.1
Band-Wise Separability Index Table for Crops

$$SI_x = \frac{|\mu_i - \mu_j|}{\sigma_i + \sigma_j}$$

Band Information	Rice		Sugarcane		Fodder		SI_1	SI_2	SI_3
	μ_1	σ_1	μ_2	σ_2	μ_3	σ_3			
Blue	1376.14	41.98	1282.54	21.27	1282.38	27.86	1.479	0.01	1.34
Green	1337.43	58.62	1201.09	28.39	1220.74	35.09	1.566	0.31	1.24
Red	1205.85	90.35	969.08	50.46	1001.69	82.88	1.68	0.24	1.18
NIR	2286.17	117.19	2835.15	126.43	3114.73	226.99	2.25	0.79	2.01
SWIR-1	11.33	0.33	11.70	0.12	12.33	0.15	0.82	2.33	2.08
SWIR-2	1788.55	136.33	1750.00	84.54	2064.436	86.323	0.17	1.84	1.24
NDVI	0.314	0.065	0.473	0.052	0.4905	0.068	1.36	0.15	2.327
VH	0.0255	0.0044	0.0255	0.0044	0.0192	0.0031	0	0.84	0.84
VV	0.0856	0.0142	0.0856	0.0142	0.0944	0.0145	0	0.31	0.30
VH/VV	3.478	0.657	3.4783	0.657	5.042	0.874	0.01	1.02	1.02

2014). Adaptive thresholding techniques dynamically change the image's threshold value, unlike the traditional thresholding techniques, which use a global threshold for all pixels (Wang & Dong 2007). Many approaches can be used in determining an image's adaptive threshold value (Wang & Dong 2007). In the past, a contrast-dependent threshold method-based approach was developed by comparing the local mean and standard deviation of the adjacent pixels within the local frame (Niblack 1985). Another dynamic threshold selection-based adaptive thresholding approach was introduced by Otsu (Otsu 1979). This approach suggested maximizing the weighted total of the preliminary and background pixel interclass variances to determine the best threshold. In this thresholding approach, the image was divided into two classes W_1 and W_2 at gray levels T such that $W_1 = \{0,1,2,...,T\}$ and $W_2 = \{T+1,T+2,...,L-1\}$. Here, the total number of gray levels of the image were represented by L (Otsu 1979). Let the number of pixels at i gray level be n_i, and $N = \sum_{i=0}^{L-1} n_i$ be the total number of pixels in each image. The probability of occurrence of gray level i is defined as given in Equation 2.3:

$$p_i = \frac{n_i}{N}, p_i \geq 0, \sum_{i=0}^{L-1} p_i = 1 - P_{w1}. \tag{2.3}$$

The means of the classes W_1 and W_2 can be computed as shown in Equation 2.4:

$$\mu_{w1} = \sum_{i=0}^{T} \frac{i * p_i}{P_{w1}}, \qquad \mu_{w2} = \sum_{i=T+1}^{L-1} \frac{i * p_i}{P_{w2}} \tag{2.4}$$

So the equivalent formula is given as shown in Equation 2.5:

$$\sigma^2(T) = P_{w1} P_{w2} (\mu_{w1} - \mu_{w2})^2 \tag{2.5}$$

By optimizing the interclass variance, the optimum threshold T^* is reached as given in Equation 2.6:

$$T^* = Arg\,max\,\sigma^2(T) \tag{2.6}$$

The Otsu method is a simple and stable adaptive thresholding method. It is extensively used in image segmentation and plays an important role in automatic threshold selection (Otsu 1979). However, this Otsu thresholding technique is extremely susceptible to the noise and size of the target. After experiments with several images, it succeeds in an image with only one peak variance (Zhan & Zhang 2019). The Otsu method's thresholds appear to be nearer to the class where the gap between the two intraclass variances is high with more significant interclass variance. This means more pixels in this class are categorized into another class. Hence it is important to improve the segmentation results (Wang & Dong 2007). When the distribution is distorted or heavy-tailed for class $C_k (k = 0\,or\,1)$, the medium value is well known to be a very stable measurement value relative to the average (Wang & Dong 2007). It is noticed that a medium-value substitute can achieve an average t^* that is highly correct in comparison to those thresholds chosen by the Otsu algorithm in the presence of a strongly tailed distribution C_k.

Hence, an improved Otsu algorithm is introduced by replacing the total mean μ_t with the total median level m_r of all points in the entire gray level image (Wang & Dong 2007). Similarly, to the whole image mean value μ_t, the mean value μ_0 and μ_1 can also be replaced by the median gray level m_0 and m_1 of the foreground part C_0 and the background part C_1 respectively. The between-class variance σ_B^2 of the two parts C_0 and C_1 can be rewritten as:

$$\sigma_B^2 = \omega_0 \left(m_0 - m_T \right)^2 + \omega_1 \left(m_1 - m_T \right)^2 \tag{2.7}$$

And the threshold t^* is chosen by maximizing σ_B^2

$$t^* = argarg\, \sigma_B^2 \tag{2.8}$$

FLOW DIAGRAM FOR THE PROPOSED MODEL

In Figure 2.4, the flowchart indicating the workflow of the proposed modeling approach has been shown. The multispectral bands from the Sentinel-2 datasets (Blue, Green, Red, NIR, SWIR-1 and SWIR-2), the NDVI, and the backscatter from the polarization bands of Sentinel-1 datasets in HH and VV polarization channels along with the backscatter ratio $(\sigma^0{}_{VH} / \sigma^0{}_{VV})$ are used. Utilizing the separability index, the bands are selected for the classification of the crops. The selected multispectral bands are fused with the Sentinel-1 polarization channels to induce better susceptibility toward agriculture crop classification. Further, the Otsu-based thresholding (OT) is applied in Sentinel-2 multispectral datasets for the classification of the agriculture crops, based on the inputs from the UAV datasets. The calculated values of $OT_1 = 22.96$, $OT_2 = 46.51$, and $OT_3 = 30.23$ for rice, sugarcane, and fodder, respectively.

RESULTS AND DISCUSSION

This section shows the implementation of the proposed approach developed in this study to precisely classify the agricultural crops. The Sentinel-1, Sentinel-2, and UAV-derived datasets are used for the implementation of the proposed methodology given above. The RGB representation and the Normalized Differential Vegetation Index maps derived from UAV and Sentinel-2 multispectral datasets have been shown, and the specific crop parcels related to the different crops considered in this study have been indicated. The retrieved output from the precise classification of the agricultural crops done using the multispectral crop information from the UAV and satellite datasets were also described above.

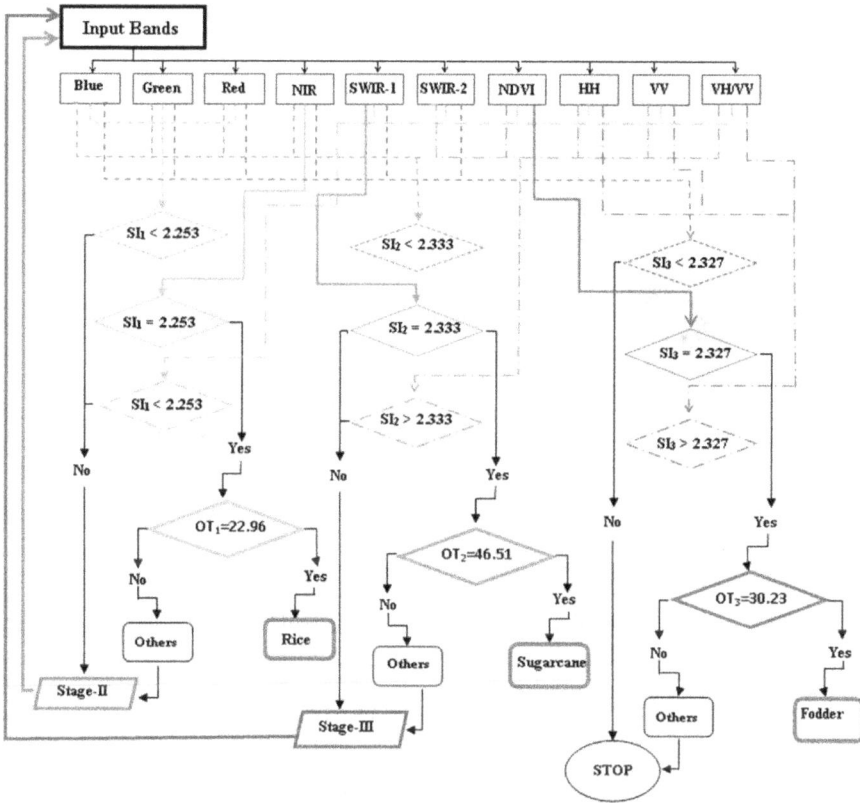

FIGURE 2.4 Flowchart for using the decision tree-based algorithm

RGB REPRESENTATION OF UAV AND SENTINEL-2 DATASETS

In Figure 2.5 (a) and Figure 2.5 (b), the RGB representation of crop parcels using UAV and Sentinel-2-derived datasets has been done, sugarcane, rice, and fodder crop parcels were considered for this study. The total crop area of these three crops was 3.48 ha, 1.78 ha, and 1.28 ha, respectively, with an average parcel area of 0.29 ha, 0.2219 ha, and 0.425 ha, respectively. The corresponding parcels in both datasets are identified together. The utilization of the multi-band information is done for analyzing the spectral variation of the multi-crop pixels. The spectral information derived from these pixels is again used for the classification of the crops. The band-wise analysis of the target pixels using Blue band (B-2), Green band (B-3), Red band (B-4), NIR band (B-8), SWIR-1 band (B-11), SWIR-2 (B-12) bands are used for the spectral analysis and classification of the agricultural crops.

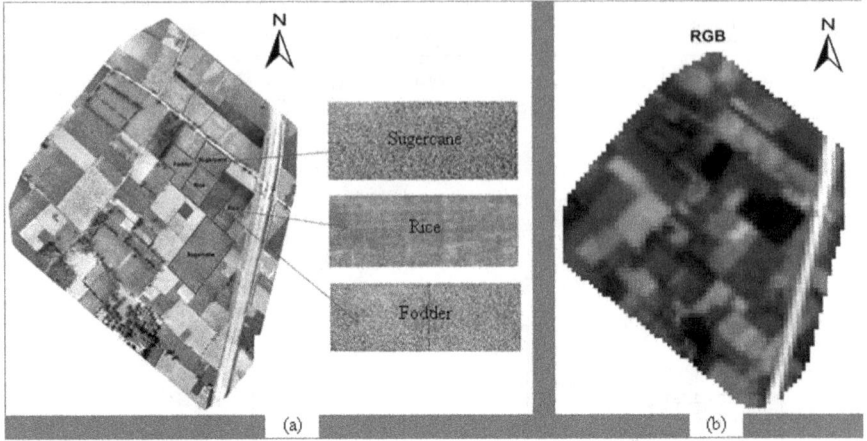

FIGURE 2.5   (a) Crop parcels using UAV datasets, (b) Crop parcels using Sentinel-2 datasets

NDVI MAPS FROM UAV AND SENTINEL-2 DATASETS

The spectral reflectance of the vegetation varies across different bands measured by the sensor. This variation in the spectra is utilized as an indicator of the presence of vegetation. The mathematical combination of two or more spectral bands is used to enhance the contrast between vegetation (having high reflectance) and bare soil. Normalized Difference Vegetation Index quantifies the vegetation targets by measuring the differences between the spectral reflectivity of the near-infrared (NIR) band and the Red band (D'Odorico et al. 2013). Figures 2.6 (a) and 2.6 (b) show the NDVI maps derived from UAV Sentinel-2 datasets, respectively.

These vegetation targets strongly reflect the electromagnetic waves in the near-infrared regions and absorb the electromagnetic waves in the Red band. This spectral reflectance and absorption are because of the chlorophyll pigment present in the vegetation (Gonçalves et al. 2012). The chlorophyll pigment in a healthy plant absorbs most of the visible red light, while the plant's cell structure reflects most of the near-infrared light. The chlorophyll pigment concentration varies in different crops, resulting in the variation of NDVI values across different vegetation.

CRO-BASED PARCEL INFORMATION

In Figure 2.7, the representation of the specific crop parcels has been done. In Figure 2.7 (a), the sugarcane parcels are shown. The parcels show the sugarcane in the matured stage. Figure 2.7 (b) shows the representation of the parcels related to fodder crops. The fodder crops are at different maturity stages in these parcels due to the different sowing dates. In Figure 2.7 (c), the cultivated rice parcels are represented. The rice parcels shown on the map are at the preliminary vegetative stage, germination to panicle initiation. The parcels related to other vegetation and

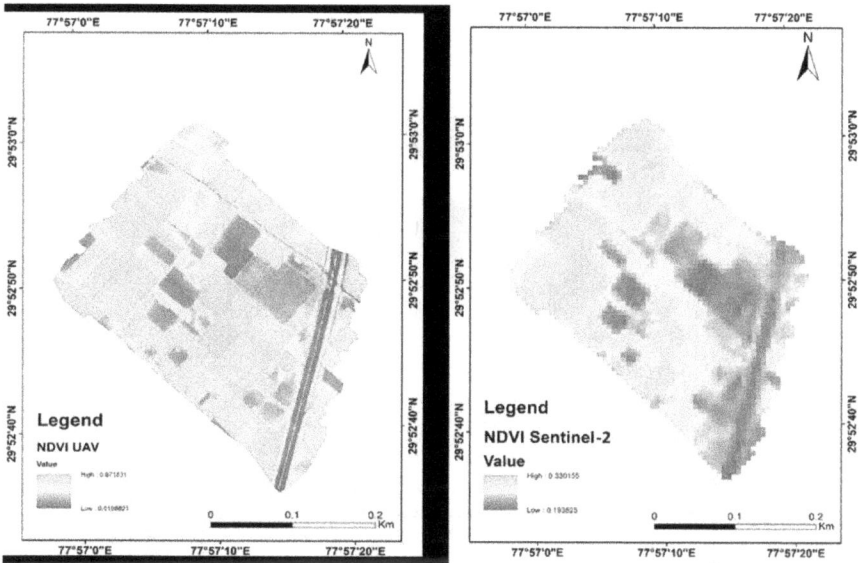

FIGURE 2.6 (a) NDVI maps derived from UAV datasets, (b) NDVI maps derived from Sentinel-2 datasets

the fallow lands have also been specified and labeled as others. The combined representation of all the crops and the classified crop map has been shown in the further section.

CLASSIFIED CROP TYPE MAP

In this section, the output result after implementing the proposed crop classification model is shown. In Figure 2.8 (a), the representation of the labeled parcel-based crop information on the UAV-acquired RGB datasets is done. The three crops considered here in this work are sugarcane, fodder, and rice. The output classified crop type map has been generated and is shown in Figure 2.8 (b). sugarcane is the most dominant crop with maximum crop coverage area pixels, followed by fodder and rice. The sugarcane class is classified as fodder for the few pixels because of somewhat similar properties. The sugarcane crop is toward attaining maturity and has somewhat similar structural properties to fodder. The Sentinel-2 pixels are classified based on the overall dominating spectral mechanism. Hence, the part of specific pixels comes in different classes. An overall high correlation has been found between the labeled crop image and the classified crop type output map in Figure 2.8 (a) and Figure 2.8 (b), respectively. It signifies a very high efficiency of the developed model during the classification of these crops.

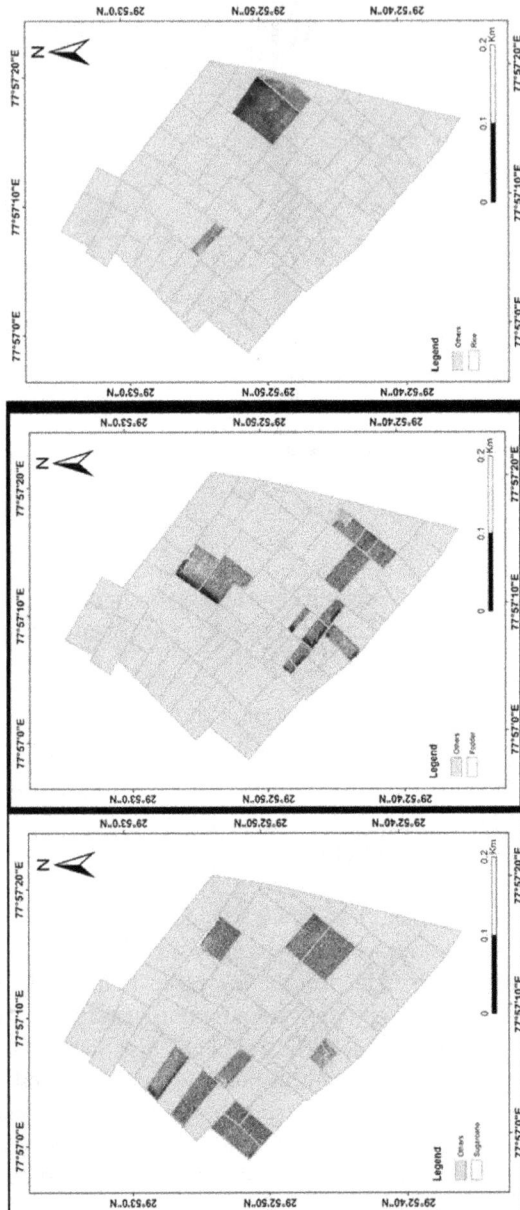

FIGURE 2.7 (a) Sugarcane parcels representation, (b) Fodder parcels representation, (c) Rice parcels representation

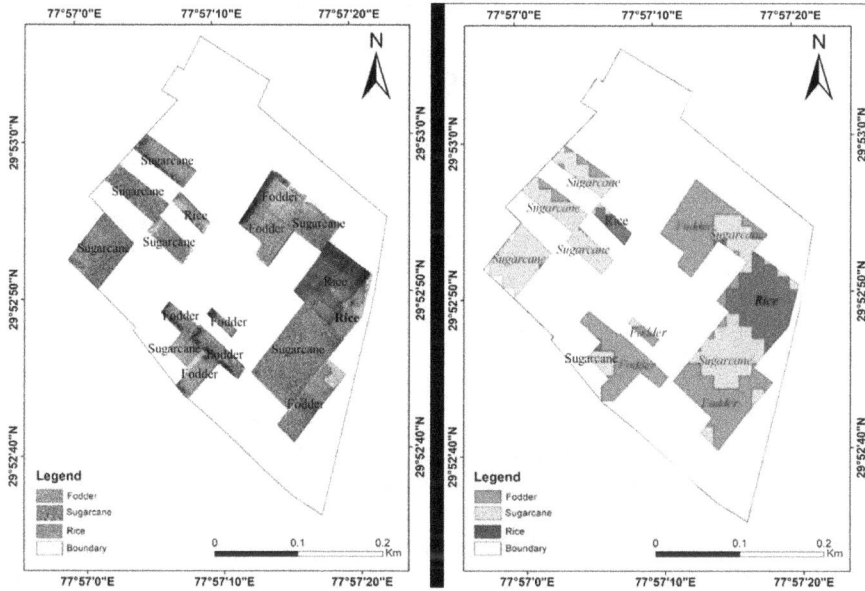

FIGURE 2.8 (a) Labeled crop image from UAV RGB dataset, (b) Classified crop type output map

ACCURACY ASSESSMENT

The efficiency of the proposed crop classification model is tested by pixel-wise accuracy assessment of the classified outputs. Table 2.2 shows the classification accuracy assessment using a confusion matrix and Table 2.3 portrays the user accuracy and producer accuracy for sugarcane, fodder and rice crops using the proposed classification approach. The classified crop maps generated by applying the proposed crop classification model are used as the testing dataset. The reference pixels of sugarcane, fodder, and rice crops are retrieved using the high-resolution UAV datasets. The confusion matrix is generated for the accuracy assessment of the developed model utilizing total classified and total ground truth pixels. The overall classification accuracy of the proposed model is coming as 82.14%, with a Kappa coefficient of 0.72. The proposed model performs well for separating these three crops with high accuracy.

TABLE 2.2
Classification Accuracy Assessment Using a Confusion Matrix

	Sugarcane	Fodder	Rice	Total Classified Pixels
Sugarcane	57	14	2	73
Fodder	9	52	2	63
Rice	2	1	29	32
Total Ground Truth Pixel	68	67	33	168

TABLE 2.3
User Accuracy and Producer Accuracy

	User Accuracy	Producer Accuracy
Sugarcane	57/73 = 90.5%	57/68= 83.9%
Fodder	52/63 = 82.5%	52/67=77.6%
Rice	29/32 = 90.6%	29/33= 87.9%
Overall Accuracy:	128/168= 82.14%	
Kappa Coefficient:	12,943/17,983 = 0.72	

CONCLUSION

This study presents a modeling approach for the agriculture crop type classification using Sentinel-1 and Sentinel-2 along with UAV datasets. The proposed modeling approach will help reduce the repetitive use of UAVs for agriculture monitoring by utilizing the satellite datasets for the classification of agricultural crops aiming toward a large scale at large-scale precision agriculture. The utilization of the UAV-acquired datasets as an add-on to the satellite datasets will provide precise information on the agricultural crops over a large area. The total crop area of these three crops sugarcane, rice, and fodder was 3.48 ha, 1.78 ha, and 1.28 ha, respectively. The proposed approach utilized the multispectral Blue band (B-2), Green band (B-3), Red band (B-4), NIR band (B-8), SWIR-1 band (B-11), SWIR-2 (B-12) bands, and NDVI derived from Sentinel-2 datasets, along with the backscatter information σ^0_{HH}, σ^0_{VH} bands and backscatter ratio ($\sigma^0_{VH} / \sigma^0_{VV}$) were used from the Sentinel-1 datasets. Further, these bands were used to estimate the crop-wise separability index for identifying the best band to separate the individual crop types. During the model development, individual crop information was retrieved using the high-resolution UAV-acquired RGB datasets. The modified Otsu-based adaptive thresholding approach was used to segment various crops based on their spectral and backscatter information. The developed model showed a good performance during the crop classification with an overall accuracy of 79.81 percentage and with the Kappa coefficient of 0.72. The estimated user's accuracy for the sugarcane is 90.5%, fodder is 82.5%, and rice is 90.6%. and the retrieved producer's accuracy for the sugarcane is 83.9%, fodder is 77.6%, and rice is 87.9%. This proposed crop classification model can be generalized for other agricultural crops by adjusting the adaptive thresholding according to the specific crops. Still, the proposed approach requires highly synchronized UAV and satellite data acquisitions. The developed hybrid approach methodology can also be utilized for future satellite missions and is valuable for estimating precise crop information. Hence, it can be effectively utilized in precision agriculture for attaining food security and crop area estimation. In this work, since the UAV imagery was acquired for a single date only, in the future scope, the approach can be further improved by incorporating multi-temporal datasets at different crop stages and corresponding crop growth data at the BBCH scale to identify the phenological development stages of crops. Another improvement in this approach is the UAV datasets acquired over a large region, which can be fused with the satellite datasets for better accuracy and more precision agriculture parameters. The present adaptive thresholding approach is far better than the approaches using absolute thresholding. Still, the study recommends the development of automatic approaches for crop type classification at very high precision using satellite and UAV datasets.

ACKNOWLEDGMENT

The authors of this paper acknowledge ESA (European Space Agency) for providing Sentinel-1 and Sentinel-2 datasets and the SNAP platform for data processing. We also thank Sutapa for providing essential inputs during the development of the methodology in this work. The authors would also like to thank Shivam Madan, and Ajeet Patel for providing assistance during UAV data acquisition.

FUNDING STATEMENT

No funding was used in this study. Also, the authors received no financial support for the research publication of this article.

REFERENCES

Adrian J, Sagan V, Maimaitijiang M. 2021. Sentinel SAR-optical fusion for crop type mapping using deep learning and Google earth engine. *ISPRS J Photogramm Remote Sens* 175:215–235.

Alam Iqbal B. 2018. *Indian Agriculture: Issues and Challenges.* http://www.imedpub.com/.

Awasthi S, Jain K, Bhattacharjee S, Gupta V, Varade D, Singh H, Narayan AB, Buddilon A. 2022. Analyzing urbanization induced groundwater stress and land deformation using time-series Sentinel-1 datasets applying PSInSAR approach. *Sci Total Environ*:157103.

Awasthi S, Jain K, Mishra V, Kumar A. 2020. An approach for multi-dimensional land subsidence velocity estimation using time-series Sentinel-1 SAR datasets by applying persistent scatterer interferometry technique. *Geocarto Int*:1–32. https://www.tandfonline.com/doi/full/10.1080/10106049.2020.1831624.

Awasthi S, Jain K, Pandey A. 2019. Psinsar based land deformation based disaster monitoring using Sentinel-1 datasets. In: *IGARSS 2019–2019 International Geoscience and Remote Sensing Symposium*: IEEE; pp. 1713–1716.

Awasthi S, Kumar S, Thakur PK, Jain K, Kumar A, Snehmani. 2021. Snow depth retrieval in North-Western Himalayan region using pursuit-monostatic TanDEM-X datasets applying polarimetric synthetic aperture radar interferometry based inversion Modelling. *Int J Remote Sens* 42(8):2872–2897. https://www.tandfonline.com/doi/full/10.1080/01431161.2020.1862439.

Awasthi S, Thakur PK, Kumar S, Kumar A, Jain K, Mani S. 2020. Snow density retrieval using Hybrid polarimetric RISAT-1 datasets. *IEEE J Sel Top Appl Earth Obs Remote Sens* 1.

Awasthi S, Varade D, Bhattacharjee S, Singh H, Shahab S, Kamal J. 2022. Assessment of land deformation and the associated causes along a rapidly developing Himalayan foothill region using. *Land* 11:1–22.

Awasthi S, Varade D, Thakur PK, Kumar A, Singh H, Jain K. 2022. Development of a novel approach for snow wetness estimation using hybrid polarimetric RISAT-1 SAR datasets in North-Western Himalayan region. *J Hydrol* 612:128252.

Bawa A, Samanta S, Himanshu SK, Singh J, Kim J, Zhang T., ... Ale S. 2023. A support vector machine and image processing based approach for counting open cotton bolls and estimating lint yield from UAV imagery. *Smart Agric Technol* 3:100140.

Chauhan S, Darvishzadeh R, Boschetti M, Nelson A. 2020. Discriminant analysis for lodging severity classification in wheat using RADARSAT-2 and Sentinel-1 data. *ISPRS J Photogramm Remote Sens.* 164:138–151.

Clevers JGPW, Gitelson AA. 2013. Remote estimation of crop and grass chlorophyll and nitrogen content using red-edge bands on Sentinel-2 and-3. *Int J Appl Earth Obs Geoinf* 23(1):344–351.

Colomina I, Molina P. 2014. Unmanned aerial systems for photogrammetry and remote sensing: A review. *ISPRS J Photogramm Remote Sens* 92:79–97. http://doi.org/10.1016/j.isprsjprs.2014.02.013.

D'Odorico P, Gonsamo A, Damm A, Schaepman ME. 2013. Experimental evaluation of Sentinel-2 spectral response functions for NDVI time-series continuity. *IEEE Trans Geosci Remote Sensing* 51(3):1336–1348.

Daponte P, De Vito L, Glielmo L, Iannelli L, Liuzza D, Picariello F, Silano G. 2019. A review on the use of drones for precision agriculture. In: *IOP Conference Series: Earth and Environmental Science*, Vol. 275. IOP Publishing; p. 12022.

De Bernardis CG, Vicente-Guijalba F, Martinez-Marin T, Lopez-Sanchez JM. 2015. Estimation of key dates and stages in rice crops using dual-polarization SAR time series and a particle filtering approach. *IEEE J Sel Top Appl Earth Obs Remote Sens* 8(3):1008–1018.

Dedeoğlu M, Başayiğit L, Yüksel M, Kaya F. 2020. Assessment of the vegetation indices on Sentinel-2A images for predicting the soil productivity potential in Bursa, Turkey. *Environ Monit Assess* 192(1):1–16. https://doi.org/10.1007/s10661-019-7989-8.

Defourny P, Bontemps S, Bellemans N, Cara C, Dedieu G, Guzzonato E, Hagolle O, Inglada J, Nicola L, Rabaute T, et al. 2019. Near real-time agriculture monitoring at national scale at parcel resolution: Performance assessment of the Sen2-Agri automated system in various cropping systems around the world. *Remote Sens Environ* 221:551–568.

Delegido J, Verrelst J, Alonso L, Moreno J. 2011. Evaluation of Sentinel-2 red-edge bands for empirical estimation of green LAI and chlorophyll content. *Sensors* 11(7):7063–7081.

Dhawan V. 2017. Water and agriculture in India. In: *Background Paper for the South Asia Expert Panel During the Global Forum for Food and Agriculture*, Vol. 28. [place unknown].

Dimov D, Löw F, Ibrakhimov M, Stulina G, Conrad C. 2017. SAR and optical time series for crop classification. In: *IEEE International Geoscience and Remote Sensing Symposium* [place unknown]. IEEE; pp. 811–814.

Drusch M, Del Bello U, Carlier S, Colin O, Fernandez V, Gascon F, Hoersch B, Isola C, Laberinti P, Martimort P, et al. 2012. Sentinel-2: ESA's optical high-resolution mission for GMES operational services. *Remote Sens Environ* 120:25–36.

Frampton WJ, Dash J, Watmough G, Milton EJ. 2013. Evaluating the capabilities of Sentinel-2 for quantitative estimation of biophysical variables in vegetation. *ISPRS J Photogramm Remote Sens* 82:83–92.

Gago J, Douthe C, Florez-Sarasa I, Escalona JM, Galmes J, Fernie AR, Flexas J, Medrano H. 2014. Opportunities for improving leaf water use efficiency under climate change conditions. *Plant Sci* 226:108–119.

Gonçalves RRV, Zullo J, Romani LAS, Nascimento CR, Traina AJM. 2012. Analysis of NDVI time series using cross-correlation and forecasting methods for monitoring sugarcane

fields in Brazil. *Int J Remote Sens* 33(15):4653–4672. https://www.tandfonline.com/doi
/full/10.1080/01431161.2011.638334.

Hedayati P, Bargiel D. 2018. Fusion of Sentinel-1 and Sentinel-2 images for classification of agricultural areas using a novel classification approach. In: *International Geoscience and Remote Sensing Symposium*, Vol. 2018-July [place unknown]. Institute of Electrical and Electronics Engineers Inc.; pp. 6643–6646.

Heupel K, Spengler D, Itzerott S. 2018. A progressive crop-type classification using multitem-poral remote sensing data and phenological information. *PFG J Photogramm Remote Sens Geoinf Sci* 86286(2):53–69. https://link.springer.com/article/10.1007/s41064-018 -0050-7.

Hong G, Abd El-Hamid HT. 2020. Hyperspectral imaging using multivariate analysis for simulation and prediction of agricultural crops in Ningxia, China. *Comput Electron Agric* 172:105355.

Husak GJ, Marshall MT, Michaelsen J, Pedreros D, Funk C, Galu G. 2008. Crop area estimation using high and medium resolution satellite imagery in areas with complex topography. *J Geophys Res* 113(D14):D14112. http://doi.wiley.com/10.1029 /2007JD009175.

Jain K, Pandey A. 2020. Calibration of satellite imagery with multispectral UAV imagery. *J Indian Soc Remote Sens*:1–12.

Kobayashi N, Tani H, Wang X, Sonobe R. 2020. Crop classification using spectral indices derived from Sentinel-2A imagery. *J Inf Telecommun* 4(1):67–90.

Kolecka N, Ginzler C, Pazur R, Price B, Verburg PH. 2018. Regional scale mapping of grass-land mowing frequency with Sentinel-2 time series. *Remote Sens* 10(8):1221.

Lottes P, Khanna R, Pfeifer J, Siegwart R, Stachniss C. 2017. UAV-based crop and weed clas-sification for smart farming. In: *2017 IEEE International Conference on Robotics and Automation*. IEEE; pp. 3024–3031.

Maha MM, Bhuiyan S, Masuduzzaman M. 2019. Smart board for precision farming using wireless sensor network. In: *International Conference on Robotics, Electrical and Signal Processing Techniques*. Institute of Electrical and Electronics Engineers Inc.; pp. 445–450.

Manakos I, Kordelas GA, Marini K. 2019. Fusion of Sentinel-1 data with Sentinel-2 products to overcome non-favourable atmospheric conditions for the delineation of inundation maps. *Eur J Remote Sens*:1–14.

Mascolo L, Lopez-Sanchez JM, Vicente-Guijalba F, Nunziata F, Migliaccio M, Mazzarella G. 2016. A complete procedure for crop phenology estimation with PolSAR data based on the complex Wishart classifier. *IEEE Trans Geosci Remote Sensing* 54(11):6505–6515.

Mishra P, Singh D. 2014. A statistical-measure-based adaptive land cover classification algo-rithm by efficient utilization of Polarimetric SAR observables. *IEEE Trans Geosci Remote Sensing* 52(5):2889–2900.

Mitchell JJ, Glenn NF, Anderson MO, Hruska RC, Charlie AH. 2012. Unmanned Aerial Vehicle (UAV) hyperspectral remote sensing for dryland vegetation monitoring hyper-spectral image and signal sensing. *4th Work Hyperspectral Image Signal Process.*

Momm HG, ElKadiri R, Porter W. 2020. Crop-type classification for long-term modeling: An integrated remote sensing and machine learning approach. *Remote Sens* 12(3):449.

Moumni A, Lahrouni A. 2021. Machine learning-based classification for crop-type map-ping using the fusion of high-resolution satellite imagery in a semiarid area. *Scientifica (Cairo)* 2021.

Niblack W. 1985. *An Introduction to Digital Image Processing*. Strandberg. https://books .google.co.in/books?id=Lcg8PgAACAAJ.

Orynbaikyzy A, Gessner U, Conrad C. 2019. Crop type classification using a combination of optical and radar remote sensing data: A review. *Int J Remote Sens* 40(17):6553–6595. https://www.tandfonline.com/doi/full/10.1080/01431161.2019.1569791.

Orynbaikyzy A, Gessner U, Mack B, Conrad C. 2020. Crop type classification using fusion of Sentinel-1 and Sentinel-2 data: Assessing the impact of feature selection, optical data availability, and parcel sizes on the accuracies. *Remote Sens* 12(17):2779.

Otsu N. 1979. A Threshold selection method from gray-level histograms. *IEEE Trans. Syst Man Cybern. SMC* 9(1):62–66.

Palanisami K, Kakumanu KR, Nagothu US, Ranganathan CR. 2019. *Climate Change and Future Rice Production in India: A Cross Country Study of Major Rice Growing States of India*. https://doi.org/10.1007/978-981-13-8363-2.

Pandey A, Jain K. 2022. An intelligent system for crop identification and classification from UAV images using conjugated dense convolutional neural network. *Comput Electron Agric* 192:106543.

Pandey A Jain K. 2022a. A robust deep attention dense convolutional neural network for plant leaf disease identification and classification from smart phone captured real world images. *Ecol Inform*. 70(June). https://doi.org/10.1016/j.ecoinf.2022.101725

Pandey A, Jain K. 2022b. Plant leaf disease classification using deep attention residual network optimized by opposition-based symbiotic organisms search algorithm. *Neural Comput Appl* [Internet]. 34(23):21049–21066. https://doi.org/10.1007/s00521-022 -07587-6

Park H, Choi J, Park N, Choi S. 2017. Sharpening the VNIR and SWIR bands of Sentinel-2A imagery through modified selected and synthesized band schemes. *Remote Sens* 9(10):1080.

Park JK, Jonghwa, P. 2015 *Crop Classification Using Imagery of Drone*:91–94. https://www .atlantis-press.com/proceedings/eers-15/25839154.

Phung H-P, Nguyen L-D, Thong N-H, Thuy L-T, Apan AA. 2020. Monitoring rice growth status in the Mekong Delta, Vietnam using multitemporal Sentinel-1 data. *J Appl Remote Sens* 14(01):1.

PIB Delhi Ministry of Finance. 2021, 29 January. Recent agricultural reforms a remedy, not a Malady says. *Econ Surv*. https://pib.gov.in/PressReleasePage.aspx?PRID=1693205.

Piedelobo L, Hernández-López D, Ballesteros R, Chakhar A, Del Pozo S, González-Aguilera D, Moreno MA. 2019. Scalable pixel-based crop classification combining Sentinel-2 and Landsat-8 data time series: Case study of the Duero river basin. *Agric Syst* 171:36–50.

Praseartkul P, Taota K, Pipatsitee P, Tisarum R, Sakulleerungroj K, Sotesaritkul T., … Cha-um S. 2022. Unmanned aerial vehicle-based vegetation monitoring of aboveground and belowground traits of the turmeric plant (Curcuma longa L.). *Int J Environ Sci Technol*:1–14.

Primicerio J, Di Gennaro SF, Fiorillo E, Genesio L, Lugato E, Matese A, Vaccari FP. 2012. A flexible unmanned aerial vehicle for precision agriculture. *Precis Agric* 13(4):517–523.

Puri V, Nayyar A, Raja L. 2017. Agriculture drones: A modern breakthrough in precision agriculture. *J Stat Manag Syst* 20(4):507–518.

Radoglou-Grammatikis P, Sarigiannidis P, Lagkas T, Moscholios I. 2020. A compilation of UAV applications for precision agriculture. *Comput Netw* 172:107148.

Roy P, Dutta S, Dey N, Dey G, Chakraborty S, Ray R. 2014. Adaptive thresholding: A comparative study. In: *International Conference on Control, Instrumentation, Communication and Computational Technologies*. Institute of Electrical and Electronics Engineers Inc.; pp. 1182–1186.

Seelan SK, Laguette S, Casady GM, Seielstad GA. 2003. Remote sensing applications for precision agriculture: A learning community approach. *Remote Sens Environ* 88(1–2):157–169.

Sharma BR, Rao KV, Vittal KPR, Ramakrishna YS, Amarasinghe U. 2010. Estimating the potential of rainfed agriculture in India: Prospects for water productivity improvements. *Agric Water Manag* 97(1):23–30.

Sonobe R, Yamaya Y, Tani H, Wang X, Kobayashi N, Mochizuki K. 2018. Crop classification from Sentinel-2-derived vegetation indices using ensemble learning. *J Appl Remote Sens* 12(02):1.

Stafford JV. 2000. Implementing precision agriculture in the 21st century. *J Agric Eng Res* 76(3):267–275.

Sun C, Li J, Cao L, Liu Y, Jin S, Zhao B. 2020. Evaluation of vegetation index-based curve fitting models for accurate classification of salt marsh vegetation using Sentinel-2 time-series. *Sensors* 20(19):5551.

Théau J, Gavelle E, Ménard P. 2020. Crop scouting using UAV imagery: A case study for potatoes. *J Unmanned Veh Syst* 8(2):99–118.

Tiwari A, Jaga PK. 2012. Precision farming in India – A review. *Outlook Agric* 41(2):139–143. http://journals.sagepub.com/doi/10.5367/oa.2012.0082.

Tiwari A, Suresh M, Jain K. Dynamics for Quantifying the Changing Pattern of. 37(2): 399–411.

Townshend J, Justice C, Li W, Gurney C, McManus J. 1991. Global land cover classification by remote sensing: Present capabilities and future possibilities. *Remote Sens Environ* 35(2–3):243–255.

Tsouros DC, Bibi S, Sarigiannidis PG. 2019. A review on UAV-based applications for precision agriculture. *Information* 10(11):349. https://www.mdpi.com/2078-2489/10/11/349.

Ustuner M, Sanli FB, Abdikan S, Esetlili MT, Kurucu Y. 2014. Crop type classification using vegetation indices of rapideye imagery. In: *International Archives of the Photogrammetry, Remote Sensing and Spatial Information Sciences*, Vol. 40. International Society for Photogrammetry and Remote Sensing; pp. 195–198.

Varade DM, Maurya AK, Dikshit O. 2019. Development of spectral indexes in hyperspectral imagery for land cover assessment. *IETE Tech Rev* 36(5):475–483. https://www.tandfonline.com/doi/full/10.1080/02564602.2018.1503569.

Vicente-Guijalba F, Martinez-Marin T, Lopez-Sanchez JM. 2014. Crop phenology estimation using a multitemporal model and a kalman filtering strategy. *IEEE Geosci Remote Sens Lett* 11(6):1081–1085.

Waldner F, Lambert MJ, Li W, Weiss M, Demarez V, Morin D, Marais-Sicre C, Hagolle O, Baret F, Defourny P. 2015. Land cover and crop type classification along the season based on biophysical variables retrieved from multi-sensor high-resolution time series. *Remote Sens* 7(8):10400–10424. http://www.mdpi.com/2072-4292/7/8/10400.

Wang F, Wang F, Zhang Y, Hu J, Huang J, Xie J. 2019. Rice yield estimation using parcel-level relative spectral variables from UAV-based hyperspectral imagery. *Front Plant Sci* 10:453. https://www.frontiersin.org/article/10.3389/fpls.2019.00453/full.

Wang H, Dong Y. 2007. An improved image segmentation algorithm based on OTSU method. In: Zhou L, editor. *International Symposium on Photo-electronic Detection and Imaging 2007 Related Technologies Applications*, Vol. 6625; p. 66250I. http://proceedings.spiedigitallibrary.org/proceeding.aspx?doi=10.1117/12.790781.

Yadav A. 2022. RESEARCH ARTICLE A Novel Change Point Detection Approach for Analysis of Time- Ordered Satellite Imagery. 5.

Yadav A, Jain K, Pandey A, Ranyal E, Majumdar J. 2022. An Optimal RetinaNet Model For Automatic Satellite Image-Based Missile Site Detection. *Def Sci J.* 72(5):753–761. https://doi.org/10.14429/dsj.72.18215

Zhan Y, Zhang G. 2019. An improved OTSU algorithm using histogram accumulation moment for ore segmentation. *Symmetry (Basel)* 11(3):431. https://www.mdpi.com/2073-8994/11/3/431.

3 IoT and Smart Sensor Applications in Nutrient and Irrigation Water Management

Ranjan Kumar, Anil Kumar, and Sushil Kumar Himanshu

INTRODUCTION

Agriculture is an important sector that forms the basis for a country to build a strong economy. As the global population continues to grow, so does the requirement for food production. World cereal production needs to increase by 3 billion tons per year to meet the needs of the growing global population by 2050 (Trendov et al., 2019). Crop production and productivity must be increased, and advanced technologies must be adopted in agriculture (Ayoub Shaikh et al., 2022). To meet challenges arising from population growth, climate change, resource depletion, and pollution (Gupta, 2020a,b; Mahajan et al., 2022), farmers and agribusinesses are turning to new technologies, such as the IoT and smart sensors, to improve efficiency, reduce costs, and increase yields (Farooq et al., 2019; Kulkarni et al., 2020; Parmar & Kumar, 2022; Said Mohamed et al., 2021; Wolfert et al., 2017; Xu et al., 2022). The IoT and smart sensors have the potential to revolutionize nutrient and irrigation water management in agriculture by providing real-time data on soil moisture, temperature, pH levels, nutrient levels, and weather parameters (Said Mohamed et al., 2021; Wolfert et al., 2017). This data can be used to inform irrigation and fertilization decisions, enabling more precise and efficient practices. In addition, IoT and smart sensors can be used to monitor and control irrigation systems and equipment, improving efficiency and reducing the need for manual labor (Krishnan & Swarna, 2020; Kulkarni et al., 2020; Pavón-Pulido et al., 2017).

However, adopting the IoT and smart sensors in agriculture is challenging (Centenaro et al., 2021). Various sensors and IoT platforms are available on the market, and selecting the right one for a specific application can be a complex task. In addition, there are concerns about the cost, reliability, and maintenance of these technologies, and the need for robust cybersecurity measures to protect against data breaches and cyber-attacks. Furthermore, the IoT and smart sensors can provide

DOI: 10.1201/9781003441175-3

real-time information to farmers, allowing them to make more informed harvesting decisions. Moreover, farmers can now monitor their crops with the help of IoT sensors, which eliminate the need to rely on un-scientific forecasts. IoT sensors are not just valuable for agriculture; they can be used in various other sectors, saving workers from having to spend their days diligently monitoring data such as temperature and humidity levels. The information gathered by IoT sensors has the potential to present a picture of what is happening in the field in real time. As a result, farmers will have a better idea of irrigation scheduling, state of soil nutrient status, requirements of inputs such as fertilizers, pesticides, and herbicides, and harvesting-related information.

This chapter aims to provide an overview of the current state of the IoT and smart sensor applications for nutrient and irrigation water management in agriculture. It discusses various types of available sensors and IoT platforms and the challenges and opportunities of these technologies. It also outlines some of the key benefits and limitations of using IoT and smart sensors in agriculture and suggests future research and development directions. The chapter provides an in-depth understanding of the IoT and smart sensor applications in the field of nutrient and irrigation management as well as how these approaches contribute to fulfilling the Sustainable Development Goals (SDGs). The chapter employs a qualitative method to provide information on these crucial topics from the fields of agricultural sciences, information technology, computer sciences, and related fields. The information is documented through an exhaustive review of related journal articles, reports, and chapters through a snowballing process.

THE INTERNET OF THINGS

The IoT is an innovative field of study and development that has the potential to provide reliable and effective solutions for various industries (Burton et al., 2018; Muangprathub et al., 2019; Xu et al., 2022a). To reduce the amount of work required on the part of humans, IoT-based solutions are being created to automate the maintenance and monitoring of farms (Farooq et al., 2019). The IoT allows various devices and sensors to transmit data through the internet in a manner that closely resembles real-time processes. Information technology systems can analyze this real-time data and promptly notify users to take corrective action when abnormal conditions are detected, reducing the need for constant human monitoring (Burton et al., 2018; Muangprathub et al., 2019a).

In agriculture, the IoT refers to a network that connects various physical elements in the agricultural system, such as animals, plants, environmental factors, production equipment, and virtual elements, to the internet using specialized agricultural information gathering equipment and predetermined protocols. This enables information exchange and communication (Xu et al., 2022a). Furthermore, the IoT has ushered in a revolutionary change in the agricultural landscape by analyzing many of the complexities and difficulties inherent to farming. The advent of the IoT in agriculture has resulted in some novel developments in the sector. Not only it boosts agricultural productivity, but it also has the potential to successfully improve the quality of farm produce, lower the expenses of labor, and increase farmers' incomes (Xu et al., 2022a). Additionally, it can genuinely achieve agricultural modernity and intelligence. Controlling the parameters that can be coordinated by humans, such as the soil pH, temperature, humidity, soil moisture, and rate of application of nutrients, can be accomplished with the help of the IoT, which enables the production of the highest-quality produce possible (Farooq et al., 2019; Pang et al., 2015).

Furthermore, efficient freshwater management, including precision irrigation in agriculture, is necessary to optimize crop productivity while minimizing costs and maintaining environmental sustainability (Kamienski et al., 2019; Himanshu et al., 2021; Kumar Roy et al., 2021; Fan et al., 2022). Using the IoT is a no-brainer for the next generation of smart water management and irrigation management systems. The widespread adoption of the IoT for precision irrigation has some obstacles. First, there is still a lack of fully automated software for IoT-based smart applications like agricultural irrigation (Kamienski et al., 2017). Additionally, the lack of advanced IoT software platforms hinders the automation of certain processes and the integration of various technologies, such as the IoT, big data analytics, cloud computing, and fog computing, which are necessary for implementing pilot smart water management applications. There is also a need for appropriate standards and information models to support the integration of diverse, advanced sensors.

SMART FARMING

"Smart farming" refers to a management philosophy that centers on equipping farms with the tools to take advantage of technological advances in the food production sector. Technology like big data, the cloud, and the IoT can be used to monitor and analyze farming processes in real time and with greater precision. Smart farming, often called precision agriculture, relies on computer algorithms and sensor data to optimize yields. Moreover, due to the growing global population, desire for increased agricultural yields, need to manage natural resources efficiently, use and sophistication of information and communication technology and need for climate-sensitive agriculture, smart farming is becoming more important. The integration of these tools paves the way for information gathered via machine-to-machine (M2M) interactions. This data is fed into a decision support system to provide farmers with a more granular level of insight into what is occurring than was previously possible. The tools that may be incorporated into "smart farming" are illustrated in Figure 3.1.

SENSORS

This section provides a detailed discussion of what IoT sensors are and how they function. As a result of the proliferation of IoT technology in agricultural settings, which was spurred on by advances in digital technology, sensing technology, and internet connectivity, sensors built with newly developed technologies are constantly evolving and progressing in the direction of becoming integrated, intelligent, embedded, and miniaturized. Currently, the United States, Japan, and Germany are the major countries in sensor technology and manufacturing processes, and they

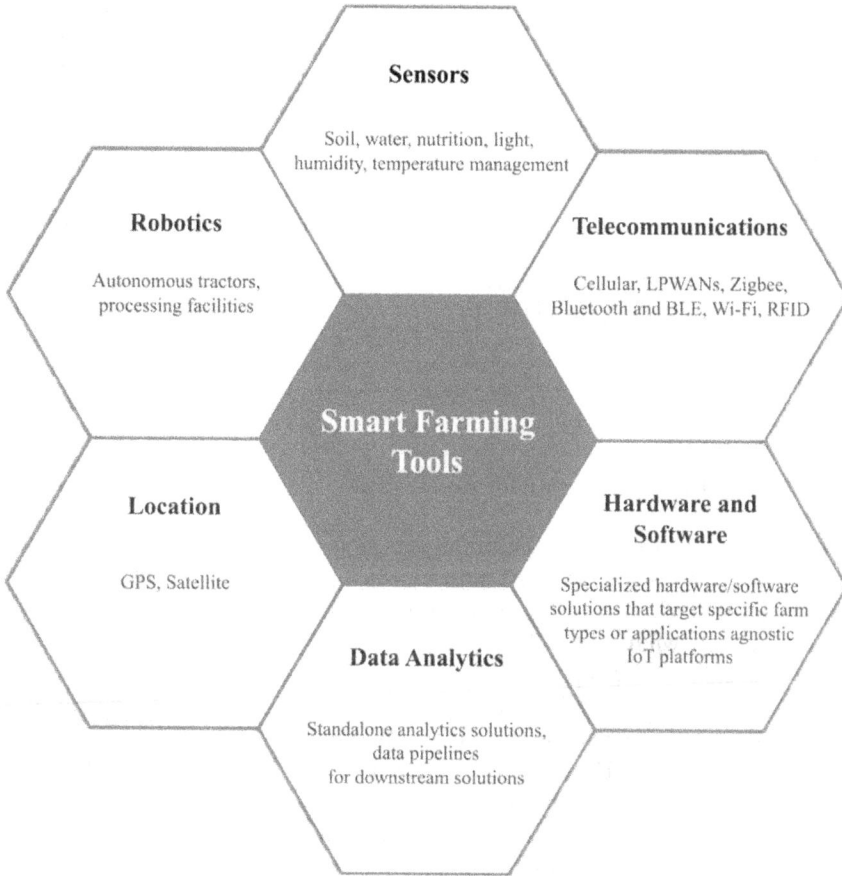

FIGURE 3.1 Smart farming tools

have a strong position worldwide (Xu et al., 2022a). More and more types of sensors, such as those that monitor soil moisture, nutrients, weather, water, air humidity, air temperature and plants, are being developed for agriculture. The data gathered from these object-detection sensors is a tremendous improvement in agricultural output.

In the existing literature, most of the attention paid by researchers is concentrated on water-saving techniques for water management in agricultural fields. This is accomplished with the assistance of an automated irrigation system that makes use of Wireless Sensor Networks (WSNs), in which the irrigation condition is a predetermined static value for the total timespan of the crop (Kumar Roy et al., 2021). WSNs installed in the field have increased farmers' productivity and efficiency in this type of agriculture optimization (Capello et al., 2016; Kodali et al., 2014; Parmar & Kumar, 2022). The IoT can be used to analyze field variables such as soil conditions, atmospheric conditions, and plant or animal biomass (Pang et al., 2015). In addition, WSN can be used to keep tabs on and adjust variables that impact agricultural

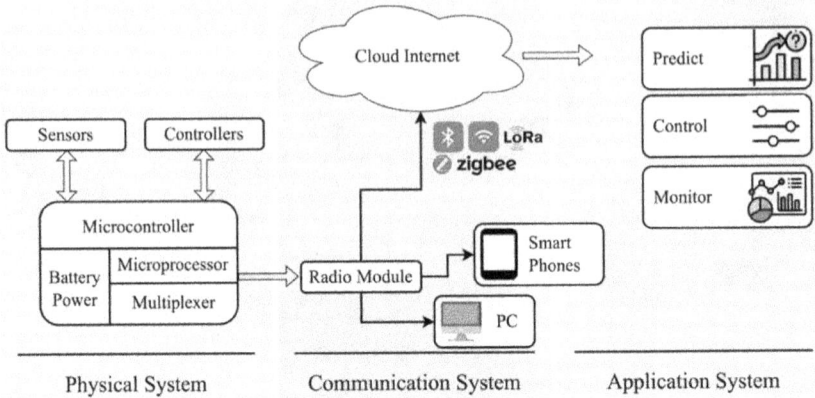

FIGURE 3.2   Diagram of a typical IoT sensor system

output. They can also be utilized in the detection of diseases, the management of equipment, and the selection of the most suitable farmers for a certain set of circumstances (Ndzi et al., 2014). A diagram of a typical IoT sensor system for real-time monitoring is shown in Figure 3.2.

Researchers have attempted to develop in-situ soil nutrient sensors for real-time monitoring using optical and electrochemical methods (Laskar & Mukherjee, 2016; Viscarra Rossel & Bouma, 2016). It is worth noting that only some of these sensor systems are currently available for purchase. As can be seen in Table 3.1, the research conducted thus far has uncovered a wide variety of soil nutrient sensors, some of which are already in the market while others are still in the research and development phase. These sensors are versatile enough to be used in map-based or

TABLE 3.1

Types of Soil Nutrient Sensors

Sensor concept	Status of development	Current results	References
Vis-NIR	Laboratory/field	Soil pH and nutrients	(Christy et al., 2003; Viscarra Rossel et al., 2006)
Vis-MIR	Laboratory	Soil mineral nitrogen	(Ehsani et al., 1999, 2000)
ATR Spec	Laboratory/field	Soil nutrients	(Christy et al., 2003)
Raman	Laboratory/field	Soil nutrients	(Jahn et al., 2006)
ISE	Laboratory/field	Soil pH and nutrients	(Tyszczuk-Rotko & Jędruchniewicz, 2019; van Staden et al., 2018)
ISFET	Laboratory/field	Soil pH and nutrients	(Adamchuk et al., 2005; Sudduth et al., 1997)

real-time systems. In-situ soil sensing typically uses optical and/or electrochemical technologies out of all the available approaches.

TELECOMMUNICATIONS

Telecommunications are essential for the IoT because they facilitate the exchange of data and information between devices and systems, enabling the automation and optimization of various agricultural processes. IoT agricultural networks are made up of both long-range and short-range networks to facilitate communication. These networks use IoT technologies to create sensors and devices for monitoring crops and fields. Protocols for conveying information are the foundation of IoT agricultural networks and applications. They are accustomed to trading all farm-related data or information across the web (Al-Sarawi et al., 2017a; Navulur & Prasad, 2017). Telecommunication is used to connect sensors, actuators, and other IoT devices that are deployed in fields and greenhouses. These devices can collect and transmit data on factors such as soil moisture, temperature, pH levels, and crop health, which can then be analyzed and used to inform irrigation, fertilization, and pest management decisions. In addition, telecommunications can be used to connect farm equipment such as tractors, combines, and irrigation systems, allowing them to be operated and monitored remotely. This can improve efficiency and reduce the need for manual labor. Overall, telecommunications are a key enabler of IoT in agriculture, as they provide the necessary connectivity and communication infrastructure for exchanging data and information between devices and systems.

HARDWARE AND SOFTWARE

Many different hardware and software components can be used in IoT systems in agriculture. Some examples include:

Hardware:

- Sensors: These are devices that can be used to measure various environmental and soil conditions, such as temperature, moisture, pH levels, and nutrient levels.
- Actuators: These are devices that can be used to control or manipulate physical systems, such as pumps, valves, and motors (Farooq et al., 2019).
- Gateway devices: These are devices that act as a bridge between the sensor and actuator network and a larger communication network, such as the internet.

Software:

- IoT platforms: These are software platforms that provide the necessary infrastructure for collecting, storing, and analyzing data from IoT devices (Botta et al., 2016).

- Data analytics tools: These are software tools that can be used to extract insights from the data collected by IoT devices (Pavón-Pulido et al., 2017).
- Mobile apps: These are software applications that can be used to monitor and control IoT devices remotely using a smartphone or tablet.

Overall, the hardware and software components used in IoT systems in agriculture are designed to enable the automation and optimization of various agricultural processes, such as irrigation, fertilization, and pest management.

DATA ANALYTICS

IoT data analytics in agriculture refers to using data analytics tools and techniques to extract insights from the data collected by IoT devices in the agriculture sector. This data is used to inform a wide range of decisions and actions, such as optimizing irrigation and fertilization schedules, detecting crop pests and diseases early, and improving the efficiency of farm equipment (Alahi et al., 2017; Balaji et al., 2019). Overall, IoT data analytics in agriculture aims to help farmers and agribusinesses make more informed, data-driven decisions that can improve efficiency. Many different types of data analytics tools and techniques can be used in agriculture, depending on the specific needs and goals of the operation. Some examples include:

- Statistical analysis: This involves using statistical techniques to identify patterns and trends in the data.
- Machine learning: This involves using algorithms to automatically learn and make predictions based on the data.
- Data visualization: This involves using charts, graphs, and other visual aids to help make sense of the data and identify key trends and patterns.

LOCATION

The use of satellite technology can help to improve the accuracy and timeliness of data collection in agriculture, and enable farmers and agribusinesses to make more informed, data-driven decisions that can improve efficiency, reduce costs, and increase yields. Satellite technology can be used in various ways to support agricultural IoT applications. Some examples include:

- Data collection: Satellites can be used to collect data on various environmental and soil conditions, such as temperature, moisture, pH levels, and nutrient levels. This data can be used to inform irrigation, fertilization, and pest management decisions (Farooq et al., 2019).
- Remote sensing: Satellites can be equipped with sensors that can detect and measure various factors relevant to agriculture, such as crop health, soil moisture levels, and pest infestations (Pallavi et al., 2017).
- Communication: Satellites can provide connectivity and communication infrastructure for IoT devices and systems deployed in remote or hard-to-reach areas (Centenaro et al., 2021).

Robotics

IoT robotics in agriculture refers to using robotic systems connected to the internet and equipped with sensors, actuators, and other IoT devices. These systems can be used to automate and optimize various agricultural processes, such as planting, watering, fertilizing, and harvesting (Krishnan & Swarna, 2020; Kulkarni et al., 2020). Moreover, using IoT robotics in agriculture can help to improve efficiency, reduce the need for manual labor, and increase yields. It can also help to reduce the environmental impact of farming by enabling more precise application of fertilizers and pesticides. Many different types of IoT robotics systems can be used in agriculture, depending on the specific needs and goals of the operation. Some examples include:

- Autonomous tractors: Tractors that are equipped with sensors and navigation systems can be programmed to perform tasks such as plowing, planting, and fertilizing.
- Drones: These are unmanned aerial vehicles that can be equipped with sensors and cameras, and can be used to survey crops and fields, and to apply pesticides and fertilizers.
- Harvesting robots: These are robots that are designed to pick and harvest crops, such as fruits and vegetables.

SMART FARMING USING IOT

The ability to extract information from sensors and transmit it through the internet is the foundation of the IoT (Al-Sarawi et al., 2017b). IoT devices deployed on a farm repeatedly gather and analyze data to optimize the efficiency of agricultural processes. This allows farmers to respond more rapidly to developing problems and shifts in the surrounding environment. High precision and continuous control are the outcomes of this automated smart farming process. These outcomes eventually lead to significant cost reductions across the board for all key resources used, including water, energy, fertilizers, and the amount of time spent by strategic people as well as less qualified human resources (Said Mohamed et al., 2021; Wolfert et al., 2017). Figure 3.3 shows the key steps in smart farming.

APPLICATION OF IOT IN NUTRIENT AND IRRIGATION WATER MANAGEMENT

Decision-making in modern agricultural practices has become easier with data available from different sensors and related processes that aid in better operational performance and resource optimization (Obaideen et al., 2022). The IoT is emerging as a promising field in different domains of agriculture that provide support for effective decision-making (Gandhi et al., 2020; Kundalia et al., 2020; Sharma et al., 2022). The solution provided by the IoT provides better crop yield with fewer inputs because of higher system efficiency. The crop yield is better in terms of productivity and quality (Hemalatha & Sujatha, 2015). The IoT in agriculture is used in different

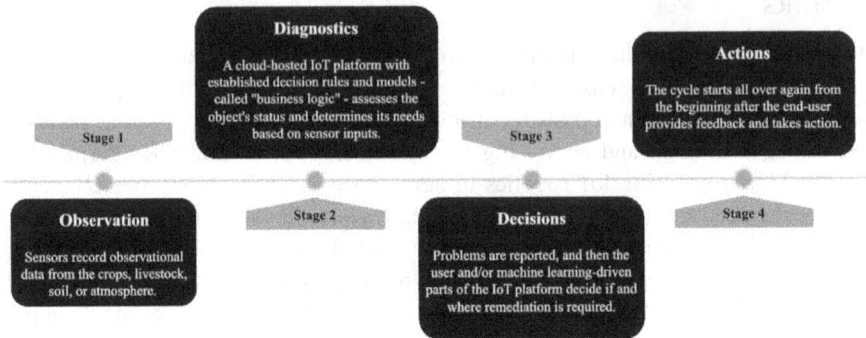

FIGURE 3.3   Smart farming steps

domains. Some of the common application areas of IoT in agriculture comprise soil and nutrition management, precision irrigation, crop monitoring, livestock monitoring, supply chain tracking, weather forecasting, pest and disease control, resource management, and greenhouse management. The recent developments in the IoT are also employed in agriculture for critical aspects of precision farming, nutrient management, weed detection, crop quality monitoring, and weed detection (Liakos et al., 2018; Sharma et al., 2022). Figure 3.4 shows the application of the IoT in different areas of agriculture. Farmers, for instance, can improve the efficiency of pesticides and fertilizers by measuring differences within a field and adjusting their strategy accordingly. Similarly, utilizing modern agricultural practices enables farmers to monitor livestock and their dietary requirements and disease surveillance. Figure 3.4 shows the application of IoT tools in agricultural farms.

The IoT, Smart Sensors and Irrigation Water Management

Irrigation is a vital process in agriculture. As we move forward, it is important to understand some critical terminologies used interchangeably in irrigation water management. The first and foremost is to understand irrigation. The process of applying water in soil to meet the needs of growing plants is known as irrigation. Applying water to plants adequately and efficiently helps in enhancing crop production. Irrigation management is the practice of applying a specific amount of irrigation water using a specific method at a designated time to achieve specific agronomic and economic goals (Evett et al., 2014). Another important terminology used in modern agriculture science is irrigation automation. In simple terms, irrigation automation is defined as

> use of in situ, remotely sensed, or near-surface remotely sensed crop, soil, and micro-meteorological properties sensed by a supervisory control and data acquisition system as inputs to a decision-making algorithm in the system, which then applies defined amounts of water at defined times automatically through control of an integrated irrigation application system.
>
> **(Evett et al., 2014)**

FIGURE 3.4   Applications of IoT in agriculture

Figure 3.5 shows IoT tools application in an agricultural farm.

Irrigation for plants can come from various sources, including rivers, lakes, aquifers, gravity, canals, ditches, pipes, and natural streams (Stolojescu-Crisan et al., 2022). Different irrigation methods have been developed to meet crops' irrigation needs. Irrigation methods also depend on the geographical and climatic conditions of the region. These days, water is scarce in some areas. Supplying water to scarce areas is a challenging and costly affair. The science of watering plants has progressed many folds. With the advancement of IoT and sensors, smart irrigation is becoming a critical system using data-intensive methods for efficient agricultural productivity, minimizing losses and therefore reducing environmental impacts.

Smart irrigation is a system of using science and technology to save water for irrigation purposes. The system contains soil sensors, weather sensors, and other controllers. The soil sensor monitors the actual ground moisture condition, and weather sensors monitor the current weather conditions and further facilitate future predictions. The controllers control the water valve to open or close based on the inputs from the sensors. The system integrates IoT, remote sensing and Geographical

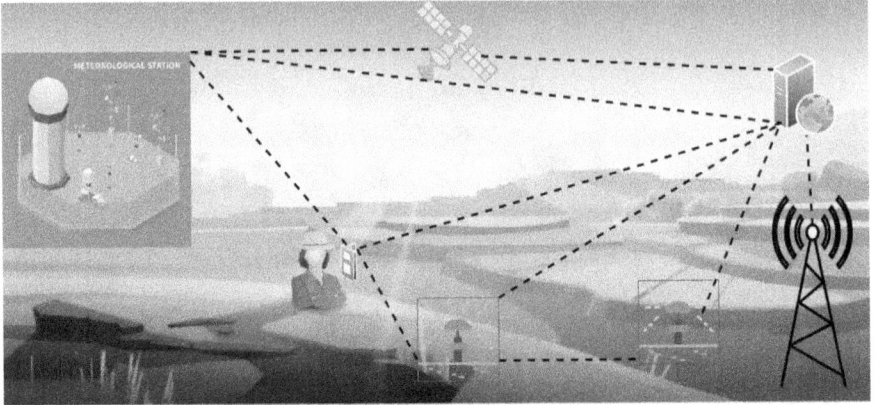

FIGURE 3.5   Application of IoT tools in agricultural farm

Information Systems (GIS), LoRa, LoRaWAN, Wi-Fi, and mobile internet. These integrated technologies provide the right amount of water according to soil and weather conditions to optimize water use.

IoT, Sensors and Nutrient Management

Plants need nutrients to grow. Nitrogen, phosphate, and potash (NPK) are three essential plant nutrients. While commercial fertilizers are the primary source of nutrients, manure and organic materials also contribute nutrients to plants. If nutrients are applied in access quantity, it can harm the environment. As the cost of fertilizers continues to grow, minimizing loss and optimizing the use of fertilizers is important. Therefore, effective nutrient management for the crop is crucial. Nutrient management is the science of achieving optimum nutrient efficiency for crop yield, quality, and minimum environmental harm. IoT and sensors play a crucial role in farm nutrient management, considering different parameters such as soil nutrient content, pH value, moisture content, and other environmental factors (Sun et al., 2020). The system reduces the environmental impact of fertilizer use by conserving energy and raw materials needed to produce fertilizers and by protecting the environment from pollution (Bacenetti et al., 2020).

The IoT and sensors-based nutrient management uses several smart tools, including remote sensing, soil nutrient sensors, robotics, and other IoT-based methods. Remote sensing and GIS-based approaches use plant and soil parameters obtained from ground sensors and through remote sensing for decision-making. Moreover, the vegetation index (VI) generated from the multi-spectral data aids in soil nutrient statuses such as NPK and crop growth (Osco et al., 2019). The information obtained from these sources further aids in appropriate fertilizer application in soil. The Variable Rate Applicator (VRA), which is installed on a GPS-enabled tractor, delivers NPK fertilizer to the appropriate location on the farm. It is achieved using advanced sensors, cameras, GPS, and satellite images (Maes & Steppe, 2019).

Components of Smart Nutrient and Irrigation Water Management Systems

- *Automated Weather Monitoring System*

  An automated weather monitoring system (AWMS) is a system of integrated components and processes that automatically measure, record, and transmit various weather parameters, such as temperature, rainfall, wind speed, wind direction, solar radiation, humidity, and atmospheric pressure (Ahmad et al., 2022).

- *Soil Moisture Monitoring*

  Soil moisture monitoring consists of sensors that directly measure soil moisture content. It measures the amount of water in the soil and is generally accurate. The sensor is usually placed at the same depth as the plant's roots. Soil moisture sensors work with the irrigation controller to manage water flow. The control platform sets high or low alarm values for soil sensors, depending on the soil type and weather pattern. Depending on the soil type, sensors are selected. However, it also depends on the system compatibility (Muangprathub et al., 2019b).

- *Smart Fertigation System*

  Fertigation is the application of fertilizer to plants, incorporated with irrigation water. Fertigation delivers the right combination of water and nutrients directly to the roots of the plant as per the crop development cycle. It is a highly efficient method of providing fertilizers and irrigation together directly at the roots of the plant. The method is commonly used in horticulture and floriculture in automated systems such as greenhouses, polyhouses, hydroponics, and agricultural fields (Ahmad et al., 2022).

- *Smart Irrigation System*

  Smart irrigation systems use advanced sensors and control mechanisms to optimize water use. Compared to traditional irrigation, smart irrigation systems can reduce water wastage by 20–40%. The system can scientifically estimate the water requirement of plants based on soil moisture and local weather parameters. Weather monitoring systems provide information on current and future weather parameters. The information is fed into a central system which gets information on other crop parameters such as soil health, crop health, and weather information.

- *Evaporation and Leaf Wetness Sensor*

  Evaporation sensors are part of the weather monitoring system. It helps in estimating the evaporation of soil. Information gathered from this device helps in adjusting irrigation duration. The leaf wetness sensor estimates the evaporation of plant leaves. These sensors resemble a leaf, as they emulate the characteristics of leaves.

ADVANTAGES AND DISADVANTAGES OF SMART IRRIGATION AND NUTRIENT MANAGEMENT SYSTEMS

As with any other system, IoT-enabled irrigation and nutrient applications have both advantages and disadvantages, which are discussed in subsequent sections.

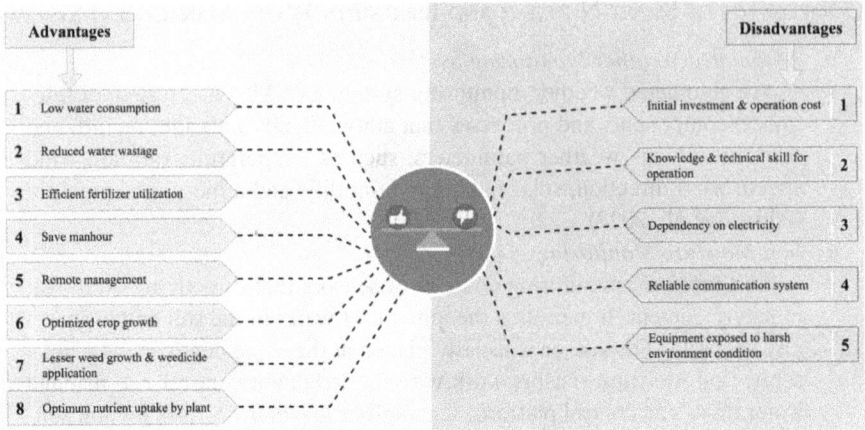

FIGURE 3.6   Advantages and disadvantages of smart irrigation and nutrient management system

Advantages

Smart IoT-based systems offer several benefits to farmers and farming communities. It also offers benefits to agriculture and allied sectors as a whole through optimizing resources. The scientific basis of input utilization by monitoring various agroclimatic factors aids in cost minimization reduces wastage, and offers environmental benefits. Some of the key advantages of the system are listed in the Figure 3.6 and summarized below:

- *Low Water Consumption*
    Smart irrigation system is designed to reduce water consumption by efficient water utilization. It utilizes soil moisture data and weather forecast data to water plants. Depending on soil moisture conditions and weather forecasts, water scheduling is performed by the system. The method saves more water as compared to other irrigation methods such as flooding, furrowing, and other non-smart irrigation systems.
- *Reduced Water Wastage*
    One of the main advantages of a smart irrigation system is that it reduces water wastage. The system uses a precision method for water use; therefore, only the required amount of water is released near plant roots. The method provides optimum water at plant roots after calculating factors influencing plant watering, reducing water wastage.
- *Efficient Fertilizer Utilization*
    Different fertilizer application methods include broadcast, placement, foliar, and aerial. Fertilizer use efficiency is low in all these methods. A lot of fertilizer gets wasted during application. When fertilizer is applied through irrigation water, it is known as fertigation. The fertigation process enhances fertilizer use efficiency. Fertigation, using sprinkler or drip

irrigation with smart irrigation tools, sensors, and the IoT increases fertilizer utilization.

- *Saves Manhours*

  The system is highly dependent on automatic machines. The system is based on the automatic method managed by sensors, control panels, and other machines with less intervention or effort from human beings. Automation minimizes human intervention. The installation process requires human intervention at the beginning. However, once the installation process is complete, the system can perform its function in automatic or semi-automatic mode. Automating the system reduces human intervention, saving the workforce or manhours in the processes involved.

- *Remote management*

  The entire process can be managed remotely without physically being present in the field. The main controller manages the system, which receives, processes, and gives necessary commands for water release. The controller is situated near to farm or away from the farm. Once the water, fertilizer, pesticide, and herbicide scheduling is done, the system continues running. The operator can monitor the situation with their handheld device.

- *Optimized crop growth*

  For optimal plant growth, the optimal condition of inputs is essential. Smart irrigation provides adequate water based on the local weather and soil conditions. When integrated with a fertilizer distribution system, popularly known as fertigation, it further optimizes crop growth.

- *Lesser weed growth and weedicide application*

  Since water is released at the plant roots, less or no water is available for weeds near the plant, reducing weed growth in agricultural fields. With less weed growth in the agricultural field, weedicide application is also less. Lesser weedicide application further improves soil health conditions.

- *Optimum nutrient uptake by plant*

  Plant growth and development are influenced by the combination and concentration of nutrients in the soil (Morgan & Connolly, 2013). Deficiencies in the supply of nutrients and excess supply of nutrients can hamper plant growth adversely. Availability of nutrients depends on soil characteristics as well as artificial application. The smart system, with available information on soil nutrient availability, soil moisture availability, and other related parameters, provides regulated water and nutrients to plants which help in optimum nutrient and water uptake.

DISADVANTAGES

The system's disadvantages are depicted in Figure 3.6 and are summarized below.

- *Initial Investment and Operation Cost*

  The smart irrigation system is expensive and requires an initial investment in establishing sensors, equipment, controllers, cameras, and farm machines. The cost of initial investment depends on the farm size. Furthermore, the system also requires maintenance from time to time for running the system. The maintenance cost also contributes significantly to the system.

- *Knowledge and Technical Skills for Operation*

  The system requires technical know-how for operation. Basic understanding to operate equipment and read different weather and environment parameters is essential for farmers managing fields. Operators with less knowledge need skill upgradation to understand the technical aspects of the operation.

- *Dependency on Electricity*

  An automatic system that raises water at heights in the fields requires energy to pump water. The energy is used from electricity or batteries. In the absence of an electricity supply, the system cannot function. Therefore, electricity makes the system dependent and adds to the operation cost.

- *Reliable Communication System*

  Establishing a reliable communication system is a prerequisite for smart farming. The communication system connects and establishes data flow among smart farming components. The system performs at its optimum level without a reliable communication system. Therefore, robust communication is required for the functioning of the system.

- *Equipment Exposed to Harsh Environment Condition*

  The system comprises several farm equipment that are exposed to field conditions. Soil moisture sensors, weather sensors, nutrient sensors, and communication sets are exposed to water, sunshine, wind, and chemicals that cause corrosion and aberration in equipment (Obaideen et al., 2022).

CONTRIBUTION OF SMART NUTRIENT AND IRRIGATION WATER MANAGEMENT TO SUSTAINABLE DEVELOPMENT GOALS

Food and water are two of the most crucial commodities in the world. Any stress on the food and water supply system can lead to a delipidated effect on society. Water scarcity led by climate change puts stress on the agricultural system. Smart irrigation and nutrient management systems are proven effective in efficient production, minimizing loss of resources, and damage to the environment. These effects have a direct and indirect impact on several of the United Nations Sustainable Development Goals. SDG 6 aims to provide access to sustainable water and sanitation management for everyone. The smart irrigation system based on IoT also contributes to achieving SDG. Farmers need to adapt to practices that are sustainable to boost productivity. Smart irrigation and nutrient management contribute to attaining SDGs for food security, reducing poverty, good health, and well-being of communities (Obaideen et al., 2022). Table 3.2 shows the contribution of nutrient and irrigation water management to the SDGs.

TABLE 3.2

Contribution of Smart Irrigation and Nutrient Management to the SDG

SDGs		Contribution of Smart Irrigation and Nutrient Management to the SDGs
SDG 1	No Poverty	Brings economic benefits through better agricultural productivity by efficient utilization of resources
SDG 2	Zero Hunger	Improves farm production and productivity, contributing to fight against hunger
SDG 3	Good Health and Well-Being	Reduced application of chemical fertilizers and pesticides, promotes good health and well-being
SDG 6	Clean Water and Sanitation	Less fertilizer and pesticide use resulting lesser soil, water, and air pollution
SDG 7	Affordable and Clean Energy	When used with renewable sources such as solar and wind energy in farms contributes to clean energy
SDG 8	Decent Work and Economic Growth	Contributes to agriculture and other related sectors leading to economic growth
SDG 9	Industry, Innovation, and Infrastructure	Supporting innovation and industries
SDG 10	Reduced Inequalities	Promotes economic and social development through better agricultural practices
SDG 11	Sustainable Cities and Communities	Promotes sustainable use of inputs, optimized crop protection and sustained food supply for rural and urban communities
SDG 12	Responsible Consumption and Production	Inputs such as water, fertilizer and pesticides are consumed judiciously and efficiently leading to optimized production
SDG 13	Climate Action	Provision of efficient water and nutrient supply for dry regions as adaptation measures
SDG 14	Life Below Water	Less groundwater extraction and lesser water pollution
SDG 15	Life on Land	Support sustainable food production and crop growth
SDG 17	Partnerships for the Goals	Technical collaboration for research and development among institutions for advanced system deployment

CONCLUSION

The agriculture sector is noticing a significant intervention from modern usage of IoT and sensors. Smart solutions provide a great opportunity for the farming community and other stakeholders engaged directly or indirectly with agriculture. In conclusion, the IoT and smart sensors have the potential to revolutionize nutrient and irrigation water management in agriculture. By collecting data on soil moisture, temperature, pH levels, and nutrient levels and transmitting this data in real time to farmers and agribusinesses, the IoT and smart sensors can enable more precise and efficient irrigation and fertilization practices. This can help to reduce costs, improve crop yields, and conserve water resources. However, adopting the IoT and smart sensors in agriculture is not without challenges. Various sensors and IoT platforms are available on the market, and selecting the right ones for a specific application can be a complex task. In addition, there are concerns about the cost, reliability, and maintenance of these technologies and the need for robust cybersecurity measures to protect against data breaches and cyber-attacks. To fully realize the potential of IoT and smart sensors in agriculture, it will be important to address these challenges and continue to develop and improve these technologies. This will require a combination of research and development efforts and collaboration between farmers, agribusinesses, and technology companies to create a more sustainable and productive agricultural sector for the 21st century. It is not hard to imagine how IoT sensors could be helpful in farming situations as we continue to transition toward a more interconnected society with more intelligent devices in every household. They can be set up in greenhouses or along irrigation lines to keep track of environmental conditions and crop health, allowing farmers to prevent problems rather than fix them as they arise. There will be shifts in smart farming as these technologies evolve and become more accessible.

REFERENCES

Adamchuk, V. I., Lund, E. D., Sethuramasamyraja, B., Morgan, M. T., Dobermann, A., & Marx, D. B. (2005). Direct measurement of soil chemical properties on-the-go using ion-selective electrodes. *Computers and Electronics in Agriculture*, *48*(3), 272–294. https://doi.org/10.1016/j.compag.2005.05.001.

Ahmad, U., Alvino, A., & Marino, S. (2022). Solar fertigation: A sustainable and smart IoT-based irrigation and fertilization system for efficient water and nutrient management. *Agronomy*, *12*(5). https://doi.org/10.3390/agronomy12051012.

Alahi, M. E. E., Xie, L., Mukhopadhyay, S., & Burkitt, L. (2017). A temperature compensated smart nitrate-sensor for agricultural industry. *IEEE Transactions on Industrial Electronics*, *64*(9), 7333–7341.

Al-Sarawi, S., Anbar, M., Alieyan, K., & Alzubaidi, M. (2017a). Internet of Things (IoT) communication protocols. In *2017 8th International Conference on Information Technology (ICIT)*, 685–690.

Al-Sarawi, S., Anbar, M., Alieyan, K., & Alzubaidi, M. (2017b). Internet of Things (IoT) communication protocols. In *2017 8th International Conference on Information Technology (ICIT)*, 685–690.

Ayoub Shaikh, T., Rasool, T., & Rasheed Lone, F. (2022). Towards leveraging the role of machine learning and artificial intelligence in precision agriculture and smart farming. *Computers and Electronics in Agriculture*, *198*, 107119. https://doi.org/10.1016/J.COMPAG.2022.107119.

Bacenetti, J., Paleari, L., Tartarini, S., Vesely, F. M., Foi, M., Movedi, E., Ravasi, R. A., Bellopede, V., Durello, S., & Ceravolo, C. (2020). May smart technologies reduce the environmental impact of nitrogen fertilization. A case study for paddy rice. *Science of the Total Environment*, *715*, 136956.

Balaji, S., Nathani, K., & Santhakumar, R. (2019). IoT technology, applications and challenges: A contemporary survey. *Wireless Personal Communications*, *108*(1), 363–388.

Botta, A., de Donato, W., Persico, V., & Pescapé, A. (2016). Integration of cloud computing and internet of things: A survey. *Future Generation Computer Systems*, *56*, 684–700.

Burton, L., Dave, N., Fernandez, R. E., Jayachandran, K., & Bhansali, S. (2018). Smart gardening IoT soil sheets for real-time nutrient analysis. *Journal of the Electrochemical Society*, *165*(8), B3157–B3162. https://doi.org/10.1149/2.0201808jes.

Capello, F., Toja, M., & Trapani, N. (2016). A real-time monitoring service based on industrial internet of things to manage agrifood logistics. In *6th International Conference on Information Systems, Logistics and Supply Chain*, 1–8.

Centenaro, M., Costa, C. E., Granelli, F., Sacchi, C., & Vangelista, L. (2021). A survey on technologies, standards and open challenges in satellite IoT. *IEEE Communications Surveys and Tutorials, 23*(3), 1693–1720.

Christy, C. D., Drummond, P., & Laird, D. A. (2003). An on-the-go spectral reflectance sensor for soil. In *2003 ASAE Annual Meeting*, 1.

Ehsani, M. R., Upadhyaya, S. K., Slaughter, D., Protsailo, L. V., & Fawcett, W. R. (2000). Quantitative measurement of soil nitrate content using mid-infrared diffuse reflectance spectroscopy. ASAE Annual International Meeting, Milwaukee, Wisconsin, USA, 9-12 July 2000 pp. 1–15 ref.9.

Ehsani, M. R., Upadhyaya, S. K., Slaughter, D., Shafii, S., & Pelletier, M. (1999). A NIR technique for rapid determination of soil mineral nitrogen. *Precision Agriculture, 1*(2), 219–236.

Evett, S. R., Colaizzi, P. D., O'Shaughnessy, S. A., Hunsaker, D. J., & Evans, R. G. (2014). *Irrigation Management BT - Encyclopedia of Remote Sensing* (E. G. Njoku, Ed.), pp. 291–302. New York: Springer. https://doi.org/10.1007/978-0-387-36699-9_73.

Fan, Y., Himanshu, S. K., Ale, S., DeLaune, P. B., Zhang, T., Park, S. C., ... Baumhardt, R. L. (2022). The synergy between water conservation and economic profitability of adopting alternative irrigation systems for cotton production in the Texas high plains. *Agricultural Water Management, 262*, 107386. https://doi.org/10.1016/j.agwat.2021.107386.

Farooq, M. S., Riaz, S., Abid, A., Abid, K., & Naeem, M. A. (2019). A survey on the role of IoT in agriculture for the implementation of smart farming. *IEEE Access, 7*, 156237–156271. https://doi.org/10.1109/ACCESS.2019.2949703.

Gandhi, M., Shahani, S., & Vyas, N. (2020). Knowledge and opinions of postgraduate resident doctors regarding promotional drug literature: A cross-sectional study. *National Journal of Physiology, Pharmacy and Pharmacology, 10*, 1. https://doi.org/10.5455/njppp.2020.10.07198202028072020.

Gupta, P. K. (2020a). Pollution load on Indian soil-water systems and associated health hazards: a review. *Journal of Environmental Engineering, 146*(5), 03120004. https://doi.org/10.1061/(ASCE)EE.1943-7870.0001693.

Gupta, P. K. (2020b). Fate, transport, and bioremediation of biodiesel and blended biodiesel in subsurface environment: a review. *Journal of Environmental Engineering, 146*(1), 03119001. https://doi.org/10.1061/(ASCE)EE.1943-7870.0001619.

Hemalatha, T., & Sujatha, B. (2015). Sensor based autonomous field monitoring agriculture robot providing data acquisition & wireless transmission. *International Journal of Innovative Research in Computer and Communication Engineering, 3*(8), 7651–7657.

Himanshu, S. K., Ale, S., Bordovsky, J. P., Kim, J., Samanta, S., Omani, N., & Barnes, E. M. (2021). Assessing the impacts of irrigation termination periods on cotton productivity under strategic deficit irrigation regimes. *Scientific Reports, 11*(1), 1–16. https://doi.org/10.1038/s41598-021-99472-w.

Jahn, B. R., Linker, R., Upadhyaya, S. K., Shaviv, A., Slaughter, D. C., & Shmulevich, I. (2006). Mid-infrared spectroscopic determination of soil nitrate content. *Biosystems Engineering, 94*(4), 505–515. https://doi.org/10.1016/j.biosystemseng.2006.05.011.

Kamienski, C., Jentsch, M., Eisenhauer, M., Kiljander, J., Ferrera, E., Rosengren, P., Thestrup, J., Souto, E., Andrade, W. S., & Sadok, D. (2017). Application development for the Internet of Things: A context-aware mixed criticality systems development platform. *Computer Communications, 104*, 1–16. https://doi.org/10.1016/j.comcom.2016.09.014.

Kamienski, C., Soininen, J. P., Taumberger, M., Dantas, R., Toscano, A., Cinotti, T. S., Maia, R. F., & Neto, A. T. (2019). Smart water management platform: IoT-based precision irrigation for agriculture. *Sensors (Switzerland)*, *19*(2). https://doi.org/10.3390/s19020276.

Kodali, R. K., Rawat, N., & Boppana, L. (2014). WSN sensors for precision agriculture. In *IEEE TENSYMP 2014 - 2014 IEEE Region 10 Symposium*, 651–656. https://doi.org/10.1109/tenconspring.2014.6863114.

Krishnan, A., & Swarna, S. (2020). Robotics, IoT, and AI in the automation of agricultural industry: A review. In *2020 IEEE Bangalore Humanitarian Technology Conference (B-HTC)*, 1–6.

Kulkarni, A. A., Dhanush, P., Chetan, B. S., Gowda, C. S. T., & Shrivastava, P. K. (2020). Applications of automation and robotics in agriculture industries; a review. *IOP Conference Series: Materials Science and Engineering*, *748*(1), 012002.

Kumar Roy, S., Member, S., Misra, S., Member, S., Singh Raghuwanshi, N., & Das, S. K. (2021). AgriSens: IoT-based dynamic irrigation scheduling system for water management of irrigated crops. *IEEE Internet of Things Journal*, *8*(6). https://doi.org/10.1109/JIOT.2020.

Kundalia, K., Patel, Y., & Shah, M. (2020). Multi-label movie genre detection from a movie poster using knowledge transfer learning. *Augmented Human Research*, *5*(1), 1–9.

Laskar, S., & Mukherjee, S. (2016). Optical sensing methods for assessment of soil macronutrients and other properties for application in precision agriculture: A review. *ADBU Journal of Engineering Technology*, *4*, 206–210.

Liakos, K. G., Busato, P., Moshou, D., Pearson, S., & Bochtis, D. (2018). Machine learning in agriculture: A review. *Sensors*, *18*(8), 2674.

Maes, W. H., & Steppe, K. (2019). Perspectives for remote sensing with unmanned aerial vehicles in precision agriculture. *Trends in Plant Science*, *24*(2), 152–164.

Mahajan, M., Gupta, P. K., Singh, A., Vaish, B., Singh, P., Kothari, R., & Singh, R. P. (2022). A comprehensive study on aquatic chemistry, health risk and remediation techniques of cadmium in groundwater. *Science of The Total Environment*, *818*, 151784. https://doi.org/10.1016/j.scitotenv.2021.151784.

Morgan, J. B., & Connolly, E. L. (2013). Plant-soil interactions: Nutrient uptake I Learn science at Scitable. *Nature Education Knowledge*, *4*(8), 2.

Muangprathub, J., Boonnam, N., Kajornkasirat, S., Lekbangpong, N., Wanichsombat, A., & Nillaor, P. (2019a). IoT and agriculture data analysis for smart farm. *Computers and Electronics in Agriculture*, *156*, 467–474. https://doi.org/10.1016/j.compag.2018.12.011.

Muangprathub, J., Boonnam, N., Kajornkasirat, S., Lekbangpong, N., Wanichsombat, A., & Nillaor, P. (2019b). IoT and agriculture data analysis for smart farm. *Computers and Electronics in Agriculture*, *156*, 467–474. https://doi.org/10.1016/j.compag.2018.12.011.

Navulur, S., & Prasad, M. N. G. (2017). Agricultural management through wireless sensors and internet of things. *International Journal of Electrical and Computer Engineering*, *7*(6), 3492.

Ndzi, D. L., Harun, A., Ramli, F. M., Kamarudin, M. L., Zakaria, A., Shakaff, A. Y. M., Jaafar, M. N., Zhou, S., & Farook, R. S. (2014). Wireless sensor network coverage measurement and planning in mixed crop farming. *Computers and Electronics in Agriculture*, *105*, 83–94. https://doi.org/10.1016/j.compag.2014.04.012.

Obaideen, K., Yousef, B. A. A., AlMallahi, M. N., Tan, Y. C., Mahmoud, M., Jaber, H., & Ramadan, M. (2022). An overview of smart irrigation systems using IoT. *Energy Nexus*, *7*, 100124. https://doi.org/10.1016/j.nexus.2022.100124.

Osco, L. P., Ramos, A. P. M., Moriya, É. A. S., de Souza, M., Junior, J. M., Matsubara, E. T., Imai, N. N., & Creste, J. E. (2019). Improvement of leaf nitrogen content inference in Valencia-orange trees applying spectral analysis algorithms in UAV mounted-sensor images. *International Journal of Applied Earth Observation and Geoinformation*, *83*, 101907.

Pallavi, S., Mallapur, J. D., & Bendigeri, K. Y. (2017). Remote sensing and controlling of greenhouse agriculture parameters based on IoT. In *2017 International Conference on Big Data, IoT and Data Science (BID)*, 44–48.

Pang, Z., Chen, Q., Han, W., & Zheng, L. (2015). Value-centric design of the internet-of-things solution for food supply chain: Value creation, sensor portfolio and information fusion. *Information Systems Frontiers*, *17*(2), 289–319. https://doi.org/10.1007/s10796-012-9374-9.

Parmar, M., & Kumar, R. (2022). Overview of IoT in the agroecosystem. *Advanced Series in Management*, *27*, 111–122. https://doi.org/10.1108/S1877-636120220000027008.

Pavón-Pulido, N., López-Riquelme, J. A., Torres, R., Morais, R., & Pastor, J. A. (2017). New trends in precision agriculture: A novel cloud-based system for enabling data storage and agricultural task planning and automation. *Precision Agriculture*, *18*(6), 1038–1068.

Said Mohamed, E., Belal, A. A., Kotb Abd-Elmabod, S., El-Shirbeny, M. A., Gad, A., & Zahran, M. B. (2021). Smart farming for improving agricultural management. *Egyptian Journal of Remote Sensing and Space Science*, *24*(3), 971–981. https://doi.org/10.1016/j.ejrs.2021.08.007.

Sharma, A., Georgi, M., Tregubenko, M., Tselykh, A., & Tselykh, A. (2022). Enabling smart agriculture by implementing artificial intelligence and embedded sensing. *Computers and Industrial Engineering*, *165*, 107936. https://doi.org/10.1016/J.CIE.2022.107936.

Stolojescu-Crisan, C., Butunoi, B. P., & Crisan, C. (2022). An IoT based smart irrigation system. *IEEE Consumer Electronics Magazine*, *11*(3), 50–58. https://doi.org/10.1109/MCE.2021.3084123.

Sudduth, K. A., Hummel, J. W., & Birrell, S. J. (1997). Sensors for site-specific management. In *The State of Site-Specific Management for Agriculture* (F. J. Pierce & E. J. Sadler Eds) pp. 183–210. ASA/CSSA/SSSA: Madison, WI, USA.

Sun, J., Abdulghani, A. M., Imran, M. A., & Abbasi, Q. H. (2020). IoT enabled smart fertilization and irrigation aid for agricultural purposes. In *Proceedings of the 2020 International Conference on Computing, Networks and Internet of Things*, 71–75.

Trendov, M., Varas, S., & Zeng, M. (2019). Digital technologies in agriculture and rural areas: Status report. Food and Agriculture Organization of the United Nations Rome, 2019. https://www.fao.org/3/ca4985en/ca4985en.pdf.

Tyszczuk-Rotko, K., & Jędruchniewicz, K. (2019). Ultrasensitive sensor for uranium monitoring in water ecosystems. *Journal of the Electrochemical Society*, *166*(10), B837.

van Staden, J. K. F., Nuta, R.-G., & Tatu, G.-L. (2018). Determination of nitrite from water catchment areas using graphite based electrodes. *Journal of the Electrochemical Society*, *165*(13), B565.

Viscarra Rossel, R. A., & Bouma, J. (2016). Soil sensing: A new paradigm for agriculture. *Agricultural Systems*, *148*, 71–74. https://doi.org/10.1016/j.agsy.2016.07.001.

Viscarra Rossel, R. A., Walvoort, D. J. J., McBratney, A. B., Janik, L. J., & Skjemstad, J. O. (2006). Visible, near infrared, mid infrared or combined diffuse reflectance spectroscopy for simultaneous assessment of various soil properties. *Geoderma*, *131*(1–2), 59–75. https://doi.org/10.1016/j.geoderma.2005.03.007.

Wolfert, S., Ge, L., Verdouw, C., & Bogaardt, M. J. (2017). Big Data in Smart Farming – A review. *Agricultural Systems*, *153*. https://doi.org/10.1016/j.agsy.2017.01.023.

Xu, J., Gu, B., & Tian, G. (2022). Review of agricultural IoT technology. *Artificial Intelligence in Agriculture*, *6*. https://doi.org/10.1016/j.aiia.2022.01.001.

4 Assessing Nutrient Variability Across Irrigation Management Zones Using Unsupervised Learning and Mixed Models

Hemendra Kumar, Brenda V. Ortiz,
Puneet Srivastava, and Jasmeet Lamba

INTRODUCTION

The world population is projected to increase from 8.27 billion in 2030 to 9.322 billion in 2050 with a population growth rate of 0.64% annually. Increasing human population demands increasing food production per unit area over the growing seasons. Extreme precipitation events are becoming common in the southeastern states causing micro droughts or floods over the growing seasons (Takhellambam et al., 2023, 2022). Due to the lack of precipitation during the growing season in the southeastern US (Kumar et al., 2022a, 2022b), farmers are increasingly adopting irrigation. Irrigation adoption is essential to meet crop water demands not only in arid regions (Ale et al., 2021; Himanshu et al., 2021, 2023; Mauget et al., 2021), but also in humid regions (Kumar et al., 2021b). The amount of irrigated land has expanded by 5% in the last 15 years in the United States (USDA-NASS, 2019). Irrigation adoption has been increasing to help farmers improve water use efficiency, avoid water stress, and maintain desirable soil moisture (Kumar et al., 2021a). Adopting appropriate irrigation practices can also help with improving nutrient availability to plants, reducing pressure on water resource systems, and reducing contamination of surface water due to surface runoff during the growing season (Kumar et al., 2021a; Yost et al., 2019; Mahajan et al., 2022; Himanshu et al., 2022; Singh et al., 2023).

Phosphorus (P) and nitrogen (N) are essential nutrients for plant growth and can impact crop yield. The excess application of nutrients in agricultural fields results in contamination of surface and groundwater (Hanrahan et al., 2019). A number of

DOI: 10.1201/9781003441175-4

factors are responsible for nutrient loss from agricultural fields during a growing season. For example, studies have found a decrease in the nutrient loss in the runoff with an increase in the time between fertilizer or manure application and the first precipitation event (Schröder et al., 2004; Sharpley, 1997; Smith et al., 2007). Based on previous studies, to reduce nutrient loss in surface runoff, it is recommended not to irrigate soon after fertilizer application or not to apply fertilizer or manure in any form before an impending rain event.

Variations in soil nutrients depend on various environmental factors including extrinsic (land uses and farming practices) and intrinsic factors (topography, soil, climate, etc.) (Gao et al., 2019). The best management of soil nutrients in agricultural fields is a crucial topic and challenging to study (Su et al., 2018) for precision agricultural management. Current studies have focused on the nutrient variability from plot scale to watershed scale; however, there is a gap in studies related to site-specific management with crop yield implications due to the limited applicability of controlled experimental studies. The nutrient variability across the crop field at varying crop stages during a growing season is important to improve nutrient use efficiency and decrease the loss in crop yield. The loss in crop yield and nutrient variability across the field suggests optimizing the agricultural system for precision agricultural management to meet food demands and improve yield management (Jiang et al., 2021; McEntee et al., 2019). Since field management is essential to avoid nutrient loss and improve yield management, studies are needed to identify the significance of prevailing factors influencing nutrient variability. This study aimed to investigate nutrient (P and N) variability across the crop field at different soil depths and its impact on crop yield. To achieve this goal, the field was delineated into different irrigation management zones (MZs) using unsupervised learning to evaluate the spatial and temporal variability in soil P and N at various crop stages and depths in a crop-growing period. We hypothesized that large differences in crop yield can occur across the field with variability in nutrient concentrations during a growing season.

The study was conducted in an agricultural field of 120 ha sown with corn. An unsupervised fuzzy c-means classification and cluster analysis based on the Euclidean distance was adopted to delineate the field into three zones using topography, soil texture, and historical yield data of ten years. Based on historical yield data collected by farmers, we found substantial yield variability across the field (Kumar, 2022).

METHODS

Soil moisture sensors were installed to measure soil water potential at 15 cm, 30 cm, and 60 cm soil depths in each MZ. Irrigation was triggered using the soil water depletion approach and was scheduled when soil water depletion based on two consecutive days showed approximately 2.54 cm as irrigation depth in any MZ. The soil samples were collected at multiple locations within each zone with three replicates of each location for nutrient analysis (Figure 4.1). The rationale behind the multiple locations was to capture maximum variability and reduce the uncertainty in soil nutrient data. Above-ground plant samples were collected for nutrient analysis at different stages of corn in each MZ. These undisturbed soil samples were collected

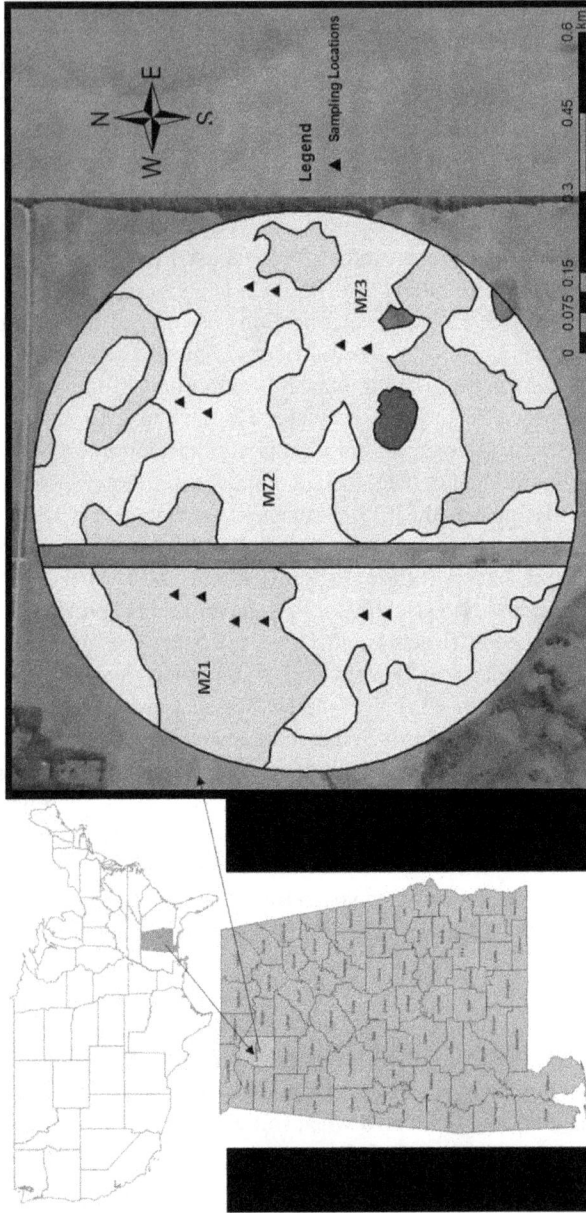

FIGURE 4.1   Management zones showing soil sampling locations for nutrient analysis during the growing season

TABLE 4.1

Soil Properties, Sand (%), Silt (%), Clay (%), Soil Organic Matter (SOM, %), Bulk Density (gcm^{-3}), and pH in Each Zone. (Within each zone, mean values (± standard deviation) of various soil properties followed by the identical letter are not significantly different at the 0.05 significant level)

Zone	Sand (%)	Silt (%)	Clay (%)	Soil texture	SOM (%)	Bulk density (gcm^{-3})	pH
MZ1	16±9.06a	47.93±10.65a	36.07±15a	Silty clay loam	3.94±0.40a	1.60±08a	5.82±0.60a
MZ2	22.31±6.55a	40.67±8.06a	37.02±9.52a	Clay loam	4.18±0.69a	1.60±06a	5.13±0.55b
MZ3	30 36±7.86b	28.87±9.27b	40.78±8.78a	Clay	4.50±0.78b	1.67±08a	5.35±0.68b

to generate the soil water retention curves. For a detailed procedure for developing soil water retention curves, readers can follow Lena et al. (2021, 2022).

The collected replicates of soil samples were composited to homogenize individual depth at each location. The soil samples were oven-dried and ground using a soil grinder and passed through a 2-mm sieve for chemical analysis. The soil samples were analyzed for total P, total N, ortho-phosphorus (OP), and soil organic matter (SOM). The SOM was determined using the loss-on-ignition method (Wright et al., 2008) in the Department of Biosystems Engineering at Auburn University. An ammonium molybdate-ascorbic acid colorimetry method was used for OP analysis (Kumar et al., 2022b). Under acidic conditions, OP ions react with ammonium molybdate and antimony potassium tartrate to form a molybdate-phosphate complex during the analysis. Total P and total N in soil samples were analyzed with acid digestion using a persulfate Kjeldahl digesting solution (Dayton et al., 2017; American Public Health Association, 1998). They kept the same matrix across the soil samples for nutrient analysis, all the standard and blanks were also digested. The nutrient analysis of soil samples was performed using a Lachat automated flow injection analyzer (Hach Company, Loveland, CO, USA). We used a protocol to run standard and blank solutions after every 15 samples to ensure the accuracy of the analysis. Total P and total N in plant samples were analyzed using ICAP-Open vessel wet digestion Digi Block 3000 and LECO-Nitrogen gas analyzer, respectively, at Waters Agricultural Laboratories, Inc. (Camilla, GA).

Field hydrologic characteristics are critically important in nutrient variability across the MZs within the crop field. A low Topographic Wetness Index (TWI) value represents a low water accumulation and high slope area, and a high TWI value represents a high water accumulation and flat area with a low slope. For detailed information, readers can follow Kumar et al. (2023) about management zone variation and impact of hydrological characteristics. Areas that allow water accumulation without runoff and waterlogging exhibit large TWI values. The drainage lines (Figure 4.2b) were also delineated to understand the direction of the water flow.

STATISTICAL ANALYSIS

Any two sample locations were at least 100 m apart. It is assumed that the soil samples are independent of each other at the same depth across the field. Moran's I test also supports this assumption of independent sampling. A common symbol, C, was used as a notation for nutrient concentration in this research. The nutrient concentration C_{ijl} at each sampling location i, each day j, and each depth l was used to calculate different statistical features to understand the variability in nutrients during the growing season. The nutrient data were normalized based on the average concentration to understand the variation at different sampling locations and at different sampling times in the cropland during the growing season. The relative difference, $\delta\left(C_{ijl}\right)$, for each location, and each time was defined as:

$$\delta\left(C_{ijl}\right) = \frac{\left(C_{ijl} - \bar{C}_{jl}\right)}{\bar{C}_{jl}} \tag{4.1}$$

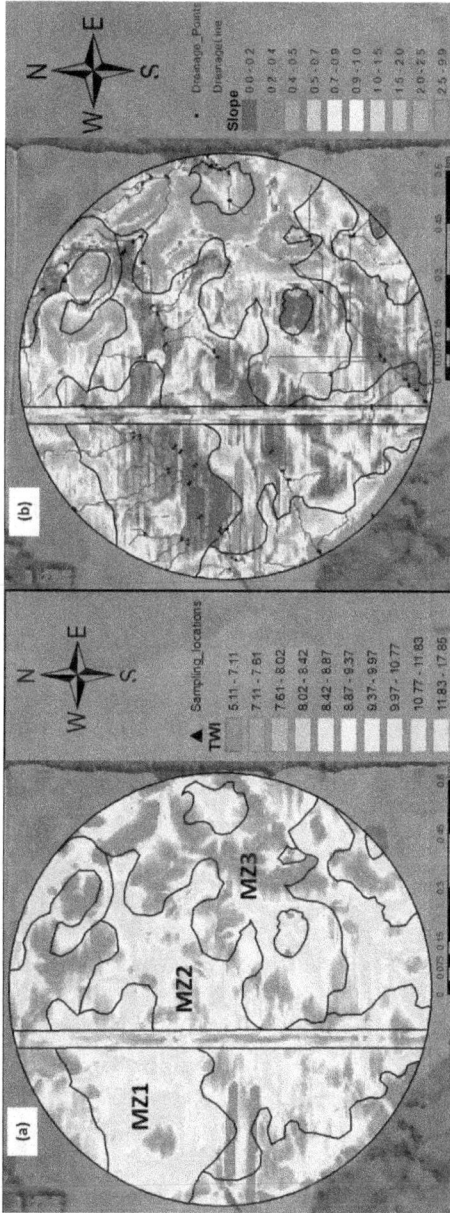

FIGURE 4.2 Hydrologic characteristics showing (a) Topographic Wetness Index (TWI) variation and (b) Slope and drainage pattern in the corn field

where $\bar{C}_{jl}$ is spatial mean of nutrient concentration Based on the relative difference, the mean relative difference $(\bar{\delta}_{c,il})$ and standard deviation of relative difference $\left(\sigma(\bar{\delta}_{c,il})\right)$ for each location (i) and each depth (l) can be given by

$$\bar{\delta}_{c,il} = \frac{\sum_{j=1}^{N} \delta\left(C_{ijl}\right)}{N} \tag{4.2}$$

$$\sigma\left(\bar{\delta}_{c,il}\right) = \sqrt{\frac{1}{N-1}\sum_{j=1}^{N}\left(\delta\left(C_{ijl}\right)-\bar{\delta}_{c,il}\right)^2} \tag{4.3}$$

The results were clustered in sampling MZs (indexed by k in the equations) to understand the nutrient variation across the zones during the growing season. The mean relative difference measures the variation of a nutrient in an MZ from the average nutrient concentration within the field and distinguishes whether the MZ was greater or lower than the average concentration of a nutrient. The independent analyses were performed for each depth (l).

Finally, a mixed-linear model with nested random effects with repeated observations was considered to determine nutrient variation between MZs and depths. Since all effects are factors, the model can be written as:

$$C_{ijkl} = \beta_0 + \beta_j + \beta_k + \beta_l + \beta_{kl} + \gamma_k + \gamma_{k(i)} + \gamma_{k(i(l))} + \varepsilon_{ijkl} \tag{4.4}$$

where β_m is the coefficient of the fixed effect at a level m, γ_m is the random effect at a level m, and ϵ is the error term. For example, β_{kl} explains the fixed effect interaction between MZ and sampling depth while $\gamma_{k(i)}$ and $\gamma_{k(i(l))}$ explain the random effect of location, respectively (nested in MZ k) and sampling depth (nested in the zone k and location i). Aside from taking into account the sampling time through the coefficient β_j, this mixed-linear model aims to account for and isolate the dependence in the data due to the repeated measurements over time for each depth, location, and zone. Using such a mixed model approach, more precise and reliable testing is achieved for the fixed effects of interest (time, zone, and depth) on the nutrient concentration. We will use nutrient symbols instead of a general notation C in the following sections of this chapter (we will use TP for soil total P instead of C for explanation).

RESULTS AND DISCUSSIONS

SOIL NUTRIENT VARIABILITY DURING THE GROWING SEASON

Based on the mixed-linear model and relative difference approach, the spatiotemporal patterns in total P between the MZs across the field were significantly different. For total P in soil, $\bar{\delta}_{TP}$, was a positive value in MZ1 and MZ2, and a negative value in MZ3 at studied soil profiles. It represents the adequate condition in MZ1 and MZ2 of soil total P and inadequate conditions of soil total P in the MZ3. This pattern of total P variability indicated that the MZ3 zone was low in soil total P relative to the average total P concentration in the field. The mixed model showed significant differences in total P between MZ1 and MZ3, and MZ2 and MZ3. However, there was no significant difference between MZ1 and MZ2 during the growing season. The effects of depth, time, MZs, and interaction between MZ and depth were significant for soil total P in the study.

According to the above-mentioned total P data, the total P variability in different MZs led to the investigation of different factors (e.g., topography, soil properties, and hydrology of the field) that caused total P variation within the field. A similar pattern of the mean relative difference in soil total P was also observed in soil OP. MZ1 had maximum and MZ3 had minimum soil OP concentration than the average soil OP concentration across the field. However, the MZ2 had soil OP between MZ1 and MZ2. It indicated the adequate soil OP concentration in MZ1 and MZ2 and inadequate soil OP concentration in the MZ3 during the growing season. The mixed model also indicated a similar pattern of variability in soil OP. MZ1 had significantly greater soil OP concentration as compared to MZ3. The effects of MZs, depth, and time were significant in soil OP; however, the soil depth did not show any interaction effect with the MZ during the growing season. An increasing pattern of mean relative difference in soil P (total P and OP) with soil depths indicated the difference in soil P concentration increased with increasing soil depth in MZ1 as compared to an average concentration of soil P in MZ1 during the growing season. However, soil P concentrations in MZ3 were always below the average concentration during the growing season. It was due to the greater saturated hydraulic conductivity of 63 cm/day in MZ1 as compared to MZ3, which had a saturated hydraulic conductivity of 9 cm/day. The applied fertilizers and manures were washed away due to the precipitation or rainfall events during the growing season and erosion patterns were observed in MZ3, it can also be seen Figure 4.2 for drainage lines and terrain attributes within each MZ. The P concentration is critical for plant height, stalk strength, growth, and crop yield. The plant heights at the silking stage of corn were 225 cm and 195 cm in MZ1 and MZ3, respectively, which explained the soil P inadequacy in the MZ3.

SOIL N VARIABILITY DURING THE GROWING SEASON

The mean relative difference of soil total N was positive for 0–15 cm, 15–30 cm, and 30–60 cm soil depths in MZ1 and MZ2, and a negative value was in MZ3 during the growing season. A positive $\bar{\delta}_{TN}$ value described adequate soil total N conditions in MZ1 and MZ2, and a negative $\bar{\delta}_{TN}$ value indicated inadequate soil total N conditions in MZ3. Based on the linear-mixed model and relative difference method for soil total N, MZ1 had higher concentrations and MZ3 had lower concentrations as compared to the average soil total N concentration in the field. The linear-mixed model showed that deeper soil profiles (15–30 cm and 30–60 cm) had significantly lower soil total N than the 0–15 soil profile with no interaction between MZs and soil depths during the growing season. The effects of the MZ, time, and soil depth were significant for the mixed-linear model of soil total N in the field. The investigation showed that soil total N was not uniformly available throughout the field during the growing season. MZ1 had a higher proportion available as compared to MZ3, which revealed that N also had the corresponding spatial variability in these MZs similar to P.

The sampling time of post-plant, silking, and pre-harvest as a fixed effect had a significant effect on soil total N during the growing season. It was due to supplemental N concentration applied by farmers as they apply traditionally during the growing season. A positive increasing trend of $\bar{\delta}_{TN}$ and a negative increasing trend of $\bar{\delta}_{TN}$ with increasing soil depths were observed in MZ1 and MZ3, respectively, during the growing season, which clarifies the greater and lower soil total N in MZ1 and MZ3 relating the average soil total N at the subsequent soil depths in the field.

FACTORS RESPONSIBLE FOR SITE-SPECIFIC NUTRIENT VARIABILITY

Based on TWI (Figure 4.2a), lower values of TWI in MZ3 also had lower nutrient concentrations of total P, OP, and total N. The lower TWI affects the mineralization influenced due to biotic and abiotic factors, such as organic P substrate availability and the geochemical properties of soil (Shaw & Cleveland, 2020). The slow process of mineralization in the field resulted in low concentrations of P and N in MZ3. Spearman's correlation between TWI and the temporal average concentration of soil nutrients was significant in the different soil layers during the growing season (Figure 4.3). A higher TWI in MZ1 had higher soil nutrient concentrations in MZ1, and a lower value of TWI in MZ3 had lower soil nutrient concentrations in MZ3. The soil P had A strong correlation between soil P and TWI in the upper soil layers explaining the strong soil P dependency on TWI at the upper soil layer than the deep soil layers. However, soil total N had a stronger correlation in the deep soil layers than in the upper soil layers. It implied the strong soil N dependency on TWI in deep soil layers.

Erosion gullies observed in MZ3 during the growing season and after harvesting showed nutrient losses due to runoff and topographical variability in the field. A high slope area increases the chance of lateral water movement and reduces the infiltration in deeper soil layers (Huat et al., 2006). MZ3 had a higher elevation and slope than MZ1. The surface runoff was likely one of the prevailing factors to lower the soil nutrient concentration in MZ3 (Plach et al., 2018; Smith et al., 2007), as it can be seen with higher slope and lower TWI in MZ3 as compared to MZ1. MZ1 acted as a pool for nutrients since no runoff loss and erosion was observed during the rainfall or irrigation events over the growing season.

The Spearman's correlation between slope and soil nutrients was significant for all nutrients except soil OP at the deepest soil layer (Figure 4.3). A negative correlation between slope and soil nutrients indicated a higher nutrient concentration in MZ1 with a lower slope, and a lower nutrient concentration in MZ3 with a higher slope. Soil texture can impact nutrient loss in surface runoff. The weak binding of

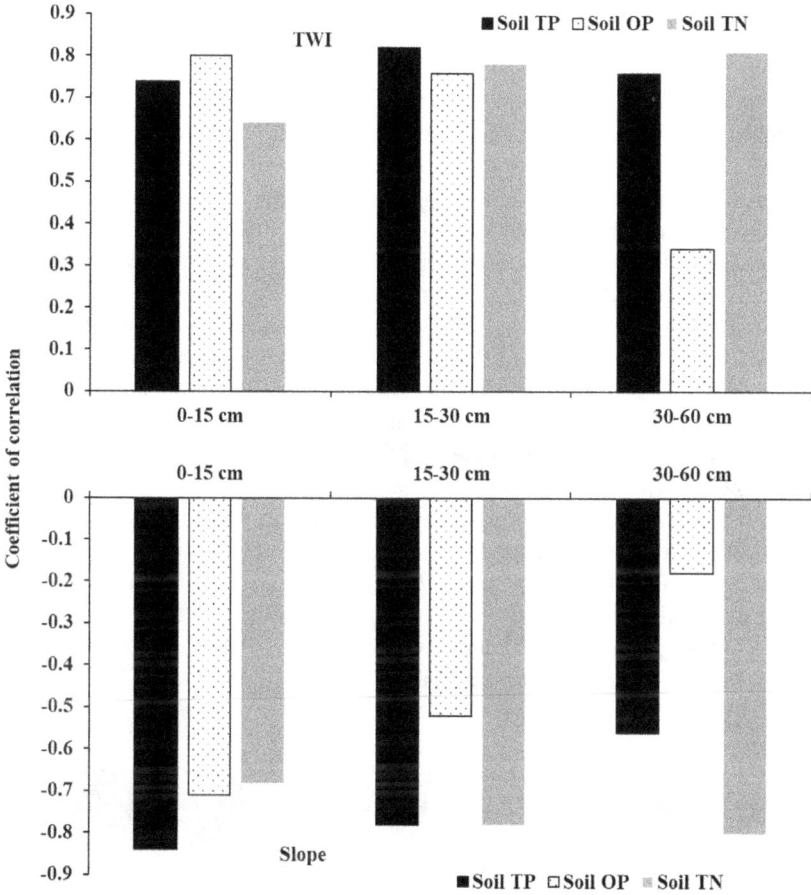

FIGURE 4.3 Correlation coefficients between terrain attributes and soil nutrients in the cornfield

nutrients with soils due to soil content heterogeneity in MZ3 could have enhanced losses.

A soil profile with higher hydraulic conductivity has a high infiltration rate and reduced runoff, which was found in this study that lower hydraulic conductivity, higher slope, and lower TWI in MZ3 as compared to MZ1. Based on water input data, the entire field received the same precipitation events and MZ3 received the least irrigation; however, no significant differences were found in irrigation amount at the end of the growing season. The Spearman's correlation between soil OP and soil total P had a better correlation in MZ1 than MZ3. Low water accumulation, an outward drainage pattern, surface runoff, and lower hydraulic conductivity limited the flow of nutrients within the soil layers in MZ3. The inward drainage pattern along with other hydrologic characteristics caused the development of a nutrient pool in MZ1, which resulted in nutrient leaching than MZ3 (Figure 4.2b).

IMPACT OF NUTRIENT VARIABILITY ON CROP YIELD

The spatial variation in crop yield had a similar pattern as of nutrient variability across the field. The yield produced in the MZs in 2019 varied from 3,139 kg/ha to 18,829 kg/ha. A significant difference was recorded in the crop yield in MZ1 and MZ3 across the field. The least yield was produced in MZ3 (red) (Figure 4.4). However, a higher yield was produced in the other two zones (MZ1 and MZ2) of the field (green) (Figure 4.4). During the soil sampling at different times in the season, we noticed that the plants were smaller in MZ3 as compared to MZ1 and MZ2. The P and N constraints can limit plant production during the growing season. Less availability of nutrients due to loss can be one of the main reasons for the crop yield variability in these zones within the field. This can also be proved with the plant heights and leaf area index data. The plant heights were 225 cm in MZ1 and 195 cm in MZ3 with corresponding LAI of 4.13 cm^2/cm^2 and 3.34 cm^2/cm^2, respectively.

Soil nutrients provide major guidelines for fertilizer management in current and future growing seasons. However, the tracking of nutrients' status retained in the plants can be a complementary guideline to adopt BMPs for precision nutrient management. We investigated the plant total N and plant total P retained in the plants at different corn stages during the growing season. Based on the mean relative difference in plant total N and plant total P, plants in MZ1 and MZ2 had adequate nutrients and inadequate nutrients in MZ3 during the growing season. The plant nutrients had a similar pattern of soil nutrient variability obtained across the field. This showed

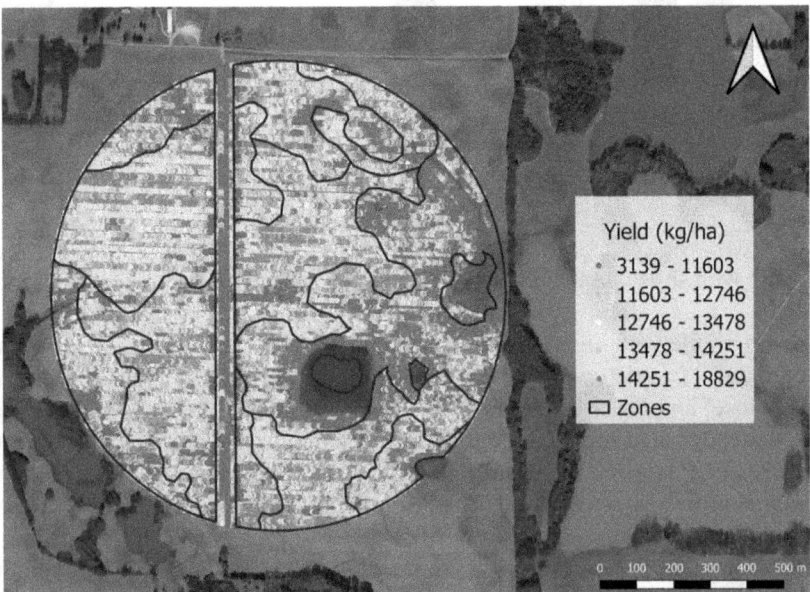

FIGURE 4.4 Crop yield variation in the cornfield during the 2019 growing season. The corn grain yield is adjusted to 15.5% moisture

the negative impact of nutrient loss in MZ3 and on crop yield. Therefore, the results of this study showed that yield in MZ1 can be related to the adequate soil and plant nutrients in MZ1, and yield in MZ3 can be related to inadequate soil and plant nutrients in the MZ3.

The Spearman correlation between plant total P and yield indicated that increasing P can increase crop yield at the silking stage. The plant total N showed a positive significant correlation of 0.85 ($p < 0.0001$) at the tasseling stage, however, a negative correlation of -0.63 ($p < 0.0001$) was at the V4 stage of the corn. It can be explained that yield had a dependency on the nutrients obtained by the plants during the growing season. Spearman's correlation coefficient between yield and plant nutrients demonstrated that higher nutrients retained in the plants at tasseling and silking stages can increase yield. A strong Spearman correlation coefficient between the TWI and crop yield suggested a strong dependency of crop yield on the topographical wetness of the field. A lower TWI leads to low crop yield, and in contrast, a higher TWI leads to higher crop yield in the field. A strong negative Spearman correlation between slope and crop yield suggested a higher slope in the field led to lower crop yield of corn.

SUMMARY AND CONCLUSION

Soil nutrient loss and nutrient uptake in plants did not occur uniformly throughout the cornfield. Nutrient variability had an impact on plant growth, which resulted in crop yield variability. MZ1 had greater nutrient concentrations in soil and plants than average nutrients within the field and also had a higher crop yield as compared to MZ3, which had lower nutrient concentrations in soil and plants than the average nutrients within the field. Consistent management practices with variable soil physical properties, field hydrologic characteristics, and surface runoff generation impact plant growth and result in a loss in crop yield in some areas of the field. Considering nutrient variability in management zone delineation can be an effective way to improve precision agriculture adoption during the growing season. Mainly, two extreme clusters of nutrient variability (MZ1 and MZ3) in soil and plants were identified, which play an important role in the site-specific management of nutrients to improve crop yield. Adoption of data data-driven approach for nutrient management can improve productivity across the crop fields and benefit farmers. Therefore, it is of utmost importance to study nutrient variability based on the management zone concept using machine learning approaches to increase crop yield, reduce loss of nutrients, and reduce negative environmental aspects.

REFERENCES

Ale, S., Himanshu, S. K., Mauget, S. A., Hudson, D., Goebel, T. S., Liu, B., … Gitz III, D. C. (2021). Simulated dryland cotton yield response to selected scenario factors associated with soil health. *Frontiers in Sustainable Food Systems*, 4, 617509.

American Public Health Association. (1998). *Standard Methods for the Examination of Water and Wastewater*. Washington, DC: American Public Health Association, American Water Works Association, and Water Environment Federation.

Dayton, E. A., Whitacre, S., & Holloman, C. (2017). Comparison of three persulfate digestion methods for total phosphorus analysis and estimation of suspended sediments. *Applied Geochemistry*, *78*, 357–362. https://doi.org/10.1016/j.apgeochem.2017.01.011

Gao, X., Xiao, Y., Deng, L., Li, Q., Quan, W. C., Li, B., Deng, O. P., & Zeng, M. (2019). Spatial variability of soil total nitrogen, phosphorus and potassium in Renshou County of Sichuan Basin, China. *Journal of Integrative Agriculture*, *18*(2), 279–289. https://doi .org/10.1016/S2095-3119(18)62069-6

Hanrahan, B. R., King, K. W., Williams, M. R., Duncan, E. W., Pease, L. A., & LaBarge, G. A. (2019). Nutrient balances influence hydrologic losses of nitrogen and phosphorus across agricultural fields in northwestern Ohio. *Nutrient Cycling in Agroecosystems*, *113*(3), 231–245. https://doi.org/10.1007/s10705-019-09981-4

Himanshu, S. K., Ale, S., Bordovsky, J. P., Kim, J., Samanta, S., Omani, N., & Barnes, E. M. (2021). Assessing the impacts of irrigation termination periods on cotton productivity under strategic deficit irrigation regimes. *Scientific Reports*, *11*(1), 1–16.

Himanshu, S. K., Ale, S., DeLaune, P. B., Singh, J., Mauget, S. A., & Barnes, E. M. (2022). Assessing the effects of a winter wheat cover crop on soil water use, cotton yield, and soil organic carbon in no-till cotton production systems. *Journal of the ASABE*, *65*(5), 1163–1177. https://doi.org/10.13031/ja.15181

Himanshu, S. K., Ale, S., Bell, J., Fan, Y., Samanta, S., Bordovsky, J. P., ... & Brauer, D. K. (2023). Evaluation of growth-stage-based variable deficit irrigation strategies for cotton production in the Texas High Plains. *Agricultural Water Management*, *280*, 108222. https://doi.org/10.1016/j.agwat.2023.108222.

Huat, B. B. K., Ali, F. H. J., & Low, T. H. (2006). Water infiltration characteristics of unsaturated soil slope and its effect on suction and stability. *Geotechnical and Geological Engineering*, *24*(5), 1293–1306. https://doi.org/10.1007/s10706-005-1881-8

Jiang, G., Grafton, M., Pearson, D., Bretherton, M., & Holmes, A. (2021). Predicting spatiotemporal yield variability to aid arable precision agriculture in New Zealand: A case study of maize-grain crop production in the Waikato region. *New Zealand Journal of Crop and Horticultural Science*, *49*(1), 41–62. https://doi.org/10.1080/01140671.2020 .1865413

Kumar, H. (2022). *Quantifying Within-Field Variability in Soil Moisture and Nutrients and Scheduling Site-Specific Irrigation Using Numerical Modeling*. Auburn University, Auburn.

Kumar, H., Srivastava, P., Lamba, J., Diamantopoulos, E., Ortiz, B., Morata, G., Takhellambam, B., & Bondesan, L. (2022a). Site-specific irrigation scheduling using one-layer soil hydraulic properties and inverse modeling. *Agricultural Water Management*, *273*, 107877. https://doi.org/10.1016/J.AGWAT.2022.107877

Kumar, H., Srivastava, P., Lamba, J., Lena, B., Diamantopoulos, E., Ortiz, B., Takhellambam, B., Morata, G., & Bondesan, L. (2023). A methodology to optimize site-specific field capacity and irrigation thresholds. *Agricultural Water Management*, *286*, 108385. https://doi.org/10.1016/j.agwat.2023.108385

Kumar, H., Srivastava, P., Lamba, J., Ortiz, B. V., Way, T. R., Sangha, L., Takhellambam, B. S., & Morata, G. (2021a). Phosphorus variability in the irrigated cropland during a growing season. *ASABE Annual International Virtual Meeting*, *4*, 1. American Society of Agricultural and Biological Engineers. https://doi.org/10.13031/AIM.202100886

Kumar, H., Srivastava, P., Lamba, J., Ortiz, B. V., Way, T. R., Sangha, L., Singh Takhellambam, B., Morata, G., & Molinari, R. (2022b). Within-field variability in nutrients for site-specific agricultural management in irrigated cornfield. *Journal of the ASABE*. https:// doi.org/10.13031/JA.15042

Kumar, H., Srivastava, P., Ortiz, B. V., Morata, G., Takhellambam, B. S., Lamba, J., & Bondesan, L. (2021b). Field-scale spatial and temporal soil water variability in irrigated croplands. *Transactions of the ASABE, 64*(4), 1277–1294. https://doi.org/10.13031/TRANS.14335

Lena, B. P., Bondesan, L., Ortiz, B. V., Pinheiro, E. A. R., Morata, G. T., & Kumar, H., (2021). Evaluation of different negligible drainage flux for field capacity estimation and its implication on irrigation depth for major soil types in Alabama, USA. *American Society of Agricultural and Biological Engineers Annual International Meeting, ASABE 2021.* https://doi.org/10.13031/AIM.202100415

Lena, B. P., Bondesan, L., Pinheiro, E. A. R., Ortiz, B. V., Morata, G. T., & Kumar, H. (2022). Determination of irrigation scheduling thresholds based on HYDRUS-1D simulations of field capacity for multilayered agronomic soils in Alabama, USA. *Agricultural Water Management, 259,* 107234. https://doi.org/10.1016/J.AGWAT.2021.107234

Mahajan, M., Gupta, P. K., Singh, A., Vaish, B., Singh, P., Kothari, R., & Singh, R. P. (2022). A comprehensive study on aquatic chemistry, health risk and remediation techniques of cadmium in groundwater. *Science of The Total Environment, 818,* 151784. https://doi.org/10.1016/j.scitotenv.2021.151784.

Mauget, S. A., Himanshu, S. K., Goebel, T. S., Ale, S., Lascano, R. J., & Gitz III, D. C. (2021). Soil and soil organic carbon effects on simulated southern high plains dryland cotton production. *Soil and Tillage Research, 212,* 105040

McEntee, P. J., Bennett, S. J., & Belford, R. K. (2019). Mapping the spatial and temporal stability of production in mixed farming systems: An index that integrates crop and pasture productivity to assist in the management of variability. *Precision Agriculture, 21*(1), 77–106. https://doi.org/10.1007/S11119-019-09658-6

Plach, J. M., Macrae, M. L., Ali, G. A., Brunke, R. R., English, M. C., Ferguson, G., Lam, W. V., Lozier, T. M., McKague, K., O'Halloran, I. P., Opolko, G., & Van Esbroeck, C. J. (2018). Supply and transport limitations on phosphorus losses from agricultural fields in the lower Great Lakes region, Canada. *Journal of Environmental Quality, 47*(1), 96–105. https://doi.org/10.2134/jeq2017.06.0234

Schröder, J. J., Scholefield, D., Cabral, F., & Hofman, G. (2004). The effects of nutrient losses from agriculture on ground and surface water quality: The position of science in developing indicators for regulation. *Environmental Science and Policy, 7*(1), 15–23. https://doi.org/10.1016/j.envsci.2003.10.006

Sharpley, A. N. (1997). Rainfall frequency and nitrogen and phosphorus runoff from soil amended with poultry litter. *Journal of Environmental Quality, 26*(4), 1127–1132. https://doi.org/10.2134/jeq1997.00472425002600040026x

Shaw, A. N., & Cleveland, C. C. (2020). The effects of temperature on soil phosphorus availability and phosphatase enzyme activities: A cross-ecosystem study from the tropics to the Arctic. *Biogeochemistry, 151*(2–3), 113–125. https://doi.org/10.1007/S10533-020-00710-6/TABLES/3

Singh, R. P., Mahajan, M., Gandhi, K., Gupta, P. K., Singh, A., Singh, P., ... & Kidwai, M. K. (2023). A holistic review on trend, occurrence, factors affecting pesticide concentration, and ecological risk assessment. *Environmental Monitoring and Assessment, 195*(4), 451. https://doi.org/10.1007/s10661-023-11005-2.

Smith, D. R., Owens, P. R., Leytem, A. B., & Warnemuende, E. A. (2007). Nutrient losses from manure and fertilizer applications as impacted by time to first runoff event. *Environmental Pollution, 147*(1), 131–137. https://doi.org/10.1016/j.cnvpol.2006.08.021

Su, B., Zhao, G., & Dong, C. (2018). Spatiotemporal variability of soil nutrients and the responses of growth during growth stages of winter wheat in northern China. *PLoS One, 13*(12), e0203509. https://doi.org/10.1371/journal.pone.0203509

Takhellambam, B. S., Srivastava, P., Lamba, J., McGehee, R. P., Kumar, H., & Tian, D. (2023). Projected mid-century rainfall erosivity under climate change over the southeastern United States. *Science of the Total Environment, 865*, 161119. https://doi.org/10.1016/J.SCITOTENV.2022.161119

Takhellambam, B. S., Srivastava, P., Lamba, J., Zhao, W., Kumar, H., & Tian, D. (2022, July 17–20). Assessment of projected change in Intensity-duration-frequency (IDF) curves for Southeastern, United States using artificial neural networks. *Annual International Meeting, 1.* https://doi.org/10.13031/AIM.202200175

USDA-NASS. (2019). *2017 Census of Agriculture.* Washington, DC: United States Department of Agriculture, National Agriculture Statistics Service.

Wright, A. L., Wang, Y., & Reddy, K. R. (2008). Loss-on-ignition method to assess soil organic carbon in calcareous Everglades wetlands. *Communications in Soil Science and Plant Analysis, 39*(19–20), 3074–3083. https://doi.org/10.1080/00103620802432931

Yost, J. L., Huang, J., & Hartemink, A. E. (2019, March). Spatial-temporal analysis of soil water storage and deep drainage under irrigated potatoes in the Central Sands of Wisconsin, USA. *Agricultural Water Management, 217*, 226–235. https://doi.org/10.1016/j.agwat.2019.02.045

5 Comparative Study on Estimation of Sediment Yield Index Using GIS and Remote Sensing for Soil Erosion Prediction

*Bareerah Khalid, Pema Wangmo,
Prakash Subedi, and Sakron Vilavan*

INTRODUCTION

Soil is one of the essential natural resources that control the economic conditions of a country and that help sustain life on Earth (Pandey et al., 2016). Although the process of soil formation takes many decades, rainwater erosion can reverse this in just a few severe storms, leaving deteriorated soil residues that lead to lower yields (Mauget et al., 2021). Ecosystems and sustainable agriculture are both threatened by the major global issue of soil erosion (Jain et al., 2010). The relocation of nutrients, soils, and geochemical components is predicted by the sediment transfer ratio (Förstner et al., 2004; Owens et al., 2004).

Even though erosion has existed for as long as agriculture, its magnitude has increased considerably (Pimentel et al., 1995; Mishra and Rai., 2013). According to the FAO-led Global Soil Partnership (GSP, 2017), a huge loss of about 400 billion USD per year results from annual soil erosion of 75 billion tons (Pg). The principal causes of soil erosion and nutrient loss are overland flow and rainfall, which result in decreased production, soil sterility, and ecological deterioration (Adimassu et al., 2016; Ma et al., 2016). There are various types of soil erosion, i.e., water erosion, landslides, sheet erosion, terrace failure, mass erosion, pollution (Mahajan et al., 2022; Yatoo et al., 2022; Singh et al., 2023), and many others. Human-induced landslide intensification has been facilitated by the development of buildings and roads, mining, hydropower project construction, and clearing vegetation (Bhattacharyya et al., 2015).

Soil erosion has drastic effects on water quality, agricultural production, and soil fertility, and also influences hydrology and environmental sustainability (Lal,

DOI: 10.1201/9781003441175-5

1998; Lal, 2015). Nutrient loss is associated with soil erosion. Globally, poor land use practices, increased deforestation, overgrazing, and wildfires are human-induced soil erosion factors (UNEP/ISRIC, 1990; Stefanidis, 2011; Efthimiou et al., 2020). Soil erosion is a global issue, but developing nations are particularly affected since their farming populations cannot replenish the nutrients and soil that have been lost (Erenstein, 1999). Therefore, to protect the productivity potential of their land, sustainable land management approaches are critical.

The usage of the Geographical Information System (GIS) helps us in the classification of a river basin. According to studies (Morgan et al., 1998; Jain et al., 2001; Van Rompaey et al., 2001; Behera et al., 2005; Mishra et al., 2007; Machiwal et al., 2010; Sharma et al., 2010; Biswas, 2012) on GIS and Revised Universal Soil Loss Equation (RUSLE) approaches, it is possible to estimate soil erosion and its spatial distribution with more accuracy and fewer parameters than physical-based models in larger regions (Palmate et al., 2022).

We can calculate the amount of sediment loss from our study area by integrating the GIS with the RUSLE and Remote sensing. In this study, we have used ArcGIS software to estimate soil loss by calculating the Sediment Yield Index of non-point sources. The comparative study of erosion intensity and priority ranking for two basins of hilly and plain areas are being discussed.

Soil erosion occurs mainly when dirt particles are left exposed to hard rains, strong winds, and flowing water. In some cases, human activities, particularly farming, land clearing, overgrazing, intensive agriculture, and deforestation increase the rate of soil erosion by up to 1000 times . In Thailand, the introduction of cash crops in the upland has caused major soil erosion not only upland but also in the rain-fed paddy of the lowland (Lorsirirat & Maita, 2006). Bhutan has limited productive land, less than 8% of its total area. Water-induced degradation such as landslides, gullies, ravine formations, and local flooding is a major cause of soil erosion. This chapter provides a comparative study on the estimation of Sediment Yield Index in Thailand and Bhutan using GIS and remote sensing for soil erosion prediction. This will help policymakers to make wise decisions to support sustainable development planning and appropriate mitigation interventions.

STUDY AREA

This study was carried out in two river basins, namely, the Drangme Chhu River Basin (DCRB) in Bhutan and the Lower Chao Phraya River Basin (LCPRB) in Thailand. The climate of the Drangme Chhu River is extremely diverse, ranging from hot, humid subtropical conditions in the south to cold, dry alpine conditions in the north. From May to October, the southwest monsoon brings heavy rainfall of more than 4000 millimeters in the southern part – and there is a marked dry season in winter. The climate of the Lower Chao Phraya River Basin ranges from hot to humid, mostly affected by an Asian tropical monsoon in the north and mild throughout the year in the south which is faced with a marine climate. The topography of the Drangme Chhu River Basin is hilly and mountainous with balance in the plains whereas the Lower Chao Phraya River Basin is mainly an alluvial plain formed by the river system.

The Drangme Chhu River Basin is located in the temperate zone of Bhutan, at a latitude of 26°56′N to 28°04′N and longitude of 90°51′E to 91°59′E, covering approximately an area of 7025 square kilometers. The basin covers the eastern part of Bhutan and is one of the sub-watersheds to the biggest river basin of the country – Manas Basin. The main tributaries, Kuri Chhu and Gong Chhu, form the main DRCB basin, both of which originate from Tibet in the north and exit in the south to join the Manas and Brahmaputra rivers in India. Almost 93% of the basin is covered by vegetation and has an annual rainfall range from 720–1900 mm.

The Lower Chao Phraya River Basin covers approximately 4300 square kilometers of the total land area of Thailand. It is located at the latitude of 15°20′N to 16°0′N and longitude of 99°06′E to 100°4′E, forming the lower part of the main Chao Phraya. This lower basin is formed by four main rivers, namely, Ping, Wang, Yom, and Nan, which follow from the north to combine at Nakorn Sawan province to form the main Chao Phraya and then flow south to the Gulf of Thailand. Almost 91% of the basin area is used under cultivation. All parts of the basin receive an almost uniform annual precipitation of 1162–1185 mm.

DATA USED

The spatial datasets of the digital elevation model (DEM), Landsat 8, FAO soil, and precipitation data from the Climate Research Unit were the main sources of data for this study.

DIGITAL ELEVATION MODEL (DEM)

The DEM datasets were obtained from the Earth Explorer website (https://earthexplorer.usgs.gov/) at 30 m spatial resolution. The datasets are used for the delineation of watersheds and sub-basins. Drangme Chhu River Basin is divided into six sub-watersheds and the Lower Chao Phraya River Basin is into five sub-watersheds (Figure 5.1). DEM was also used to derive the slope for the study areas.

LAND USE LAND COVER CLASSIFICATION

The Landsat 8 data obtained from the Earth Explorer website for the time span of January 2020 to January 2021 (https://earthexplorer.usgs.gov/) on August 23, 2022, was used for land use classifications of the study areas. The land use land cover classes were extracted using the ArcGIS ISO unsupervised classification method. We evaluated the accuracy, Kappa Coefficient (T), and found an 85% accuracy assessment image classification. The extracted land use is classified into four different classes (Figure 5.2). Table 5.1 shows the land use classes for two different basins. Vegetation is the major land cover for the Drangme Chhu River Basin with more than 93% of coverage while cultivation is the major land use land cover for the

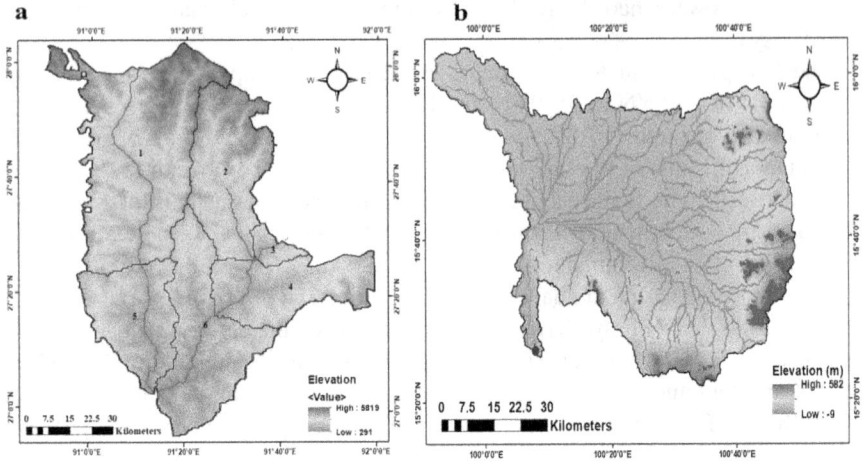

FIGURE 5.1   Elevation map: Drangme Chhu River Basin (a); (b) Lower Chao Phraya River Basin

Lower Chao Phraya River Basin covering about 91% considering only the agricultural work.

SOIL CLASSIFICATION

Soil texture and soil type are some of the main parameters which determine the erosion intensity of an area. For this study, the soil data have been obtained from the Digital Soil Map of the World (DSMW), FAO website (https://www.fao.org/soils

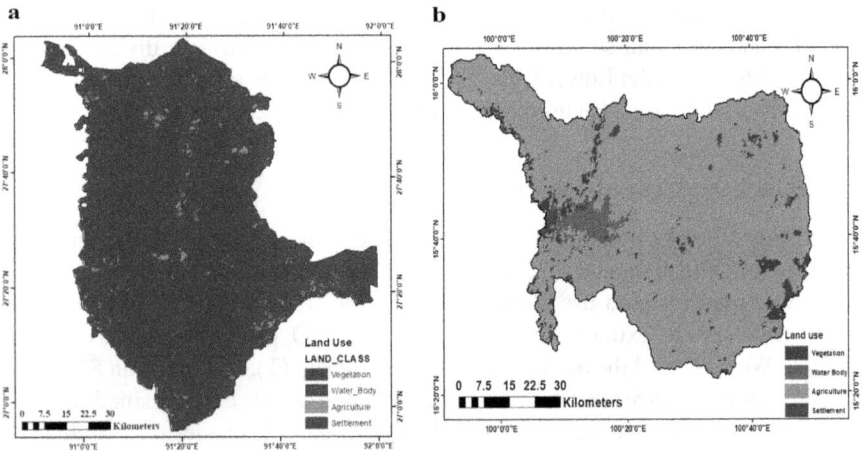

FIGURE 5.2   Basin-wise land use land cover classification

TABLE 5.1
Area-Wise Land Use Classification for Study Areas

		Area (sq. km)		Area (%)	
Sl. No	Land Use Class	DCRB	LCPRB	DCRB	LCPRB
1	Vegetation	6538.75	193.19	93.08	4.39
2	Water Body	236.37	150.91	3.36	3.43
3	Cultivation	92.74	4009.61	1.32	91.15
4	Settlement	156.79	45.29	2.23	1.03
Total		7024.65	4399.00	100	100

-portal/data-hub/soil-maps-and-databases/en/). Loamy soil is the most dominant soil cover in the Drangme Chhu River Basin of which orthic acrisols soil texture type is prominent, covering about 66% followed by lithosols with 33% of the basin area (Table 5.2, A). On the other side, silt can be observed as the major soil cover in the Nakorn Sawan Basin, covering about 42% and silty loam texture with 37% of the basin soil type (Table 5.2, B) (Figure 5.3).

RAINFALL DATA

The Drangme Chhu River flows through Bhutan in the southwest direction between two ranges of the Lower Himalayas in V-shaped gorges and enters Assam in India into the south-central foothills of the Himalayas. The valley opens up in the foot-hills; marked by the formation of swamps and marshes in the plains. The upper

TABLE 5.2
Area-Wise Soil Classification

A. Drangme Chhu River Basin

Sl. No	Soil class	Area (sq. km)	Area (%)
1	Loamy	2378	33.81
2	Loamy Sand	4637.5	65.93
3	Water	18.5	0.26
Total		7034	100

B. Lower Chao Phraya River Basin

Sl. No	Soil class	Area (sq. km)	Area (%)
1	Silt loam	1680.40	37.87
2	Silt	1867.30	42.08
3	Silt Clay	745.08	16.79
4	Water	144.34	3.25
Total		4437.13	100

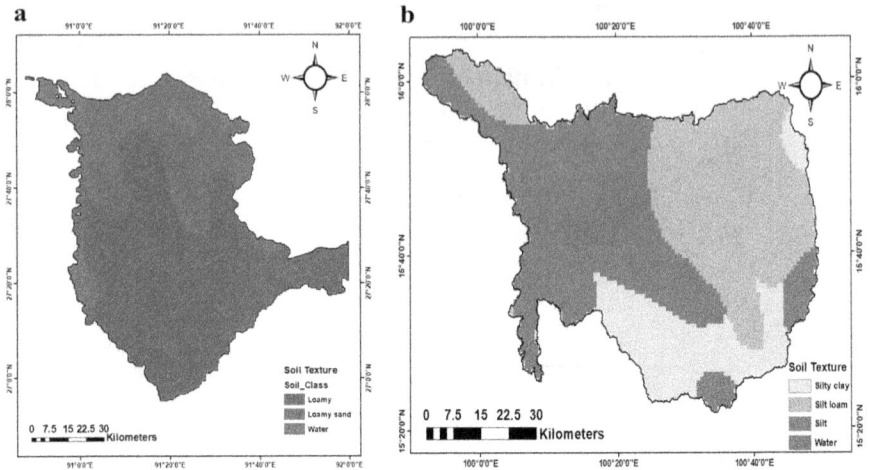

FIGURE 5.3   Basin-wise soil classification

catchment is snowbound while the middle and lower catchments are thickly forested. However, we do not have any literature to quantify how much water glacier melting generates the water flows.

Soil erosion is mainly dependent on the amount and intensity of rainfall for an area. For these study areas, rainfall data for the year 2022 time period were obtained from high-resolution gridded datasets of the Climate Research Unit (https://crudata .uea.ac.uk/cru/data/hrg/). The minimum rainfall received by the Drangme Chhu River Basin is 720 mm and the maximum rainfall is up to 1900 mm. The areas at lower elevations, near the outlet have received maximum rainfall for that particular year whereas areas at higher altitudes have received less rainfall compared to other parts of the basin.

As the Lower Chao Phraya River Basin has almost uniform elevation, the rainfall received is the same throughout the basin, ranging from 1142–1185 mm (Figure 5.4).

SLOPE

The rate of soil loss is directly linked with the slope of an area as most areas with high slopes are prone to more soil loss compared to gentle slopes. The slope for both the study areas has been generated from DEM (Figure 5.5). These slopes are further divided into subcategories to determine the land slope variations in particular study areas. Table 5.3 shows that almost 28% of the part of the basin is classified under a moderately steep slope in the Drangme Chhu River Basin while the majority of the area in the Nakorn Sawan Basin falls under the gentle slope category, covering 79% of the area.

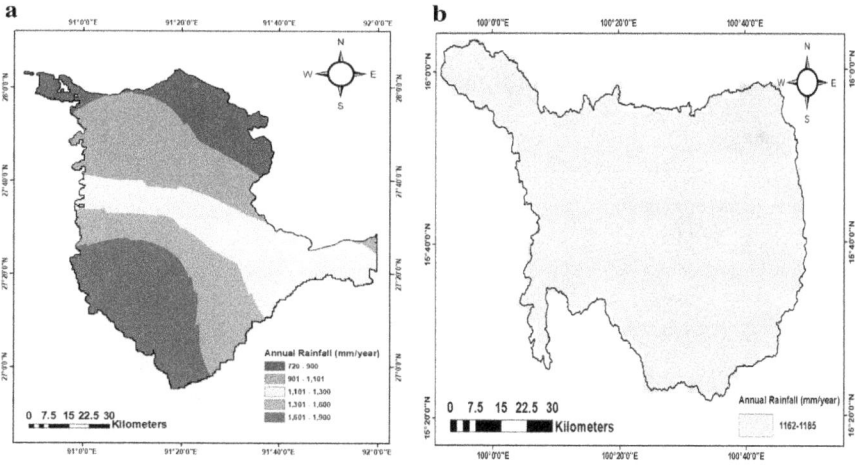

FIGURE 5.4   Basin-wise rainfall distribution

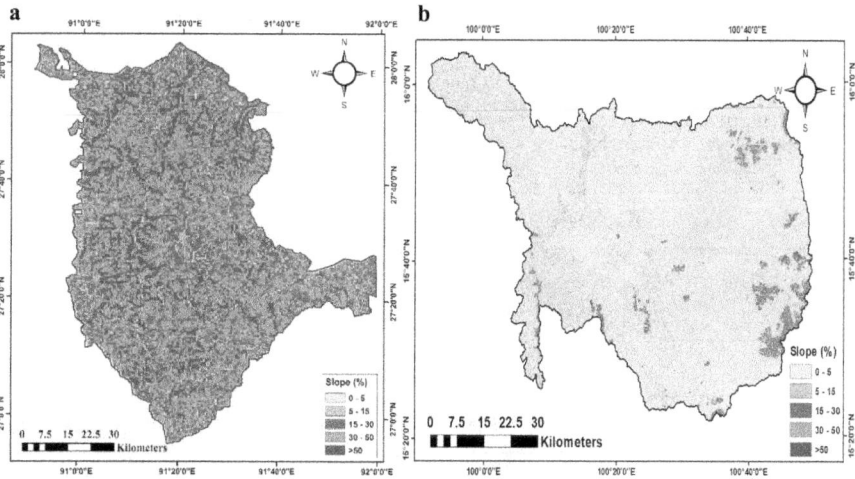

FIGURE 5.5   Basin-wise slope category in percent rise

METHODS

The rate of soil erosion and sediment flow from an area is generally dependent upon climatic variables such as rainfall and runoff as well as the land characteristics of soil and vegetation. The parameters like slope, land use, and soil types are some of the driving forces that influence the rate of sediment flow and deposition in an area. However, these characteristics are found to differ considerably within the different

TABLE 5.3
Basin-Wise Slope Category and Classification

Slope category (%)	Slope classification	Area (sq. km) DCRB	LCPRB	Area (%) DCRB	LCPRB
0–5	Gentle	877.00	3488.55	12.77	79.54
5–15	Moderate	1582.75	742.55	23.00	16.93
15–30	Moderately Steep	1975.75	72.81	27.92	1.66
30–50	Steep	1638.25	61.35	23.68	1.40
> 50	Very Steep	879.50	20.91	12.63	0.48
Total		6953.25	4386.17	100	100

sub-basins of a catchment which therefore needs to be discretized for soil loss computation. A grid-based discretization is found to be the best-suited procedure in each process-based model as well as in other simple models (Beven, 1996; Kothyari & Jain, 1997). The overall methodology for this study is described in Figure 5.6.

ASSIGNING DELIVERY RATIO

The sediment eroded and transferred from any area gets deposited within a catchment. The sediment transported and deposited in a catchment embraces the determination of the rate of soil erosion. The amount of sediment removed per unit area from

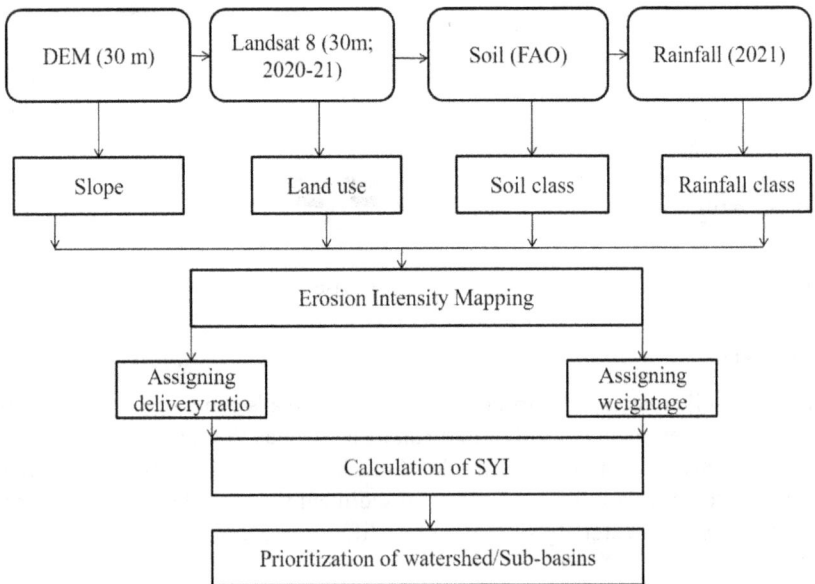

FIGURE 5.6   Overall methodological framework

TABLE 5.4
Stream Distance and Delivery Ratio

Stream distance (km)	Delivery ratio
0–1	1
1–10	0.9
10–30	0.8
> 30	0.7

a catchment by different parameters is termed the sediment yield. The delivery ratio is the ratio of sediment yield to the total surface area (Jain & Kothyari, 2000). The delivery ratio is assigned based on the nearest river network distance from the outlet. The values are assigned based on the length of the stream (Table 5.4).

ASSIGNING WEIGHTAGE

The weightage for a different land characteristic is assigned based on the effect of sediment detachment from upstream and deposition within the catchment (Table 5.5). This study classified different forest cover as one under vegetation with the lowest weightage. The highest weightage value is assigned to parameters that can cause or correspond to a high erosion rate.

COMPUTATION OF SEDIMENT YIELD INDEX

The map layer for each parameter of the slope, land use land cover, soil, type, and rainfall was prepared followed by assigning the weightage for the determined classes or categories to calculate the Sediment Yield Index using the following equation.

$$SYI = \frac{\Sigma\left(A_i \times W_i \times D_i\right) \times 100}{Aw},$$

Where, $i = 1\text{-}N$
A_i = Area of ith basin unit
W_i = Weighted value of ith mapping unit
D_i = Delivery ratio
Aw = Total area of a sub-basin
N = Numbers of mapping units

RESULTS AND DISCUSSIONS

EROSION INTENSITY CLASSIFICATION

The erosion intensity for both catchments was classified under four classes which are negligible, slight, moderate, and severe (Figure 5.7). The analysis showed slight erosion in 58.7% of the area and negligible in 24.8% of the area under the Drangme

TABLE 5.5
Assigned Weightage to the Key Parameters Used for Erosion Intensity Mapping

Sl. No	Parameters	Category/ Class	Weightage
1	Land Use Land Cover	Vegetation	1
		Water body	2
		Agriculture	3
		Settlement	4
2	Soil Class	Silt loam	4
		Silt	3
		Silty clay	3
		Loamy Sand	3
		Loam	2
		Water	1
3	Slope (percent rise)	0–5	1
		5–15	2
		15–30	3
		30–50	4
		>50	5
4	Rainfall	720–900	1
		901–1100	2
		1101–1300	3
		1301–1600	4
		1601–1900	5

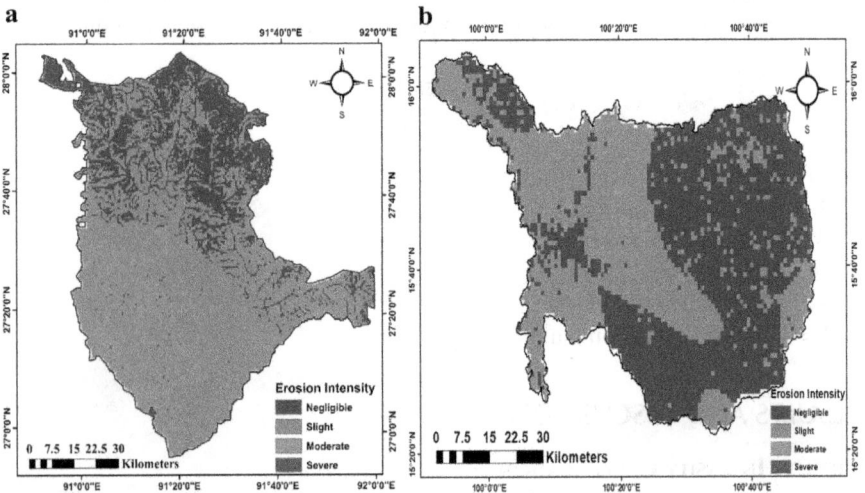

FIGURE 5.7 Basin-wise classification of erosion intensity

TABLE 5.6

Basin-Wise Erosion Intensity Classification

A. Drangme Chhu River Basin

Sl. No	Sum of weightage	Erosion intensity class	Area (sq. km)	Area (%)
1	4.0–7	Negligible	1686.50	24.84
2	8.0–11	Slight	3986.25	58.72
3	12.0–14	Moderate	1097.25	16.16
4	15.0–17	Severe	18.75	0.28
Total			6788.75	100.00

B. Lower Chao Phraya River Basin

Sl. No	Sum of weightage	Erosion intensity class	Area (sq. km)	Area (%)
1	3–5	Negligible	2301.90	53.38
2	6–8	Slight	1939.05	44.96
3	9–10	Moderate	66.50	1.54
4	11–12	Severe	4.93	0.11
Total			4312.37	100.00

Chhu River Basin. While 16% of the area showed moderate erosion intensity, less than 1% of the area under this river basin is prone to erosion (Table 5.6 A).

More than 53% of the area in the Lower Chao Phraya River Basin showed negligible erosion intensity while 44.9% of the area is prone to slight erosion. Moderate and severe erosion intensities were less than 2% in this catchment area (Table 5.6 B). The result revealed these two catchments are not at risk of severe erosion due to the high vegetation cover of the Drangme Chhu River Basin and the lower slope rise of the Lower Chao Phraya River Basin.

PRIORITIZATION OF SUB-BASINS

The Sediment Yield Index was calculated using ArcGIS software environments for each of the sub-basins. The priority classes are further divided into four groups of low, medium, high, and very high rank (Table 5.7 and Figure 5.8). The sub-basins having SYI value of more than 600 were categorized as very high-priority classes. This means the areas under this are prone to severe erosion and need to be taken under conservation treatments.

The analysis shows that 33% of the total Drangme Chhu River Basin falls under a very high-priority rating while 50% of the area falls under a low to medium priority rank. On the other hand, the Lower Chao Phraya River Basin has a total of approximately 64% of the area under low- to medium priority rating and less than 2% of the total area falls under high priority.

TABLE 5.7
Basin-Wise Priority Ranking Based on SYI and Area Coverage

Sl. No	SYI values	Area (sq. km)		Area (%)		Priority rank
		DCRB	LCPRB	DCRB	LCPRB	
1	200–270	1158.75	1444.84	17.07	33.50	Low
2	271–400	3387.75	2745.20	49.90	63.66	Medium
3	401–600	0.00	122.32	0.00	2.84	High
4	> 601	2242.25	0.00	33.03	0.00	Very high
	Total	6788.75	4312.37	100	100	

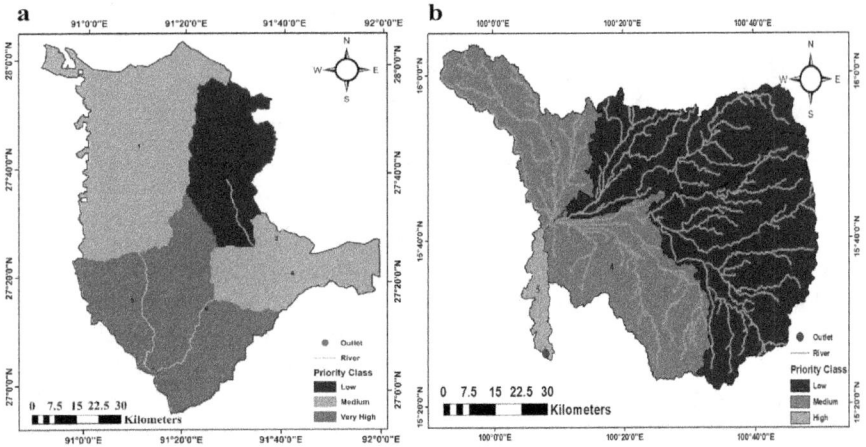

FIGURE 5.8 Basin-wise priority ranking of sub-basins

In Table 5.8 A, two of the sub-basins under the Drangme Chhu River Basins are categorized into very high-priority ratings, and therefore conservation treatment is necessary as the erosion rate can be very high. The table also shows three sub-basins in medium and one sub-basin under low-priority classes.

Table 5.8 B shows one sub-basin of the Lower Chao Phraya River Basin under high and low-priority classes whereas three sub-basins are shown under medium priority rating class. The areas under low and medium priority classes don't need conservation treatment as the erosion rate is either very low or negligible.

TABLE 5.8

Sub-Basin-Wise Priority Ranking as per Calculated SYI Values

A. Drangme Chhu River Basin

Sub-basin	Area (sq. km)	SYI	Priority	Rank
1	2345.75	309.33	Medium	5
2	1158.75	268.66	Low	6
3	128	326.21	Medium	4
4	914	387.18	Medium	3
5	905.25	616.32	Very high	2
6	1337	819.13	Very high	1

B. Lower Chao Phraya River Basin

Sub-basin	Area (sq. km)	SYI	Priority	Rank
1	709.29	359.07	Medium	2
2	1023.71	274.78	Medium	4
3	1444.84	252.32	Low	5
4	1012.21	324.09	Medium	3
5	122.32	490.60	High	1

CONCLUSION

This comparative study investigated the soil erosion prediction using the Sediment Yield Index using GIS tools by analyzing spatial data information on slope, land use, soil classes, and rainfall from different sources for two catchments, which are, the Drangme Chhu and the Lower Chao Phraya River basins. Each of these basins was subdivided into sub-basins for erosion priority ranking to determine conservation prioritization. In the Drangme Chhu River, sub-basins 5 and 6 showed high SYI and whereas sub-basin 5 showed high SYI in the Lower Chao Phraya River Basin which requires higher priority. Sub-basins 1, 3, and 4 are under the medium priority of the Drangme Chhu River Basin while sub-basins 1, 2, and 4 are under the Lower Chao Phraya River Basin. Sub-basins 2 and sub-basins 3 of Drangme Chhu and Lower Chao Phraya River basins fall under the lower priority.

REFERENCES

Adimassu, Z.; Langan, S.; Johnston, R.; Mekuria, W.; Amede, T. Impacts of soil and water conservation practices on crop yield, run-off, soil loss and nutrient loss in Ethiopia: Review and synthesis. *Environ. Manag.* 2016, 59(1), 87–101.

Behera, P.; Rao, K.H.V.D.; Das, K.K. Soil erosion modeling using MMF model—A remote sensing and GIS perspective. *J. Indian Soc. Remote Sens.* 2005, 33(1), 165–176.

Beven, K.J. (1996) A discussion of distributed modeling. In *Distributed Hydrological Modelling* (ed. M. B. Abbott; J. C. Refsgaard), pp. 255–278. Dordrecht: Kluwer.

Bhattacharyya, R.; Ghosh, B.N.; Mishra, P.K.; Mandal, B.; Rao, C.S.; Sarkar, D.; Das, K.; Anil, K.S.; Lalitha, M.; Hati, K.M.; et al. Soil degradation in India: Challenges and potential solutions. *Sustainability* 2015, 7(4), 3528–3570.

Biswas, S. Estimation of soil erosion using remote sensing and GIS and prioritization of catchments. *Int. J. Emerg. Technol. Adv. Eng.* 2012, 2, 124–128.

Efthimiou, N.; Psomiadis, E.; Panagos, P. Fire severity and soil erosion susceptibility mapping using multi-temporal earth observation data: The case of Mati fatal wildfire in Eastern Attica, Greece. *Catena* 2020, 187, 104320.

Erenstein, O.C.A. (1999). *The Economics of Soil Conservation in Developing Countries: The Case of Crop Residue Mulching.* Wageningen: Wageningen University.

Förstner, U.; Heise, S.; Schwartz, R.; Westrich, B.J.; Ahlf, W. Historical contaminated sediments and soils at the river basin scale: Examples from the Elbe River catchment area. *J. Soils Sediments* 2004, 4(4), 247–260.

GSP Global Soil Partnership endorses guidelines on sustainable soil management 2017. Available online: http://www.fao.org/global-soilpartnership/resources/highlights/detail/en/c/416516/ (accessed on 15 October 2021).

Jain, M.K.; Kothyari, U.C. Estimation of soil erosion and sediment yield using GIS. *Hydrol. Sci. J.* 2000, 45(5), 771–786.

Jain, M.; Mishra, K.; Surendra, K.; Shah, R. Estimation of sediment yield and areas vulnerable to soil erosion and deposition in Himalayan watershed using GIS. *Curr. Sci.* 2010, 98(2), 25.

Jain, S.K.; Kumar, S.; Varghese, J. Estimation of soil erosion for a Himalayan watershed using GIS technique. *Water Resour. Manag.* 2001, 15(1), 41–54.

Kothyari, U.C.; Jain, S.K. Sediment yield estimation using GIS. *Hydrol. Sci. J.* 1997, 42(6), 833–843.

Lal, R. Restoring soil quality to mitigate soil degradation. *Sustainability* 2015, 7(5), 5875–5895.

Lal, R. Soil erosion impact on agronomic productivity and environment quality. *Crit. Rev. Plant Sci.* 1998, 17(4), 319–464.

Lorsirirat, K.; Maita, H. (2006). Soil erosion problems in Northeast Thailand: A case study from the view of agricultural development in a rural community near Khon Kaen. *Disaster Mitigation of Debris Flows, Slope Failures and Landslides*, 675.

Ma, X.; Li, Y.; Li, B.; Han, W.; Liu, D.; Gan, X. Nitrogen and phosphorus losses by runoff erosion: Field data monitored under natural rainfall in three gorges reservoir area, China. *CATENA* 2016, 147, 797–808.

Machiwal, D.; Srivastava, S.K.; Jain, S. Estimation of sediment yield and selection of suitable sites for soil conservation measures in Ahar river basin of Udaipur, Rajasthan using RS and GIS techniques. *J. Indian Soc. Remote Sens.* 2010, 38(4), 696–707.

Mahajan, M., Gupta, P. K., Singh, A., Vaish, B., Singh, P., Kothari, R., & Singh, R. P. (2022). A comprehensive study on aquatic chemistry, health risk and remediation techniques of cadmium in groundwater. *Science of The Total Environment, 818*, 151784. https://doi.org/10.1016/j.scitotenv.2021.151784.

Mauget, S.A.; Himanshu, S.K.; Goebel, T.S.; Ale, S.; Lascano, R.J.; Gitz III, D.C. Soil and soil organic carbon effects on simulated Southern High Plains dryland Cotton production. *Soil Till. Res.* 2021, 212, 105040.

Mishra, A.; Kar, S.; Singh, V.P. Prioritizing structural management by quantifying the effect of land use and land cover on watershed runoff and sediment yield. *Water Resour. Manag.* 2007, 21(11), 1899–1913.

Mishra, P.K.; Rai, S.C. Use of indigenous soil and water conservation practices among farmers in Sikkim Himalaya. *Indian J. Tradit. Knowl.* 2013, 12, 454–464.

Morgan, R.P.C.; Quinton, J.N.; Smith, R.E.; Govers, G.; Poesen, J.W.A.; Auerswald, K.; Styczen, M.E. The European Soil Erosion Model (EUROSEM): A dynamic approach for predicting sediment transport from fields and small catchments. *Earth Surf. Process. Landf. J. Br.* 1998, 23, 527–544.

Owens, P.N.; Apitz, S.; Batalla, R.; Collins, A.; Eisma, M.; Glindemann, H.; Hoonstra, S.; Köthe, H.; Quinton, J. K., Westrich, B, White, S, Wilkinson, H Sediment management at the river basin scale: Synthesis of SedNet Working Group 2 outcomes. *J. Soils Sediments* 2004, 4, 219–222.

Palmate, S.S.; Amrit, K.; Jadhao, V.G.; Dayal, D.; Himanshu, S.K. (2022) Prioritization of erosion prone areas based on a sediment yield index for conservation treatments: A case study of the upper Tapi River basin. In *Advances in Remediation Techniques for Polluted Soils and Groundwater*, Edited by Pankaj Kumar Gupta, Basant Yadav and Sushil Kumar Himanshu, pp. 291–307. Elsevier.

Pandey, A.; Himanshu, S.K.; Mishra, S.K.; Singh, V.P. Physically based soil erosion and sediment yield models revisited. *Catena*, 147, 595–620.

Pimentel, D.; Harvey, C.; Resosudarmo, P.; Sinclair, K.; Kurz, D.; McNair, M.; Crist, S.; Shpritz, L.; Fitton, L.; Saffouri, R.; et al. Environmental and economic costs of soil erosion and conservation benefits. *Science* 1995, 267(5201), 1117–1123.

Sharma, A.; Tiwari, K.N.; Bhadoria, P.B.S. Effect of land use land cover change on soil erosion potential in an agricultural watershed. *Environ. Monit. Assess.* 2010, 173(1–4), 789–801.

Singh, R. P., Mahajan, M., Gandhi, K., Gupta, P. K., Singh, A., Singh, P., ... & Kidwai, M. K. (2023). A holistic review on trend, occurrence, factors affecting pesticide concentration, and ecological risk assessment. *Environmental Monitoring and Assessment*, *195*(4), 451. https://doi.org/10.1007/s10661-023-11005-2.

Stefanidis, S. Estimation of the mean annual sediment discharge in fire affected watersheds. *Silva Balc.* 2011, 12, 91–96.

UNEP; ISRIC. (1990). *World Map on Status of Human Induced Soil Degradation*. Nairobi: UNEP.

Van Rompaey, A.; Verstraeten, G.; Van Oost, K.; Govers, G.; Poesen, J. Modelling mean annual sediment yield using a distributed approach. *Earth Surf. Process Landf.* 2001, 26(11), 1221–1236.

Yatoo, A.M., Ali, M.N., Zaheen, Z., Baba, Z.A., Ali, S., Rasool, S., Sheikh, T.A., Sillanpää, M., Gupta, P.K., Hamid, B. and Hamid, B., (2022). Assessment of pesticide toxicity on earthworms using multiple biomarkers: a review. *Environmental Chemistry Letters*, *20*(4), 2573–2596. https://doi.org/10.1007/s10311-022-01386-0.

6 Soil Moisture Flows Modeling for Micro-Irrigation and Nutrient Load Management

Mohammad Abdul Kader and Md. Shariot-Ullah

INTRODUCTION

Irrigation is necessary to ensure crop production sustainability and has become an indispensable element in landscape architecture (Incrocci et al., 2020). The increasing demand for water and the effects of global warming have made water management a critical issue (Himanshu et al., 2022). Soil moisture is essential for irrigation management, as it affects the growth and productivity of crops, the movement of water and nutrients, and the overall water balance of an ecosystem (Gülser & Candemir, 2014). It can be influenced by rainfall, evaporation, and anthropogenic activities such as irrigation, drainage, and land use change.

Crop productivity is impacted by limited access to water and inefficient irrigation practices in the farmer fields (Ale et al., 2020; Fan et al., 2022; Singh et al., 2022). Although several water-saving technologies are available, the installation cost is very high and limited in developing countries (Viswanathan et al., 2016). Thus, finding alternative irrigation water management techniques to improve sustainable crop water productivity is crucial (Himanshu et al., 2021, 2023). In contrast, optimal irrigation water management may require a thorough understanding of plant water demands, as well as water and nutrient dynamics in the soil (Kumar et al., 2022). Therefore, the temporal and spatial variability of soil water dynamics in the soil–plant–atmospheric continuum should be represented by decision-support technologies. Water flow simulation models are needed to accurately capture crop growth, soil physical processes, and water balance components. In contrast, proper understanding and knowledge of soil physical properties are essential for sustainable irrigation and water management within the soil root zone (Scott et al., 2000). Advances in irrigation water storage and distribution have enabled improved application methods, such as low-level sprinklers, microjets, and drip irrigation. A well-designed fertigation and irrigation system can minimize water and fertilizer application costs while supplying nutrients to the wetted irrigation zone around the plant in precise

DOI: 10.1201/9781003441175-6

and consistent amounts. Applying a small amount of nutrients/fertilizers to trees at regular intervals during the growing season provides advantages over traditional fertilizer approaches.

Monitoring soil moisture in agricultural cropland is critical for irrigation planning and understanding the relationship between soil water potential, crop water uptake, and irrigation water volume (Kumar et al., 2021; Bwambale et al., 2022). Until now, soil moisture has been measured in several ways, including gravimetric methods, time domain reflectometry (TDR), neutron probes, and soil water potential using different sensors/probes. The dynamics of soil moisture and modeling studies provide accurate information to forecast the water and solute amount in the vadose zone (Gupta et al., 2023), which can support saving water and predict crop yield. In order to maximize irrigation and water management strategies, it is crucial to monitor soil moisture and understand how different components affect it.

Understanding and managing soil moisture dynamics are essential for sustainable agriculture and environmental conservation, which is a critical aspect that influences crop growth, yield, and soil health. Micro-irrigation systems have become increasingly popular for their potential to save water, improve crop yields, and reduce nutrient loss. This chapter discusses different approaches of soil moisture flow models, the role of micro-irrigation in managing soil water and nutrient flows, and their potential for saving water in agriculture and environmental conservation.

WATER FLOW MODELING APPROACHES

Unsaturated hydraulic conductivity and soil hydraulic head gradients are responsible for explaining the spatiotemporal dynamics of soil moisture (Vereecken et al., 2016). The predominant techniques employed in modeling soil water dynamics involve the utilization of the Richard equation methods, which integrate the Darcy equation and the continuity equation, to facilitate the soil water balance (Hunink et al., 2011). Soil water flow modeling simulates water movement through the soil using mathematical equations and algorithms (Šimůnek et al., 2007). This simulation can help predict the movement of water and solutes in the soil and can be used for various applications, such as irrigation and drainage design, crop water uptake, and environmental impact assessments. Moreover, modeling approaches help to minimize water and nutrient losses due to runoff or leaching, which can have negative ecological impacts (Tu et al., 2021). Water movement in the soil and its effects on crop growth and water use are crucially essential in agricultural water management. Various models can be used to predict soil water flows, including analytical solutions, numerical models, and physically based models. Each model needs specific data to fit the parameters based on the objective functions.

Predicting soil water flux in the vadose zone, taking into account water uptake by plant roots, is essential to achieve optimal agricultural water use and prevent transport of dissolved soil matter to surface and groundwater reservoirs (Xu et al., 2017). The modeling of soil water is essential for saving irrigation because it allows for predicting water movement in the soil. By understanding soil moisture dynamics, farmers can optimize irrigation schedules to ensure that crops receive enough water to grow. However, the soil moisture flow variation is not the same under different soil layers. The amount of soil moisture dynamics depends on soil types and climatic conditions. The top layers contain lower soil moisture than the deeper soil layer due to high evaporation demand from the surface layer, followed by deep layers. The subsurface drip system is prejudiced by various factors including the crop type, soil type, initial soil water content, evaporation rates, emitter discharge, water application rates, drip spacing, and lateral depth (Kader et al., 2019). In order to optimize water application to the soil root zone, the placement depth and spacing of the emitter must be determined to minimize seepage and evaporation. However, soil water distribution models can be divided into analytical, numerical and empirical models.

Numerical models simulate water movement in the soil using finite differences, finite elements, and finite volumes. The choice of data types for the preceding analysis is particularly significant because soil models frequently incorporate a large number of state variables (such as soil moisture, soil matric potential, transpiration, and soil heat flux). Different data types carry various information about the individual compartments and their respective processes (Vereecken et al., 2008). They have become increasingly popular due to the limitations of analytical models based on numerous assumptions. Various numerical models can simulate soil water movements, such as HYDRUS, SWAT, SWAP, RZWQM, and RMS. These models use different mathematical equations and algorithms to simulate water movement in the soil and can consider factors such as soil properties, plant growth, micro-irrigation and fertilization practices, and climatic conditions. The output of these models can be used to optimize irrigation schedules and fertilization rates and to improve the efficiency of irrigation systems as well as enhance crop yield and quality. It is crucial to remember that the accuracy of soil water simulation relies on the quality of input data, the complexity of the model, and the model's ability to replicate the processes in the field.

THE HYDRUS MODEL

The HYDRUS model is a commonly utilized tool for the simulation of water and solute movement within soil (Šimůnek et al., 2013). It uses the finite element method to solve the Richard equation, which describes water flow in unsaturated soils. The software can simulate the movement of water, heat, and solutes in one- and two-dimensional systems. In numerous research, HYDRUS-1D has been extensively utilized to model water movement under drip irrigation. The model has been widely and successfully used to simulate water movement under various conditions due to its adaptability to accommodate distinct types of boundary conditions, the root uptake of water and nutrients, and its ease of use due to its graphical, user-friendly interface. One of the crucial components in improving micro-irrigation management is the ability to assess irrigation performance by modeling soil evaporation using HYDRUS in conjunction with irrigation performance indicators like water supply, relative irrigation, depleted fraction, and overall consumed ratio (Zhou & Zhao, 2019).

HYDRUS especially version 2D/3D is a versatile model that can be used for a broad range of applications, including unsaturated zone models, enabling users to evaluate various irrigation schedules, root water uptake studies, and leaching of nutrients and contaminants in agricultural applications (Morianou et al., 2023). Additionally, it has been used extensively to analyze the effects of different irrigation and fertigation systems (Phogat et al., 2014). One of the significant advantages of HYDRUS is that it has no limitations on spatial or temporal scales, making it a powerful tool for modeling complex systems. Phogat et al. (2014) utilized HYDRUS-2D to enhance fertigation efficiency and prevent groundwater contamination by optimizing drip irrigation timescales of horticultural crops. Mguidiche et al. (2015) simulated soil moisture content in potatoes using subsurface drip irrigation and demonstrated a close correlation between the simulated and measured data within the soil root zone. Provenzano (2007) demonstrated the appropriateness of using HYDRUS for modeling water contents around a buried emitter. A comprehensive study of the mechanisms taking place in a subsurface drip irrigated soil profile, including root growth, distribution, and plant water uptake, is crucial for creating accurate models of soil water dynamics.

THE AQUACROP MODEL

AquaCrop is a crop growth model that simulates crops' water, nutrient, and radiation balance (Steduto et al., 2009). AquaCrop can be used to optimize irrigation schedules and predict crop yields under different water and nutrient management scenarios. Based on a modification of Ritchie's equations, the water balance method incorporates the dividing of soil evaporation and plant transpiration. Under water-limited circumstances, the AquaCrop model can also forecast crop productivity, water needs, and use of water efficiency. Previous research on micro-irrigated and fertigated crops using the AquaCrop model with proper calibration accurately predicted yield, nutrient uptake, and increased agricultural productivity (Plauborg et al., 2022). Furthermore, there is evidence that micro-irrigation and nutrient control based on models saved water and fertilizer, particularly in subsurface drip irrigation systems, compared to surface drip irrigation systems and sprinkler irrigation with split nitrogen application which significantly improved crop productivity (Plauborg et al., 2022; Tan et al., 2018).

HYDRUS and AquaCrop are popular numerical models for simulating soil water dynamics. Although HYDRUS is a powerful model for simulating water and solute transport in the soil, AquaCrop is a crop growth model that simulates water and nutrient uptake by crops. Both models have advantages and can be used for different purposes depending on the available data and the specific application. Both models have their strengths and weaknesses, and the selection of which model to use depends on the specific application and the data available.

SWAP (SOIL–WATER–ATMOSPHERE–PLANT)

The dynamics of water and solutes in the soil, plant, and atmosphere continuum are simulated by the SWAP computer model (Van Dam et al., 1997). This model solves the Richard equation with the finite difference technique. SWAP has the capacity to manage complex boundary conditions and simulate the effects of drainage and irrigation. Yields of crops and water productivity were simulated using a SWAP agro-hydrological model with planned sprinkler and surface irrigation systems (Xue & Ren, 2016). Sabzzadeh & Alimohammadi (2023) used to SWAP model to simulate the regional distributions of salinity, phosphate, and nitrate in an irrigation district in west Iran.

RZWQM (Root Zone Water Quality Model)

RZWQM is a computer model that simulates the water, heat, and solute dynamics in the soil–plant–atmosphere continuum. It uses the finite difference method to solve the Richards equation and can simulate water flow in one- and two-dimensional systems. It also considers plant water and nutrient uptake, soil organic matter dynamics, and pollutants. The RZWQM assessed how various surface drip fertigation management affected soil volumetric water content, leaf area index, plant height, and maize yield (Cheng et al., 2022).

RMS (Resource Management System)

The RMS software package simulates water, heat, and solute transport in porous media developed by USDA. It uses the finite element method to solve the Richards equation and a variety of applications including irrigation scheduling, water management, and crop yield prediction (Kersebaum & Janssen, 2012). RMS can manage complex boundary conditions and simulate irrigation and drainage effects in soil-plant-atmosphere systems and its ability to integrate diverse datasets, including remote sensing data, and for its user-friendly interface. It has been used extensively in agricultural and environmental research and has been applied in different countries around the world.

The SWAT (Soil and Water Tool Assessment)

The Soil and Water Assessment Tool is a hydrological model that operates at the scale of small watersheds to river basins. It is designed to forecast the ecological consequences of land use, land management practices, and shifts in climatic conditions (Arnold, 1994). SWAT is commonly used to analyze the prevention of soil erosion, control of nonpoint source pollution and regional watershed management. It was first developed in the early 1990s to simulate the effects of hydrology and/or water quality. It is commonly used to analyze soil erosion prevention, control, and regional watershed management. SWAT can be modified to simulate the watershed processes including the water flow, nonpoint sources predictions, groundwater flow modeling, and drainage losses (Wei et al., 2018). Moreover, Chen et al. (2018) reported the wide application of SWAT for predicting the subsurface drip irrigation and simulating the water and heat flows by scenario analysis.

THE SOIL WATER BALANCE METHOD

The water balance method calculates the water input and output in a specific area over time. Changes in soil water storage are calculated using precipitation, irrigation, evapotranspiration, and drainage. The water balance equation is usually represented as follows:

Precipitation + Irrigation – Evapotranspiration – Drainage = Change in soil water storage

The utilization of soil water flow dynamics prediction has the potential to enhance water management practices and augment crop production consequences. The water balance equation is based on certain assumptions and is susceptible to measurement errors and uncertainties in the input and output. To assess the potential impact of these uncertainties on the predictions, sensitivity analysis should be performed. The precise simulation of the water budget process in a cropland system has emerged as a prominent area of investigation and gained significant interest within the hydrology discipline (Tu et al., 2021). In the context of soil modeling, it is imperative to consider remote sensing flux measurements across the plant-soil-atmosphere continuum. The aforementioned measurements may be obtained by utilizing surface-characteristic parameters such as latent and sensible heat, as well as actual evapotranspiration. These parameters can be acquired through the integration of visible to thermal infrared data (Mu et al., 2011).

THE PLANT–WATER STRESS MODEL

Plant stress models can be employed to predict plant water status in response to various environmental and plant physiological components. These models aim to understand the relationship between plant water uptake, water transport, water utilization, and environmental conditions. The models consider factors such as soil water content, air temperature, vapor pressure, and wind speed to predict the water status of the plant and its response to water stress (Noun et al., 2022). The ultimate goal of these models is to provide information for effective irrigation scheduling and improve crop production. However, implementing suitable irrigation scheduling methods requires early water stress detection in crops before it causes irreversible damage and yield losses (Ihuoma & Madramootoo, 2017). Crop water stress is a water supply deficiency caused by soil water content or plants' physiological responses (Ihuoma & Madramootoo, 2017). FAO-56 guideline explains the most popular and useful technique for calculating crop water demand and the water balance of soil (Allen et al., 1998). In addition, water movement is continuous in the soil–plant–atmosphere continuum (SPAC), a unified concept of water potential better coupling changes in underlying surface temperature, humidity, and evapotranspiration to water movement (García-Tejera et al., 2017). The measurement accuracy used to determine the water deficit affects the efficiency of plant-based irrigation scheduling. However, plant-based monitoring can be classified into two categories: direct measurement of leaf, stem, and water potential (plant–water-status monitoring) and measurements of crop physiology such as plant stomatal conductance, thermal sensing, sap flow, and cavitation of xylem (Bwambale et al., 2022). Water stress has an impact on all aspects of plant development, resulting in decreased plant size, leaf area, and yield of crops (Simbeye et al., 2023). The onset of plant water deficit occurs when the water demand of the crop surpasses the available supply. Plant water stress can be investigated using the approaches of stomatal conductance, leaf water potential, sap flow, Infrared thermometry, and crop water stress index (CWSI). By using these models, farmers can optimize irrigation scheduling to minimize water use while maximizing plant growth and yield. The SPAC model has developed several modules in recent decades, including the first root water absorption function, which described root water uptake, and the effective root density and soil solute transport model, which physically simulated soil and root interactions (Yang et al., 2018). The numerous plant characteristics that directly or indirectly provide crop water-status information and can be used to help plan irrigation have been extensively discussed by several researchers (Jones, 2004; Fernández, 2014).

NUTRIENT LOAD MANAGEMENT

Nutrient load management focuses on controlling the movement of N and P. Nutrients are essential for plant growth, but excess nutrients can lead to various problems. These include pollution of nearby groundwater and surface water sources, eutrophication of water bodies, and air pollution. Nutrient load management is the process of managing the amount of nutrients, such as nitrogen and phosphorus, added to an ecosystem to optimize plant growth.

There are several ways to manage nutrient loads, including fertilizer management:

- Application time: Applying the right amount of fertilizer at the right time, according to crop requirements and soil conditions.
- Crop rotation: Alternating crops with high nutrient requirements with lower needs or fixing nutrients from the air.
- Conservation tillage: reducing soil disturbance, which can reduce the amount of nutrients lost through erosion.
- Manure management: Proper handling and storage of manure to reduce nutrient losses to the environment.
- Cover cropping: planting cover crops that fix nutrients from the air and add organic matter to the soil.
- Irrigation management: Optimizing the irrigation schedule and water management to reduce nutrient leaching.
- Nutrient budgeting: keeping track of the nutrient inputs and outputs to balance the nutrient cycle and reduce the nutrient surplus.

Numerous research on nutrient cycling have concentrated on a single element, which is most frequently N. Although nitrogen is the nutrient that must be consumed in the greatest amounts, other factors can regulate how much nitrogen gets into and out of plants. Micro-nutrients are rarely taken into consideration in ecosystem-scale models; however, P is gradually being used in such models. When applying manures and composts that include N and P nutrients, both must be taken into account. Farmers are concerned about nitrogen losses because, if they are not controlled effectively, a massive portion of applied N fertilizer may be lost rather than utilized by crops. N issues that affect the environment include nitrate leaching from soil into groundwater, too much N in runoff, and nitrous oxide losses (Figure 6.1). According to Magdoff (1993), the crucial issues with P are the losses to freshwater bodies caused by runoff and leaching into tile drains.

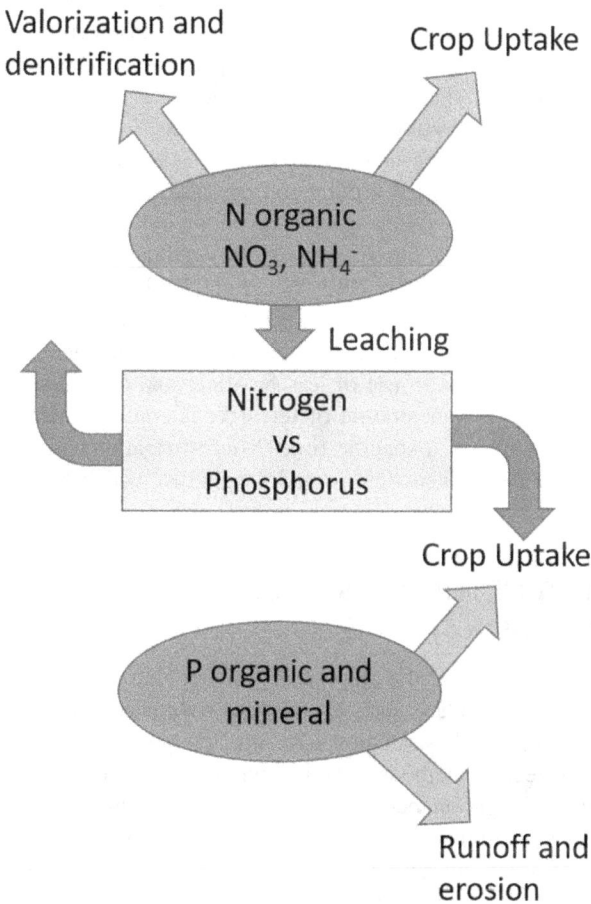

FIGURE 6.1 Different pathways for nitrogen and phosphorus losses from soils (Magdoff, 1993)

NITROGEN AND PHOSPHORUS LOAD MANAGEMENT

Although N is necessary for plant growth, too much nitrogen can cause a number of complications including slowing down plant growth (Magdoff, 1993). By mineralizing soil organic matter as well as external organic and inorganic sources, nitrogen is extracted from the soil. Nitrogen load management controls the amount of nitrogen added to an ecosystem to optimize plant growth and minimize negative environmental impacts. It is a complex process requiring a comprehensive approach considering many elements, such as soil type, crop type, meteorological, and environmental. Traditional studies have focused on the predominant processes affecting the dynamics of soil organic matter, particularly mineralization/immobilization turnover.

Eutrophication of phosphorus (P) in aquatic systems is a significant issue having detrimental effects on numerous sectors and ecosystems. Both point and nonpoint sources contribute to P, with nonpoint sources, particularly agricultural ones, being more challenging to control. The main sources of P translocation into the aquatic system include runoff and soil erosion, which promotes excessive algal development, anoxic conditions, changes in the composition of plant species, disturbance of food webs, and deterioration of recreational areas. Controlling nutrient loads and ecosystem restoration are methods to mitigate this. Restructuring the industrial layout can be used to control point sources, whereas catchment management can control nonpoint sources. Ecosystem restoration can be achieved through phytoremediation, restoration of riparian areas and wetlands, and river maintenance and restoration (Ngatia & Taylor, 2019; Wu et al., 2012). Micro-irrigation systems can apply fertilizer and chemicals to plants by spraying directly on the plant roots. Fertigation is the process of adding fertilizers to soil to enhance plant growth and development which ensures to use of less fertilizer and overall saving up to 50% of the total cost. The proper amount of fertilizer is applied to the plants through micro-irrigation devices at a specific time. Micro-irrigation systems may also be used to apply herbicides, insecticides, and fungicides, which help to increase the yields of crops.

MANAGEMENT OF SOIL MOISTURE IN MICRO-IRRIGATION SYSTEMS

There are two ways to lower the amount of water required for agriculture: supply-side management approaches, such as creating watersheds and developing water resources, and demand-side strategies, like enhancing water management technologies. The most effective methods for conserving water and raising the production of crops are micro-irrigation techniques like drip and sprinkler irrigation. Evidence suggests that a well-designed and operated micro-irrigation system may preserve up to 40% to 80% of water and raise water use efficiency by up to 100% as compared to 30–40% under conventional practice (Kaur et al., 2023). Soil water management is the process of controlling the amount of water applied to plants by using micro-irrigation. Micro-irrigation systems are designed to deliver water and nutrients directly to the root zone of plants, which can help to improve water use efficiency and reduce water losses due to runoff or evaporation. The attention in irrigation management is shifting from the supply side of the water to a system where the yield and economic benefits to water use are maximized. There are several ways to manage soil water in micro-irrigation systems, including:

- Irrigation scheduling: applying water demand on crop water requirements and soil moisture conditions.
- Soil moisture monitoring: using sensors to measure soil moisture and adjust irrigation schedules accordingly.
- Water budgeting: keeping track of the water inputs and outputs to balance the water cycle and reduce water surplus.
- Crop coefficient: using reference evapotranspiration and crop coefficient to calculate the water requirement for the specific crop.
- Pressure regulation: controlling the water pressure in the system to ensure that water is supplied at the correct rate and to the right place.

It is recommended to refrain from applying water and nutrients in regions where roots are not present or in quantities that exceed the capacity of roots for uptake. A popular concept in micro-irrigation is the drip fertigation method, which combines the application of water and fertilizers by drip-tube irrigation. Drip fertigation can also reduce water usage by applying water and nutrients directly to the roots (Kaur et al., 2023). It is a form of micro-irrigation that allows water to drip slowly to the roots of plants. It can improve water efficiency and reduce water losses through evaporation and runoff because it applies water directly to the root zone of the crop rather than flooding the entire field. A better understanding of the relationship between irrigation techniques, soil composition, root dispersion, water and nutrient absorption rates can facilitate the development of more effective and efficient micro-irrigation water management strategies (Hopmans & Bristow, 2002).

Micro-irrigation is a water conservation technique that can be employed in regions with limited water resources to alleviate the effects of climate change, which include reduced rainfall and increased temperatures. It allows precise control of water application, which can be especially beneficial for crops sensitive to water stress. However, if water is applied too frequently or at too high a rate, it can lead to waterlogging and oxygen deprivation in the root zone, leading to poor crop growth or even crop death. The plant root water uptake profiles of different irrigation methods are illustrated in Figure 6.2. Compared to the conventional irrigation approach, drip irrigation, which wets the whole root zone of the plant instead of just a portion of it, may significantly increase production by meeting the water requirement of crops with high evaporation demands (Figure 6.2) (Kader et al., 2017; Li et al., 2018). It can be said that micro-irrigation can be a valuable tool for managing soil water dynamics in different crops and therefore proper design, operation, and maintenance are essential to ensure optimal water and nutrient efficiency and crop growth.

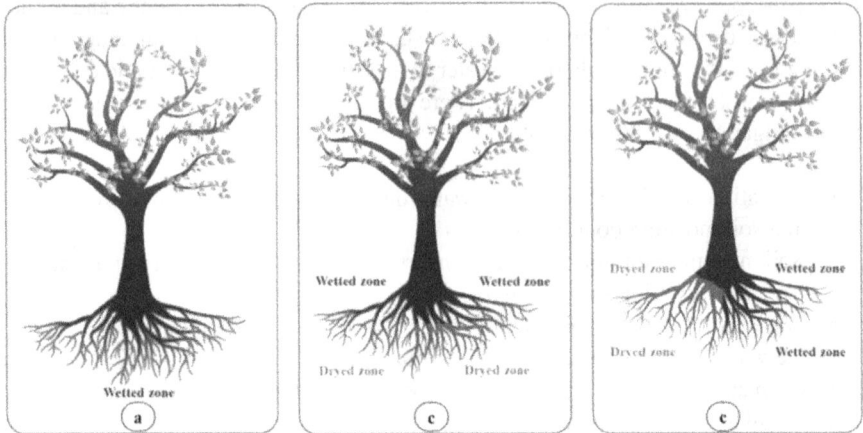

FIGURE 6.2 Root water uptake profiles under different types of irrigation: (a) conventional, (b) deficit, (c) partial deficit (Slamini et al., 2022)

DRIP AND SPRINKLER IRRIGATION

Drip and sprinkler are the two most advanced irrigation methods in agriculture. Micro-irrigation through drip tubes delivers water and nutrients directly to the root zone of crops by a network of small-diameter tubes or emitters, offering many advantages over traditional irrigation methods (Rodríguez Díaz, 2020). Micro-irrigation system has the advantage of saving irrigation water and electricity which reduces the cultivation cost besides enhancing the crop yield with high quality. It can also decrease the amount of water applied to the soil by up to 70% compared to flood irrigation (Ravikumar, 2023). Drip irrigation is being used more frequently to control the timing, frequency, and quantity of water applied to crops due to their high application efficiency which ranges from approximately 80–95% (Madramootoo & Morrison, 2013). A layout of the drip system with the necessary component of micro-irrigation is illustrated in Figure 6.3.

Drip and sprinkler irrigation are both methods used to provide water and nutrients to crops, but they have different effects on soil water and nutrient movement (Ravikumar, 2023). On the other hand, a micro sprinkler kit is suitable for farmers with access to pressurized water for groundnuts, vegetables, nurseries, and home gardens. By using an emitter 15 to 30 cm away from the soil surface, water is supplied to the soil surface using micro-spray irrigation. This technique can also be used for fertigation. In drip irrigation systems, water is supplied by drip tubes with tiny holes (emitters) placed at regular intervals to enable the water to enter the soil. Improved water management in drip irrigation systems reduces disease and weed growth which significantly enhances crop yields. Additionally, it can decrease the amount of water by up to 70% compared to flood irrigation. As a result, water is used more efficiently, and the water balance of the soil is improved (Hedley et al., 2014; Sokol et al., 2019).

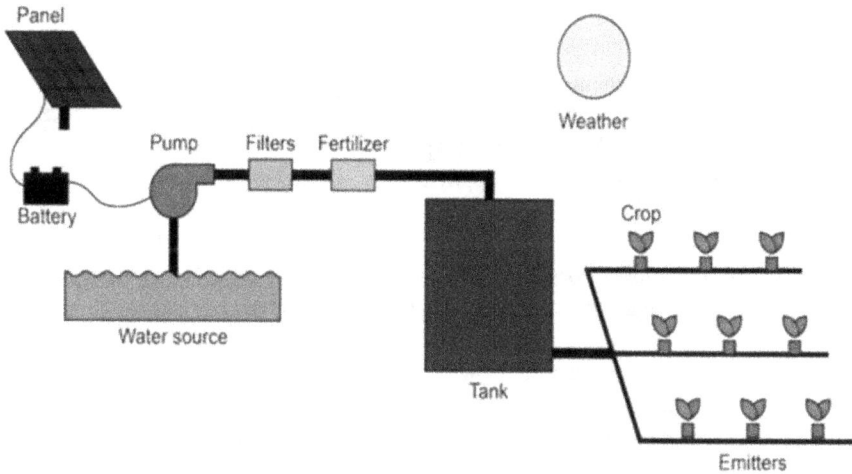

FIGURE 6.3 Layout of drip irrigation system (Source: MIT GEAR Lab)

Sprinkler irrigation, on the other hand, involves applying water to soil through a network of sprinklers that release water in the form of droplets. This method is considered less efficient than drip irrigation because it irrigates the entire field and not just the plant's root zone. A layout of sprinkler irrigation is given in Figure 6.4.

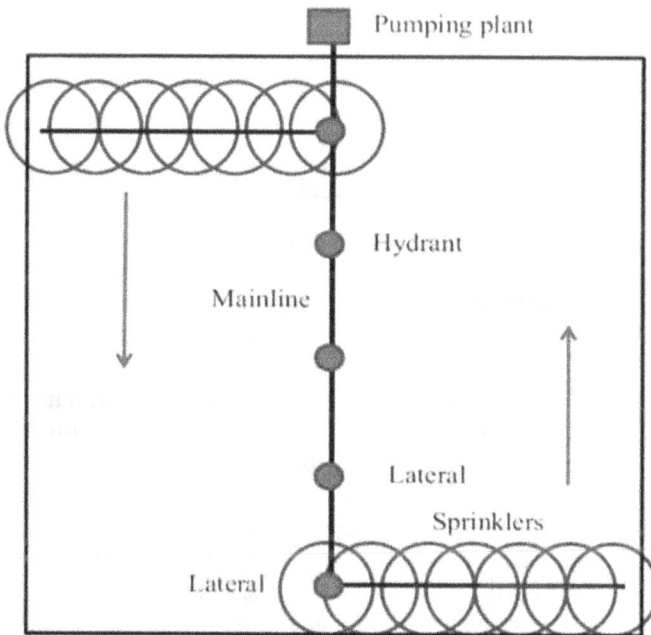

FIGURE 6.4 Layout of sprinkler irrigation system. Source: Rodríguez Díaz, 2020

Additionally, water loss through evaporation and runoff is higher with sprinkler irrigation than drip irrigation. As a result, the water balance of the soil is affected. In terms of nutrient movement, drip irrigation allows for more precise control of the nutrient application, as it can apply small amounts of water and nutrients directly to the plant root zone. On the other hand, sprinkler irrigation applies nutrients to the entire field, which can lead to overapplication and potential nutrient loss. It can be reported that, drip irrigation is considered more efficient and effective for managing soil water and nutrient movement than sprinkler irrigation. It allows for more precise water and nutrient application control, leading to better crop growth and yield.

Whether drip irrigation or sprinkler irrigation is the better option for farmers' fields highly depends on various factors, such as the crop being grown, the soil type, the climate, and the available water sources. Several studies show that drip irrigation can be more water-efficient than sprinkler irrigation, as it delivers water directly to the plant roots where it is needed and can be especially effective in areas with limited water resources or water restrictions. Drip irrigation can also be less labor intensive than sprinkler irrigation, as it requires less frequent watering and can be easily automated. On the other hand, sprinkler irrigation can be more effective in larger areas or crops requiring more water, such as field crops or turfgrass. Sprinkler irrigation can also help cool plants and reduce heat stress in hot weather, which can benefit some climates. However, sprinkler irrigation can be less water-efficient than drip irrigation, as some water can be lost due to evaporation and wind drift.

In summary, drip irrigation can be a good option for water-scarce areas or crops which require less water, while sprinkler irrigation can be more appropriate for larger areas or crops that require more water. Ultimately, the choice between drip and sprinkler irrigation will depend on the specific conditions and requirements of each situation.

FURTHER RESEARCH ON SOIL WATER MODELING

One potential area for further research on soil water modeling under micro-irrigation could be the examination of the effects of distinct types of micro-irrigation systems (e.g., drip irrigation vs. subsurface drip irrigation) on soil water movement and nutrient distribution for various soil and climatic conditions. Moreover, field studies also require measuring soil moisture and nutrient levels at different depths and distances from the irrigation source, as well as using numerical models to simulate and compare the water and nutrient movement under different micro-irrigation scenarios. Additionally, the combined effects of drip and fertigation on soil moisture and nutrient distributions could be a future research topic to optimize irrigation use efficiency. To address the intimate connections between plants and soil, future models must better integrate soil physical knowledge with agronomic and plant physiological information. Further investigation may be conducted to identify an efficient approach for attaining nutrient dynamics and accessibility within soils through the examination of hydrological and biogeochemical processes within the soil-plant-atmospheric continuum. On the other hand, simulating root growth and soil water intake as well as plant transpiration in response to varied soil conditions will provide new research ideas for soil moisture modeling (Vereecken et al., 2016).

CONCLUSIONS

This chapter comprehensively overviews the state-of-the-art water and nutrient flow models in different micro-irrigation methods. It explains the applications of different numerical models such as HYDRUS, SWAP, and AquaCrop to predict the movement of water and nutrients in the soil by considering several factors such as soil properties, plant growth, irrigation, and fertilization practices. It also investigates the potential environmental benefits of using soil moisture flow modeling, such as reducing water and nutrient losses and minimizing negative impacts on water bodies. Proper design, operation, and maintenance are essential to ensure optimal water and nutrient efficiency and crop growth in micro-irrigation systems. To better understand soil water management in micro-irrigation systems and maximize the effectiveness of these systems, further research is needed.

REFERENCES

Ale, S., Omani, N., Himanshu, S. K., Bordovsky, J. P., Thorp, K. R., & Barnes, E. M. (2020). Determining optimum irrigation termination periods for cotton production in the Texas High Plains. *Transactions of the ASABE, 63*(1), 105–115. https://doi.org/10.13031/trans .13483

Allen, R., Pereira, L. S., Raes, D., & Smith, M. (1998). Crop evapotranspiration: Guidelines for computing crop requirements. *Irrigation and Drainage Paper No. 56, FAO, 56,* 300. https://doi.org/10.1016/j.eja.2010.12.001

Arnold, J. (1994). *SWAT-Soil and Water Assessment Tool.* US Department of Agriculture, Agricultural Research Service, Grassland, Soil and Water Research Laboratory, Temple, TX.

Bwambale, E., Abagale, F. K., & Anornu, G. K. (2022). Smart irrigation monitoring and control strategies for improving water use efficiency in precision agriculture: A review. *Agricultural Water Management, 260*(July 2021), 107324. https://doi.org/10.1016/j .agwat.2021.107324

Chen, Y., Marek, G. W., Marek, T. H., Brauer, D. K., & Srinivasan, R. (2018). Improving SWAT auto-irrigation functions for simulating agricultural irrigation management using long-term lysimeter field data. *Environmental Modelling and Software, 99,* 25–38.

Cheng, H., Yu, Q., Abdalhi, M. A., Li, F., Qi, Z., Zhu, T., … Feng, S. (2022). RZWQM2 simulated drip fertigation management to improve water and nitrogen use efficiency of maize in a solar greenhouse. *Agriculture, 12*(5), 672.

Fan, Y., Himanshu, S. K., Ale, S., DeLaune, P. B., Zhang, T., Park, S. C., … Baumhardt, R. L. (2022). The synergy between water conservation and economic profitability of adopting alternative irrigation systems for cotton production in the Texas High Plains. *Agricultural Water Management, 262,* 107386. https://doi.org/10.1016/j.agwat.2021 .107386

Fernández, J. E. (2014). Plant-based sensing to monitor water stress: Applicability to commercial orchards. *Agricultural Water Management, 142,* 99–109. https://doi.org/10.1016/j .agwat.2014.04.017

García-Tejera, O., López-Bernal, Á., Testi, L., & Villalobos, F. J. (2017). A soil-plant-atmosphere continuum (SPAC) model for simulating tree transpiration with a soil multi-compartment solution. *Plant and Soil, 412*(1–2), 215–233. https://doi.org/10.1007/s11104-016-3049-0

Gülser, C., & Candemir, F. (2014). Using soil moisture constants and physical properties to predict saturated hydraulic conductivity. *Eurasian Journal of Soil Science, 3*(1), 77–81.

Gupta, P. K., Gharedaghloo, B., & Price, J. S. (2023). Multiphase flow behavior of diesel in bog, fen, and swamp peats. *Journal of Contaminant Hydrology, 255*, 104162.

Hedley, C. B., Knox, J. W., Raine, S. R., & Smith, R. (2014). Water: Advanced irrigation technologies. In *Encyclopedia of Agriculture and Food Systems* (pp. 378–406). Elsevier. https://doi.org/10.1016/B978-0-444-52512-3.00087-5

Himanshu, S. K., Ale, S., Bordovsky, J. P., Kim, J., Samanta, S., Omani, N., & Barnes, E. M. (2021). Assessing the impacts of irrigation termination periods on cotton productivity under strategic deficit irrigation regimes. *Scientific Reports, 11*(1), 1–16. https://doi.org/10.1038/s41598-021-99472-w

Himanshu, S. K., Ale, S., DeLaune, P. B., Singh, J., Mauget, S. A., & Barnes, E. M. (2022). Assessing the effects of a winter wheat cover crop on soil water use, cotton yield, and soil organic carbon in no-till cotton production systems. *Journal of the ASABE, 65*(5), 1163–1177.

Himanshu, S. K., Ale, S., Bell, J., Fan, Y., Samanta, S., Bordovsky, J. P., … Brauer, D. K. (2023). Evaluation of growth-stage-based variable deficit irrigation strategies for cotton production in the Texas High Plains. *Agricultural Water Management, 280*, 108222. https://doi.org/10.1016/j.agwat.2023.108222

Hunink, J., Vila, M., & Baille, A. (2011). REDSIM: Approach to soil water modelling- Tools and data considerations to provide relevant soil water information for deficit irrigation (May), 33. http://www.futurewater.nl/wp-content/uploads/2012/04/Approach-to-soil-water-modelling-for-REDSIM.pdf

Hopmans, J. W., & Bristow, K. L. (2002). Current capabilities and future needs of root water and nutrient uptake modeling. *Advances in Agronomy, 77*, 103–183.

Ihuoma, S. O., & Madramootoo, C. A. (2017). Recent advances in crop water stress detection. *Computers and Electronics in Agriculture, 141*, 267–275. https://doi.org/10.1016/j.compag.2017.07.026

Incrocci, L., Thompson, R. B., Fernandez-Fernandez, M. D., De Pascale, S., Pardossi, A., Stanghellini, C., … Gallardo, M. (2020). Irrigation management of European greenhouse vegetable crops. *Agricultural Water Management, 242*(April), 106393.

Jones, H. G. (2004). Irrigation scheduling: Advantages and pitfalls of plant-based methods. *Journal of Experimental Botany, 55*(407), 2427–2436. https://doi.org/10.1093/jxb/erh213. https://doi.org/10.1016/j.agwat.2020.106393

Kader, M. A., Nakamura, K., Senge, M., Mojid, M. A., & Kawashima, S. (2019). Numerical simulation of water- and heat-flow regimes of mulched soil in rain-fed soybean field in central Japan. *Soil and Tillage Research, 191*(December 2018), 142–155. https://doi.org/10.1016/j.still.2019.04.006

Kader, M. A., Senge, M., Mojid, M. A., & Ito, K. (2017). Recent advances in mulching materials and methods for modifying soil environment. *Soil and Tillage Research, 168*, 155–166. https://doi.org/10.1016/j.still.2017.01.001

Kaur, A., Singh, K. B., Gupta, R. K., Alataway, A., & Dewidar, A. Z. (2023). Interactive effects of nitrogen application and irrigation on water use, growth and tuber yield of potato under subsurface drip irrigation, *Agronomy, 13*(1), 1–19. https://doi.org/10.3390/agronomy13010011

Kersebaum, K. C., & Janssen, S. J. (Eds.) (2012). *Environmental Software Systems. Frameworks of Environment.* Springer Science & Business Media.

Kumar, H., Srivastava, P., Lamba, J., Diamantopoulos, E., Ortiz, B., Morata, G., ... Bondesan, L. (2022). Site-specific irrigation scheduling using one-layer soil hydraulic properties and inverse modeling. *Agricultural Water Management, 273*, 107877. https://doi.org/10.1016/j.agwat.2022.107877

Kumar, H., Srivastava, P., Ortiz, B. V., Morata, G., Takhellambam, B. S., Lamba, J., & Bondesan, L. (2021). Field-scale spatial and temporal soil water variability in irrigated croplands. *Transactions of the ASABE, 64*(4), 1277–1294. https://doi.org/10.13031/trans.14335

Li, Q., Li, H., Zhang, L., Zhang, S., & Chen, Y. (2018). Mulching improves yield and water-use efficiency of potato cropping in China: A meta-analysis. *Field Crops Research, 221*, 50–60. https://doi.org/10.1016/j.fcr.2018.02.017

Madramootoo, C. A., & Morrison, J. (2013). Advances and challenges with micro-irrigation. *Irrigation and Drainage, 62*(3), 255–261. https://doi.org/10.1002/ird.1704

Magdoff, F. (1993). Building soils for better crops. *Soil Science, 156*. https://doi.org/10.1097/00010694-199311000-00014

Mguidiche, A., Provenzano, G., Douh, B., Khila, S., Rallo, G., & Boujelben, A. (2015). Assessing Hydrus-2D to simulate soil water content (swc) and salt accumulation under an sdi system: Application to a potato crop in a semi-arid area of central Tunisia. *Irrigation and Drainage, 64*(2), 263–274.

Morianou, G., Kourgialas, N. N., & Karatzas, G. P. (2023). A review of HYDRUS 2D/3D applications for simulations of water dynamics, root uptake and solute transport in tree crops under drip irrigation. *Water, 15*(4), 741.

Mu, Q., Zhao, M., & Running, S. W. (2011). Improvements and evaluations of the MODIS global evapotranspiration algorithm. *Remote Sensing of Environment, 115*(8), 1781–1800.

Ngatia, L., & Taylor, R. (2019). Phosphorus eutrophication and mitigation strategies. In *Phosphorus - Recovery and Recycling*. https://doi.org/10.5772/intechopen.79173

Noun, G., Lo Cascio, M., Spano, D., Marras, S., & Sirca, C. (2022). Plant-based methodologies and approaches for estimating plant water status of Mediterranean tree species: A semi-systematic review. *Agronomy, 12*(9). https://doi.org/10.3390/agronomy12092127

Plauborg, F., Motarjemi, S. K., Nagy, D., & Zhou, Z. (2022). Analysing potato response to subsurface drip irrigation and nitrogen fertigation regimes in a temperate environment using the Daisy model. *Field Crops Research, 276*, 108367.

Phogat, V., Skewes, M. A., Cox, J. W., Sanderson, G., Alam, J., & Šimůnek, J. (2014). Seasonal simulation of water, salinity and nitrate dynamics under drip irrigated mandarin (Citrus reticulata) and assessing management options for drainage and nitrate leaching. *Journal of Hydrology, 513*, 504–516.

Provenzano, G. (2007). Using HYDRUS-2D simulation model to evaluate wetted soil volume in subsurface drip irrigation systems. *Journal of Irrigation and Drainage Engineering, 133*(4), 342–349.

Ravikumar, V. (2023). Sprinkler and drip irrigation. *Sprinkler and Drip Irrigation*. https://doi.org/10.1007/978-981-19-2775-1

Rodríguez Díaz, J. A. (2020). Concepts of sprinkler irrigation : Type, design, assessment, implementation and operation. Department of Agronomy, University of Córdoba, Campus Rabanales, Edif. da Vinci, 14071 Córdoba, Spain

Sabzzadeh, I., & Alimohammadi, S. (2023). Spatiotemporal simulation of nitrate, phosphate, and salinity in the unsaturated zone for an irrigation district west of Iran using SWAP-animo model. *Journal of Hydrologic Engineering, 28*(1), 04022037.

Scott, R. L., Shuttleworth, W. J., Keefer, T. O., & Warrick, A. W. (2000). Modeling multi-year observations of soil moisture recharge in the semiarid American Southwest. *Water Resources Research, 36*(8), 2233–2247. https://doi.org/10.1029/2000wr900116

Šimůnek, J., Šejna, M., & Genuchten, M. V. (2007). HYDRUS (2D/3D). *Software Package for Simulating Two-and Three-Dimensional Movement of Water, Heat, and Multiple Solutes in Variably-Saturated Media, User Manual, Version, 1*, 203.

Šimůnek, J., Šejna, M. A., Saito, H., Sakai, M., & Genuchten, M. T. Van (2013). The HYDRUS-1D software package for simulating the movement of water, Heat, and multiple solutes in variably saturated media, version 4.17, HYDRUS Software Series 3 (June), 343.

Slamini, M., Sbaa, M., Arabi, M., & Darmous, A. (2022). Review on partial root-zone drying irrigation: Impact on crop yield, soil and water pollution. *Agricultural Water Management*, *271*(June), 107807. https://doi.org/10.1016/j.agwat.2022.107807

Sokol, J., Amrose, S., Nangia, V., Talozi, S., Brownell, E., Montanaro, G., … Winter, A. G. (2019). Energy reduction and uniformity of low-pressure online drip irrigation emitters in field tests. *Water (Switzerland)*, *11*(6). https://doi.org/10.3390/w11061195

Steduto, P., Hsiao, T. C., Raes, D., & Fereres, E. (2009). AquaCrop—The FAO crop model to simulate yield response to water: I. Concepts and underlying principles. *Agronomy Journal*, *101*(3), 426–437.

Simbeye, D. S., Mkiramweni, M. E., Karaman, B., & Taskin, S. (2023). Plant water stress monitoring and control system. *Smart Agricultural Technology*, *3*, 100066.

Singh, J., Ale, S., DeLaune, P. B., Himanshu, S. K., & Barnes, E. M. (2022). Modeling the impacts of cover crops and no-tillage on soil health and cotton yield in an irrigated cropping system of the Texas rolling plains. *Field Crops Research*, *287*, 108661.

Tan, S., Wang, Q., Zhang, J., Chen, Y., Shan, Y., & Xu, D. (2018). Performance of AquaCrop model for cotton growth simulation under film-mulched drip irrigation in southern Xinjiang, China. *Agricultural Water Management*, *196*, 99–113.

Tu, A., Xie, S., Mo, M., Song, Y., & Li, Y. (2021). Water budget components estimation for a mature citrus orchard of southern China based on HYDRUS-1D model. *Agricultural Water Management*, *243*(April 2020), 106426. https://doi.org/10.1016/j.agwat.2020.106426

Van Dam, J. C., Huygen, J., Wesseling, J. G., Feddes, R. A., Kabat, P., Van Walsum, P. E. V., … Van Diepen, C. A. (1997). *Theory of SWAP Version 2.0; Simulation of Water Flow, Solute Transport and Plant Growth in the Soil-Water-Atmosphere-Plant Environment (No. 71)*. DLO Winand Staring Centre.

Vereecken, H., Huisman, J. A., Bogena, H., Vanderborght, J., Vrugt, J. A., & Hopmans, J. W. (2008). On the value of soil moisture measurements in vadose zone hydrology: A review. *Water Resources Research*, *44*(1–21), W00D06. doi:10.1029/2008WR006829.

Vereecken, H., Schnepf, A., Hopmans, J. W., Javaux, M., Or, D., Roose, T., … Young, I. M. (2016). Modeling soil processes: Review, key challenges, and new perspectives. *Vadose Zone Journal*, *15*(5), vzj2015.09.0131. https://doi.org/10.2136/vzj2015.09.0131

Viswanathan, P. K., Kumar, M. D., & Narayanamoorthy, A. (Eds.) (2016). *Micro Irrigation Systems in India: Emergence, Status and Impacts*. http://libproxy.wustl.edu/login?url=http://search.ebscohost.com/login.aspx?direct=true&db=eoh&AN=1628089&site=ehost-live&scope=site

Wu, H., Yuan, Z., Zhang, L., & Bi, J. (2012). Eutrophication mitigation strategies: Perspectives from the quantification of phosphorus flows in socioeconomic system of Feixi, Central China. *Journal of Cleaner Production*, *23*(1), 122–137. https://doi.org/10.1016/j.jclepro.2011.10.019

Wei, Z., Zhang, B., Liu, Y., & Xu, D. (2018). The application of a modified version of the SWAT model at the daily temporal scale and the hydrological response unit spatial scale: A case study covering an irrigation district in the Hei River Basin. *Water*, *10*(8), 1064.

Xu, B., Shao, D., Tan, X., Yang, X., Gu, W., & Li, H. (2017). Evaluation of soil water percolation under different irrigation practices, antecedent moisture and groundwater depths in paddy fields. *Agricultural Water Management, 192*, 149–158. https://doi.org/10.1016/j.agwat.2017.06.002

Xue, J., & Ren, L. (2016). Evaluation of crop water productivity under sprinkler irrigation regime using a distributed agro-hydrological model in an irrigation district of China. *Agricultural Water Management, 178*, 350–365.

Yang, W., Mao, X., Yang, J., Ji, M., & Adeloye, A. J. (2018). A coupled model for simulating water and heat transfer in soil-plant-atmosphere continuum with crop growth. *Water (Switzerland), 11*(1). https://doi.org/10.3390/w11010047

Zhou, H., & Zhao, W. Z. (2019). Modeling soil water balance and irrigation strategies in a flood-irrigated wheat-maize rotation system. A case in dry climate, China. *Agricultural Water Management, 221*(May), 286–302. https://doi.org/10.1016/j.agwat.2019.05.011

7 Applications of Remote Sensing and GIS Techniques in the Monitoring of Ecosystem Services

*Deeksha Nayak, Anoop Kumar Shukla,
and Nandineni Rama Devi*

INTRODUCTION

Multiple aspects influence the biophysical status of an ecosystem, as does people's capacity to enjoy the services (MEA, 2005). Furthermore, both human and non-human actions can alter the biological and chemical cycles, and the balance energy of the earth, resulting in atmospheric warming and climate change. Rapid urbanisation, on the other hand, decreases ecological services. Gretchen (1997) points out in his study that people's lifestyles may be jeopardising the flourishing biological biodiversity at the expense of their descendants. Many writers define ecosystem services as the mechanism that helps support human existence through interaction between the natural ecosystem and the species when discussing ecological biodiversity (Costanza et al., 1997; Daily Gretchen, 1997). Globally, human systems are sustained by nature's contribution, i.e. ecosystem services, which asserts that human-induced land use and land cover changes (LULC) have grown over the previous three decades, resulting in changes in a natural environment to human-conquered territory. Furthermore, according to Gómez-Baggethun and Barton (2013), more than half of the world's population lives in urban regions that may benefit from ecosystem services; the projected prediction of people living in urban areas is 66% by 2050. The primary goal of this research is to understand the applications of Remote Sensing (RS) and Geographic Information Systems (GIS) in studying ecosystem services.

Ecosystem services (ES) are the advantages that people receive from nature (Costanza et al., 1997; Daily Gretchen, 1997). To build a bridge between nature and society, ES combines human well-being and natural systems for ecological and economic growth. Changes in land use and land cover have a significant influence on

DOI: 10.1201/9781003441175-7

ecosystem services, leading to their deterioration (Deeksha & Shukla, 2022; Kumar Shukla et al., 2018; Shukla et al., 2020). As a result, evaluating ecosystem services has been a focus of academic research for many years. Recent interventions also demonstrate the study's preparedness to advise policymakers in making critical decisions in the policymaking process, as well as the integration of ecology, geography, and economy.

ESs play an important role in ensuring an individual's well-being by providing security, satisfying fundamental necessities for day-to-day living, and maintaining strong social relationships. As half of the world's population lives in cities, urban ecosystems remain a major topic of ecosystem services study. According to MEA (2005), over 60% of worldwide ESs have been threatened or inappropriately exploited, and a similar trend is projected to continue for the foreseeable future. As a result, ES is now widely regarded as a critical component of land-use planning, ecological environmental planning, and management.

The interaction between the ESs might be of two forms. The first is trade-offs, in which an increase in the influence of one ES leads to a decrease in the effect of another ES. Second, there are synergies, which occur when the action of one ES increases the effect of another ES (Han et al., 2020). Understanding this link is crucial because it focuses on the interaction between ESs by focusing on intrinsic bundles (Cord et al., 2017) rather than separate ESs when similar relationships arise again throughout place and time. According to studies, trade-offs and synergies are created by interactions between multiple ecosystems, hence ecosystem services cannot be regarded as autonomous. Researchers conclude that studying multiple ecosystem services is difficult (Braat & de Groot, 2012).

Several more paradigms for ES investigations have recently developed (Li et al., 2021). To account for natural capital, the Common International Classification of Ecosystem Services (CICES) includes numerous ES criteria. The framework designed by the Intergovernmental Science-Policy Platform on Biodiversity and Ecosystem Services (IPBES) closely captures themes that link nature's contribution to humanity. To better comprehend ES, researchers developed a foundation framework for worldwide ES research; hence, ES may be divided into four categories (Figure 7.1): provisioning ecosystem services, regulating ecosystem services, cultural ecosystem services, and supporting ecosystem services.

Ecosystem services are described as a commodity that may be directly derived from nature and consumed and have a commercial value. Water, food, timber, and biofuels are all examples of provisioning ecosystem services (Gupta, 2020a,b; Mahajan et al., 2022; Gupta et al., 2020; Gupta et al., 2022; Yatoo et al., 2022; Singh et al., 2023). Similarly, for the freshwater supply ecosystem service to be provided there needs to be an environment that is working in good condition (Abera et al., 2021). Precipitation, evaporation, and climatic fluctuation are major components that influence the region's water output (Zhang et al., 2018). Water yield has a favourable relationship with evapotranspiration and soil conservation, as well as other components such as food production and wood, among others.

Regulating ecosystem services may be described as the advantages derived from the ecosystem's process that affects the current state of affairs. Climate change,

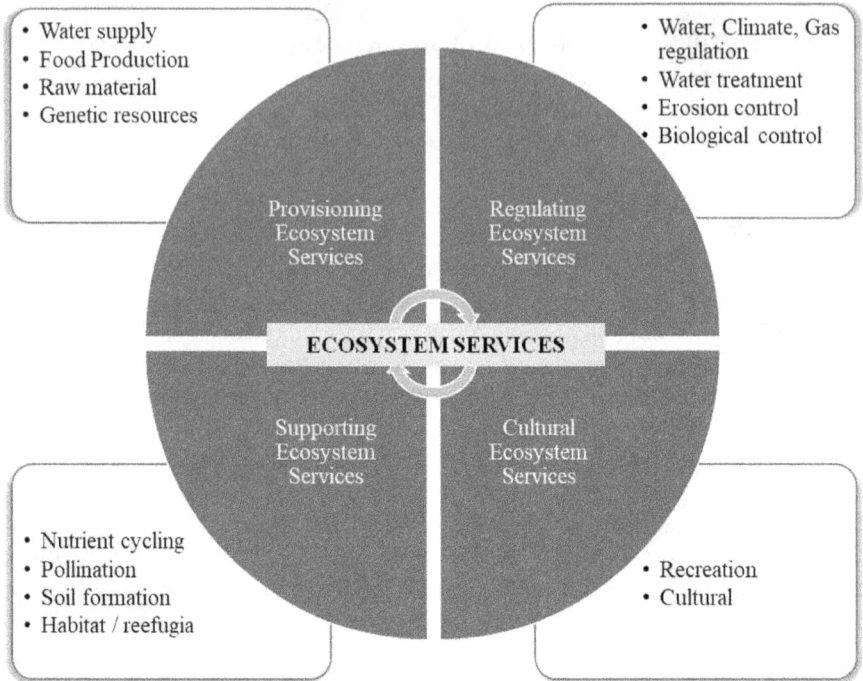

- Water supply
- Food Production
- Raw material
- Genetic resources

- Water, Climate, Gas regulation
- Water treatment
- Erosion control
- Biological control

Provisioning Ecosystem Services

Regulating Ecosystem Services

ECOSYSTEM SERVICES

Supporting Ecosystem Services

Cultural Ecosystem Services

- Nutrient cycling
- Pollination
- Soil formation
- Habitat / reefugia

- Recreation
- Cultural

FIGURE 7.1   Classification of ecosystem services

carbon storage, soil fertility, floods, and other examples will be given. The study's researcher (Bennett et al., 2009) stresses the interaction between regulating ecosystems and other ESs in terms of trade-offs and/or synergies, as a result of which regulating ESs may be regarded as one of the essential metrics for assessing ecological resilience. Managing one ES component improves synergies with other ES parameters, particularly carbon storage, low flow, biodiversity, and so forth (Carter Berry et al., 2020). Carbon storage is a critical component in the worldwide service of climate management (Gómez-Baggethun & Barton, 2013). Because of the ineffectiveness of non-interpreting scientific criteria, the practical application of carbon sequestration knowledge will take a step back in public governance. Carbon sequestration is a critical factor in global climate management. Because carbon is held in the terrestrial environment, researchers (Delibas et al., 2021) investigated the significance of soil in avoiding climate change by sequestering carbon. As a result, the carbon in the soil has a significant influence on both geographical and non-spatial data. As a result, modelling and comparing various LULC scenarios can provide crucial information to policymakers in the decision-making process. By linking various rationales to economic possibilities and regulatory legislation, this process may be re-scaled globally, regionally, and locally. By linking various rationales to economic possibilities and regulatory legislation, this process may be re-scaled globally, regionally, and locally.

Researchers introduced cultural ecosystem services, examined them, and characterised them as "the non-material" or intangible advantages humans gain from the ecosystem either spiritually, via cognitive growth, recreation, self-reflection, or by enjoying aesthetically (Gómez-Baggethun & Barton, 2013; MEA, 2005). Some of the services in this ES, such as recreation, have monetary value, whilst others do not. The many methods in which cultural ecosystems are incorporated into the research are functions that fulfil life information functioning (de Groot et al., 2012). Furthermore, simplify the term by linking human socio-cultural behaviours to psychological growth (Sen & Guchhait, 2021). CES is also linked to the intangible advantages that individuals gain from nature as a result of engagement. Several kinds of research on cultural ecosystem services focus on nature-based and aesthetic recreation services (TEEB, 2010), with little attention paid to the spiritual value of landscape due to modelling limitations (Nelson et al., 2013).

Supporting ecosystem services are the underlying processes of the ecosystem that sustain life, such as photosynthesis, the nutrient cycle, and evolution; this is a critical service that the ecosystem offers, allowing the rest of the ecosystem services to be delivered.

The scenario includes evaluating the potential of RS for protecting urban ecological integrity. As a result, the current study is meant to highlight the relevance and potential of Remote Sensing in measuring ES. It examines attitudes regarding RS in current research on ecosystem services.

RS AND GIS IMPLEMENTED IN ES STUDIES

Remote Sensing is a basic form of science that helps us acquire and analyse information about any object or phenomenon from a certain distance through the help of satellites. Today, we find a varied application of Remote Sensing in ground water analysis, mining and mineral exploration, biophysical mapping, geological planning and mapping, landslide hazard analysis, earthquake analysis, geomorphology studies, etc. Whereas the Geographic Information System data model represents how we intend to look at the world as well as gives an understanding of how the world works (Signorello et al., 2020; Shukla et al., 2020; Deeksha & Shukla, 2022). GIS data gives new representation to one or more aspects of the real world. The models created may be of the static type where the input to the model and the processed output from the model pertains to a specific study area for the same period. These studies provide general indicators, impact factors, and soft points towards the current scenario study area like soil exploration, water contamination estimates, Land use land cover (LULC) change studies, ecosystem service studies, etc. In this section, we shall understand the application of RS and GIS in studying ecosystem services.

Ecosystems may be investigated at several scales, including global, regional, and local. Global-level studies are conducted globally, but researchers recommend examining the services at the local level, which provides us with a better knowledge of the issue and aids in the implementation of suitable mitigation methods at the regional level. This contributes to the achievement of global sustainable goals (Jia et al., 2014). Although several research investigations have shed light on the interdependence of diverse ecosystem services in recent years, the merger of our existing knowledge reservoir and gaps remains insufficient (Lee & Lautenbach, 2016).

Remote Sensing and Geographic Information System implementations in ecosystem surveillance nowadays are increasingly widespread, overcoming the limits of old approaches and allowing observation of multiple spatiotemporal scales in a repeated and unbiased way (Lausch et al., 2016; Soubry et al., 2021). The inherent properties of GIS and RS allow for the synchronisation of synoptic, geographically linked, and periodic observations. The data is used to understand man–environment interactions in a specific metropolitan area (Luederitz et al., 2015). RS-based evaluations are gaining popularity as a valid method for modelling, mapping, and evaluating ecosystems and their products (Pettorelli et al., 2014). The scenario includes evaluating the potential of RS for protecting urban ecological integrity. As a result, the current study is meant to highlight the relevance and potential of Remote Sensing in measuring ES. It examines attitudes regarding RS in current research on ecosystem services.

The ES model assists the researcher in quantifying, physically locating, and perhaps evaluating economic patterns. According to Daily, 1997, this knowledge is critical in helping urban planners, urban designers, and legislators understand the impact of urban growth on ES. There is a profusion of models and technologies available now that assist us in mapping and accessing ESs and likewise (Signorello et al., 2020).

A growing body of research has tried to comprehend land-use developments and their influence on ESs at the same time, which has assisted designers and policymakers in taking proper actions to address the issue. Satellite photos have been widely utilised as the most accurate technique for monitoring LULC and ES changes (Shukla et al., 2020; Verma & Raghubanshi, 2019). Models are used to examine the interactions of ESs (such as trade-offs, synergies, bundles/clusters, and flows) and to consciously bring forth advantages that humans experience for their well-being (Dang et al., 2021). Enhancing ecosystem service management by objectively assessing interactions among multiple ES is highly valued.

The Integrated Valuation of Environmental Services and Trade-offs (InVEST) tool was created inside the Natural Capital Project (Signorello et al., 2020). The InVEST model is capable of displaying a spatially displayed map of ESs. When compared to other models, the InVEST model requires little experience; it provides a virtually exact evaluation with a minimum number of data input criteria and is useful in comprehending areas dealing with ecological processes. The InVEST model is a helpful tool for evaluating small-scale and local research that produces meaningful and trustworthy results for LULC and ES. Using user-defined base settings like land usage, land cover, and climate variations, the InVEST toolbox is used to calculate roughly 14 ES for supply changes.

The SWAT (Soil and Water Assessment Tool) is widely used to simulate hydrological processes (Notter et al., 2012; Himanshu et al., 2023). Furthermore, the model offers spatial discretisation flexibility, allowing it to assess space locally, regionally, and globally. To measure ES changes, a variety of models are utilised, including ARIES, LUCI, CA-Markov, SLEUTH, CLUES, and others.

Ecosystem services are studied quantitatively using mapping and modelling tools. Researchers also employed a mix of models to analyse ESs, such as a combination of models mapping ESs (such as InVEST, SWAT, and ARIES), models mapping urban expansion, and statistical modelling. LUSD-urban (Land Use Scenario Dynamics-urban) is an urban expansion model that helps in a multi-scale simulation of urban development. LUSD-urban, together with Cellular Automata (CA) and system dynamics models, represents micro-scale evolutionary factors and macro-scale resource restrictions. This model has gone through several versions in recent years, with improved accuracy and a kappa index on average (Xie et al., 2018). SLEUTH (Slope, Land use, Exclusion, Urban extent, Transportation, and Hill shade) and CLUES are the other models (the Conversion of Land Use and its effects at a small regional extent). Correlation analysis, regression analysis, and root mean square deviation are examples of statistical models. This model combination is effective for creating correlations among a few variables, but it is not regarded as practically feasible. InVEST, ARIES, and SWAT are the most well-known models.

LULC, soil data, topographical data, and hydrological data are the most widely utilised base data. This provides a comprehensive picture of several criteria such as habitats, soil types, vegetation classes, and biomes. According to the investigator adding to the same data, additional types of data utilised for ES evaluation include census data, meteorological data such as precipitation data, which is used for water yield assessment, and digital elevation model (DEM), which is employed for Hydrology evaluation.

InVEST, SWAT, ARIES (Artificial Intelligence for Ecosystem Services), LUCI, and other models are utilised to obtain spatiotemporal ESs. According to the publishing trend, the InVEST model is widely utilised due to its input data criterion; it uses publicly available open-source data and has a mapping/modelling scale of 30 x 30 m. This strategy enables us to have access to a variety of ecological services (water quality, soil erosion, carbon sequestration, biodiversity conservation, nutrients, agricultural produce, etc).

A few instances employ the InVEST tool, which is a software suite that merges GIS and RS data to map and quantify environmental services. Researchers used the InVEST programme to map Colombian ecosystem services such as carbon storage, water availability, and biodiversity. They discovered that the tool is successful at identifying locations where ecosystem services are threatened and that it may be used to influence land-use planning choices. The review paper written by (Zaman-Ul-haq et al., 2022) offers an overview of the use of RS for mapping and monitoring ecosystem services. The authors highlight the promise of RS for delivering spatially explicit information on ecosystem services whilst also identifying important obstacles, such as the need for standardised techniques and integrating RS with other data sources. The study maps and models ecosystem services in European cities and peri-urban regions using RS and GIS. The authors discovered that the technique may give useful information for decision-making in urban planning and management, as well as emphasise the need for multidisciplinary collaboration in ecosystem service research. These works show the promise of RS and GIS for mapping and quantifying environmental services, as well as the need for standardised procedures and multi-disciplinary collaboration in this sector.

RESULTS AND DISCUSSION

According to the study, we find an intensive amount of work conducted in highly industrialised countries, whilst the portion from emerging countries was fairly small (Zaman-Ul-haq et al., 2022).

We discover the use of long-term multispectral data obtained from Landsat, Sentinel, MODIS, and other satellites. Images from hyperspectral sensors are also employed in certain research, followed by Unmanned Aerial Vehicle (UAV) and aerial images, and finally LiDAR and radar data. Previously, we employed UAV and aerial images, as well as radar sensors, to analyse ES; but, in the early 2000s, multispectral, hyperspectral, and LiDAR data entered the picture. Multispectral sensors are utilised extensively in regional studies, whereas hyperspectral data, aerial photos, and LiDAR are employed extensively in local studies. The most regularly used sensors for assessing ecosystem services are Landsat and MODIS.

Talking about the spatial resolution of the sensors, we find that the 30 m long-term available Landsat Satellite is used, followed by 250 m and 1 km, which is acquired from MODIS. Then we find a 10 m resolution sensor connected to Sentinel-2. The sensor of Sentinel-2 delivers one of the finest spatial resolution products accessible for free usage beginning in 2015. These are the most used spatial resolutions employed to study ecosystem services.

Aside from remote sensors and indices, many forms of GIS data are utilised. The majority of them may be obtained from RS; however, because they are employed directly without the usage of RS, this is categorised separately. This can be categorised into spatial data and non-spatial data. In spatial data, we have toposheets of local, regional, and global scale, along with a digital elevation model, contour maps, land-use maps, soil maps, etc. Non-spatial data consists of precipitation, temperature, wind speed, relative humidity, solar radiation, etc. on various timelines (monthly, daily), this can be acquired from the regional meteorological department.

The methodological framework (Figure 7.2) used to assess ecosystem services starts with image acquisitions from earth observation satellites. Followed by spatial map preparation, where LULC maps are created using satellite images and ERDAS Imagine image processing software. The map can be prepared using Pixel Based Image Analysis (PBIA) or Object-Based Image Analysis (OBIA) method (Shukla et al., 2018). In the PIBA framework, we have two types of map classification, i.e. Supervised classification and Unsupervised classification.

OBIA and Nearest Neighbor (NN) classification are used to create land-use land cover (LULC) maps. The categorisation indices for OBIA are as follows: 1) the Normalized Difference Vegetation Index (NDVI), which estimates the density of green cover on a specific land parcel, 2) the Soil Adjustment Vegetation Index (SAVI), which corrects NDVI for the impact of soil brightness on the given land parcel and is employed where vegetation cover is poor. 3) Normalized Difference Water Index (MNDWI) modification aids in the development of open water features

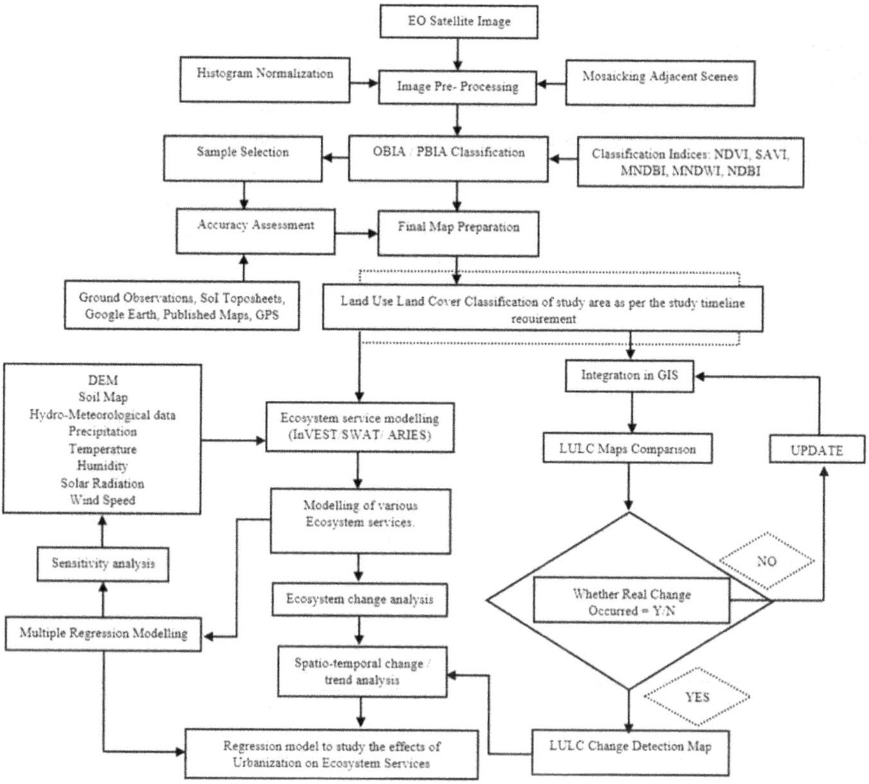

FIGURE 7.2   Methodological framework used to study ESs using RS and GIS

by reducing built-up, vegetation, and soil land sounds. 4) The Normalized Difference Built-Up Index (NDBI) is a tool for mapping the urban built-up area.

The majority of studies relied on free "open data sources", according to key findings. For data collecting, the Landsat, MODIS, and Sentinel satellites series were the ones most commonly used. The bulk of research relied on open and free data repositories for this purpose, such as aerial photos and Google Earth imagery. The data is then processed, analysed, and visualised using a GIS environment.

CONCLUSION

The urban environment is becoming stressed as a result of urbanisation and changes in metropolitan areas. To maintain the resilience of urban life, holistic assessments of simultaneous consequences on the ES are required. The main findings of this examination validate the researchers' arguments that accessibility, availability, and affordability of data and analysis are critical elements to consider. According to the findings, the focus of modern urban studies is shifting from interpreting to information acquisition for efficient decision-making. The use of RS and GIS for this goal is becoming more popular. For evaluating ESs, the LULC-based evaluation approach is often used. The analyses confirm that the developed countries account for a disproportionately higher part of ES research. For examining ES, researchers from both developed and developing nations favour free or open data sources. Nevertheless, information barriers to access, a lack of RS and GIS-based software, and a lack of capacity-building initiatives in poor nations are hindering any research efforts. As a result, dependence on remotely sensed data is dwindling. It calls for extraordinary solutions to stimulate RS-based ES research in developing economies. The availability of open-source technologies such as Google Earth Engine and free access to databases, together with other capacity-building efforts, is critical.

REFERENCE

Abera, W., Tamene, L., Kassawmar, T., Mulatu, K., Kassa, H., Verchot, L., & Quintero, M. (2021). Impacts of land use and land cover dynamics on ecosystem services in the Yayo coffee forest biosphere reserve, southwestern Ethiopia. *Ecosystem Services*, 50. https://doi.org/10.1016/j.ecoser.2021.101338

Bennett, E. M., Peterson, G. D., & Gordon, L. J. (2009). Understanding relationships among multiple ecosystem services. *Ecology Letters*, 12(12), 1394–1404. https://doi.org/10.1111/j.1461-0248.2009.01387.x

Braat, L. C., & de Groot, R. (2012). The ecosystem services agenda: bridging the worlds of natural science and economics, conservation and development, and public and private policy. *Ecosystem Services*, 1(1), 4–15. https://doi.org/10.1016/j.ecoser.2012.07.011

Carter Berry, Z., Jones, K. W., Gomez Aguilar, L. R., Congalton, R. G., Holwerda, F., Kolka, R., Looker, N., Lopez Ramirez, S. M., Manson, R., Mayer, A., Muñoz-Villers, L., Ortiz Colin, P., Romero-Uribe, H., Saenz, L., von Thaden, J. J., Vizcaíno Bravo, M. Q., Williams-Linera, G., & Asbjornsen, H. (2020). Evaluating ecosystem service trade-offs along a land-use intensification gradient in central Veracruz, Mexico. *Ecosystem Services*, 45. https://doi.org/10.1016/j.ecoser.2020.101181

Cord, A. F., Bartkowski, B., Beckmann, M., Dittrich, A., Hermans-Neumann, K., Kaim, A., Lienhoop, N., Locher-Krause, K., Priess, J., Schröter-Schlaack, C., Schwarz, N., Seppelt, R., Strauch, M., Václavík, T., & Volk, M. (2017). Towards systematic analyses of ecosystem service trade-offs and synergies: Main concepts, methods and the road ahead. *Ecosystem Services*, 28. https://doi.org/10.1016/j.ecoser.2017.07.012

Costanza, R., d'Arge, R., de Groot, R., Farber, S., Grasso, M., Hannon, B., Limburg, K., Naeem, S., O'Neill, R. V., Paruelo, J., Robert, G., Raskin, P. S., & van den Belt, M. (1997). The value of the world's ecosystem services and natural capital. *Nature*, 387, 253–260.

Daily Gretchen, C. (1997). Daily_1997_Natures-services-chapter-1. In G. C. Daily (Ed.) *Societal Dependence on Natural Ecosystems* (Nature's Services.). Island Press, Washington, DC., pp. xx–392.

Dang, A. N., Jackson, B. M., Benavidez, R., & Tomscha, S. A. (2021). Review of ecosystem service assessments: Pathways for policy integration in Southeast Asia. *Ecosystem Services*, 49. https://doi.org/10.1016/j.ecoser.2021.101266

Deeksha, & Shukla, A. K. (2022). Ecosystem services: A systematic literature review and future dimension in freshwater ecosystems. *Applied Sciences (Switzerland)*, 12(17). https://doi.org/10.3390/app12178518

de Groot, R., Brander, L., van der Ploeg, S., Costanza, R., Bernard, F., Braat, L., Christie, M., Crossman, N., Ghermandi, A., Hein, L., Hussain, S., Kumar, P., McVittie, A., Portela, R., Rodriguez, L. C., ten Brink, P., & van Beukering, P. (2012). Global estimates of the value of ecosystems and their services in monetary units. *Ecosystem Services*, *1*(1), 50–61. https://doi.org/10.1016/j.ecoser.2012.07.005

Delibas, M., Tezer, A., & Kuzniecow Bacchin, T. (2021). Towards embedding soil ecosystem services in spatial planning. *Cities*, *113*. https://doi.org/10.1016/j.cities.2021.103150

Gómez-Baggethun, E., & Barton, D. N. (2013). Classifying and valuing ecosystem services for urban planning. *Ecological Economics*, *86*, 235–245. https://doi.org/10.1016/j.eco-lecon.2012.08.019

Gupta, P. K. (2020a). Pollution load on Indian soil-water systems and associated health hazards: a review. *Journal of Environmental Engineering*, *146*(5), 03120004. https://doi.org/10.1061/(ASCE)EE.1943-7870.0001693.

Gupta, P. K. (2020b). Fate, transport, and bioremediation of biodiesel and blended biodiesel in subsurface environment: a review. *Journal of Environmental Engineering*, *146*(1), 03119001. https://doi.org/10.1061/(ASCE)EE.1943-7870.0001619.

Gupta, P. K., Gharedaghloo, B., Lynch, M., Cheng, J., Strack, M., Charles, T. C., & Price, J. S. (2020). Dynamics of microbial populations and diversity in NAPL contaminated peat soil under varying water table conditions. *Environmental Research*, *191*, 110167. https://doi.org/10.1016/j.envres.2020.110167.

Gupta, P. K., Mustapha, H. I., Singh, B., & Sharma, Y. C. (2022). Bioremediation of petroleum contaminated soil-water resources using neat biodiesel: A review. *Sustainable Energy Technologies and Assessments*, *53*, 102703. https://doi.org/10.1016/j.seta.2022.102703.

Han, H. Q., Liu, Y., Gao, H. J., Zhang, Y. J., Wang, Z., & Chen, X. Q. (2020). Tradeoffs and synergies between ecosystem services: A comparison of the karst and non-karst area. *Journal of Mountain Science*, *17*(5), 1221–1234. https://doi.org/10.1007/s11629-019-5667-5

Himanshu, S. K., Pandey, A., Madolli, M. J., Palmate, S. S., Kumar, A., Patidar, N., & Yadav, B. (2023). An ensemble hydrologic modeling system for runoff and evapotranspiration evaluation over an agricultural watershed. *Journal of the Indian Society of Remote Sensing*, *51*(1), 177–196.https://doi.org/10.1007/s12524-022-01634-4

Jia, X., Fu, B., Feng, X., Hou, G., Liu, Y., & Wang, X. (2014). The tradeoff and synergy between ecosystem services in the Grain-for-Green areas in Northern Shaanxi, China. *Ecological Indicators*, *43*, 103–113. https://doi.org/10.1016/j.ecolind.2014.02.028

Shukla, A.K., Ojha, C. S. P., Mijic, A., Buytaert, W., Pathak, S., Dev Garg, R., & Shukla, S. (2018). Population growth, land use and land cover transformations, and water quality nexus in the Upper Ganga River basin. *Hydrology and Earth System Sciences*, *22*(9), 4745–4770. https://doi.org/10.5194/hess-22-4745-2018

Lausch, A., Erasmi, S., King, D. J., Magdon, P., & Heurich, M. (2016). Understanding forest health with remote sensing-Part I-A review of spectral traits, processes and remote-sensing characteristics. *Remote Sensing*, *8*(12). https://doi.org/10.3390/rs8121029

Lee, H., & Lautenbach, S. (2016). A quantitative review of relationships between ecosystem services. *Ecological Indicators*, *66*. https://doi.org/10.1016/j.ecolind.2016.02.004

Li, J., Zhang, C., & Zhu, S. (2021). Relative contributions of climate and land-use change to ecosystem services in arid inland basins. *Journal of Cleaner Production*, *298*. https://doi.org/10.1016/j.jclepro.2021.126844

Luederitz, C., Brink, E., Gralla, F., Hermelingmeier, V., Meyer, M., Niven, L., Panzer, L., Partelow, S., Rau, A. L., Sasaki, R., Abson, D. J., Lang, D. J., Wamsler, C., & von Wehrden, H. (2015). A review of urban ecosystem services: Six key challenges for future research. *Ecosystem Services*, *14*. https://doi.org/10.1016/j.ecoser.2015.05.001

Mahajan, M., Gupta, P. K., Singh, A., Vaish, B., Singh, P., Kothari, R., & Singh, R. P. (2022). A comprehensive study on aquatic chemistry, health risk and remediation techniques of cadmium in groundwater. *Science of The Total Environment*, *818*, 151784. https://doi.org/10.1016/j.scitotenv.2021.151784.

MEA. (2005). *Ecosystems and Human Well-Being : Synthesis*. Island Press.

Nelson, E., Bhagabati, N., Ennaanay, D., Lonsdorf, E., Pennington, D., & Sharma, M. (2013). Modeling terrestrial ecosystem services. In *Encyclopedia of Biodiversity* (2nd ed., pp. 347–361). Elsevier Inc. https://doi.org/10.1016/B978-0-12-384719-5.00427-5

Notter, B., Hurni, H., Wiesmann, U., & Abbaspour, K. C. (2012). Modelling water provision as an ecosystem service in a large East African river basin. *Hydrology and Earth System Sciences*, *16*(1), 69–86. https://doi.org/10.5194/hess-16-69-2012

Pettorelli, N., Laurance, W. F., O'Brien, T. G., Wegmann, M., Nagendra, H., & Turner, W. (2014). Satellite remote sensing for applied ecologists: Opportunities and challenges. *Journal of Applied Ecology*, *51*(4), 839–848. https://doi.org/10.1111/1365-2664.12261

Sen, S., & Guchhait, S. K. (2021). Urban green space in India: Perception of cultural ecosystem services and psychology of situatedness and connectedness. *Ecological Indicators*, *123*. https://doi.org/10.1016/j.ecolind.2021.107338

Shukla, A. K., Ojha, C. S. P., Garg, R. D., Shukla, S., & Pal, L. (2020). Influence of spatial urbanization on hydrological components of the Upper Ganga River Basin, India. *Journal of Hazardous, Toxic, and Radioactive Waste*, *24*(4). https://doi.org/10.1061/(asce)hz.2153-5515.0000508

Signorello, G., Marzo, A., Prato, C., Sturiale, G., & de Salvo, M. (2020). Assessing the hidden impacts of hypothetical eruption events at Mount Etna. *Environmental and Sustainability Indicators*, *8*. https://doi.org/10.1016/j.indic.2020.100056

Singh, R. P., Mahajan, M., Gandhi, K., Gupta, P. K., Singh, A., Singh, P., ... & Kidwai, M. K. (2023). A holistic review on trend, occurrence, factors affecting pesticide concentration, and ecological risk assessment. *Environmental Monitoring and Assessment*, *195*(4), 451. https://doi.org/10.1007/s10661-023-11005-2.

Soubry, I., Doan, T., Chu, T., & Guo, X. (2021). A systematic review on the integration of remote sensing and gis to forest and grassland ecosystem health attributes, indicators, and measures. *Remote Sensing*, *13*(16). https://doi.org/10.3390/rs13163262

TEEB. (2010). *The Economics of Ecosystems and Biodiversity*.

Verma, P., & Raghubanshi, A. S. (2019). Rural development and land use land cover change in a rapidly developing agrarian South Asian landscape. *Remote Sensing Applications: Society and Environment*, *14*, 138–147. https://doi.org/10.1016/j.rsase.2019.03.002

Xie, W., Huang, Q., He, C., & Zhao, X. (2018). Projecting the impacts of urban expansion on simultaneous losses of ecosystem services: A case study in Beijing, China. *Ecological Indicators*, *84*, 183–193. https://doi.org/10.1016/j.ecolind.2017.08.055

Yatoo, A.M., Ali, M.N., Zaheen, Z., Baba, Z.A., Ali, S., Rasool, S., Sheikh, T.A., Sillanpää, M., Gupta, P.K., Hamid, B. and Hamid, B., (2022). Assessment of pesticide toxicity on earthworms using multiple biomarkers: a review. *Environmental Chemistry Letters*, *20*(4), 2573–2596. https://doi.org/10.1007/s10311-022-01386-0.

Zaman-Ul-haq, M., Saqib, Z., Kanwal, A., Naseer, S., Shafiq, M., Akhtar, N., Bokhari, S. A., Irshad, A., & Hamam, H. (2022). The trajectories, trends, and opportunities for assessing urban ecosystem services: A systematic review of geospatial methods. *Sustainability (Switzerland)*, *14*(3). https://doi.org/10.3390/su14031471

Zhang, L., Cheng, L., Chiew, F., & Fu, B. (2018). Understanding the impacts of climate and landuse change on water yield. *Current Opinion in Environmental Sustainability*, *33*. https://doi.org/10.1016/j.cosust.2018.04.017

8 Evaluation of Machine Learning Algorithms in Soil Water Content Prediction at Multiple Depths

*Shubham Jain, Sayantan Samanta,
and Sushil Kumar Himanshu*

INTRODUCTION

Efficient irrigation management is critical to optimize crop production while reducing water usage and associated costs (Fan et al., 2022; Himanshu et al., 2022). With increasing demands for food supply and ever-fluctuating and limited freshwater availability, soil water predictions can greatly aid the farmer's irrigation decision-making process. Typically, growers use experience-based scheduling (Fereres et al., 2003) and tend to over-irrigate crops to prevent production from being limited by water shortages (Thompson et al., 2007). On the other hand, low soil moisture due to under-irrigation can lead to crop stress and lower yields. To ensure that the limited water resource is used in the most efficient way, monitoring soil water content (SWC) is extremely important. Based on the crop type, SWC can indicate soil water thresholds (SWT), which are used to determine the time and quantity of required irrigation. The chance of yield reduction or even failure is high if a crop encounters subsequent days with SWC below the permanent wilting point (water stress) during its critical growth stages, such as the peak bloom stage in cotton (Himanshu et al., 2019, 2023) or tasseling and ear formation stages in corn (Cakir, 2004). In the era of precision agriculture, on-farm SWC monitoring has become very popular in the form of Time Domain Reflectometry (Ledieu et al., 1986). However, the cost of setup can be cumbersome for the growers, especially in Asia, where the farms are relatively smaller (Otsuka et al., 2016).

An alternative monitoring approach is using remote sensing products to derive theoretical and statistical models for predicting soil moisture at a point scale (Ahmad et al., 2010; Gupta et al. 2023). These datasets provide information about the earth's surface, including features such as land cover, vegetation, and soil moisture. Recent advancements in remote sensing techniques coupled with data science methods can help improve large-scale water content predictions at lower costs. The resolution of these remote sensing products can range from a few centimeters (high-resolution aerial imagery) to several kilometers (satellite imagery). High-resolution aerial imagery obtained from either handheld or Unpiloted Aerial Systems (UAS) mounted multispectral, hyperspectral, and thermal sensors have been used to estimate SWC at multiple depths in the soil profile (Ge et al., 2021; Li et al., 2022). Commonly used satellite imagery products used to estimate soil moisture status are obtained from the Soil Moisture Active Passive (SMAP) mission (Entekhabi, Njoku, et al., 2010) and the Soil Moisture and Ocean Salinity (SMOS) mission (Kerr et al., 2001). These satellites use a combination of active radar and passive microwave sensors to measure SWC at shallow depths (5 cm). Due to the poor resolution (40 km and 50 km, respectively) and estimations of SWC at shallow depths, spatially interpolated high-resolution products have also been developed (Abbaszadeh et al., 2021; Das et al., 2019). Overall, the use of remote sensing datasets in physical and empirical models can lead to improved decision-making and more effective management of natural resources.

Machine learning (ML) algorithms can potentially improve the accuracy of SWC prediction by learning the relationships between a large set of independent characteristics and soil moisture. These methods can outperform the traditional Multiple Linear Regression (MLR) methods as they can learn the non-linear relationships between these variables (Acharya et al., 2021). Machine learning models can also outperform process-based models such as HYDRUS-2D (Simunek et al., 1999), especially when there is a lack of available data for appropriate calibration and validation (Karandish & Šimůnek, 2016). Multiple ML algorithms have been previously introduced for prediction (Adab et al., 2020; Cai et al., 2019), forecasting (Gill et al., 2006), and gap-filling (Mao et al., 2019; Sun & Xu, 2021) of SWC. Some commonly used ML models include decision trees, random forests, support vector machines, and artificial Neural Networks. Each of these algorithms has its benefits and drawbacks, and the optimal algorithm for a given application will depend on the data and prediction task (Benos et al., 2021).

The selection of appropriate explanatory variables for use in ML regression models is a crucial and challenging task as it helps maintain the model's complexity and prevent overfitting. A common approach to feature selection is knowledge-based, which typically uses existing literature to understand the relationships between the target variable and various climatological and geological characteristics. Soil water content is often linked to soil physical properties (Gupta et al. 2023), climate, and topographical features of a region (Karthikeyan & Mishra, 2021). Celik et al. (2022) used the Sentinal-1 and Sentinal-2 backscatter coefficients, Soil Moisture Active Passive (SMAP), and climate data as dynamic features and topographic and soil texture data as static data to fit on long short-term memory (LSTM) model. Nguyen et al. (2022) used a total of 52 variables, including vegetation indices, soil indices, water index, and topographic features, to predict soil moisture at 10-meter spatial resolution using a combination of extreme gradient boosting regression and genetic algorithm. Liu et al. (2020) compared the performance of six machine-learning algorithms for spatial downscaling of soil moisture using land surface temperature, vegetation indices, and DEM data as explanatory variables. Yamaç et al. (2020) found that ML models outperformed pedotransfer functions for estimating field capacity and permanent wilting point based on a subset of four soil characteristics. However, using more variables such as climate forcings and vegetation characteristics, may also influence the model accuracy and increase its generalizability (Szabó et al., 2019). When selecting variables, it is also important to consider the correlation between them to eliminate redundant features and reduce the complexity of the model. Another approach to variable selection is to evaluate the performance of each variable in predicting soil moisture based on the specific model and available training dataset. Various variable selection methods seek to determine the global importance of each variable, such as Permutation Feature Importance (Molnar, 2019), shapely additive explanations (Lundberg & Lee, 2017), and partial dependence plots (Greenwell, 2017).

This study presents a comparison of multiple machine learning approaches for the prediction and selection of independent features for regression modeling of soil moisture at a point scale. The specific objectives are to 1) compare the ML models based on their predictive performance on SWC at a daily time scale and at five soil depths using a range of independent features and 2) identify suitable features that have a significant effect on predicting SWC at multiple depths using the feature importance metrics.

METHODOLOGY

STUDY AREA AND DATASETS

This study was conducted in the Texas-Gulf region located in the southern United States. The region boundary was obtained from the USGS watershed boundary dataset (Simley & Carswell, 2009). The western region experiences water stress and primarily consists of arid and semi-arid climates. The eastern part of the region experiences a more humid subtropical climate and high precipitation events as it is located close to the Gulf of Mexico. The entire region spans approximately 471,000 km^2 and receives a mean annual rainfall of 807.8 mm and a mean annual temperature of 20.12°C (Wu et al., 2021).

The Soil and Climate Analysis Network (SCAN; Schaefer et al., 2007) is a continental-scale network that records soil water content in North America primarily using the Hydra Probe sensors. A total of 13 sites (Figure 8.1) were selected within the Texas-Gulf region from the SCAN network for this study. These sites are evenly spread in the region and have sufficient available soil moisture data between 2016 and 2021 for training and testing the ML models are various soil depths. These sites are situated in different climatic regions with varying precipitation trends, which would allow the model to learn different water stress situations.

Explanatory variables used in the ML regressions to predict SWC are listed in Table 8.1. The SCAN also records temperature and precipitation data, which was used in the models, along with the static variables such as elevation, latitude, and longitude, which added spatial and geographical complexities to the models. Antecedent

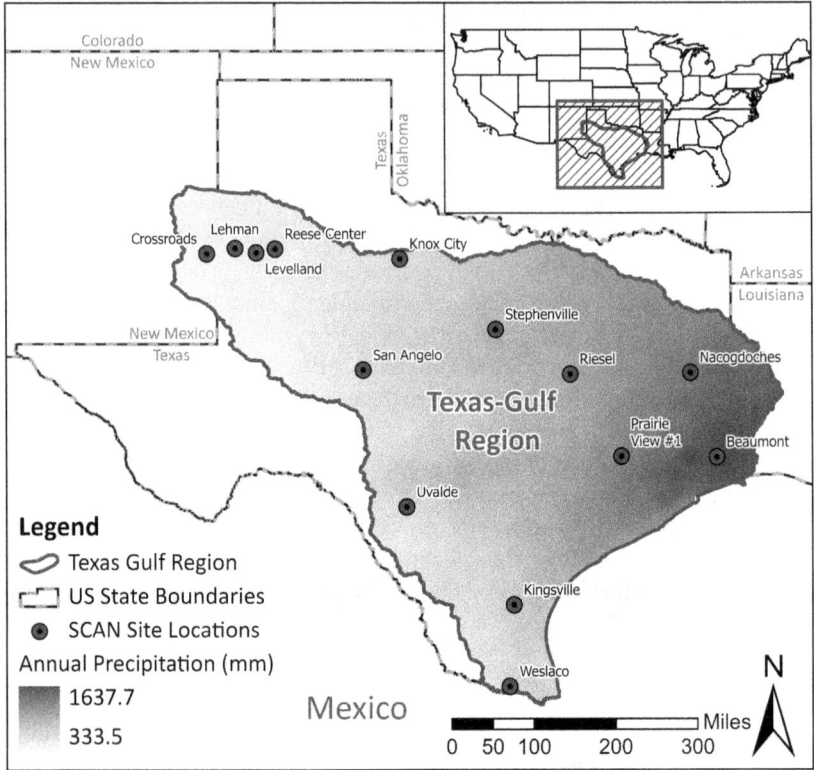

FIGURE 8.1   Map showing the location of 13 Soil and Climate Analysis Network (SCAN) sites used in this study overlaying on the long-term annual precipitation (1981–2010) in the Texas-Gulf region obtained from the PRISM dataset

precipitation conditions at three, five, and seven days were also estimated based on daily precipitation data. The Soil Survey Geographic Database (SSURGO) database was used to obtain the soil profile and characteristics, including soil texture, bulk density, and saturation hydraulic conductivity. Gridded surface SWC at a resolution of 9 km was obtained from the SMAP data (Reichle et al., 2022) at each site for surface and root zone water content. Land surface temperatures (daytime and night-time) were obtained from NASA's Moderate Resolution Imaging Spectroradiometer (MODIS/Terra). Normalized Daily Vegetation Index (NDVI; Equation 8.1; Rouse et al., 1973), Enhanced Vegetation Index (EVI; Equation 8.2; Huete et al., 2002), and Normalized Multi-band Drought Index (NMDI; Equation 8.3; Wang and Qu, 2007) were used as vegetation and water indices in this study. The NMDI was approximated using the MODIS/Terra bands, as shown in Equation 8.3.

$$NDVI = \frac{(NIR - Red)}{(NIR + Red)}; \tag{8.1}$$

TABLE 8.1

List of Features Used in the Machine Learning (ML) Regression Models Along with a Description of the Data Sources and Temporal Scale for the 13 Soil and Climate Analysis Network (SCAN) Sites

Feature	Description	Dataset	Scale	Unit
WATER DAY	Day of the year	Soil Climate Analysis Networks (SCAN)	Daily	–
TAVG	Average daily temperature		Daily	Celsius
TMIN	Minimum daily temperature		Daily	Celsius
TMAX	Maximum daily temperature		Daily	Celsius
PRCP	Daily precipitation		Daily	Inches
PRCP 3DAY	three-day antecedent precipitation		Daily	Inches
PRCP 5DAY	five-day antecedent precipitation		Daily	Inches
PRCP 7DAY	seven-day antecedent precipitation		Daily	Inches
ELEVATION	Elevation		Static	Feet
LATITUDE	Latitude coordinates		Static	Degree
LONGITUDE	Longitude coordinates		Static	Degree
SMAP RZ SM	SMAP root zone soil moisture (0–100 cm)	SMAP level 4 (SPL4SMGP)	Daily	m^3/m^3
SMAP SURF SM	SMAP surface soil moisture (0–5 cm)		Daily	m^3/m^3
NDVI	Normalized daily vegetation index	MODIS/Terra (MOD09GA V006)	Daily	–
EVI	Enhanced Vegetation Index		Daily	–
NMDI	Normalized Multi-band Drought Index (NMDI)		Daily	–
LST DAY	Daytime Land surface temperature	MODIS/Terra (MOD11A1 V006)	Daily	Kelvin
LST NIGHT	Nighttime Land surface temperature		Daily	Kelvin
KSAT	Saturated hydraulic conductivity	SSURGO	Static	μm/s
PERCLAY	% Clay soil		Static	%
PERSAND	% Sand		Static	%
PERSILT	% Silt		Static	%
BULK DENSITY	Bulk density		Static	g/cm^3

$$EVI = 2.5 \times \frac{(NIR - Red)}{(NIR + 6 \times Red - 7.5 \times Blue + 1)} \; ; \text{ and} \qquad (8.2)$$

$$NMDI_{broadband} = \frac{NIR - (SWIR_{1640} - SWIR_{2130})}{NIR + (SWIR_{1640} - SWIR_{2130})} ; \qquad (8.3)$$

where NIR is the Near Infrared band (841–876 nm), Red is the visible red band (620–670 nm), Blue is the visible blue band (459–479 nm), and $SWIR_{1640}$ (1,628–1,642 nm) and $SWIR_{2130}$ (2,105–2,155 nm) are shortwave infrared bands. The missing values in these datasets due to cloud cover were interpolated at a daily scale.

A pairwise correlation analysis was performed on the explanatory variables to identify the redundant explanatory variables using Pearson Correlation. Although the ML models were trained using all the explanatory variables listed in Table 8.1, the correlation metrics provided a check for the ML models' ability to disregard multiple redundant variables.

MACHINE LEARNING MODELS

Random Forest

Random Forest (RF) is a tree-based ensemble learning method comprised of binary partitioning trees that partition the predictor space using a series of feature splits. It is an extension of the Bootstrap Aggregation (Bagging) technique in which the bootstrap trees are decorrelated by selecting only a subset of features for each tree split. Due to recursive partitioning and local model fitting, random forests are also resistant to outliers and noise (Breiman, 2001).

The RF model uses the out-of-bag error, which is calculated by estimating the prediction of the ith observation x_i by aggregating the predictors built on bootstrap samples excluding the ith observation (Genuer & Poggi, 2020). The out-of-bag error is calculated as follows (Equation 8.4):

$$OOB \; Error = \frac{1}{N} \sum_{i=1}^{N} (Y_i - \widehat{Y}_i) ; \qquad (8.4)$$

where, Y_i is the observed value and $\widehat{Y}_i$ is the predicted value for the ith observation.

The hyperparameter "mtry" for the regression trees was determined for each soil depth using the out-of-bag error. The mtry parameter determined the number of variables randomly sampled as possible candidates during each split in constructing the random forest tree (Figure 8.2a). The model selects the best split that minimizes the error during each of these splits. One-third of the total number of variables is a typical mtry value for random forests. However, the mtry values for each model at different soil depths used in the analysis were different based on out-of-bag error. All models were built using 1,000 trees, which was sufficient to produce minimal out-of-bag errors for all soil depths and was computationally feasible.

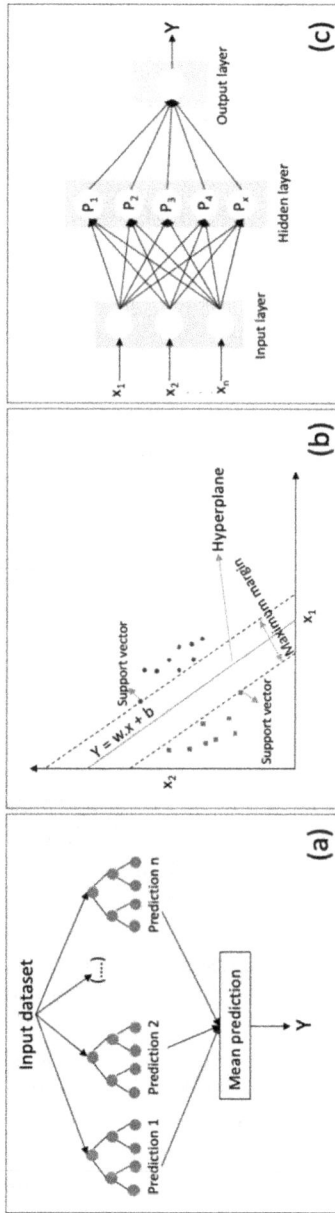

FIGURE 8.2 A basic structure of (a) Random Forest (RF) algorithm, (b) Support Vector Regression algorithm with a linear kernel, and (c) Neural Network (NN) algorithm with a single hidden layer

Support Vector Regression

Support Vector Regression (SVR) is simply a term used for applying support vector machines in the case of regression. The support vector algorithm and have been very successful in modeling non-linear classification and regression problems in its present form (Cortes and Vapnik, 1995; Vapnik, 1963). For a case of a linear SVR and a training dataset $\{(x_1, y_1), \ldots, (x_k, y_k)\}$, we need to obtain a function ($f(x) = w.x + b$), that has a maximum of e deviation from y_i (Figure 8.2b) and is also as flat as possible (i.e., small w) (Smola & Schölkopf, 2004; Trafalis & Ince, 2000).

In the case of non-linear classification and regression, a kernel function is required to transform the data into a high-dimensional feature space. For this study, a radial basis function kernel (RBF; Equation 8.5; Scholkopf et al., 1997) performed best for the input dataset compared to other generally used kernels (i.e., linear and polynomial kernels).

$$K(x, y) = \exp\left(-\frac{(x-y)^2}{2\sigma^2}\right);$$
(8.5)

where, x and y are input vectors, and σ is a measure of the width of the kernel function.

Selection of an optimal value for parameter σ is crucial as a lower σ can cause overfitting. SVR models were simulated for each soil depth individually using an RBF kernel. The σ and a cost parameter (C) that determines the width of the margin of tolerance within which the error is acceptable, were selected using a grid search algorithm for each model.

Neural Networks

The Neural Networks (NN) consist of a series of layers that build incrementally complex representations of data which are very effective for learning non-linear patterns in the data and constructing predictive models. They are usually composed of an input layer, hidden layers, and an output layer. The layers comprise several computational units or nodes with weighted input and output connections (Figure 8.2c). A network also requires an appropriate loss function and an optimizer to update the weights assigned to connections between nodes of the network during each training run using stochastic gradient descent (Chollet, 2018).

Training a NN requires tuning multiple parameters of the network, such as the number of layers, the number of nodes within each layer, the learning rate, dropout rate for each dense layer, which makes the model complex and thus requires enough training data to not cause overfitting. Most hydrologic predictive models involve few observations due to the expense of measuring data, and therefore the implementation of NN for hydrologic response variables is very challenging.

In this study, a NN with a single hidden layer is built using the "nnet" package in R (Ripley et al., 2016) for each soil depth individually. The number of units in the hidden layer and the weight decay parameter were estimated using grid search.

The weight decay is a regularization method that reduces the model complexity and allows the model to prevent overfitting of the predictive results.

EVALUATION METRICS

The models were evaluated for their performance of predicting SWC using three evaluation metrics: Nash-Sutcliffe efficiency (NSE; Equation 8.6), unbiased Root Mean Squared Error (ubRMSE; Equation 8.9; Entekhabi et al., 2010), and Percent Bias (% Bias; Equation 8.10).

$$NSE = 1 - \frac{\sum_{i=1}^{N}\left(SM_{obs,i} - SM_{pred,i}\right)^2}{\sum_{i=1}^{N}(SM_{obs,i} - \overline{SM}_{obs})} ; \qquad (8.6)$$

$$MD = \frac{\sum_{i=1}^{N} SM_{pred,i} - SM_{obs,i}}{N} ; \qquad (8.7)$$

$$RMSE = \sqrt{\frac{1}{N}\sum_{i=1}^{N}\left(SM_{pred,i} - SM_{obs,i}\right)^2} ; \qquad (8.8)$$

$$ubRMSE = \sqrt[2]{(RMSE)^2 - (MD)^2} ; \qquad \text{and (8.9)}$$

$$\% BIAS = 100 \times \frac{\sum_{i=1}^{N} SM_{pred,i} - SM_{obs,i}}{\sum_{i=1}^{N} SM_{obs,i}} ; \qquad (8.10)$$

where N is the number of quantiles, SM_{pred} is the predicted SW, SM_{obs} is the observed SWC, $\overline{SM}_{obs}$ is the mean of predicted SWC, and $\overline{SM}_{pred}$ is the mean of predicted SWC.

FEATURE IMPORTANCE

The Permuted Feature Importance (PFI) was calculated by randomly shuffling one variable at a time and comparing the increase in model performance error induced by the shuffling (Cutler et al., 2012; Ishwaran & Lu, 2019). A larger increase in model error implies a higher importance of that variable. The feature importance is model-agnostic and thus can be applied to all the ML regression models used in the study to compare the variable importance in each model and soil depth. The significance of a variable is computed using the subsequent steps:

 i) Fit the model to the training data and calculate the model's error using the original training data.
 ii) Permute the explanatory variable whose importance is to be determined by shuffling its values.
 iii) Refit the model to the new data containing the randomized variable.
 iv) Recompute the error in the model and record the error's percent change from the original model.
 v) Repeat steps ii through iv for each feature in the dataset and rank each explanatory variable according to the change in model error.

RESULTS AND DISCUSSION

CORRELATION ANALYSIS

Based on the pairwise correlation between the explanatory variables utilized in this study, a strong correlation was noted between the soil texture variables and latitude and longitude. A strong negative correlation was seen between elevation and longitude (Figure 8.3). The SMAP surface and root zone soil moisture and soil texture variables were positively correlated, suggesting soil texture's strong influence on soil moisture. The day of the year was also positively correlated with surface and air temperatures as they follow similar seasonal trends.

Although correlation analysis was not used to perform variable selection in this study, it helped to examine the models' ability to efficiently eliminate multiple redundant features (Fouad & Loáiciga, 2020). It is to be noted that the pairwise correlation analysis did not account for multicollinearity in ML models and is beyond the scope of this investigation.

ML PREDICTIVE PERFORMANCE

The RF, SVR, and NN algorithms were set up for each of the five soil depths (5 cm, 10 cm, 20 cm, 51 cm, and 102 cm) individually, resulting in 15 regression models. The performance of these models on the test dataset based on NSE, ubRMSE, and % Bias is shown in Figure 8.4.

The three ML algorithms performed comparably and adequately across all soil depths. Support vector regression performed marginally worse across most soil depths compared to other methods. The overall performance was highest for the 5 cm depth with an ubRMSE of 0.06 m^3/m^3 and then decreased substantially at the 10 cm depth with an ubRMSE of 0.11 m^3/m^3. All three ML algorithms performed poorly at 10 cm depth (Table A1; Table A2). The low performance at the 10 cm depth can be attributed to the lack of availability of training data compared to other soil depths and the lower temporal variability in SWC at deeper soil layers. Figure 8.5 shows the comparison between the simulated and measured SWC at the Stephenville SCAN site using the RF model at 5 and 20 cm depths. Although the RF model performed well at higher SWC, the model over-predicted at lower SWCs suggesting that the wilting point may not be accurately captured by the model.

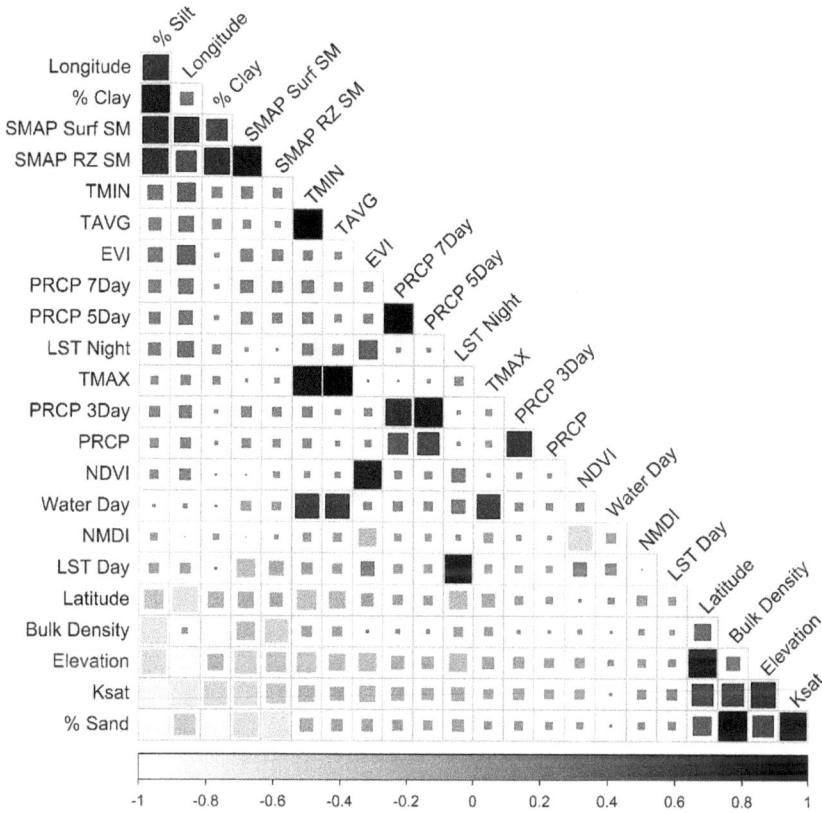

FIGURE 8.3 Correlation plot for explanatory variables. Soil characteristics in this figure correspond to 5 cm depth. Variable descriptions are provided in Table 8.1

This is one of the limitations of this study, as the ML models are not bound by physical conditions.

The prediction performance varied substantially between SCAN sites. The lowest ubRMSE obtained from all three ML algorithms was for the Lehman site at 102 cm (Table 8.3; Table A1–A2). Most sites showed an Root Mean Squared Error (RMSE) between 0.03 m^3/m^3 to 0.1 m^3/m^3, which was comparable with the previous studies (Huang et al., 2022; Q. Li et al., 2021; Senyurek et al., 2020). Karthikeyan and Mishra (2021) reported unRMSE of less than 0.04 m^3/m^3 between ML simulated and measured SWC suggesting that ML algorithms are effective in predicting SWC based on remote sensing products. Among all the test sites and depths, the highest ubRMSE was noted at the Stephenville site at 51 cm depth (0.128 m^3/m^3; Table 8.3) for the RF model. Interestingly, the model was able to predict the soil moisture adequately at other depths during the same test period. The SWC predictions at the Riesel site also had an ubRMSE higher than 0.1 m^3/m^3 for all the soil depths, which can be attributed to the missing measured SWC retrievals in the training and test periods.

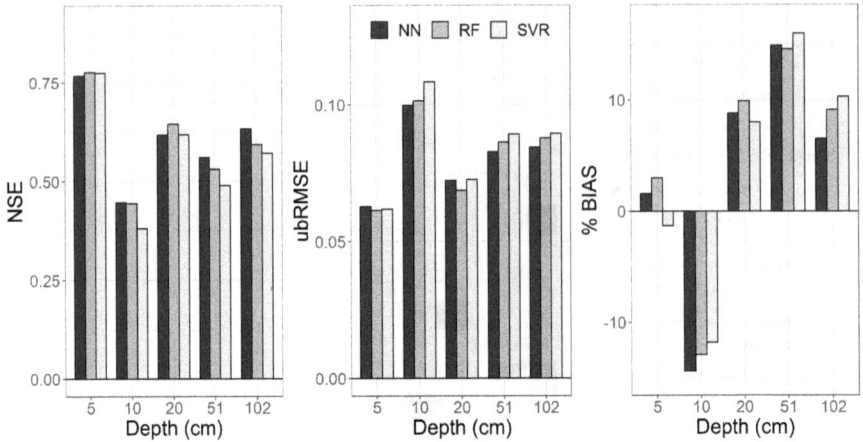

FIGURE 8.4 Performance comparison of the machine learning models (NN: Neural Network, RF: Random Forest, and SVR: Support Vector Regression) based on the Nash-Sutcliffe efficiency (NSE), Root Mean Square Error (ubRMSE), and Percent Bias (% BIAS)

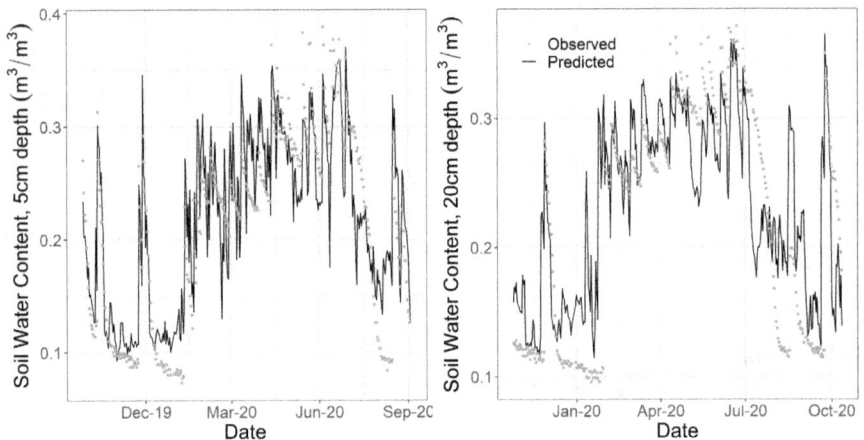

FIGURE 8.5 Plots of predicted and observed soil water content (SWC) at the Stephenville site using the Random Forest model at 5 cm and 20 cm depths

Variable Importance

The variable importance allows us to perform variable selection and reduce the model complexity by removing the features that do not have a significant impact on the model performance. It also allows us to visualize the correlation between various climatological and geophysical features with the soil moisture and gain some physical understanding of the model's working, which is otherwise a black-box model. The representative plot for PFI from RF, SVR, and NN at 5 cm depth shows the

TABLE 8.3

Unbiased Root Mean Square Error (ubRMSE) for Soil Moisture Estimates Using the Random Forest (RF) Model During the Test Period for Each Soil and Climate Analysis Network (SCAN) Station in the Study

	ubRMSE (m³/m³)				
Site	5 cm	10 cm	20 cm	51 cm	102 cm
Beaumont	0.082	0.052	0.037	0.025	0.108
Crossroads	0.031	0.030	0.055	0.014	0.051
Kingsville	0.029	0.040	0.031	0.075	0.020
Knox City	0.036	0.080	0.034	0.042	0.031
Lehman	0.034	0.024	0.034	0.010	0.003
Levelland	0.030	0.053	0.048	0.082	0.010
Prairie View	0.080	0.063	0.062	0.066	0.072
Reese Center	0.031	0.109	0.027	0.033	0.043
Riesel	0.112	0.107	0.111	0.123	0.100
San Angelo	0.055	0.065	0.090	0.083	0.111
Stephenville	0.040	0.046	0.056	0.129	0.085
Uvalde	0.061	0.033	0.054	0.091	0.055
Weslaco	0.037	0.030	0.028	0.037	0.066

percent increase in ubRMSE in the model when the feature is randomly permuted (Figure 8.6). While both SMAP surface and root zone soil moisture were used as predictors in the models, all the models selected root zone soil moisture as an important covariate at all soil depths and highly influenced the temporal variability at each site. The feature importance varies significantly between models and soil depths for

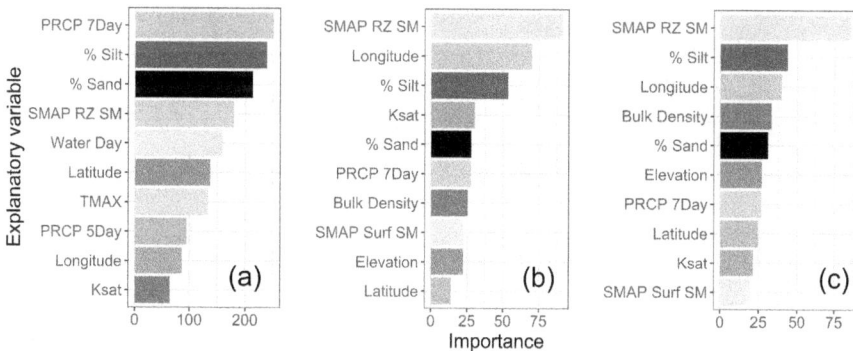

FIGURE 8.6 Permutation Feature Importance (PFI) of top-ten explanatory variables from (a) Random Forest (RF), (b) Support Vector Regression (SVR), (c) Neural Network (NN) models at 5 cm depth based on the percent increase in Root Mean Squared Error (RMSE) as shown on the x-axis of each bar plot

TABLE 8.4

Top-Five Explanatory Variables Selected by Random Forest (RF), Support Vector Regression (SVR), and Neural Network (NN) Algorithms at 5 cm, 10 cm, 20 cm, 51 cm, and 102 cm Depths Based on Permutation Feature Importance

Depth	RF	SVR	NN
5 cm	PRCP 7Day	SMAP RZ SM	SMAP RZ SM
	% Silt	Longitude	% Silt
	% Sand	% Silt	Longitude
	SMAP RZ SM	Ksat	Bulk Density
	Water Day	% Sand	% Sand
10 cm	Longitude	Elevation	Elevation
	SMAP RZ SM	SMAP RZ SM	Bulk Density
	% Sand	Latitude	Longitude
	PRCP 7Day	Longitude	SMAP RZ SM
	Water Day	% Silt	% Clay
20 cm	SMAP RZ SM	SMAP RZ SM	Longitude
	% Clay	Elevation	Elevation
	PRCP 7Day	Latitude	Ksat
	Water Day	% Sand	Bulk Density
	% Silt	Longitude	SMAP RZ SM
51 cm	% Sand	SMAP RZ SM	Longitude
	SMAP RZ SM	Latitude	Bulk Density
	% Clay	Elevation	Elevation
	Water Day	Longitude	Latitude
	Elevation	Bulk Density	SMAP RZ SM
102 cm	SMAP RZ SM	SMAP RZ SM	Longitude
	% Silt	Longitude	Bulk Density
	Water Day	Bulk Density	% Silt
	% Sand	% Silt	SMAP RZ SM
	Longitude	Elevation	Latitude

all the explanatory variables. Feature importance at depths 10 cm, 20 cm, 51 cm, and 102 cm are shown in Figures A1–A4.

The static site descriptors, such as the soil textures, site location, and elevation, had a high importance at all soil depths (Table 8.4). The daily precipitation was not selected as an important covariate, but the five-day and seven-day total antecedent precipitation had higher importance indicating the influence of prior-day precipitation on soil moisture. The vegetation and water indices from the MODIS/Terra product had very little significance in predicting soil moisture in the models. This can be attributed to these satellite products having a lower spatial resolution and thus not having a high correlation with the in-situ soil moisture. The land surface temperatures at day and night were also found to be less significant for all soil depths.

However, the nighttime land surface temperature had a greater importance than the daytime land surface temperature except at 10 cm depth. The water day had some significance in the Random Forests model but was not used in the other models. Water day can be significant in determining the seasonality in the model, but its effect may have been suppressed by surface and air temperature variables due to a high correlation between them.

The RF model was more efficient in ignoring the redundant explanatory variables observed in the correlation analysis compared to the SVR and NN models. The highly correlated variables such as soil texture, elevation, latitude, and longitude were more often selected together as important variables in the SVR and NN models suggesting that the RF model could be more advantageous in performing variable selection.

The vegetation and water indices obtained from satellite data were found to have little significance in predicting soil moisture in this study. Also, antecedent precipitation could provide more information about soil moisture than same-day precipitation and thus was ranked higher in variable importance by all models (Figures A1–A4). Static features such as soil texture were more influential than other dynamic variables, as the spatial variation in soil moisture was higher than the temporal variation. Overall, the exploratory variables selected based on the theoretical understanding of relationships between soil moisture and climatological and geographical variables can further be filtered based on their performance in predicting soil moisture.

CONCLUSIONS

In this chapter, we examined the three different ML models (RF, NN, and SVR) to predict soil moisture at five soil depths (5 cm, 10 cm, 20 cm, 51 cm, and 102 cm) using in-situ data from SCAN sites in the Texas-Gulf region. We found that all three models had similar accuracy in predicting soil moisture. Overall, the models performed satisfactorily at all depths except for the 10 cm depth that can be attributed to lower soil moisture retrievals for training and testing at this depth. The importance ranking of explanatory variables varied between models, but soil textures, elevation, latitude, longitude, elevation, antecedent precipitation and SMAP root zone soil moisture had a strong influence in predicting soil moisture.

Machine learning models can enhance the performance in predicting and forecasting hydrological and geophysical variables by learning the non-linear relationships between these variables and other explanatory variables which are often not considered in conventional linear regression or several physically based approaches. Also, ML models are advantageous in handling high-dimensional datasets which is highly pertinent when modeling hydrological and geophysical characteristics. With the increasing availability of large-scale remote sensing data and in-situ observations, the accuracy of these ML models is expected to increase further with more available training data. However, one of the apparent limitations of using ML models is the black-box nature of these models that hinders their use in the community and by stakeholders due to limited explanations of the model outputs. The knowledge of the models' understanding of the underlying physical phenomena is crucial for the stakeholders and for the model to perform well beyond the range of training data. Future work could involve the use of local surrogate model explanation methods that allow for learning insights into model explanations by understanding the relationships between explanatory variables and the predictions at the local scale.

REFERENCES

Abbaszadeh, P., Moradkhani, H., Gavahi, K., Kumar, S., Hain, C., Zhan, X., Duan, Q., Peters-Lidard, C., & Karimiziarani, S. (2021). High-resolution SMAP satellite soil moisture product: Exploring the opportunities. *Bulletin of the American Meteorological Society*, 102(4), 309–315. https://doi.org/10.1175/BAMS-D-21-0016.1

Acharya, U., Daigh, A. L. M., & Oduor, P. G. (2021). Machine learning for predicting field soil moisture using soil, crop, and nearby weather station data in the red river valley of the north. *Soil Systems*, 5(4). https://doi.org/10.3390/soilsystems5040057

Adab, H., Morbidelli, R., Saltalippi, C., Moradian, M., & Ghalhari, G. A. F. (2020). Machine learning to estimate surface soil moisture from remote sensing data. *Water*, 12(11), 3223. https://doi.org/10.3390/w12113223

Ahmad, S., Kalra, A., & Stephen, H. (2010). Estimating soil moisture using remote sensing data: A machine learning approach. *Advances in Water Resources*, 33(1), 69–80. https://doi.org/10.1016/j.advwatres.2009.10.008

Benos, L., Tagarakis, A. C., Dolias, G., Berruto, R., Kateris, D., & Bochtis, D. (2021). Machine learning in agriculture: A comprehensive updated review. *Sensors*, 21(11), 3758. https://doi.org/10.3390/s21113758

Breiman, L. (2001). Random forests. *Machine Learning*, 45(1), 5–32. https://doi.org/10.1023/A:1010933404324

Cai, Y., Zheng, W., Zhang, X., Zhangzhong, L., & Xue, X. (2019). Research on soil moisture prediction model based on deep learning. *PLoS One*, 14(4), e0214508. https://doi.org/10.1371/journal.pone.0214508

Cakir, R. (2004). Effect of water stress at different development stages on vegetative and reproductive growth of corn. *Field Crops Research*, 89(1), 1–16. https://doi.org/10.1016/j.fcr.2004.01.005

Celik, M. F., Isik, M. S., Yuzugullu, O., Fajraoui, N., & Erten, E. (2022). Soil moisture prediction from remote sensing images coupled with climate, soil texture and topography via deep learning. *Remote Sensing*, 14(21), 5584. https://doi.org/10.3390/rs14215584

Chollet, F. (2018). *Deep learning with Python*, Vol. 361. Manning New York.

Cortes, C., & Vapnik, V. (1995). SUPPORT-VECTOR NETWORKS. *Machine Learning*, 20(3), 273–297. https://doi.org/10.1023/a:1022627411411

Cutler, A., Cutler, D. R., & Stevens, J. R. (2012). Random forests. In C. Zhang & Y. Ma (Eds.) *Ensemble Machine Learning: Methods and Applications*, pp. 157–175. Springer US. https://doi.org/10.1007/978-1-4419-9326-7_5

Das, N. N., Entekhabi, D., Dunbar, R. S., Chaubell, M. J., Colliander, A., Yueh, S., Jagdhuber, T., Chen, F., Crow, W., & O'Neill, P. E. (2019). The SMAP and Copernicus Sentinel 1A/B microwave active-passive high resolution surface soil moisture product. *Remote Sensing of Environment*, 233, 111380. https://doi.org/10.1016/j.rse.2019.111380

Entekhabi, D., Njoku, E. G., O'Neill, P. E., Kellogg, K. H., Crow, W. T., Edelstein, W. N., Entin, J. K., Goodman, S. D., Jackson, T. J., & Johnson, J. (2010). The soil moisture active passive (SMAP) mission. *Proceedings of the IEEE*, 98(5), 704–716. https://doi.org/10.1109/JPROC.2010.2043918

Entekhabi, D., Reichle, R. H., Koster, R. D., & Crow, W. T. (2010). Performance metrics for soil moisture retrievals and application requirements. *Journal of Hydrology andmeteorology*, 11(3), 832–840. https://doi.org/10.1175/2010JHM1223.1

Fan, Y., Himanshu, S. K., Ale, S., DeLaune, P. B., Zhang, T., Park, S C., & Baumhardt, R. L. (2022). The synergy between water conservation and economic profitability of adopting alternative irrigation systems for cotton production in the Texas High Plains. *Agricultural Water Management*, 262, 107386.

Fereres, E., Goldhamer, D. A., & Parsons, L. R. (2003). Irrigation water management of horticultural crops. *Hortscience*, 38(5), 1036–1042. https://doi.org/10.21273/HORTSCI .38.5.1036

Fouad, G., & Loáiciga, H. A. (2020). Independent variable selection for regression modeling of the flow duration curve for ungauged basins in the United States. *Journal of Hydrology*, 587, 124975. https://doi.org/10.1016/j.jhydrol.2020.124975

Ge, X., Ding, J., Jin, X., Wang, J., Chen, X., Li, X., Liu, J., & Xie, B. (2021). Estimating agricultural soil moisture content through UAV-based hyperspectral images in the arid region. *Remote Sensing*, 13(8), 1562. https://doi.org/10.3390/rs13081562

Genuer, R., & Poggi, J.-M. (2020). Random forests. In *Random Forests with R*, pp. 33–55. Springer. https://doi.org/10.1007/978-3-030-56485-8_3

Gill, M. K., Asefa, T., Kemblowski, M. W., & McKee, M. (2006). Soil moisture prediction using support vector machines 1. *JAWRA Journal of the American Water Resources Association*, 42(4), 1033–1046. https://doi.org/10.1111/j.1752-1688.2006.tb04512.x

Greenwell, B. M. (2017). pdp: An R package for constructing partial dependence plots. *R Journal*, 9(1), 421. https://doi.org/10.32614/RJ-2017-016

Gupta, P. K., Gharedaghloo, B., & Price, J. S. (2023). Multiphase flow behavior of diesel in bog, fen, and swamp peats. *Journal of Contaminant Hydrology*, 255, 104162.

Himanshu, S. K., Ale, S., Bell, J., Fan, Y., Samanta, S., Bordovsky, J. P., Gitz III, D. C., Lascano, R. J., & Brauer, D. K. (2023). Evaluation of growth-stage-based variable deficit irrigation strategies for cotton production in the Texas High Plains. *Agricultural Water Management*, 280, 108222. https://doi.org/10.1016/j.agwat.2023.108222

Himanshu, S. K., Ale, S., Bordovsky, J., & Darapuneni, M. (2019). Evaluation of crop-growth-stage-based deficit irrigation strategies for cotton production in the Southern High Plains. *Agricultural Water Management*, 225, 105782. https://doi.org/10.1016/j .agwat.2019.105782

Himanshu, S. K., Ale, S., DeLaune, P. B., Singh, J., Mauget, S. A., & Barnes, E. M. (2022). Assessing the effects of a winter wheat cover crop on soil water use, cotton yield, and soil organic carbon in no-till cotton production systems. *Journal of the ASABE*, 65(5), 1163–1177.

Huang, S., Zhang, X., Chen, N., Ma, H., Zeng, J., Fu, P., Nam, W.-H., & Niyogi, D. (2022). Generating high-accuracy and cloud-free surface soil moisture at 1 km resolution by point-surface data fusion over the Southwestern US. *Agricultural and Forest Meteorology*, 321, 108985. https://doi.org/10.1016/j.agrformet.2022.108985

Huete, A., Didan, K., Miura, T., Rodriguez, E. P., Gao, X., & Ferreira, L. G. (2002). Overview of the radiometric and biophysical performance of the MODIS vegetation indices. *Remote Sensing of Environment*, 83(1–2), 195–213. https://doi.org/10.1016/S0034 -4257(02)00096-2

Ishwaran, H., & Lu, M. (2019). Standard errors and confidence intervals for variable importance in random forest regression, classification, and survival. *Statistics in Medicine*, 38(4), 558–582. https://doi.org/10.1002/sim.7803

Karandish, F., & Šimůnek, J. (2016). A comparison of numerical and machine-learning modeling of soil water content with limited input data. *Journal of Hydrology*, 543, 892–909. https://doi.org/10.1016/j.jhydrol.2016.11.007

Karthikeyan, L., & Mishra, A. K. (2021). Multi-layer high-resolution soil moisture estimation using machine learning over the United States. *Remote Sensing of Environment*, 266, 112706. https://doi.org/10.1016/j.rse.2021.112706

Kerr, Y. H., Waldteufel, P., Wigneron, J.-P., Martinuzzi, J., Font, J., & Berger, M. (2001). Soil moisture retrieval from space: The Soil Moisture and Ocean Salinity (SMOS) mission. *IEEE Transactions on Geoscience and Remote Sensing*, 39(8), 1729–1735. https://doi .org/10.1109/36.942551

Ledieu, J., de Ridder, P., de Clerck, P., & Dautrebande, S. (1986). A method of measuring soil moisture by time-domain reflectometry. *Journal of Hydrology*, 88(3–4), 319–328. https://doi.org/10.1016/0022-1694(86)90097-1

Li, Q., Wang, Z., Shangguan, W., Li, L., Yao, Y., & Yu, F. (2021). Improved daily SMAP satellite soil moisture prediction over China using deep learning model with transfer learning. *Journal of Hydrology*, 600, 126698. https://doi.org/10.1016/j.jhydrol.2021 .126698

Li, W., Liu, C., Yang, Y., Awais, M., Ying, P., Ru, W., & Cheema, M. J. M. (2022). A UAV-aided prediction system of soil moisture content relying on thermal infrared remote sensing. *International Journal of Environmental Science and Technology*, 1–14. https:// doi.org/10.1007/s13762-022-03958-7

Liu, Y., Jing, W., Wang, Q., & Xia, X. (2020). Generating high-resolution daily soil moisture by using spatial downscaling techniques: A comparison of six machine learning algorithms. *Advances in Water Resources*, 141, 103601. https://doi.org/10.1016/j.advwatres .2020.103601

Lundberg, S. M., & Lee, S.-I. (2017). A unified approach to interpreting model predictions. *Advances in Neural Information Processing Systems*, 30, 4768–4777.

Mao, H., Kathuria, D., Duffield, N., & Mohanty, B. P. (2019). Gap filling of high-resolution soil moisture for SMAP/sentinel-1: A two-layer machine learning-based framework. *Water Resources Research*, 55(8), 6986–7009. https://doi.org/10.1029/2019WR024902

Molnar, C. (2019). Interpretable machine learning: a guide for making black box models explainable. https://christophm. github. io/interpretable-ml-book.

Nguyen, T. T., Ngo, H. H., Guo, W., Chang, S. W., Nguyen, D. D., Nguyen, C. T., Zhang, J., Liang, S., Bui, X. T., & Hoang, N. B. (2022). A low-cost approach for soil moisture prediction using multi-sensor data and machine learning algorithm. *Science of the Total Environment*, 833, 155066. https://doi.org/10.1016/j.scitotenv.2022.155066

Otsuka, K., Liu, Y., & Yamauchi, F. (2016). The future of small farms in Asia. *Development Policy Review*, 34(3), 441–461. https://doi.org/10.1111/dpr.12159

Reichle, R., De Lannoy, G., Koster, R. D., Crow, W. T., Kimball, J. S., & Liu, Q. (2022). *SMAP L4 Global 3-Hourly 9 km EASE-Grid Surface and Root Zone Soil Moisture Analysis Update, Version 6*. ASA National Snow and Ice Data Center Distributed Active Archive Center. https://doi.org/10.5067/6P2EV47VMYPC

Ripley, B., Venables, W., & Ripley, M. B. (2016). Package 'nnet.' *R package Version*, 7(3–12), 700.

Rouse, Jr, J. W., Haas, R. H., Schell, J. A., & Deering, D. W. (1973). *Paper a 20. Third Earth Resources Technology Satellite-1 Symposium: The Proceedings of a Symposium Held by Goddard Space Flight Center at Washington, DC On*, 351, 309.

Schaefer, G. L., Cosh, M. H., & Jackson, T. J. (2007). The USDA natural resources conservation service soil climate analysis network (SCAN). *Journal of Atmospheric and Oceanic Technology*, 24(12), 2073–2077. https://doi.org/10.1175/2007JTECHA930.1

Scholkopf, B., Sung, K.-K., Burges, C. J. C., Girosi, F., Niyogi, P., Poggio, T., & Vapnik, V. (1997). Comparing support vector machines with Gaussian kernels to radial basis function classifiers. *IEEE Transactions on Signal Processing*, 45(11), 2758–2765. https://doi .org/10.1109/78.650102

Senyurek, V., Lei, F., Boyd, D., Kurum, M., Gurbuz, A. C., & Moorhead, R. (2020). Machine learning-based CYGNSS soil moisture estimates over ISMN sites in CONUS. *Remote Sensing*, 12(7), 1168. https://doi.org/10.3390/rs12071168

Simley, J. D., & Carswell, Jr, W. J. (2009). The national map—Hydrography. *Us Geological Survey Fact Sheet*, 3054(4), 1–4.

Simunek, J., Sejna, M., & van Genuchten, M. T. (1999). *The HYDRUS-2D Software Package.* International Ground Water Modeling Center, p. 251.

Smola, A. J., & Schölkopf, B. (2004). Statistics and computing - A tutorial on support vector regression.pdf. *Statistics and Computing,* 14(3), 199–222.

Sun, H., & Xu, Q. (2021). Evaluating machine learning and geostatistical methods for spatial gap-filling of monthly ESA CCI soil moisture in China. *Remote Sensing,* 13(14), 2848. https://doi.org/10.3390/rs13142848

Szabó, B., Szatmári, G., Takács, K., Laborczi, A., Makó, A., Rajkai, K., & Pásztor, L. (2019). Mapping soil hydraulic properties using random-forest-based pedotransfer functions and geostatistics. *Hydrology and Earth System Sciences,* 23(6), 2615–2635. https://doi.org/10.5194/hess-23-2615-2019

Thompson, R. B., Gallardo, M., Valdez, L. C., & Fernández, M. D. (2007). Using plant water status to define threshold values for irrigation management of vegetable crops using soil moisture sensors. *Agricultural Water Management,* 88(1–3), 147–158. https://doi.org/10.1016/j.agwat.2006.10.007

Trafalis, T. B., & Ince, H. (2000). Support vector machine for regression and applications to financial forecasting. *Proceedings of the International Joint Conference on Neural Networks,* 6(May 2016), 348–353. https://doi.org/10.1109/ijcnn.2000.859420

Vapnik, V. (1963). Pattern recognition using generalized portrait method. *Automation and remote control,* 24, 774–780.

Wang, L., & Qu, J. J. (2007). NMDI: A normalized multi-band drought index for monitoring soil and vegetation moisture with satellite remote sensing. *Geophysical Research Letters,* 34(20). https://doi.org/10.1029/2007GL031021

Wu, W.-Y., Yang, Z.-L., & Barlage, M. (2021). The impact of Noah-MP physical parameterizations on modeling water availability during droughts in the Texas–Gulf region. *Journal of Hydrology Andmeteorology,* 22(5), 1221–1233. https://doi.org/10.1175/JHM-D-20-0189.1

Yamac, S. S., Şeker, C., & Negiş, H. (2020). Evaluation of machine learning methods to predict soil moisture constants with different combinations of soil input data for calcareous soils in a semi arid area. *Agricultural Water Management,* 234, 106121. https://doi.org/10.1016/j.agwat.2020.106121

9 Artificial Intelligence Application in Database Management for SCADA Systems

Ronald Singh and Ankit Gupta

INTRODUCTION

Agriculture is the backbone of the world's economy, and it is essential to feed the ever-growing population. However, agriculture faces many challenges, including water scarcity, soil degradation, climate change, and nutrient deficiency. Agri-tech solutions can play a significant role in addressing these challenges. In this chapter, we will discuss the advances in agri-tech approaches for nutrients and irrigation water, including precision farming, hydroponics, fertigation, and sensor-based irrigation.

Water is an indispensable resource for agriculture, and its efficient management is essential for sustainable farming. However, with increasing population growth, urbanization, and climate change, water scarcity has become a significant challenge for agriculture. According to the United Nations, water scarcity affects more than 40% of the global population, and by 2050, the world's demand for water is expected to increase by 55% (UN Water, 2021). Therefore, there is a pressing need for innovative solutions to improve water-use efficiency in agriculture and ensure food security (Himanshu et al., 2021, 2023. In recent years, advances in agri-tech have played a significant role in revolutionizing water management in agriculture. These technologies provide farmers with a range of tools and solutions to optimize water use, reduce waste, and increase yields (Fan et al., 2022). For example, precision irrigation systems enable farmers to apply water precisely and efficiently, based on crop needs and site-specific conditions. These systems use sensors, weather data, and machine learning algorithms to monitor and control water application, reducing water use while increasing crop yields (Lamm et al., 2017). Similarly, sensors that monitor soil moisture and weather conditions provide farmers with real-time information on crop water needs, enabling them to adjust irrigation scheduling and avoid water stress, improving crop health and yield (Mengel et al., 2019). Furthermore, advances in agri-tech are also helping farmers to conserve water resources and mitigate the impacts of climate change. For example, using drones and satellite imagery,

DOI: 10.1201/9781003441175-9

farmers can monitor crop growth and water use across large areas, identifying areas of over or under-irrigation and optimizing water use (Zhang et al., 2021). In addition, machine learning algorithms can predict crop water needs based on weather data and crop growth stage, enabling farmers to make informed decisions on irrigation scheduling and reduce water waste.

The potential benefits of adopting agri-tech solutions for water management in agriculture are significant. According to a report by the World Bank, adopting precision irrigation technologies can reduce water use in agriculture by up to 30%, while increasing crop yields by up to 20% (World Bank, 2019). Similarly, a study by Qin et al. (2018) showed that using a sensor-based irrigation system improved water-use efficiency and yield of maize crops in China. In this chapter, we will review some of the latest advances in agri-tech for water management and their potential to transform agriculture toward more sustainable and efficient water use. We will also discuss the challenges and opportunities associated with the adoption of these technologies and their implications for policy and decision-making.

LATEST ADVANCES IN AGRI-TECH FOR WATER MANAGEMENT

Recent advances in agri-tech have enabled farmers to adopt more sustainable and efficient water management practices in agriculture. This section discusses some of the latest technologies and solutions that have emerged in this field.

IoT-Based Sensors for Precision Irrigation

IoT-based sensors have become an essential tool for precision irrigation, allowing farmers to monitor and control water applications in real time. These sensors measure soil moisture, temperature, and other parameters, allowing farmers to optimize irrigation scheduling based on crop needs and site-specific conditions. Research shows that using IoT-based sensors for irrigation can reduce water use by up to 50% while increasing crop yields by up to 30% (Albuquerque et al., 2021).

Artificial Intelligence (AI) for Crop Water Management

AI technologies such as machine learning and deep learning algorithms are increasingly being used for crop water management. These algorithms can predict crop water needs based on weather data, soil moisture, and other variables, allowing farmers to adjust irrigation schedules accordingly. In a study by Zhang et al. (2021), an AI-based irrigation system was shown to improve water-use efficiency by 23% and increase crop yields by 11% compared to traditional irrigation methods.

Cloud-Based Irrigation Management Systems

Cloud-based irrigation management systems allow farmers to remotely monitor and control irrigation from their smartphones or computers. These systems use sensors and weather data to provide real-time information on crop water needs, allowing farmers to adjust irrigation scheduling and avoid over- or under-irrigation. A study by Adhikari et al. (2021) showed that using a cloud-based irrigation management system improved water-use efficiency by 22% and increased crop yields by 18%.

WATER REUSE AND RECYCLING TECHNOLOGIES

Water reuse and recycling technologies such as drip irrigation, subsurface drip irrigation, and wastewater treatment systems are becoming increasingly popular in agriculture. These technologies allow farmers to reuse and recycle water, reducing the need for freshwater and minimizing water waste (Mahajan et al. 2022; Yatoo et al. 2022; Singh et al. 2023). A study by Garg et al. (2020) showed that using drip irrigation and subsurface drip irrigation can reduce water use by up to 50% while increasing crop yields by up to 25%. In conclusion, advances in agri-tech are enabling farmers to adopt more sustainable and efficient water management practices in agriculture. These technologies offer significant potential to reduce water use, increase crop yields, and ensure food security in a water-scarce world.

PRECISION FARMING

Precision farming, also known as precision agriculture, is an approach to farming that uses technology to optimize the use of resources such as water, fertilizer, and pesticides. It is a data-driven, site-specific, and crop-specific approach that aims to increase productivity, reduce waste, and minimize environmental impacts. Precision farming is a rapidly growing field that is expected to transform agriculture in the coming years. One of the key technologies used in precision farming is remote sensing, which involves collecting data about crops and soils from sensors mounted on aircraft, satellites, or drones. This data can be used to create maps of fields that show variations in soil type, moisture, and other factors that affect crop growth. By analyzing these maps, farmers can apply inputs such as fertilizer and water more precisely, reducing waste and improving yields. Another important technology used in precision farming is variable-rate application (VRA), which involves applying inputs such as fertilizer and pesticides at different rates in different parts of a field, based on the data collected by remote sensors. VRA has been shown to improve yields and reduce input costs compared to uniform application methods. In addition to remote sensing and VRA, precision farming also involves the use of other technologies such as GPS-guided equipment, automated irrigation systems, and data analytics software. These technologies enable farmers to monitor and manage their crops more efficiently, reducing labor costs and improving sustainability.

One of the challenges of precision farming is the complexity of the data involved. Farmers must be able to collect, analyze, and interpret large amounts of data from multiple sources in order to make informed decisions about inputs and management practices. To address this challenge, there has been a growing interest in using AI and machine learning (ML) algorithms to analyze the data and provide recommendations to farmers. There have been several studies that have demonstrated the benefits of precision farming. For example, a study by Gao et al. (2019) showed that

VRA of nitrogen fertilizer increased maize yields by 8.3% and reduced nitrogen use by 7.7% compared to uniform application methods. Another study by Arslan et al. (2021) showed that using remote sensing data to create management zones in a wheat field led to a 10.5% increase in yield and a 26% reduction in fertilizer use compared to traditional management practices. Precision farming is a promising approach to agriculture that uses technology to optimize resource use and improve productivity. The use of remote sensing, VRA, and other technologies can help farmers make more informed decisions about inputs and management practices, leading to higher yields and reduced environmental impacts. As the field of precision farming continues to evolve, there is likely to be increased interest in using AI and ML algorithms to analyze data and provide recommendations to farmers.

HYDROPONICS

Hydroponics is a modern agricultural technique that involves growing plants in a nutrient-rich water solution, without the use of soil. It has emerged as an innovative approach to agriculture due to its many benefits over traditional soil-based systems. One of the primary advantages of hydroponics is the ability to control plant nutrition, leading to increased yields and more efficient use of resources. A study published in the *Journal of Plant Nutrition* found that hydroponic systems can provide plants with more precise nutrient balances than soil-based systems, resulting in faster growth and higher yields (Gunes et al., 2006). Hydroponics also uses significantly less water than traditional agriculture, making it a more sustainable option for water-scarce areas. A study published in Agricultural Water Management found that hydroponics can reduce water usage by up to 90% compared to soil-based systems (Li & Voorhees, 2019). This is because hydroponics recirculates water, using it repeatedly rather than allowing it to drain away. Another benefit of hydroponics is that it can be used to grow plants in areas with limited access to arable land or unfavorable soil conditions. For example, hydroponics has been used successfully in urban areas to increase food production and improve food security (Schmautz & Baeza, 2017). Additionally, hydroponics allows for the vertical growth of plants, making it a more space-efficient approach to agriculture. Overall, hydroponics is a promising agri-tech approach with numerous benefits, including increased yields, more efficient resource use, and the ability to grow plants in a wider range of environments. As technology continues to improve, hydroponics has the potential to become an increasingly important part of our global food system.

FERTIGATION

Fertigation is a modern agricultural technique that combines irrigation and fertilization. It involves the application of water-soluble fertilizers through an irrigation system, allowing for a more precise application of nutrients to plants. Fertigation is a promising agri-tech approach that offers numerous benefits over traditional fertilization methods. One of the primary benefits of fertigation is its ability to provide plants with a more precise balance of nutrients. A study published in the journal *Agronomy* found that fertigation can result in more efficient nutrient uptake by crops, leading to higher yields and better plant health (Garcia-Sanchez et al., 2019). Additionally, fertigation allows for the delivery of fertilizers in smaller, more frequent doses, reducing the risk of fertilizer runoff and leaching. Fertigation also has the potential to increase water-use efficiency in agriculture. Because fertigation delivers nutrients directly to the plant roots, it reduces the amount of fertilizer that is lost to evaporation or runoff. A study published in the journal *Agricultural Water Management* found that fertigation can reduce fertilizer use by up to 30% compared to traditional broadcast fertilization (Shen et al., 2016). Another benefit of fertigation is that it allows for more efficient use of labor and resources. Fertigation systems can be automated, reducing the need for manual labor, and allowing for more precise application of fertilizers. This can lead to cost savings and improved efficiency on the farm. Overall, fertigation is a promising agri-tech approach that offers numerous benefits over traditional fertilization methods. Its ability to deliver nutrients more efficiently, reduce fertilizer runoff, and increase water-use efficiency make it an appealing option for farmers and growers worldwide.

SENSOR-BASED IRRIGATION

Sensor-based irrigation is a modern agricultural technique that involves using sensors to monitor soil moisture levels and other environmental factors to optimize water use. By providing real-time information on soil moisture, temperature, and other conditions, sensor-based irrigation systems can help farmers and growers make more informed decisions about when and how much to water their crops. One of the primary benefits of sensor-based irrigation is its ability to reduce water usage. By providing real-time data on soil moisture levels, sensor-based irrigation systems can help farmers avoid over-watering their crops. A study published in the *Journal of Irrigation and Drainage Engineering* found that sensor-based irrigation can reduce water usage by up to 50% compared to traditional irrigation methods (Tian et al., 2014). Sensor-based irrigation also has the potential to increase crop yields and improve plant health. By providing plants with the right amount of water at the right time, sensor-based irrigation can improve nutrient uptake and reduce the risk of water stress. A study published in the *Journal of Agricultural Science* found that sensor-based irrigation can increase crop yields by up to 20% compared to traditional irrigation methods (Zhang et al., 2019). Another benefit of sensor-based irrigation is that it can help farmers and growers save time and resources. By automating the irrigation process, sensor-based systems can reduce the need for manual labor and allow for more precise and efficient use of water. This can lead to cost savings and improved efficiency on the farm. Overall, sensor-based irrigation is a promising agri-tech approach that offers numerous benefits over traditional irrigation methods. Its ability to reduce water usage, increase crop yields, and improve plant health make it an appealing option for farmers and growers worldwide.

In conclusion, agri-tech approaches can help to address the challenges faced by agriculture, including water scarcity, soil degradation, climate change, and nutrient deficiency. Precision farming, hydroponics, fertigation, and sensor-based irrigation are some of the advances in agri-tech approaches that can help to conserve resources, improve yields, and minimize environmental impacts. The adoption of these technologies can help to ensure that agriculture remains sustainable and continues to feed the world's growing population.

REFERENCES

Adhikari, P., Gurung, M., & KC, N. (2021). Cloud-based irrigation management system for improved water use efficiency and yield in maize crop. *Journal of Irrigation and Drainage Engineering*, 147(5), 04021011.

Albuquerque, T. B., de Souza, R. S., de Lima, F. F., & Ribeiro, V. (2021). An IoT-based precision irrigation system for sustainable agriculture. *IEEE Access*, 9, 40167–40176.

Arslan, M., Aydin, A., & Kayadelen, E. (2021). Precision agriculture technologies: A case study for management zones in wheat fields. *Land Use Policy*, 102, 105277.

Fan, Y., Himanshu, S. K., Ale, S., DeLaune, P. B., Zhang, T., Park, S. C., ... & Baumhardt, R. L. (2022). The synergy between water conservation and economic profitability of adopting alternative irrigation systems for cotton production in the Texas High Plains. *Agricultural Water Management*, 262, 107386.

Garcia-Sanchez, F., Garcia-Sanchez, R., Molina-Martinez, J. M., Navarro-Garcia, J., & Nicolas, E. (2019). Fertigation in vegetable crops: A review. *Agronomy*, 9(5), 245.

Garg, R. K., Kumar, S., & Kumar, R. (2020). Drip irrigation and subsurface drip irrigation for sustainable agriculture: A review. *Journal of Environmental Management*, 268, 110684.

Gao, Q., Zhao, X., Guo, Y., Li, J., Li, M., Guo, X., & Li, P. (2019). Effect of variable-rate nitrogen fertilization on maize yield and nitrogen use efficiency under drip irrigation. *Agricultural Water Management*, 213, 969–978.

Ghosh, S., & Bhowmik, A. (2021). *Precision farming technologies for sustainable agriculture*. Springer.

Gunes, A., Alpaslan, M., Inal, A., & Eraslan, F. (2006). Impact of hydroponic and soil cultivation systems on nutrient levels and growth of tomato and cucumber plants. *Journal of Plant Nutrition*, 29(5), 829–841.

Himanshu, S. K., Ale, S., Bordovsky, J. P., Kim, J., Samanta, S., Omani, N., & Barnes, E. M. (2021). Assessing the impacts of irrigation termination periods on cotton productivity under strategic deficit irrigation regimes. *Scientific reports*, 11(1), 20102.

Himanshu, S. K., Ale, S., Bell, J., Fan, Y., Samanta, S., Bordovsky, J. P., ... & Brauer, D. K. (2023). Evaluation of growth-stage-based variable deficit irrigation strategies for cotton production in the Texas High Plains. *Agricultural Water Management*, 280, 108222.

Lamm, F. R., Stone, K. C., Klocke, N. L., & Eisenhauer, D. E. (2017). Advances in precision irrigation: The future of irrigated agriculture. In *Precision agriculture for sustainability* (pp. 243–266). Springer.

Li, Y., & Voorhees, W. B. (2019). Agricultural water conservation potential of hydroponic systems: A case study of lettuce production. *Agricultural Water Management*, 217, 174–181.

Mahajan, M., Gupta, P. K., Singh, A., Vaish, B., Singh, P., Kothari, R., & Singh, R. P. (2022). A comprehensive study on aquatic chemistry, health risk and remediation techniques of cadmium in groundwater. *Science of The Total Environment*, 818, 151784. https://doi.org/10.1016/j.scitotenv.2021.151784.

Mengel, D. B., Anapalli, S. S., Gowda, P. H., Marek, G. W., Porter, D. O., & Kustas, W. P. (2019). Remote sensing for irrigation management: From science to practice. *Journal of Crop Improvement*, 33(6), 758–782.

Qin, X., Li, Y., Li, M., & Li, F. (2018). Water-saving irrigation systems improve maize yield and water use efficiency in the semi-arid region of northern China. *PLoS One*, 13(10), e0204476.

Schmautz, Z., & Baeza, J. (2017). Urban agriculture and food security in cities: A review of the literature. *Sustainability*, 9(3), 1–20.

Shen, L., Zhang, X., Wang, Q., Zhang, X., & Luo, Y. (2016). Effect of fertigation on nutrient leaching and water use efficiency of apple trees. *Agricultural Water Management*, 177, 257–266.

Singh, R. P., Mahajan, M., Gandhi, K., Gupta, P. K., Singh, A., Singh, P., ... & Kidwai, M. K. (2023). A holistic review on trend, occurrence, factors affecting pesticide concentration, and ecological risk assessment. *Environmental Monitoring and Assessment*, 195(4), 451. https://doi.org/10.1007/s10661-023-11005-2.

Tian, Y., Zhang, Y., Wang, J., & Li, J. (2014). Sensor-based irrigation scheduling for cotton in arid areas of China. *Journal of Irrigation and Drainage Engineering*, 140(2), 04013015.

UN Water. (2021). Water scarcity. Retrieved from https://www.unwater.org/water-facts/scarcity/.

World Bank. (2019). *Enhancing water use efficiency in agriculture: An evaluation of precision irrigation technologies.* Retrieved from https://openknowledge.worldbank.

Yatoo, A.M., Ali, M.N., Zaheen, Z., Baba, Z.A., Ali, S., Rasool, S., Sheikh, T.A., Sillanpää, M., Gupta, P.K., Hamid, B. and Hamid, B., (2022). Assessment of pesticide toxicity on earthworms using multiple biomarkers: a review. *Environmental Chemistry Letters*, 20(4), 2573–2596. https://doi.org/10.1007/s10311-022-01386-0.

Zhang, Z., Wang, X., Liu, L., Wang, H., Gao, Z., & Li, Y. (2019). Improved tomato yield and water use efficiency by sensor-based deficit irrigation in solar greenhouse. *Journal of Agricultural Science*, 11(10), 248–255.

Zhang, C., Yuan, J., Wang, W., Wang, H., Cao, W., & Huang, J. (2021). Improving crop yield and water use efficiency using a deep learning-based irrigation system. *Computers and Electronics in Agriculture*, 189, 106259.

10 Entropy-Weighted-Multi-Criteria Decision-Making (E-MCDM) Approach for Erosion Area Prioritization
Case Study of a Himalayan River Basin

Brijesh Kumar, Dipankar Roy, Sanoj Kumar, and Ashok Kumar

INTRODUCTION

Since the beginning of human history, the soil has governed the ability to cultivate crops and ultimately the success of civilizations. The co-existence of mankind, the earth, and food confirms soil as the foundation of agriculture (Parikh and James 2013). The formation of fertile cultivable soil takes years of inter-cultural operations such as tillage, irrigation, and manuring (Mauget et al., 2021). However, this fertile soil gets washed away because of erosion – a process of detachment of soil from its seat by means of external agents like air, water, and temperature (Ellison 1948; Pandey et al. 2016). Therefore, soil erosion is the "greatest challenge for sustainable soil management" (FAO 2019). Soil erosion poses a serious environmental issue that not only affects natural assets but also the lives of millions of people worldwide due to reduced crop production (Den Biggelaar et al. 2003; Panagos et al. 2018; Palmate et al. 2022). Therefore, strategies to preserve fertile soil are very important on a catchment scale (Keesstra et al. 2018). In the last century, significant research has been done on water erosion, but still there are unknown issues that need to be researched (García-Ruiz et al. 2017) and prioritization of erosion-prone zones is one of them.

DOI: 10.1201/9781003441175-10

Soil erosion creates serious soil degradation and ultimately reduces the production of crops by removing the precious top layer of soil (Borrelli et al. 2017; Panagos et al. 2018; Mauget et al. 2021). Thus, continuously detached sediment from the catchments fill the valley and reservoir ultimately creating serious economic and environmental issues (Dutta 2016). To assess the economic and environmental repercussions of soil erosion and to conduct erosion management plans to deal with it, estimation of soil erosion rate and prioritization of erosion-prone areas at a regional and global scale is a must (Fallah et al. 2016; Kadam et al. 2019). Prioritization of erosion-prone areas is also supplementary research for direct assessment of soil loss from watersheds (Fernández-Raga et al. 2017). The drainage basins let the eroded soil particles escape through the fluvial system which depends on the morphology of the watershed. Thus, the morphological study of the watersheds provides the basis for the assessment of soil erosions from the drainage basins (Patel et al. 2013; Singh and Singh 2018).

Watershed morphometry is the quantification of basic measurements and numerical systemization of landform elements and their dimensions, that is derived from topographical maps and other elevation data sets (Santra and Mitra 2017). The effective derivation of drainage network morphology is crucial for understanding landforms, physical properties of soil, runoff rate from the catchment, and erosion characteristics (Kumar et al. 2017b). The methodologies presented by Horton (1945), Strahler (1952a), and Miller (1953), etc. have enabled the quantitative parameterization of morphological, topographical, and dimensional characteristics of river networks and catchments. Thus, it provides a quantitative description of the physiography of the watersheds. In the recent past, studies based on the morphological analysis of drainage networks have been performed in various fields viz. natural resource and environmental hazard analysis (Arnous et al. 2011; Radice et al. 2013), groundwater potential zones mapping (Ajay Kumar et al. 2020), analysis of hydrological behaviors of watersheds (Kabite and Gessesse 2018; Sharma et al. 2018), and prioritizing watersheds for soil and water conservation (Kadam et al. 2019; Raju and Nagesh Kumar 2013).

Researchers have applied remote sensing (RS) and GIS data and techniques to effectively prioritize the erosion-prone areas in the watersheds. Drainage network-based numerical analysis has been considered a principal technique for morphological analysis of drainage basin and watershed characterization (Kumar et al. 2017b, 2017a; Patel et al. 2013). Researchers have successfully applied the multi-criteria decision-making (MCDM) approach for solving complex area prioritization problems viz. selection of suitable crop area (Seyedmohammadi et al. 2018), flood-prone zone mapping (Khosravi et al. 2019), soil erosion area prioritization (Chowdary et al. 2013; Jain and Ramsankaran 2019; Kadam et al. 2019; Vulević et al. 2015), etc. worldwide. Almost every study regarding erosion-prone area prioritization has used the Analytic Hierarchy Process (AHP) (Saaty 2004) decision-making approach for the weightage computation for different criteria (Chowdary et al. 2013; Fallah et al. 2016; Vulević et al. 2015). Most of these studies are limited to small watersheds and quite accessible for physical verification so that the modeler can decide the priority in AHP weightage computation. However, the decision on weightage for each criterion is difficult for complex and inaccessible terrains like the Himalayas. This limitation can be addressed by utilizing Entropy-Based Weighting for Multi-Objective Optimization, which calculates the weightage based on the entropy of the criteria (Rocha et al. 2015; Yue 2017).

The goal of this study was to utilize the capability of morphometric parameters for erosion-prone watershed prioritization using entropy-weighted MCDM (E-MCDM) methods; COmplex PRoportional ASsessment (COPRAS), Simple Additive Weighting (SAW), Vlse Kriterijumska Optimizacija I Kompromisno Resenje in Serbian (VIKOR), and the Technique for Order of Preference by Similarity to Ideal Solution (TOPSIS) for the complex Himalayan Gandak River Basin. These MCDM methods have been applied for erosion area prioritization in varied geographic conditions (Altaf et al. 2014; Raju and Nagesh Kumar 2013; Vulević et al. 2015). The mentioned MCDM methods have sought many benefits including pair-wise comparison, ideal and negative ideal solutions, and no compensatory decision-making (Arunachalam et al. 2015; Mardani et al. 2015; Peng 2015). Soil erosion is a serious concern for the Gandak River Basin causing the failure of dams (Kothyari 2011; Kumar and Lakshmi 2018) and muddy flooding (Acharya and Prakash 2019; Bandyopadhyay and Gyawali 1994; Kumar et al. 2017a). Therefore, an attempt has been made to prioritize the watersheds of the inaccessible Gandak River Basin using the E-MCDM technique, which can assist decision-makers in framing effective soil and water conservation practices.

MATERIAL AND METHODS

DESCRIPTION OF STUDY AREA (GANDAK RIVER BASIN)

The transboundary Himalayan Gandak River Basin (GRB), which drains water from China, the mid-hills of Nepal, and the Tarai region of Bihar, India (Kumar et al. 2016) was chosen for the study. The origin point of the basin falls at an altitude of 7620 m to the northeast of Dhaulagiri on the Nepal–China border at 29.3°N and 83.97°E and confluences with the Ganges River near Patna, Bihar (India). The basin has a vast geographical area (44,797 km²) falling between 25.6°N–29.4°N and 82.8°E–85.82°E, out of which 7620 km² falls in India and the rest in Nepal and China. High rain events exceeding intensity >124.4 mm in 24 hours, frequently occur in the mid-hills over the Nepal portion of the basin (Dahal et al. 2006; GFCC 2004), bringing massive, detached soil particles and muddy flows in Bihar, India. The location of the study area is shown in Figure 10.1.

The climate ranges from subtropical in the south to Tundra toward higher Himalayan regions. The mean daily maximum temperature ranges from 43.03 to −9.42°C, while the mean daily minimum temperature range is 29.37 to −27.22°C (Kumar and Lakshmi 2018). The precipitation is highly variable in space and time within the basin, and almost 90% of rainfall occurs during the South Asian Monsoon season. The average annual rainfall within the basin varies from 2030 mm in the northern mountainous area to 1100 mm in the southern plains region of the basin (Kumar et al. 2016).

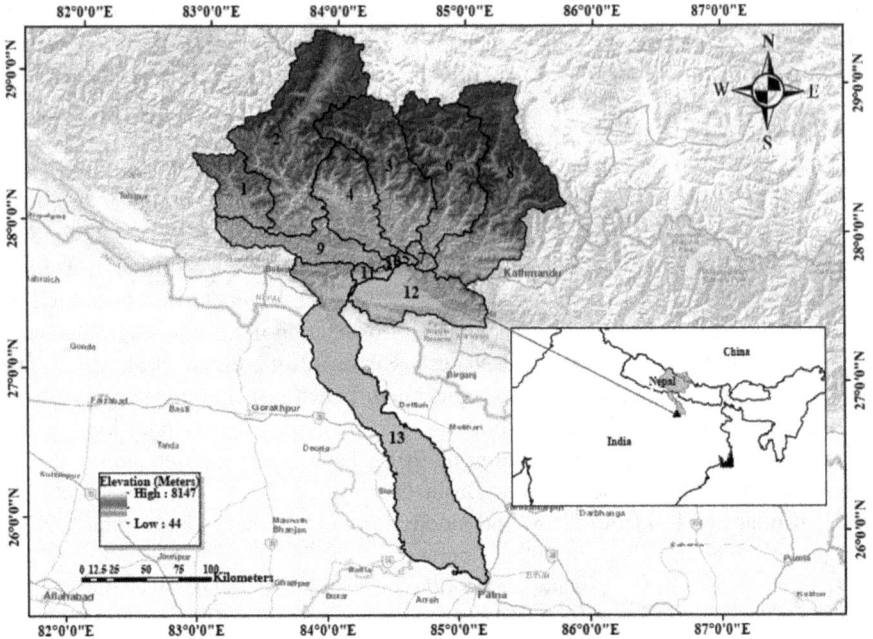

FIGURE 10.1   Location and elevation map of the study area

The Gandak River Basin contains a variety of ecosystems and biodiversity. It ranges from the alpine arid rain shadow areas in the Tibetan Plateau through the steep topography of the high mountains, including some of the world's highest points, Shivalik hills, Dhaulagiri (~8100 m from MSL), down to the flat plains (33 m from MSL) toward its confluence. The Gandak River flows for about 380 km in Nepal and China and 260 km in India before reaching Patna. It has six sub-catchments namely the Kali Gandak, Seti, Marshandi, Budhi Gadak, Trisuli, and Rapti. Out of the six, five are located in the high-altitude areas of the Himalayas (Kumar et al. 2017a, 2017b). Part of the Himalayan basin range falls in the dry alpine climate with low precipitation. The southern part, which extends from the mid-hills to the flat area, has a humid climate with relatively high rainfall. Based on the GlobeLand30 (Chen et al. 2015) land cover data 2010, GRB has 33.1% forest, 22.04% agricultural land, 20.78% grassland, 10.24% ice/snow cover, 9.4% barren land, and only 0.16% urban land [Figure 10.2(a)]. According to the FAO soil classification (2003), the basins comprise ten different types of soils dominated by various kinds of Cambisols (48% of the total basin area), Lithosols (29%), Fluvisols (9%), Glaciers (8.4%), Dystric Regosols (5%) and Rorthic Luvisols [Figure 10.2(b)].

DATA, RESOURCES, AND ERODIBILITY MAPPING

The processes followed in the erodibility prioritization for the watersheds are presented in the flowchart in Figure 10.3. In this work, hydrological units were

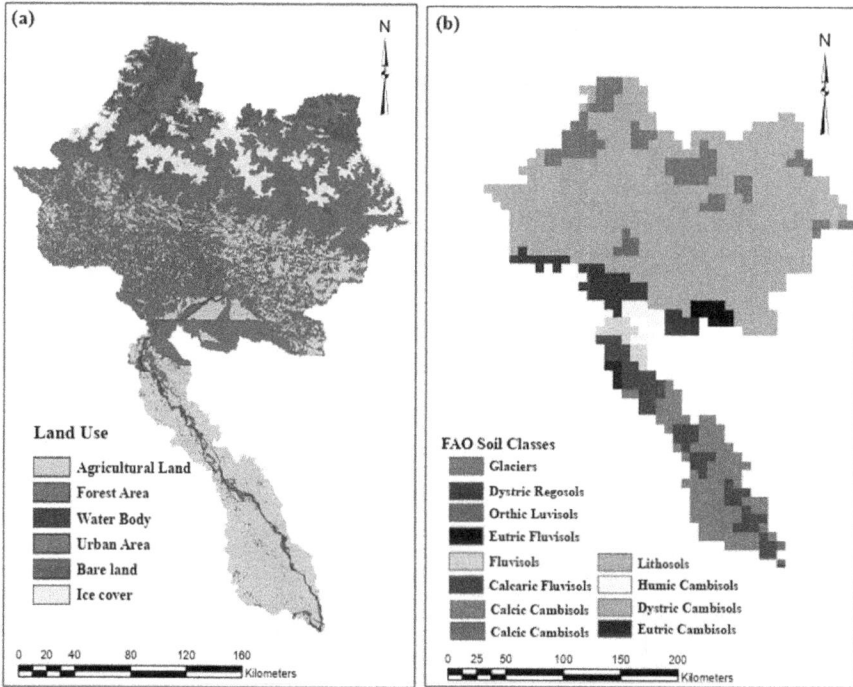

FIGURE 10.2 Land use/land cover (a) and soil map (b) for the study area

considered as a basis for studying morphometry, therefore the stream networks and watershed delineation were conducted using topographical maps of 1:250,000 scale (collected from http://legacy.lib.utexas.edu/maps/ams/india/) and the SRTM 30 m digital elevation model (DEM) (https://earthexplorer.usgs.gov/). The SRTM 30 m DEM was preferred over others as it provides the best river network delineation for the Gandak River Basin (Kumar et al. 2017) than other globally available DEMs. The watershed delineation and stream network extraction have used the D8 flow-direction method and flow accumulation for > 500 ha of the upland area to avoid very small streams. Next, the ranking of the streams was carried out as per Strahler's order (Strahler 1957) and the same is presented in Figure 10.4. In general, the 24 morphological parameters were extracted using SRTM 30 m DEM and GIS techniques as per Table 10.1. Out of the 24 parameters, six are basic topographical features (Table 10.2) which are input to derive other 16 parameters related to linear, shape, and landscape characteristics (Table 10.3). The parameters listed in Table 10.3 have been utilized as input criteria for MCDM methods. The steps followed were pre-processing (DEM reconditioning, assigning stream slope, sink evaluation, fill sinks), computation of flow direction, and flow accumulation. Next, extraction and analysis of the morphometric parameters for each watershed was done using multi-criteria decision-making methods viz. COPRAS, SAW, VIKOR and TOPSIS using the MATLAB® MCDM tool (Irik Mukhametzyanov 2020). In the end, soil erosion

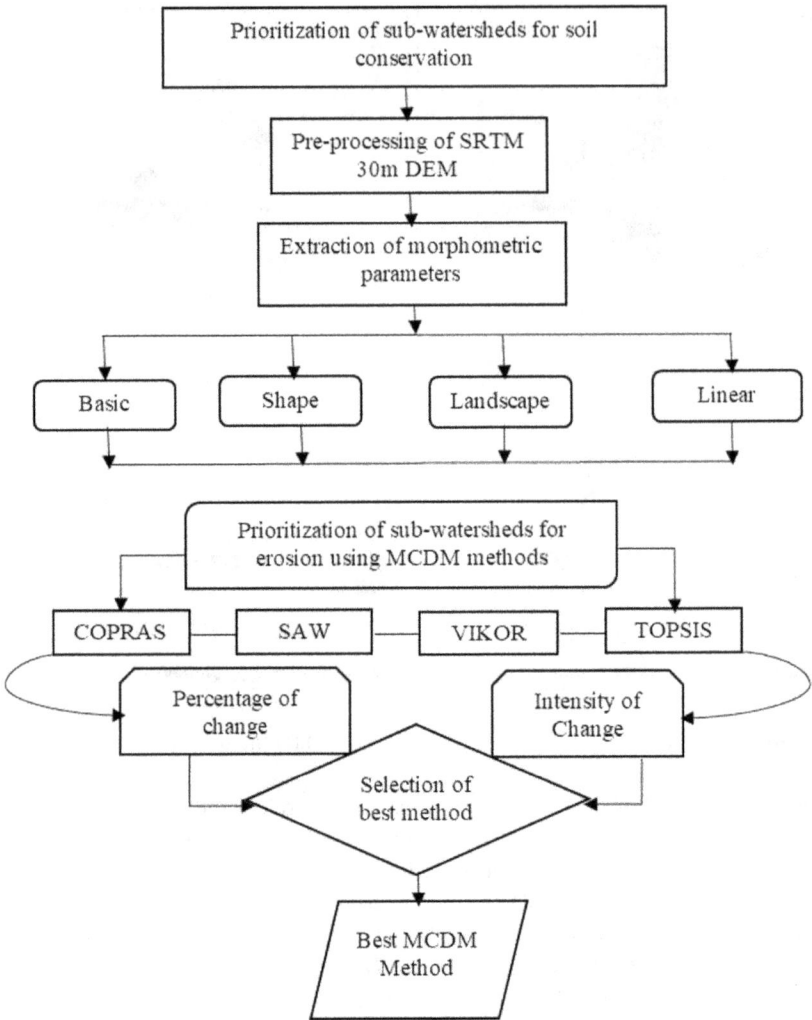

FIGURE 10.3   Flowchart of the methods used in the study

susceptibility maps were prepared for all four methods and the results were validated with the percentage of change and intensity of change methods.

DESCRIPTION OF MCDM METHODS USED FOR THE STUDY

ENTROPY-BASED WEIGHTAGE CALCULATION

In MCDM methods, the biggest challenge is assigning an accurate weightage for different criteria with respect to the score of alternatives. For this, the entropy method

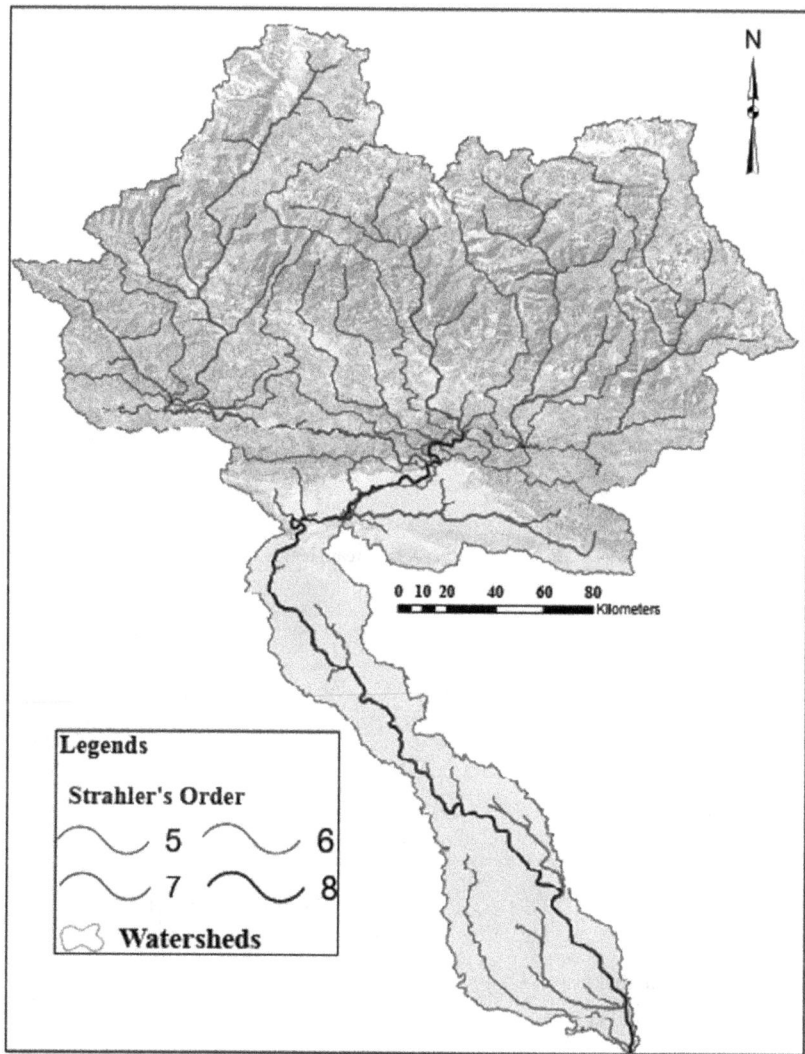

FIGURE 10.4 Strahler's order of the study area

(Shannon 1948) can be utilized. The steps involved in the weightage computation by entropy method are illustrated here

I. Formulation of decision matrix: The decision matrix of $(m \times n)$ order can be framed for m criteria and n alternatives.

$$\begin{bmatrix} a_{11} \, a_{12} \, a_{13} \ldots \ldots a_{1n} \, a_{21} \, a_{22} \, a_{23} \ldots \ldots a_{2n} \ldots \ldots \ldots \ldots \ldots a_{m1} \, a_{m2} \, a_{m3} \ldots \ldots a_{mn} \end{bmatrix} \quad (10.1)$$

TABLE 10.1

Computation of Morphological Parameters Viz. Basic, Linear, Shape, and Landscape

Aspect	Parameters	Formula	Reference
Basic	Basin Area (A)	Area of the watersheds (km^2)	(Horton 1945a)
	Perimeter of basin (P)	Perimeter of the watersheds (km)	
	Stream order (U)	Hierarchical ranking of streams	
	Stream number (N_U)	Total number of streams of all order	(Strahler 1957)
	Basin length (L_b)	Length of basin (km) $$L_b = 1.321 \times A^{0.568}$$	(Nooka Ratnam et al. 2005)
	Stream length (L_u)	Total stream length of all order (km)	(Horton 1945a)
	Mean stream length (L_{SM})	$$L_{SM} = \frac{L_u}{N_u}$$	
	Bifurcation ratio (R_b)	$$R_b = \frac{N_u}{(N_u + 1)}$$ N_u = Number of streams of order "u" $N_u + 1$ = Number of streams of the next order	(Schumm 1956)
Linear	Drainage density (D_d) (km / km^2)	$$D_d = \frac{L_u}{A}$$	(Horton 1945a)
	Stream frequency (F_u) $(no. / km^2)$	$$F_u = \frac{N_U}{A}$$	
	Mean bifurcation ratio (R_{bm})	R_{bm} =Average of bifurcation ratio of all stream orders (U)	(Strahler 1964)

(Continued)

TABLE 10.1 CONTINUED

Computation of Morphological Parameters Viz. Basic, Linear, Shape, and Landscape

Aspect	Parameters	Formula	Reference
	Texture Ratio (T) $(no./km)$	$T = \dfrac{N_U}{P}$	(Horton 1945a)
	Overland flow length $(L_o)(km)$	$L_o = \dfrac{1}{2D_d}$	
	Infiltration number (If)	$If = F_u \times D_d$	(Faniran 1968)
	Constant of channel maintenance (C)	$C = \dfrac{A}{\sum_{i=1}^{i=n} L_u}$	(Horton 1945a)
Shape	Form factor (R_f)	$R_f = \dfrac{A}{L_b^2}$	(Horton 1945a)
	Shape factor (B_s)	$B_s = \dfrac{L_b^2}{A}$	(Nooka Ratnam et al. 2005)
	Elongation ratio (R_e)	$R_e = 1.128 \times \sqrt{\dfrac{A}{L_b}}$	(Schumm 1956)
	Compactness Coefficient (C_c)	$C_c = \dfrac{P}{2\sqrt{\pi A}}$	(Horton 1945a)
	Circulatory ratio (R_c)	$R_c = 4\pi \times \left(\dfrac{A}{P^2}\right)$	(Miller 1953)
Landscape	Basin relief (B_h)	$B_h = h - h_1$ Where, $h = maximum\,height$ $h_1 = minimum\,height$	(Horton 1945a)
	Ruggedness number (R_n)	$R_n = D_d \times \left(\dfrac{B_h}{1000}\right)$	(Moore et al. 1991)
	Relief ratio $R_h)$	$R_h = \left(\dfrac{B_h}{L_b}\right)$	(Schumm 1956)
	Slope (S)	$S = \left(\dfrac{B_h}{\sqrt{A}}\right) \times 100$	(Nautiyal 1994)

TABLE 10.2
Basic Topographical Characteristics of the Gandak River Basin

Watershed	Area(A) (km2)	Perimeter(P) (km)	Total No. of Streams (Nu)	Stream Length (Lu) (km)	Basin Length (Lb) (km)	Elevation (m)		
						Max	Min	Mean
WS1	1998.09	354.45	3899	1790.07	99.00	4029	540	1868.93
WS2	7652.02	816.64	17095	7669.81	212.27	8143	540	3595.77
WS3	4789.14	559.68	11,372	4861.98	162.66	8137	286	3347.50
WS4	2949.57	416.84	6606	2885.93	123.52	7948	229	1628.65
WS5	121.19	73.83	444	122.21	20.15	1929	229	1012.11
WS6	4998.45	592.4	11,934	4959.66	166.66	8147	332	3736.41
WS7	284.66	114.06	607	266.49	32.73	1928	286	876.58
WS8	6605.88	793.88	13,114	6198.89	195.26	7408	332	3429.77
WS9	2242.04	525.06	4408	2203.91	105.70	2336	187	889.90
WS10	65.75	58.28	225	66.45	14.24	1308	187	493.20
WS11	310.72	187.35	876	392.36	34.40	1933	135	364.76
WS12	3176.78	479.65	6732	3883.06	128.83	2600	135	573.66
WS13	9455.84	1420.48	30,913	11,594.1	239.39	1842	44	125.37

TABLE 10.3

SRTM Extracted Basic, Shape and Landscape Parameters of Watersheds

Basin	Dd	Fu	Rbm	T	C	Lo	If	Re	Rc	Rf	Bs	Cc	Bh	S	Rn	Rh
WS1	0.90	1.95	20.65	11.00	1.12	0.56	1.75	0.51	0.20	0.20	4.91	2.24	4.03	9.01	3.61	0.04
WS2	1.00	2.23	1.79	20.93	1.00	0.50	2.24	0.68	0.14	0.17	5.89	2.63	8.14	9.31	8.16	0.04
WS3	1.02	2.37	2.63	20.32	0.99	0.49	2.41	0.61	0.19	0.18	5.52	2.28	8.14	11.76	8.26	0.05
WS4	0.98	2.24	2.49	15.85	1.02	0.51	2.19	0.55	0.21	0.19	5.17	2.17	7.95	14.63	7.78	0.06
WS5	1.01	3.66	9.72	6.01	0.99	0.50	3.69	0.28	0.28	0.30	3.35	1.89	1.93	17.52	1.94	0.10
WS6	0.99	2.39	2.40	20.15	1.01	0.50	2.37	0.62	0.18	0.18	5.56	2.36	8.15	11.52	8.08	0.05
WS7	0.94	2.13	24.32	5.32	1.07	0.53	2.00	0.33	0.27	0.27	3.76	1.91	1.93	11.43	1.80	0.06
WS8	0.94	1.99	38.99	16.52	1.07	0.53	1.86	0.66	0.13	0.17	5.77	2.76	7.41	9.11	6.95	0.04
WS9	0.98	1.97	2.04	8.40	1.02	0.51	1.93	0.52	0.10	0.20	4.98	3.13	2.34	4.93	2.30	0.02
WS10	1.01	3.42	2.20	3.86	0.99	0.49	3.46	0.24	0.24	0.32	3.08	2.03	1.31	16.13	1.32	0.09
WS11	1.26	2.82	2.00	4.68	0.79	0.40	3.56	0.34	0.11	0.26	3.81	3.00	1.93	10.97	2.44	0.06
WS12	1.22	2.12	8.81	14.04	0.82	0.41	2.59	0.56	0.17	0.19	5.22	2.40	2.60	4.61	3.18	0.02
WS13	1.23	3.27	47.85	21.76	0.82	0.41	4.01	0.71	0.06	0.17	6.06	4.12	1.84	1.89	2.26	0.01

II. Calculation of feature weightage $\left(P_{ij}\right)$ for the ith alternative and j^{th} criterion

$$P_{ij} = \frac{a_{ij}}{\sum_{i=1}^{m} a_{ij}^2} \qquad \left(1 \le i \le m \text{ and } 1 \le j \le n\right) \qquad (10.2)$$

III. Calculation of output entropy $\left(e_j\right)$ for the jth criteria

$$e_j = -K\left(\sum_{i=1}^{m}\left(P_{ij} \, lnP_{ij}\right)\right) \qquad (10.3)$$

$$K = \frac{1}{lnln(m)} \qquad (10.4)$$

IV. Calculation of variation coefficient $\left(g_j\right)$ for j^{th} criteria

$$g_i = \left|1 - e_j\right| \qquad (10.5)$$

V. Calculation of entropy weightage $\left(w_j\right)$ for the jth criteria

$$w_j = \frac{g_j}{\sum_{i=1}^{m} g_j} \qquad (10.6)$$

THE COPRAS MCDM METHOD

The COPRAS MCDM method was first introduced by Zavadskas and Kaklauskas (1996). The COPRAS method assumes the direct and commensurate affiliation of the level of magnitude and usefulness of alternatives in the presence of conflicting criteria (Chatterjee and Chakraborty 2013). The COPRAS method for rating the alternatives can be framed in the following steps

I. Calculation of normalized decision matrix by max-min method $\left(R_{ij}\right)$;

Decision matrix as per Equation 10.1

$$R_{ij} = \left(\frac{a_{ij} - a_j^{min}}{a_j^{max} - a_j^{min}}\right) \qquad (10.7)$$

where a_j^{min} and a_j^{max} are minimum and maximum values of j th criteria

II. Computation of weighted normalized decision matrix $\left(d_{ij}\right)$

$$d_{ij} = w_j \times R_{ij} \tag{10.8}$$

III. Computation of maximum $\left(S_j^+\right)$ and minimum $\left(S_j^+\right)$ indices for the alternatives as follows

$$S_j^+ = \sum_{j=1}^{n} y + ij \quad j = 1, 2, \ldots\ldots\ldots n \tag{10.9}$$

$$S_j^- = \sum_{j=1}^{n} y - ij \quad j = k+1, k+2, \ldots\ldots n \tag{10.10}$$

where, $y + ij$ and $y - ij$ are the weighted normalized qualities for advantageous and non-advantageous adjectives, respectively.

IV. Calculation of relative score $\left(Q_i\right)$ for each alternative

$$Q_i = S_j^+ + \frac{S_{min}^- \sum_{j=1}^{n} S_j^-}{S_j^- \sum_{j=1}^{n} \left(\dfrac{S_{min}^-}{S_j^-}\right)} = S_j^+ + \frac{\sum_{j=1}^{n} S_j^-}{S_j^- \sum_{j=1}^{n} \left(\dfrac{1}{S_j^-}\right)} \tag{10.11}$$

where S_{min}^- is the minimum value of S_j^-. S_j^+ and S_j^- are maximum and minimum indices, respectively. The best alternative is having Q_i .

THE SAW METHOD

The SAW model is the most widely used MCDM method. In this method scores of each alternative are obtained by simply aggregation of the weightage of criteria $\left(w_j\right)$ with the value of normalized alternatives $\left(R_{ij}\right)$ in such a way that the relative score $\left(Qi\right)$ can be given directly by the modeler (Sargaonkar et al. 2011). The steps to calculate the relative score for the alternatives by the SAW method are as follows:

I. Calculation of the normalized decision matrix by the max-min method $\left(R_{ij}\right)$ has been done as per Equation 10.7.

II. Calculation of the weighted normalized decision matrix $\left(d_{ij}\right)$

$$d_{ij} = w_j \times R_{ij} \tag{10.12}$$

III. The final step is to acquire the score for each alternative by data integrations as follows

$$Q_i = \Sigma R_{ij} \times w_j \tag{10.13}$$

Where Q_i is the final score of each alternative. The best alternative is having Q_i .

The VIKOR Method

The VIKOR method is a well-established MCDM method to calculate the score of alternatives, first introduced by Opricovic and Tzeng (2004). This method emphasizes on selection and ranking of alternative sets of conflicting criteria (Opricovic and Tzeng 2004). The VIKOR method for rating options can be described in the following steps

I. Calculation of normalized decision matrix by max-min method $\left(R_{ij}\right)$ as per Equation 10.7.

II. Calculation of the weighted normalized decision matrix $\left(d_{ij}\right)$

$$d_{ij} = w_j \times R_{ij} \tag{10.14}$$

III. Determination of ideal object $\left(d_j^+\right)$ and anti-ideal $\left(d_j^-\right)$ for all objectives $\left(j = 1, 2, \ldots . . j\right)$ if the ith criterion is a profit criterion, its maximum value is more relevant to the purpose, then $d_j^+ = maxd_{ij}$ and $d_j^- = mind_{ij}$

IV. Computing the values S_i (utility index) and R_i (regret index) as follows

$$S_i = L_{1,i} = \sum_{i=1}^{n} w_j \times \frac{\left(d_j^+ - d_{ij}\right)}{\left(d_j^+ - d_j^-\right)} \tag{10.15}$$

$$R_i = max\left[\sum_{i=1}^{n} w_j \times \frac{\left(d_j^+ - d_{ij}\right)}{\left(d_j^+ - d_j^-\right)}\right] \tag{10.16}$$

V. Calculation of relative score $\left(Q_i\right)$ for each alternative

$$Q_i = V * \frac{\left(S_i - S^-\right)}{\left(S^* - S^-\right)} + \left(1 - V\right) * \frac{\left(R_i - R^*\right)}{\left(R^- - R^*\right)} \tag{10.17}$$

where, $S^- = minS_i$, $S^* = S_i$, $R^- = minR_i$, $R^* = maxR_i$, and V is the weightage of the criteria. This parameter $\left(V\right)$ has ranged from 0–1 and when the $V > 0.5$, the index of Q will have the majority rule.

VI. The next step is ranking the alternatives that have been sorted by considering the values of S, R, and Q and the best alternative has been designated to one having the least value of these three parameters

VII. As per the Q parameter, the highest score is for the best alternative and needs to qualify the following two conditions

CONDITION 1:

$$Q(A2) - Q(A1) \geq \frac{1}{1-N} \tag{10.18}$$

Where N is the number of criteria and A_1 and A_2 are the alternatives having first and second rank in the list, respectively.

CONDITION 2:

The alternative which has been ranked 1 in Q, should also have rank 1 in S and R, or both of them. If condition 2 is not fulfilled, ranking of alternatives will be as per $A_1, A_2, \ldots, A_m$ where A_m is determined by as following:

$$(A_m) - Q(A_1) < \frac{1}{1-N} \tag{10.19}$$

If condition 1 has not been met, then A_1 and A_2 will be the best solution (Opricovic and Tzeng 2004).

THE TOPSIS METHOD

The TOPSIS is a distance-based MCDM introduced by Hwang and Yoon (1981). The main basis of TOPSIS is the estimation of Euclidean distance of decision-making alternatives from the positive and negative ideal solutions. In this model, the preferred alternative is one having a minimum distance from the positive ideal solution and a maximum distance from the negative ideal solution. The distance of these two (positive and negative ideal solutions) is expressed in terms of the "*closeness coefficient*". Therefore, an alternative having a higher "*closeness coefficient*" is the preferred alternative (Kannan et al. 2009).

The TOPSIS method has the following steps to reach the rating options:

I. Calculation of normalized decision matrix by max-min method $\left(R_{ij} \right)$ as per Equation 10.7

II. Calculation of the weighted normalized decision matrix $\left(d_{ij} \right)$

$$d_{ij} = w_j \times R_{ij} \tag{10.20}$$

III. Calculation of positive ideal solution $\left(x_j^+ \right)$ and negative ideal solution $\left(x_j^- \right)$

$$x_j^+ = \{ \left(if \ j \in C_j \left(max \right); d_{ij} \mid if \ j \in C_j \left(min \right) \} \right) \tag{10.21}$$

$$= \left\{ d_1^+, d_2^+, d_3^+ \ldots \ldots \ldots d_n^+ \right\}$$

$$x_j^- = \{(if \ j \in C_j(max); d_{ij} \ | \ if \ j \in C_j(min)\}) \tag{10.22}$$

$$= \{d_1^+, d_2^+, d_3^+ \ldots \ldots \ldots d_n^+\}$$

where, j and C_j are related to increasing and decreasing criteria, respectively.
IV. The calculation ideal and negative ideal solution distance is as follows

$$S_i^+ = \sqrt{\sum_{i=1}^{n}\left(d_{ij} - x_j^+\right)^2} \tag{10.23}$$

$$S_i^- = \sqrt{\sum_{i=1}^{n}\left(d_{ij} - x_j^-\right)^2} \tag{10.24}$$

V. The final step is the calculation of the *closeness coefficient* of the alternatives to the ideal solution as Equation 10.25. The value of the *closeness coefficient* varies between zero and one. The alternatives with the higher *closeness coefficient* are the superiors.

$$Q_i = \frac{S_i^-}{S_i^+ - S_i^-}; \quad 0 \le Q_i \le 1 \tag{10.25}$$

VALIDATION OF MCDM METHODS

In this study, two model validation indices were used to evaluate and compare the different MCDM methods with each other (Ameri et al. 2018).

$$\Delta P = \frac{N - N_{constant}}{N} \times 100 \tag{10.26}$$

$$I = \frac{\sum_{i=1}^{N} \frac{rank \ i(r1)}{rank \ i(r2)}}{N} \tag{10.27}$$

Where, ΔP is the percentage of change, N is the number of alternatives, $N_{canstant}$ is the number of alternatives having the same rank, ΔI is the intensity of change, $rank \ i(r1)$ is the rank of alternative in the first method, $rank \ i(r2)$ is the rank of alternative in the second method.

The percentage of change (ΔP) tells the power of individual MCDM methods to categorize the number of alternatives (N) into number of rank (r) (two alternatives may have the same rank then $r < N$ or if each alternative has a unique rank then $N =$

r). However, the intensity of change (ΔI) denotes the change in the rank (r) of alternatives between the MCDM methods. A higher value of (ΔP) and (ΔI) for a MCDM method than others, denotes that the method is able to better fragment the area into the number of classes.

RESULT AND DISCUSSION

MORPHOLOGICAL ANALYSIS

Morphological parameters provide a comprehensive description of different earth surface's systems viz. hydrology, geology, morphology, and inter-relation between them. Also, the division of precipitation into runoff and penetration water is mostly decided by drainage characteristics (Kumar et al. 2017a; Schumm and Lusby 1963). The soil erosion process (a morphological phenomenon) is directly influenced by morphological parameters namely linear, shape, and landscape. Moreover, the effect of hillslope morphology has an accelerated effect on soil erosion (Bonetti et al. 2019) which is relevant to this study. In the present study, the effect of morphometric characteristics of the drainage network on soil erosion susceptibility has been evaluated for the Gandak River Basin and its 13 watersheds (Figure 10.1). To achieve this, 23 morphological parameters (related to linear, shape, and landscape characteristics) were investigated out of which 16 parameters were used as criteria for MCDM methods for ranking of watersheds. The discussion of the morphological parameters considered as criteria for MCDM methods is given in the following sub-sections.

BASIC PARAMETERS

The quantitative value of basic morphological parameters for the watersheds of the Gandak River Basin is presented in Table 10.2. Our analysis shows that the Gandak River Basin is an eighth-order Himalayan basin with 44,650 km^2 of geographical area $\left(A\right)$ and 2633.6 km of perimeter $\left(P\right)$. It has 108,225 number of streams accumulating to 46,894.93 km of stream length over the basin. The number of watersheds, number of streams and their length were extracted with the drainage threshold of > 500 ha to avoid a very small stream network during stream network extraction from DEM. These characteristics are indicative of Horton's First Law (Horton 1945b), which states that the number of streams of different ranks in the basin tends to have an inverse geometric ratio. This geometric association has been presented between the logarithm of stream number $\left(N_u\right)$ and stream order $\left(O\right)$ (Figure 10.5). From Figure 10.5, it is evident that the number of streams is gradually decreasing with the increase in the stream order $\left(R^2 = 0.94\right)$. The basin length $\left(L_b\right)$ (Nooka Ratnam et al. 2005) ranges between 14.24–239.39 km, minimum for watershed *WS10* and maximum for *WS13*. Also, the maximum elevation ranges from 1308–8147 m for watersheds *WS10* and *WS6*, and the minimum elevation from 44–540 m for *WS13* and *WS2*. The other parameters (linear, shape, and landscape) extracted for the watersheds of the Gadak River basin have been presented in Table 10.3.

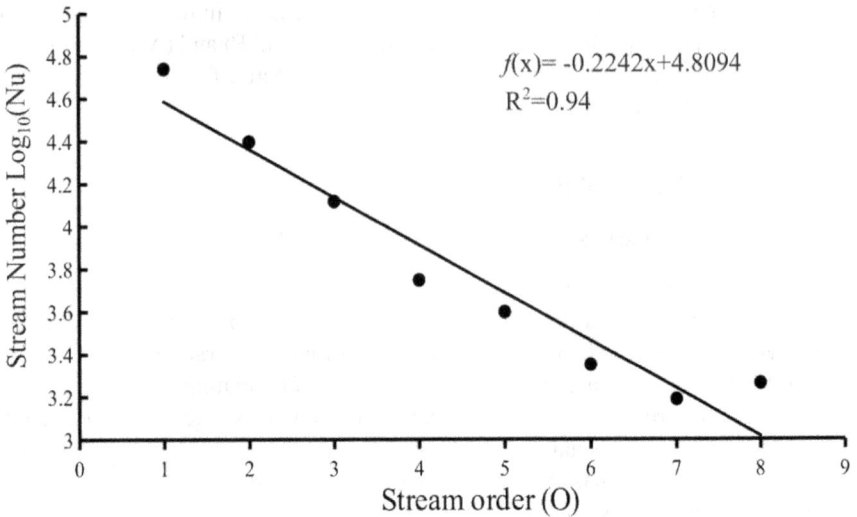

FIGURE 10.5   Stream order *V/S* stream number on a basin scale.

LINEAR PARAMETERS

Drainage Density (D_d):

Stream density represents the landscape's dissection and a basin's runoff potential. The lower drainage density shows that the basin has permeable soil, fair vegetation cover, and flat topography (Ameri et al. 2018; Prasad et al. 2008). Slope angle and relative relief are the key morphological factors that control drainage density. In the Gandak River Basin, *WS1* has the lowest $D_d (0.89)$, which means this watershed has maximum permeability among the other watersheds, or as per D_d criteria, *WS1* has resistance to erosion. On the other hand, *WS11* with the highest $D_d (1.26)$ is highly susceptible to erosion. For the Gandak River Basin, watersheds *WS7*, *WS8*, *WS4*, *WS9*, *WS6*, *WS2*, *WS5*, *WS10*, *WS3*, *WS12*, and *WS13* are in ascending order of D_d.

Stream frequency (F_u):

F_u is the ratio of the number of streams in the watershed to the watershed's area (Horton 1945b). It has a direct relation to the roughness of the watershed and an inverse relation to the permeability of the watershed's soil. High F_u value for a watershed reveals that it has a rocky surface with low permeability and high erosion (Ameri et al. 2018; Fallah et al. 2016). The F_u in the Gandak River Basin varies from 1.95–3.66 streams/km², minimum for *WS1* and maximum for *WS5*. These results reveal that *WS5* has the lowest penetration and therefore higher susceptibility to erosion, but *WS1* has the lowest erosion. As per F_u, watersheds *WS9*, *WS8*, *WS12*, *WS7*, *WS2*, *WS4*, *WS3*, *WS6*, *WS11*, *WS13*, and *WS10* are in ascending order of erosion for the Gandak River Basin.

Mean bifurcation ratio $\left(R_{bm}\right)$:

The R_{bm} is inversely correlated to infiltration alike D_d and F_u. A higher value of R_{bm} denotes the peak of the initial hydrograph during flooding, ultimately higher soil degradation (Ameri et al. 2018; Fallah et al. 2016). The value of R_{bm} in the Gandak River Basin varies from 1.78–47.84, minimum for *WS2* and maximum for *WS13*, respectively. Therefore, as per R_{bm}, *WS13* is the least permeable and highly susceptible to erosion among all watersheds of the Gandak River Basin. The ranking of soil erosion susceptibility as per R_{bm} for watersheds in ascending order are *WS11*, *WS9*, *WS10*, *WS6*, *WS4*, *WS3*, *WS12*, *WS5*, *WS1*, *WS7*, and *WS8*.

Texture ratio $\left(T\right)$:

Texture ratio $\left(T\right)$, one of the most critical parameters of morphology, depends upon physical factors such as rainfall, climate, infiltration capacity of the soil, soil type, soil evolutionary stage, rock, and vegetation. According to (Smith 1950), drainage texture is classified into four categories viz. rough $(T < 4)$, moderate $(4 < T < 10)$, soft $(10 < T < 15)$, and ultra-soft $(T > 15)$ or high lands. The high value of T for a watershed reveals the presence of soft rocks, which are more vulnerable to erosion (Ameri et al. 2018; Fallah et al. 2016). In the Gandak River Basin value T varies from 3.86–21.76, minimum for *WS10* and maximum for *WS13*. Therefore, *WS10* has the highest vulnerability for soil erosion. The watersheds *WS11*, *WS7*, *WS5*, *WS9*, *WS1*, *WS12*, *WS4*, *WS8*, *WS6*, *WS3*, and *WS2* are in ascending order T, denoting their rank of vulnerability to soil erosion. Overall, the average value of T for the basin is greater than 10, which indicates the presence of fragile soft rock mountains in the basin (Singh 2006).

Constant of channel maintenance $\left(C\right)$:

The constant channel maintenance $\left(C\right)$ indicator reflects the permeability of the land surface and control of the flow to the basin outlet. The parameter C has a direct correlation with erosion similar to the relationship of D_d to F_u (Ameri et al. 2018; Fallah et al. 2016). The value of C in the Gandak River Basin varies from 0.79–1.12, minimum for *WS11* and maximum for *WS1*. Therefore, *WS1* is more susceptible to erosion than other watersheds. The rank of watersheds for soil erosion as per the value of C are *WS13*, *WS12*, *WS3*, *WS10*, *WS5*, *WS2*, *WS6*, *WS9*, *WS4*, *WS8*, *WS7*.

Length of overland Flow $\left(L_o\right)$:

The L_o is an independent variable that affects the hydrological and physiological evolution of the basin (Strahler 1952b). This parameter is inversely related to the slope of the channel. Its value is higher for mild slopes and low on steep slope terrains (Ameri et al. 2018; Fallah et al. 2016). The value of L_o for the Gandak River Basin is much less, ranging from 0.4–0.55 reflecting a steeper slope for the basin. This finding is true since the basin has an elevation ranging from ~40–8100 m from MSL. The minimum value of L_o is seen for *WS11* and the maximum for *WS1*. Therefore, *WS1* has been found most prone to erosion as per L_o. The rank of watersheds in ascending order of L_o are *WS13, WS12, WS3, WS10, WS5, WS2, WS6, WS9, WS4, WS8,* and *WS7.*

Infiltration number (If):
The *If* is an important parameter which defines the permeability of terrain and ultimately influences the runoff potential from the watershed. The *If* and sensitivity of watersheds to erosion are inversely proportional to each other (Ameri et al. 2018; Fallah et al. 2016). The *If* value ranges from 1.74–4.00, minimum for watershed *WS1* and maximum for *WS13.* Therefore, *WS1* has the highest susceptibility to erosion. The rank of watersheds (in descending order) for susceptibility to erosion are *WS8, WS9, WS7, WS4, WS2, WS6, WS3, WS12, WS10, WS11,* and *WS5.*

SHAPE PARAMETERS

Elongation ratio (R_e):
The value of R_e in the Gandak River Basin varies from 0.24–0.70. For flat topographies, the value is close to 1 whereas a value < 0.8 denotes the area having high relief and steep ground slope (Fallah et al. 2016; Schumm 1956). In this study, *WS13* with $R_e = 0.70$ was found to be the most elongated watershed and therefore it is least susceptible to erosion. However, *WS10* with $R_e = 0.24$ was found as most susceptible to erosion. The ranks (in descending order) of susceptibility to erosion for the watersheds are *WS5, WS7, WS11, WS1, WS9, WS4, WS12, WS3, WS6, WS8,* and *WS2.*

Circularity Ratio (R_C):
The R_C is generally the function of stream length, stream frequency, roughness, slope, geological structures, and climate. The high value of R_C denotes a circular-shaped watershed with high roughness and permeability (Ameri et al. 2018; Fallah et al. 2016). However, the low R_C value represents an elongated watershed with low roughness and permeability. In the Gandak River Basin R_C value ranges from 0.058–0.279, minimum for *WS13,* and maximum for *WS5.* Therefore, *WS13* having the lowest value of R_C also has the lowest permeability. Subsequently, it has the highest sensitivity to erosion. On the other hand, WS5 with the highest value of R_C is least susceptible to erosion. The ranks of watersheds for soil erosion (in descending order) are *WS9, WS11, WS8, WS2, WS12, WS6, WS3, WS1, WS4, WS10,* and *WS7.*

Form factor (R_f):
The R_f is inversely related to soil erosion, similar to the elongation ratio. Therefore, a watershed having a lower value of R_f will be more sensitive to soil erosion (Ameri et al. 2018; Fallah et al. 2016). In the Gadak River basin the, R_f value varies from 0.16–0.32, minimum for *WS13* and maximum for *WS10.* Hence, *WS13* is the most and *WS13* is the least erosion-prone watershed among all. The ranks of susceptibility to erosion (in descending order) are *WS2, WS8, WS6, WS3, WS12, WS4, WS9, WS1, WS11, WS7,* and *WS5.*

Shape factor (B_S):

The B_S largely affects the rate of sediment and runoff along with the drainage length and roughness of the watershed. Therefore, in the case of erosion, B_S behaves similarly to the R_f (Ameri et al. 2018; Fallah et al. 2016). In the Gandak River Basin, the B_S varies from 3.08–6.06, minimum for *WS10* and maximum for *WS13*. Therefore, *WS10 is* most susceptible to erosion, whereas *WS13* is least. The ranks of watersheds (in descending order) for susceptibility to erosion are *WS5, WS7, WS11, WS1, WS9, WS4, WS12, WS3, WS6, WS8*, and *WS2*.

Compactness coefficient $\left(C_C \right)$:
The C_C is directly related to the infiltration capacity of soil for a watershed. Hence, the effect of C_C is similar to R_f and B_S for soil erosion (Ameri et al. 2018; Fallah et al. 2016). For the basin, the C_C varies from 1.89–4.12, minimum for *WS5*, and maximum for *WS13*. Therefore, as C_C *WS5* is the most and *WS13* is the least erosion-prone watershed in the basin. The ranks of the watersheds (in descending order) prone to erosion are *WS7, WS10, WS4, WS1, WS3, WS6, WS12, WS2, WS8, WS11*, and *WS9*.

LANDSCAPE PARAMETERS

Basin relief $\left(B_h \right)$:
The B_h is the difference of maximum and minimum elevation points within a watershed. It has a strong correlation with the hydrological characteristics (Schumm 1956). The B_h indicates the total slope of the watershed and therefore the intensity of soil erosion is proportional to it (Ameri et al. 2018; Fallah et al. 2016). In the Gandak River Basin, *WS10* has a minimum $B_h \left(1.30 \right)$, whereas *WS6* has a maximum (8.15). Therefore, *WS6* has the highest severity of erosion, and *WS10* has the least. The ranks of watersheds for severity of erosion in ascending order are *WS13, WS7, WS5, WS11, WS9, WS12, WS1, WS8, WS4, WS3, WS2*.

Slope $\left(S \right)$:
The slope of the watershed is a hydro-morphometric parameter that expresses the runoff volume and its concentration (Fallah et al. 2016; Mesa 2006). In the Gadak River basin, the steepest slope has been found for *WS5* (17.52) and the lowest for *WS13* (1.89). Therefore, *WS5* is most prone to soil erosion whereas *WS13* least. The ranks of watersheds for the severity of erosion in ascending order are *WS12, WS9, WS1, WS8, WS2, WS11, WS7, WS6, WS3, WS4, WS10*.

Ruggedness number $\left(Rn \right)$:
The R_n is used to estimate the flood potential streams (Strahler 1952b) and basically, it denotes the geometric features of a drainage basin. It is directly correlated with the erodibility potential of an area (Ameri et al. 2018; Fallah et al. 2016). The value of R_n in the Gandak River Basin ranges from 1.32–8.26, minimum for *WS10* and maximum for *WS3*, which means *WS3* is most susceptible to erosion whereas *WS10*

is the least. The ranks of watersheds for severity of erosion in ascending order are *WS7, WS5, WS13, WS9, WS11, WS12, WS1, WS8, WS4, WS6,* and *WS2.*

Relief ratio (Rh):
The R_h is directly related to the slope of the basin and it affects the earth's surface, subsequently the hydrological processes and erosion of the drainage basin. Like the other shape parameters, R_h is directly related to the watershed's likelihood of being exposed to erosive fluvial forces (Ameri et al. 2018; Fallah et al. 2016). In the Gadak River basin, the R_n varies from 0.007–0.095, minimum for *WS13* and maximum for *WS5.* Therefore, *WS5* is most susceptible to erosion whereas *WS13* is the least as per R_n value. The ranks of watersheds for the severity of erosion in ascending order are *WS12, WS9, WS8, WS2, WS1, WS6, WS3, WS11, WS7, WS4,* and *WS10.*

ERODIBILITY PRIORITIZATION OF WATERSHEDS USING MCDM METHODS

Various researchers have successfully used the morphological parameters in order to prioritize the erosion-prone watersheds within a basin (Ameri et al. 2018; Chowdary et al. 2013; Deshmukh et al. 2010; Jain and Ramsankaran 2019; Patel et al. 2013). Authors, in the present work, have analyzed morphological characteristics of the drainage basin (linear, shape, and landscape) that have been taken into account in order to prioritize the risk of soil erosion associated with each watershed (Ameri et al. 2018; Patel et al. 2013). Linear and landscape morphological characteristics have a positive effect on soil erosion (Nooka Ratnam et al. 2005). However, shape parameters have an inverse correlation with soil erosion (Patel et al. 2013; Patel and Dholakia 2010), and therefore the lower value of these parameters will have more effect on soil erosion. For example, linear parameter drainage density $(D_d = 1.26)$ is highest for *WS11*, and therefore, *WS11* is the highest susceptible to erosion however, *WS1* has the lowest $D_d (0.89)$, which means this watershed has the least soil erosion.

After the calculation of all morphometric parameters viz. linear, shape, and landscape (Table 10.3), a decision matrix of $(m \times n)$ was prepared using alternatives-watersheds $(i = 1, m)$ and criteria-morphometric parameters $(j = 1, n)$ (Equation 10.1). The criteria used in this study have dissimilar units (such as drainage density, stream frequency, slope, etc.), therefore the decision matrix had to be normalized before considering it for further analysis. In this research, the maximum–minimum method of normalization was used (Equation 10.7) before furnishing the data for ranking by COPRAS, SAW, VIKOR, and TOPSIS MCDM methods.

In the next step, the weightage for each criterion has been determined using the entropy method as per Equation 10.6. The weightage of the different criteria is presented in Figure 10.6. The results of the entropy weighting for morphometric parameters revealed that the $R_{bm} (0.366)$, $B_h (0.126)$, and $R_n (0.114)$ are the most effective morphometric parameters for soil erosion. In contrast, entropy weighting has shown that morphometric parameters $L_o (0.0031)$, $C(0.0032)$ and $D_d (0.0035)$

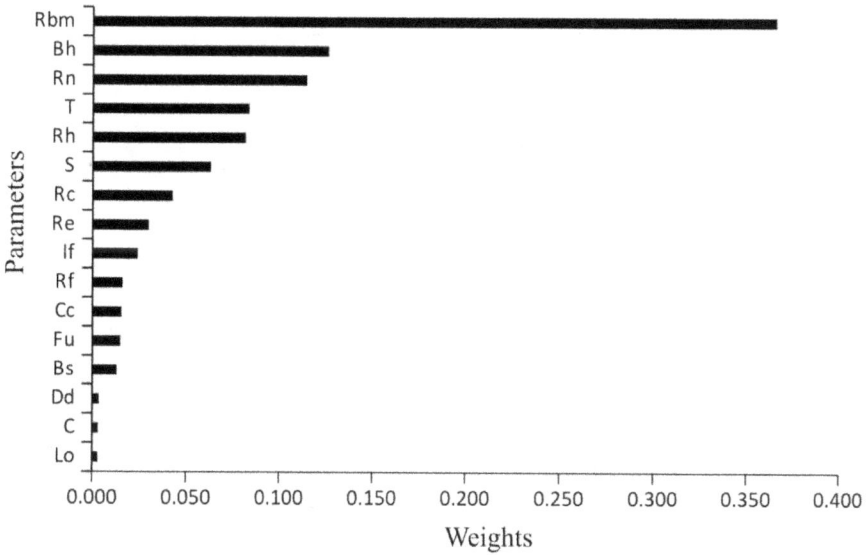

FIGURE 10.6   The relative weight of the geomorphometric parameters using the entropy method

have the least effect on soil erodibility of the watersheds. The rank of other morphometric parameters (in ascending order of erodibility effect) are B_S (0.013) , F_u (0.015), C_C (0.0153), R_f (0.016), I_f (0.024), R_e (0.03), R_C (0.042), $S(0.062)$, R_h (0.081), and $T(0.083)$. In the next step, the weighted normalized decision matrix was calculated (Equation 10.8, Equation 10.12, Equation 10.14, and Equation 10.20) in a similar fashion for all MCDM methods. After this, the final rank of each alternatives-watersheds was calculated using Equation 10.11, Equation 10.13, Equation 10.17, and Equation 10.25 for COPRAS, SAW, VIKOR, and TOPSIS, respectively. The score (Q_i) for each alternative along with their priorities of erosion susceptibility by all MCDM methods are presented in Table 10.4. The other details for the computation of elements of each MCDM method have been provided in the previous section.

The results of the watershed prioritization by the COPRAS method show that WS8, WS13, and WS3 have the highest Q_i scores, 0.694, 0.541, and 0.469, respectively, and therefore, these watersheds have been ranked as 1st, 2nd, and 3rd in severity to erosion. On the other hand, watersheds WS9, WS11, and WS10 have the lowest Q_i values 0.174, 0.204, and 0.211 respectively, and therefore, these watersheds are the least erosion-prone with ranks 13th, 12th, and 11th within the basin. The SAW method has shown similar results as the COPRAS method, watersheds WS8, WS13 and WS3 have ranked 1st, 2nd, and 3rd in the Soil Erosion Severity Index with Q_i values 0.704, 0.548, and 0.479, respectively; however, watersheds WS9, WS10, and WS11 have the lowest ranks 13th, 12th, and 11th with Q_i values 0.178, 0.213 and 0.221. In VIKOR method, watersheds WS8, WS13, and WS3 have been ranked to 1st,

TABLE 10.4
Prioritization of Watersheds Using COPRAS, SAW, VIKOR, and TOPSIS Models

Rank	COPRAS			SAW			VIKOR			TOPSIS		
	Qi	Priority	Ai	Qi	Priority	Ai	Qi	Priority	Ai	Qi	Priority	Ai
1	0.694	High	8	0.704	High	8	0.997	Very High	8	0.876	Very High	13
2	0.541	High	13	0.548	High	13	0.771	Very High	13	0.785	Very High	8
3	0.469	Moderate	3	0.479	Moderate	3	0.486	Moderate	7	0.436	Moderate	7
4	0.463	Moderate	6	0.474	Moderate	6	0.476	Moderate	1	0.409	Moderate	1
5	0.458	Moderate	4	0.469	Moderate	4	0.294	Moderate	3	0.205	Low	5
6	0.447	Moderate	2	0.458	Moderate	2	0.286	Moderate	6	0.191	Low	12
7	0.406	Moderate	1	0.416	Moderate	1	0.283	Moderate	4	0.191	Low	3
8	0.363	Moderate	7	0.374	Moderate	7	0.263	Moderate	2	0.185	Low	6
9	0.311	Moderate	5	0.319	Moderate	5	0.237	Low	5	0.177	Low	4
10	0.247	Low	12	0.258	Moderate	12	0.167	Low	12	0.171	Low	2
11	0.211	Low	10	0.221	Low	10	0.043	Low	10	0.055	Low	10
12	0.204	Low	11	0.213	Low	11	0.033	Low	11	0.048	Low	11
13	0.174	Low	9	0.178	Low	9	0	Low	9	0.047	Low	9

Qi is the final score of the method and Ai is the alternatives (watersheds)

2nd, and 3rd with Q_i values of 0.997, 0.771, and 0.486 in the soil erosion severity ranking and WS9, WS11, and WS10 have been found to be 13th, 12th, and 11th rank with Q_i values of 0, 0.033, and 0.043. At last, the TOPSIS method has examined watersheds WS13, WS8, and WS7 in rank 1st, 2nd, and 3rd with Q_i values of 0.876, 0.785, and 0.436, respectively, in Soil Erosion Severity Index and watersheds WS9, WS11, and WS10 have shown similar rank as TOPSIS with Q_i values of 0.047, 0.048, and 0.055. The Q_i value varies between 0–1 for all MCDM methods, therefore to establish the priority indexing, the Q_i range has been linearly categorized into four categories viz. low (< 0.25), moderate (0.25–0.5), high (0.5–0.75), and very high (> 0.75). The priority indexing of soil erosion severity has found categories high, moderate, and low for COPRAS and SAW methods (Figure 10.7(A) and Figure 10.7(B)). On the other hand, VIKOR and TOPSIS have shown very high, moderate, and low priority indexes similar to each other (Figure 10.7(C) and Figure 10.7(D)).

The results of watershed prioritization using MCDM methods reflect that the erosion is mostly in the lower reach of the basin (Tarai region) and western alluvial plains toward Kathmandu. The COPRAS and SAW models have ranked 30.8%, 53.8%, and 15.4% of the basin area as low, moderate, and high-priority zones, respectively, however, neither ranked any watershed as very high-priority zone. On the other side, VIKOR and TOPSIS have found 69.2%, 15.4%, and 15.4% of the basin area into low, moderate and high priority zones and none of the watersheds into high priority zones. The results are comparable with the land cover map and soil map (Figure 10.2), in which WS13 and WS8, which are ranked high (COPRAS & SAW) and very high (VIKOR & TOPSIS), have different types of Cambisols covered with agriculture, urban, and bare land cover type, which are severe soil erosion-prone characteristics. On the other hand, watersheds falling on moderate to low erosion priority zones have majority forest land cover along with ice land and bare soil. This is because the land cover type has a significant impact on drainage networks and erosion severity of the watersheds (Altaf et al. 2014; Ashraf 2020). Furthermore, the presence of a high of dense vegetation, roots, and trunk reduces the rill erosion, and it also reduces the hammering effect of the raindrops and thus splash erosion (Aber et al. 2010; Shinohara et al. 2016). The watersheds WS13 and WS8 have a high value of R_{bm} (47.85 and 38.9), B_h (1.84, and 7.41), and R_n (2.26 and 6.9), respectively, and that is why these have been ranked high-priority zones. Obviously, WS8 has a higher value of B_h and R_n than WS13 (higher-ranked watershed), but this is because of criteria I_f, R_C, R_e, etc. which implies a reducing soil erosion effect on WS8. In contrast, the watersheds ranked low priority zones (WS9, WS10, WS11) have the lower value of R_{bm}, B_h, R_n and higher value of shape parameters, which have mostly a reducing soil erosion effect (Table 10.3). These results enable the framework to identify severe erosion-prone watersheds in a data scarce river basin. In the present work, WS8 and WS13 were categorized as very high erosion-prone areas and therefore, in these two watersheds the conservation practices are most desired.

The results of the present study and the methodology used are analogous to the studies done for watershed prioritization in the country and other parts of the world. The entropy-weighting method found R_{bm}, B_h, and R_n as the first three most influential parameters which are consistent with others (Khare 2014; Patel et al. 2012). The

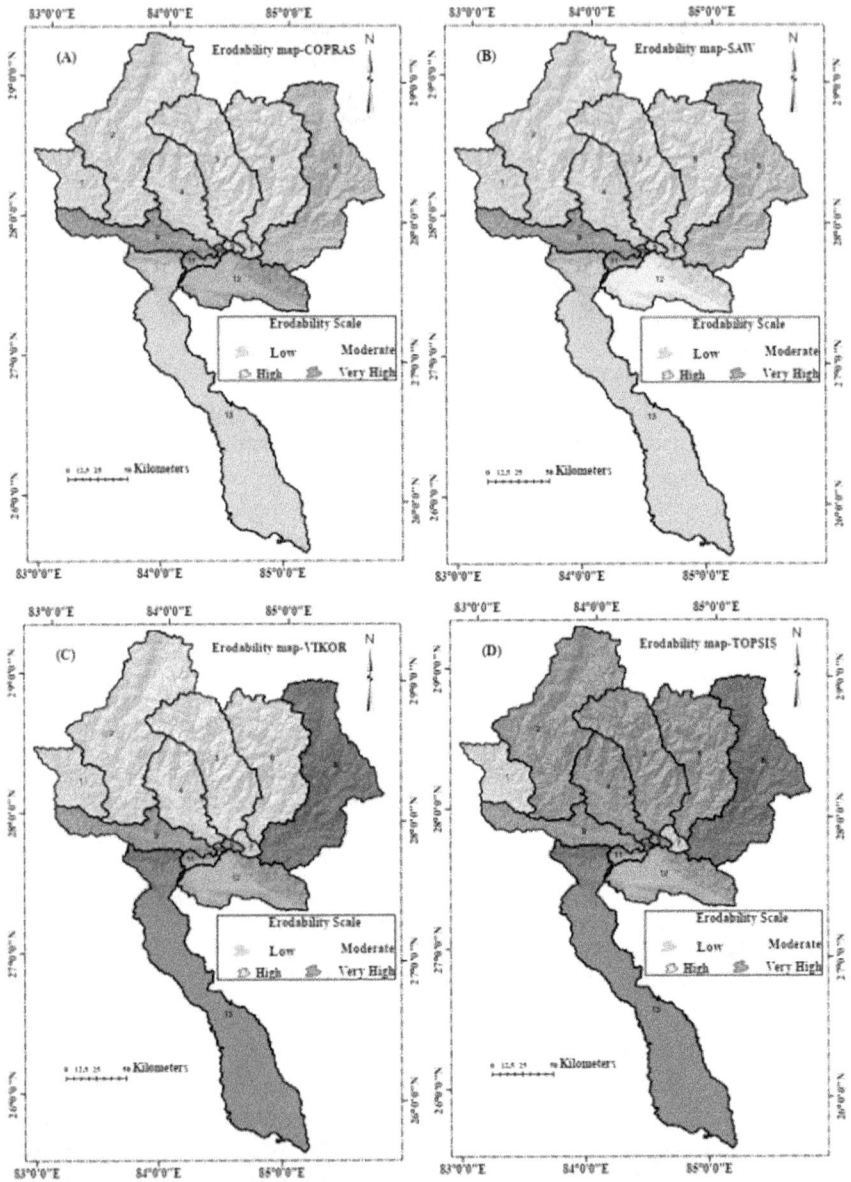

FIGURE 10.7 Erodibility prioritization of watersheds using (A) COPRAS, (B) SAW, (C) VIKOR, and (D) TOPSIS MCDM methods

watersheds having high bifurcation are found to be seriously affected by erosion and those watersheds were assigned high severity rank in studies done over surrounding Himalayan watersheds (Arabameri et al. 2018; Rather et al. 2017). Similarly, B_h and R_n also had a great influence on watershed erosion in other Himalayan catchments (Khare 2014; Patel et al. 2013). Moreover, our study found TOPIS as the best MCDM method to identify erosion-prone watersheds which is similar to the results published by other researchers (Ameri et al. 2018; Khosravi et al. 2019; Opricovic and Tzeng 2004; Vulević et al. 2015).

VALIDATION OF RESULTS FOR MCDM METHODS

The effectiveness of ranking methods (MCDM) has been examined by the percentage of change (ΔP) and intensity of change (ΔI) statistical indicators and the results for these are presented in Table 10.5 and Table 10.6. The results of ΔP found that the TOPSIS (with ΔP 215.38) is the most efficient MCDM method, which means it is able to divide the area into distinctive rank of watersheds. Followed by TOSIS, COPRAS , ($\Delta P = 123.08$), SAW (c = 123.08) and VIKOR ($\Delta P = 153.85$)in respective ranks to fragment the area into a number of sub-classes. The results of intensity of change (ΔI) show a similar agreement of superiority for the TOPSIS method with the highest ΔI 2.24. This means TOPSIS is better to categorize the alternatives for their change in rank than other MCDM methods. In the intensity of change method, COPRAS and SAW together ranked second with ΔI a value of 2.18 and VIKOR stood last with ΔI a value of 2.16. From the results of ΔI and ΔI, it can be stated that the TOPSIS MCDM method is better for fragmenting and ranking the watersheds into a number of distinctive erosion-prone areas.

After the validation, the rank of watersheds achieved through TOPSIS was plotted against morphological parameters to see the association with each other. It has been observed that three morphological parameters (R_{bm}, C_C and R_h) have direct and significant trends (Figure 10.8). The rank of the watersheds (r) is associated

TABLE 10.5
Percentages of the Change (P) Between the MCDM Methods

Methods	COPRAS	VIKOR	SAW	TOPSIS	Sum
COPRAS	0.00	46.15	0.00	76.92	123.08
VIKOR	46.15	0.00	46.15	61.54	153.85
SAW	0.00	46.15	0.00	76.92	123.08
TOPSIS	76.92	61.54	76.92	0.00	215.38

TABLE 10.6
Intensity of the Change (I) Between the MCDM Methods

Methods	COPRAS	VIKOR	SAW	TOPSIS	Sum
COPRAS	1.00	0.39	0.36	0.43	2.18
VIKOR	0.38	1.00	0.38	0.39	2.16
SAW	0.36	0.39	1.00	0.43	2.18
TOPSIS	0.43	0.39	0.43	1.00	2.24

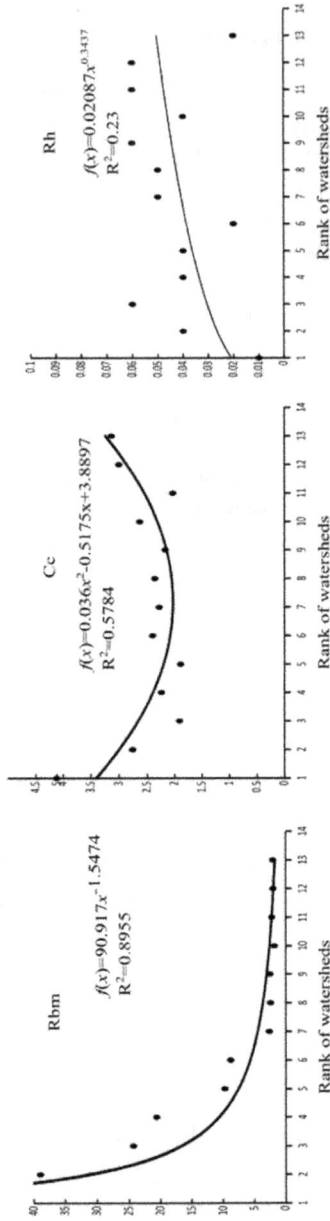

FIGURE 10.8   Relation between the erosion severity (rank of watersheds) to the morphological properties

with negative power function with mean bifurcation ratio (R_{bm}) $(R_{bm} = 90.917r^{-1.547})$. It means that the severity of the erosion for the watersheds declines with power function when the mean bifurcation ratio of the watersheds reduces. To reduce the bifurcation ratio, nala plugging and stream plugging can be the best watershed management practices in order to reduce the bifurcations of the streams. The compactness coefficient (C_c) has a second-order negative polynomial $(C_c = 0.036r^2 - 0.5175r + 3.8897)$ association with the rank of the watershed (r). It means the circular basin has the shortest time of concentration (T_c) and these watersheds are huge erosion-prone. Therefore, structural measures such as the construction of check dam and stop dam can be the best practices to reduce the time of concertation (T_c) of the flow. Similarly, the relief ratio (R_h) has a second-order positive polynomial relation $(R_h = 0.0209r0.3437)$ with rank of the watersheds (r). This shows that the watersheds having higher relief ratios are more erosion-prone and therefore in these watersheds terracing and countering practices to reduce the relief ratio can be the best practices to reduce soil erosion.

CONCLUSION

Regionalization of a severe erosion-prone area is a critical step before implementing the management plans in large and in accessible regions. Nowadays remote sensed DEMs have given the opportunity to derive the morphological parameters which are the true indicator of the land condition. However, the standalone morphological parameter cannot regionalize the erosion-prone zone and therefore the application of a criteria-based decision-making approach has found a place which can take into account the importance of the parameters based on their influence on the erosion.

In this study, morphological parameters along with the MCDM methods were used to prioritize the erosion-prone watersheds in the Gandak River Basin. The present study has extracted the morphometric characteristics using SRTM 30 m DEM and GIS techniques for the watersheds of the Himalayan Gandak River Basin. Four MCDM methods (COPRAS, SAW, VIKOR, and TOPSIS) were applied to find the weighted score (Q_i) of morphological parameters and rank the severity of the watersheds. The ranks of COPRAS, SAW, VIKOR and TOPSIS have been compared and the accuracy of these models has been evaluated using a percentage of change and intensity of change statistical index. It has been found that the TOPSIS method has more accuracy for watershed prioritization than others for the Himalayan region. In the case of the priority index, COPRAS and SAW have divided the watersheds into low, moderate and high-priority zones however, VIKOR and TOPSIS have categorized them into low, moderate and very high-priority zones.

The results of the best MCDM method TOPSIS have demarcated watersheds into low, moderate, and very high priority zones with geographical areas 69.2%, 15.4%, and 15.4%, respectively, however, none of the watersheds have been found to be high priority zones. The watersheds that have been categorized as high-priority zones are WS13 and WS8, which fall in agriculture dominated Tarai region having the most fragile calcic and dystric Cambisols soil types. Also, the study found that the rank of the watersheds (r) has a negative power function with mean bifurcation ratio, second-order negative polynomial relation with compactness coefficient and, second-order positive polynomial relation with relief ratio.

In the most affected watersheds (*WS8* and *WS13*), proper soil conservation practices need to be adopted by the government of India and the Government of Nepal in order to avoid soil erosion toward the lower trunk which ultimately confluences with the great Ganga River. The government of India may plan extensive soil conservation practices in the lower trunk of the Gandak River Basin through the National Rural Employment Guarantee Act (NREGA) especially the construction of lave on the watercourse of Gandak, Nala plugging, bunding, hush &wood gabions (Jain and Ramsankaran 2019; Tang et al. 2014).

The present study has reflected that morphological characteristics (extracted from DEMs using GIS techniques) along with E-MCDM methods are able to demarcate the prioritization of soil erosion-prone watersheds for the complex Himalayan Gandak River Basin. Therefore, these methods can be used by decision-makers and planners to make appropriate resource allocations for better soil conservation at the basin scale.

Availability of data: The data that support the findings of this study are openly available on NASA's Earth Explorer site [https://earthexplorer.usgs.gov/] and the Texas University depository [(http://legacy.lib.utexas.edu/maps/ams/india/)].

CONFLICTS OF INTEREST

"The Author(s) declare(s) that there is no conflict of interest".

FUNDING STATEMENT

No funding was available for the current research work.

REFERENCES

Aber, J. S., Marzolff, I., and Ries, J. B. (2010). "Vegetation and erosion." In *Small-Format Aerial Photography*, Elsevier, 219–228.

Acharya, A., and Prakash, A. (2019). "When the river talks to its people: Local knowledge-based flood forecasting in Gandak River basin, India." *Environmental Development*, 31, 55–67.

Ajay Kumar, V., Mondal, N. C., and Ahmed, S. (2020). "Identification of groundwater potential zones using RS, GIS and AHP techniques: A case study in a part of Deccan volcanic province (DVP), Maharashtra, India." *Journal of the Indian Society of Remote Sensing*, 48(3), 497–511.

Altaf, S., Meraj, G., and Romshoo, S. A. (2014). "Morphometry and land cover based multi-criteria analysis for assessing the soil erosion susceptibility of the western Himalayan watershed." *Environmental Monitoring and Assessment*, 186(12), 8391–8412.

Ameri, A. A., Pourghasemi, H. R., and Cerda, A. (2018). "Erodibility prioritization of sub-watersheds using morphometric parameters analysis and its mapping: A comparison among TOPSIS, VIKOR, SAW, and CF multi-criteria decision making models." *Science of the Total Environment*, 613–614, 1385–1400.

Arabameri, A., Pradhan, B., Pourghasemi, H. R., and Rezaei, K. (2018). "Identification of erosion-prone areas using different multi-criteria decision-making techniques and GIS." *Geomatics, Natural Hazards and Risk*, 9(1), 1129–1155.

Arnous, M. O., Aboulela, H. A., and Green, D. R. (2011). "Geo-environmental hazards assessment of the north western Gulf of Suez, Egypt." *Journal of Coastal Conservation*, 15(1), 37–50.

Arunachalam, A. P. S., Idapalapati, S., and Subbiah, S. (2015). "Multi-criteria decision making techniques for compliant polishing tool selection." *International Journal of Advanced Manufacturing Technology*, 79(1–4), 519–530.

Ashraf, A. (2020). "Risk modeling of soil erosion under different land use and rainfall conditions in Soan river basin, sub-Himalayan region and mitigation options." *Modeling Earth Systems and Environment*, 6(1), 417–428.

Bandyopadhyay, J., and Gyawali, D. (1994). "Himalayan water resources: Ecological and political aspects of management." *Mountain Research and Development*, 14(1), 1.

Bonetti, S., Richter, D. D., and Porporato, A. (2019). "The effect of accelerated soil erosion on hillslope morphology." *Earth Surface Processes and Landforms*, 44(15), 3007–3019.

Borrelli, P., Robinson, D. A., Fleischer, L. R., Lugato, E., Ballabio, C., Alewell, C., Meusburger, K., Modugno, S., Schütt, B., Ferro, V., Bagarello, V., Oost, K. Van, Montanarella, L., and Panagos, P. (2017). "An assessment of the global impact of 21st century land use change on soil erosion." *Nature Communications*.

Chatterjee, P., and Chakraborty, S. (2013). "Gear material selection using complex proportional assessment and additive ratio assessment-based approaches: A comparative study." *International Journal of Materials Science and Engineering*, 1(2), 104–111.

Chen, J. J., Liao, A., Cao, X., Chen, L., Chen, X., He, C., Han, G., Peng, S., Lu, M., Zhang, W., Tong, X., and Mills, J. (2015). "Global land cover mapping at 30 m resolution: A POK-based operational approach." *ISPRS Journal of Photogrammetry and Remote Sensing*, 103, 7–27.

Chowdary, V. M., Chakraborthy, D., Jeyaram, A., Murthy, Y. V. N. K., Sharma, J. R., and Dadhwal, V. K. (2013). "Multi-criteria decision making approach for watershed prioritization using analytic hierarchy process technique and GIS." *Water Resources Management*, 27(10), 3555–3571.

Dahal, R. K., Hasegawa, S., Masuda, T., and Yamanaka, M. (2006). "Roadside slope failures in Nepal during torrential rainfall and their mitigation." In *Disaster Mitigation of Debris Flows, Slope Failures and Landslides*. Universal Academy Press, Inc., 503–514.

Den Biggelaar, C., Lal, R., Wiebe, K., and Breneman, V. (2003). "The global impact of soil erosion on productivity. I. Absolute and relative erosion-induced yield losses." *Advances in Agronomy*, 1–48.

Deshmukh, D. S., Chaube, U. C., Tignath, S., and Tripathi, S. K. (2010). "Morphological analysis of Sher River Basin using GIS for identification of erosion-prone areas." *Ecohydrology and Hydrobiology*, 10(2–4), 307–313.

Dutta, S. (2016). "Soil erosion, sediment yield and sedimentation of reservoir: A review." *Modeling Earth Systems and Environment*, 2(3), 1–18.

Ellison, W. D. (1948). "Soil detachment by water in erosion processes." *Eos, Transactions American Geophysical Union*, 29(4), 499–502.

Fallah, M., Kavian, A., and Omidvar, E. (2016). "Watershed prioritization in order to implement soil and water conservation practices." *Environmental Earth Sciences*, 75(18), 1–17.

Faniran, A. (1968). "The index of drainage intensity—A provisional new drainage factor." *Australian Journal of Science*, 31, 328–330.

FAO. (2019). *Soil Erosion the Greatest Challenge for Sustainable Soil Management*. Rome. 100 pp. Licence: CC BY-NC-SA 3.0 IGO.

Fernández-Raga, M., Palencia, C., Keesstra, S., Jordán, A., Fraile, R., Angulo-Martínez, M., and Cerdà, A. (2017). "Splash erosion: A review with unanswered questions." *Earth-Science Reviews*, 171, 463–477.

García-Ruiz, M. J., Santiago, B., Noemí, L., Estela, N., and Cerdà, A. (2017). "Ongoing and emerging questions in water erosion studies." *Land Degradation and Development*, 28(1), 5–21.

GFCC. (2004). "Updated comprehensive plan of flood management of Gandak river system." In *Ganga Flood Control Commission*. Ministry of Water Resources, Government of India.

Horton, R. E. (1945a). "Erosional deviation of streams and their drainage basins; hydrophysical approach to quantitative morphology." *Geological Society of America Bulletin*, 56(3), 275.

Horton, R. E. (1945b). "Erosional development of streams and their drainage basins; Hydrophysical approach to quantitative morphology." *Bulletin of the Geological Society of America*, 56(3), 275–370.

Hwang, C.-L., and Yoon, K. (1981). "Methods for multiple attribute decision making." In Multiple Attribute Decision Making. Lecture Notes in Economics and Mathematical Systems. Springer, 186.

Irik Mukhametzyanov (2024). MCDM tools (https://www.mathworks.com/matlabcentral/fileexchange/65742-mcdm-tools), MATLAB Central File Exchange. Retrieved January 11, 2020.

Jain, P., and Ramsankaran, R. A. A. J. (2019). "GIS-based integrated multi-criteria modelling framework for watershed prioritisation in India—A demonstration in Marol watershed." *Journal of Hydrology*, 578, 124131.

Kabite, G., and Gessesse, B. (2018). "Hydro-geomorphological characterization of Dhidhessa River Basin, Ethiopia." *International Soil and Water Conservation Research*, 6(2), 175–183.

Kadam, A. K., Jaweed, T. H., Kale, S. S., Umrikar, B. N., and Sankhua, R. N. (2019). "Identification of erosion-prone areas using modified morphometric prioritization method and sediment production rate: A remote sensing and GIS approach." *Geomatics, Natural Hazards and Risk*, 10(1), 986–1006.

Kannan, G., Pokharel, S., and Kumar, P. S. (2009). "A hybrid approach using ISM and fuzzy TOPSIS for the selection of reverse logistics provider." *Resources, Conservation and Recycling*, 54(1), 28–36.

Keesstra, S., Nunes, J., Novara, A., Finger, D., Avelar, D., Kalantari, Z., and Cerdà, A. (2018). "The superior effect of nature based solutions in land management for enhancing ecosystem services." *Science of the Total Environment*, 610–611, 997–1009.

Khare, D. (2014). "Morphometric analysis for prioritization using remote sensing and GIS techniques in a hilly catchment in the state of Uttarakhand, India." *Indian Journal of Science and Technology*, 7(10), 1650–1662.

Khosravi, K., Shahabi, H., Pham, B. T., Adamowski, J., Shirzadi, A., Pradhan, B., Dou, J., Ly, H. B., Gróf, G., Ho, H. L., Hong, H., Chapi, K., and Prakash, I. (2019). "A comparative assessment of flood susceptibility modeling using multi-criteria decision-making analysis and machine learning methods." *Journal of Hydrology*, 573, 311–323.

Kothyari, U. C. (2011). "Sediment problems and sediment management in the Indian Sub-Himalayan region." *IAHS-AISH Publication*, 349(September 2009), 3–13.

Kumar, B., and Lakshmi, V. (2018). "Accessing the capability of TRMM 3B42 V7 to simulate streamflow during extreme rain events: Case study for a Himalayan River Basin." *Journal of Earth System Science*, 127(2), 1–15.

Kumar, B., Lakshmi, V., and Patra, K. C. (2017a). "Evaluating the uncertainties in the SWAT model outputs due to DEM grid size and resampling techniques in a large Himalayan river basin." *Journal of Hydrologic Engineering*, 22(9), 1–12.

Kumar, B., Patra, K. C., and Lakshmi, V. (2016). "Daily rainfall statistics of TRMM and CMORPH: A case for trans-boundary Gandak River basin." *Journal of Earth System Science*, 125(5), 919–934.

Kumar, B., Patra, K. C., and Lakshmi, V. (2017b). "Error in digital network and basin area delineation using D8 method: A case study in a sub-basin of the Ganga." *Journal of the Geological Society of India*, 89(1), 65–70.

Mardani, A., Jusoh, A., Nor, K. M. D., Khalifah, Z., Zakwan, N., and Valipour, A. (2015). "Multiple criteria decision-making techniques and their applications - A review of the literature from 2000 to 2014." *Economic Research-Ekonomska Istrazivanja*, 28(1), 516–571.

Mauget, S. A., Himanshu, S. K., Goebel, T. S., Ale, S., Lascano, R. J., and Gitz III, D. C. (2021). "Soil and soil organic carbon effects on simulated Southern High Plains dryland Cotton production." *Soil and Tillage Research*, 212, 105040.

Mesa, L. M. (2006). "Morphometric analysis of a subtropical Andean basin (Tucumán, Argentina)." *Environmental Geology*, 50(8), 1235–1242.

Miller, V. C. (1953). *A Quantitative Geomorphic Study of Drainage Basin Characteristics in the Clinch Mountain Area, Virginia and Tennessee*. Project NR 389–402. Technical Report 3, Columbia University, Department of Geology, ONR, New Yorkicago Press.

Moore, I. D., Grayson, R. B., and Ladson, A. R. (1991). "Digital terrain modelling: A review of hydrological, geomorphological, and biological applications." *Hydrological Processes*, 5(1), 3–30.

Nautiyal, M. D. (1994). "Morphometric analysis of a drainage basin using aerial photographs: A case study of Khairkuli basin, district Dehradun, U.P.." *Journal of the Indian Society of Remote Sensing*, 22(4), 252–262.

Nooka Ratnam, K., Srivastava, Y. K., Venkateswara Rao, V., Amminedu, E., and Murthy, K. S. R. (2005). "Check Dam positioning by prioritization micro-watersheds using SYI model and morphometric analysis - Remote sensing and GIS perspective." *Journal of the Indian Society of Remote Sensing*, 33(1), 25–38.

Opricovic, S., and Tzeng, G. H. (2004). "Compromise solution by MCDM methods: A comparative analysis of VIKOR and TOPSIS." *European Journal of Operational Research*, 156(2), 445–455.

Palmate, S. S., Amrit, K., Jadhao, V. G., Dayal, D., and Himanshu, S. K. (2022). Prioritization of erosion prone areas based on a sediment yield index for conservation treatments: A case study of the upper Tapi River basin. In *Advances in Remediation Techniques for Polluted Soils and Groundwater* (pp. 291–307). Elsevier.

Panagos, P., Standardi, G., Borrelli, P., Lugato, E., Montanarella, L., and Bosello, F. (2018). "Cost of agricultural productivity loss due to soil erosion in the European Union: From direct cost evaluation approaches to the use of macroeconomic models." *Land Degradation and Development*, 29(3), 471–484.

Pandey, A., Himanshu, S. K., Mishra, S. K., and Singh, V. P. (2016). "Physically based soil erosion and sediment yield models revisited." *Catena*, 147, 595–620.

Parikh, S. J., and James, B. R. (2013). "Soil: The foundation of agriculture." *Nature Education Knowledge*, 3(10), 2.

Patel, D. P., and Dholakia, M. B. (2010). "Feasible structural and non-structural measures to minimize effect of flood in lower tapi basin." *WSEAS Transactions on Fluid Mechanics*, 5(3), 104–121.

Patel, D. P., Dholakia, M. B., Naresh, N., and Srivastava, P. K. (2012). "Water harvesting structure positioning by using geo-visualization concept and prioritization of mini-watersheds through morphometric analysis in the lower Tapi Basin." *Journal of the Indian Society of Remote Sensing*, 40(2), 299–312.

Patel, D. P., Gajjar, C. A., and Srivastava, P. K. (2013). "Prioritization of Malesari mini-watersheds through morphometric analysis: A remote sensing and GIS perspective." *Environmental Earth Sciences*, 69(8), 2643–2656.

Peng, Y. (2015). "Regional earthquake vulnerability assessment using a combination of MCDM methods." *Annals of Operations Research*, 234(1), 95–110.

Prasad, R. K., Mondal, N. C., Banerjee, P., Nandakumar, M. V., and Singh, V. S. (2008). "Deciphering potential groundwater zone in hard rock through the application of GIS." *Environmental Geology*, 55(3), 467–475.

Radice, A., Rosatti, G., Ballio, F., Franzetti, S., Mauri, M., Spagnolatti, M., and Garegnani, G. (2013). "Management of flood hazard via hydro-morphological river modelling. The case of the mallero in Italian Alps." *Journal of Flood Risk Management*, 6(3), 197–209.

Raju, K. S., and Nagesh Kumar, D. (2013). "Prioritisation of micro-catchments based on morphology." *Proceedings of the Institution of Civil Engineers - Water Management*, 166(7), 367–380.

Rather, M. A., Satish Kumar, J., Farooq, M., and Rashid, H. (2017). "Assessing the influence of watershed characteristics on soil erosion susceptibility of Jhelum basin in Kashmir Himalayas." *Arabian Journal of Geosciences*, 10(3), 1–25.

Rocha, L. C. S., De Paiva, A. P., Balestrassi, P. P., Severino, G., and Rotela, P. (2015). "Entropy-based weighting for multiobjective optimization: An application on vertical turning." *Mathematical Problems in Engineering*, 2015, (Article ID 608325).

Saaty, T. L. (2004). "Decision making — The Analytic Hierarchy and Network Processes (AHP/ANP)." *Journal of Systems Science and Systems Engineering*, 13(1), 1–35.

Santra, A., and Mitra, S. S. (Eds.) (2017). "Remote sensing techniques and GIS applications in Earth and environmental studies." In *Advances in Geospatial Technologies*. IGI Global.

Sargaonkar, A. P., Rathi, B., and Baile, A. (2011). "Identifying potential sites for artificial groundwater recharge in sub-watershed of River Kanhan, India." *Environmental Earth Sciences*, 62(5), 1099–1108.

Schumm, S. A. (1956). "Evolution of drainage systems and slopes in badlands at Perth Amboy, New Jersey." *GSA Bulletin*, 67(5), 597–646.

Schumm, S. A., and Lusby, G. C. (1963). "Seasonal variation of infiltration capacity and runoff on hillslopes in western Colorado." Journal of Geophysical Research, 68(12), 3655–3666.

Seyedmohammadi, J., Sarmadian, F., Jafarzadeh, A. A., Ghorbani, M. A., and Shahbazi, F. (2018). "Application of SAW, TOPSIS and fuzzy TOPSIS models in cultivation priority planning for maize, rapeseed and soybean crops." *Geoderma*, 310, 178–190.

Shannon, C. E. (1948). "A mathematical theory of communication." *Bell System Technical Journal*, 27(3), 379–423.

Sharma, S. K., Gajbhiye, S., Tignath, S., and Patil, R. J. (2018). "Hypsometric analysis for assessing erosion status of watershed using geographical information system." In Singh, V Y. R., and Yadav, S., ed., *Hydrologic Modeling: Water Science and Technology Library*, vol. 81, 263–276.

Shinohara, Y., Otani, S., Kubota, T., Otsuki, K., and Nanko, K. (2016). "Effects of plant roots on the soil erosion rate under simulated rainfall with high kinetic energy." *Hydrological Sciences Journal*, 61(13), 2435–2442.

Singh, J. S. (2006). "Sustainable development of the Indian Himalayan region: Linking ecological and economic concerns*." *Current Science*, 90(6), 784–788.

Singh, O., and Singh, J. (2018). "Soil erosion susceptibility assessment of the lower Himachal Himalayan watershed." *Journal of the Geological Society of India*, 92(2), 157–165.

Smith, K. G. (1950). "Standards for grading texture of erosional topography." *American Journal of Science*, 248(9), 655–668.

Strahler, A. N. (1952a). "Hypsometric (Area-Altitude) analysis of erosional topograply." *Geological Society of America Bulletin*, 63(11), 1117–1142.

Strahler, A. N. (1952b). "Dynamic basis of geomorphology." *Bulletin of the Geological Society of America*, 63(9), 923–938.

Strahler, A. N. (1957). "Quantitative analysis of watershed geomorphology." *Eos, Transactions American Geophysical Union*, 380, 913–920.

Strahler, A. N. (1964). "Quantitative geomorphology of drainage basins and channel networks." In Chow, V. T. ed, *Handbook of Applied Hydrology*. McGraw-Hill.

Tang, Q., He, C., He, X., Bao, Y., Zhong, R., and Wen, A. (2014). "Farmers' sustainable strategies for soil conservation on sloping arable lands in the Upper Yangtze River Basin, China." *Sustainability (Switzerland)*, 6(8), 4795–4806.

Vulević, T., Dragović, N., Kostadinov, S., Simić, S. B., and Milovanović, I. (2015). "Prioritization of soil erosion vulnerable areas using multi-criteria analysis methods." *Polish Journal of Environmental Studies*, 24(1), 317–323.

Yue, C. (2017). "Entropy-based weights on decision makers in group decision-making setting with hybrid preference representations." *Applied Soft Computing Journal*, 60, 737–749.

Zavadskas, E., and Kaklauskas, A. (1996). "Determination of an efficient contractor by using the new method of multicriteria assessment." *International Symposium for "The Organization and Management of Construction". Shaping Theory and Practice*, 2, 94–104.

11 Role of Urea Super Granule (USG) Applicator in Efficient Management of Nitrogen Fertilizer
Case Study of Rice Farming in Bangladesh

Moniruzzaman and Md. Sadique Rahman

INTRODUCTION

Despite significant progress in cereal production, Bangladesh persists in facing food security challenges due to rising population, shrinking agricultural land, and unequal food distribution (Emran et al. 2019). In addition, fluctuations in cereal prices over the past decade have added to the burden on households (FAO 2018). Approximately 80% of Bangladesh's arable land is used to cultivate rice (*Oryza sativa*) during three cropping seasons: spring (Aus), monsoon (Aman), and winter (Boro). Rapid progress in the rice farming input management system has facilitated the adoption of modern rice varieties. Modern rice varieties are always responsive to a balanced application of fertilizer and irrigation water (Alam et al. 2011). Food security, adaptation to climate change, and the achievement of several sustainable development goals all depend on effective nutrient management (Chivenge et al. 2021).

One of the most important nutrients for rice plant growth is nitrogen (N). Improving crop growth and grain yield requires N to be managed properly. However, increased N application is arguably the most widely used strategy for boosting yield, despite the fact that over half of applied N is typically lost to the environment (Emran et al. 2019; Mauget et al., 2021; Mahajan et al. 2022; Yatoo et al. 2022; Singh et al. 2023). Urea is now widely used as a significant component of N fertilizers for rice. An estimated 80% of the urea supply goes toward producing rice (Hoque et al. 2016). However, the rice plant only takes in 15%–35% of the total applied nitrogen (Iqbal 2009; Hoque et al. 2016). Farmers frequently apply excessive amounts of nitrogen fertilizer in an attempt to maintain yield despite the inherent difficulty in estimating precise needs. However, the diminishing returns of applying too much nitrogen

DOI: 10.1201/9781003441175-11

fertilizer mean that doing so is probably pointless (Tilman et al. 2011; Yousaf et al. 2016). N fertilizer, when applied properly, can boost photosynthesis, plant resistance to biotic stress, and crop yield. Therefore, it is crucial to improve N-use efficiency in rice farming.

In flooded rice, where water depth can be deep but may also fluctuate throughout the growing season, increasing N-use efficiency is especially difficult (Ladha et al. 2005). Prilled urea, a common form of N fertilizer, is widely used because it is convenient for farmers to apply. However, urea super granules (USG) have been proposed as a means to increase N efficiency by retaining N in shallower soil layers (Huda et al. 2016). When compared to broadcast prilled urea, USG manually worked into the soil can increase N-use efficiency while also improving the environment, providing greater net benefit, and yield (Ahamed 2012; Miah et al. 2016; Huda et al. 2016). Compared to using broadcasted prilled urea, it is reported that using USG placed at a depth of 6–10 cm in a wetland rice field can save 30% of urea (Hoque et al. 2013).

However, the time and money needed to manually place USG in deep soil is significantly higher. Farmers' decisions to adopt USG-based N management are expected to be heavily influenced by labor requirements rather than gains in yield or N-use efficiency due to the fact that labor scarcity and cash availability are major constraints faced by rice farmers (USDA 2018). An effective and affordable fertilizer application machine that can adjust to the varying fertilizer depth needs of different crops is needed. The Bangladesh Agricultural Research Institute (BARI) developed a manually operated push-type fertilizer applicator for rice farms, which was then disseminated throughout the country by the Department of Agricultural Extension as a solution to the problem of USG placement by hand (Wohab et al. 2009). However, the adoption of the USG applicator is limited. Several studies have been conducted on the field-level performance of BARI-developed USG applicators (Alam et al. 2011; Hoque et al. 2013; Hoque et al. 2016). A few studies (Sarma et al. 2021; Anik et al. 2022) identified the factors influencing the adoption of manual application of USG at the farm level. This highlights the need for a socio-economic study of USG applicator users to better understand the reasons for varied adoption rates. This chapter focuses on the financial performance of the BARI-developed USG applicator at the farm level and identifies the factors responsible for the lower adoption of the USG applicator.

USG APPLICATOR

USG applicator for rice fields was developed in 2009 by BARI's Farm Machinery and the Postharvest Process Engineering (FMPE) department (Wohab et al. 2009). USG applicators typically weigh in at around 6 kilograms. Both steel and plastic were used in the construction of the applicator. The applicator's skids were set up between two rows of rice. The forward motion of the applicator is buoyed by the skids. Two 1 kg USG capacity hoppers have been installed in front of the fertilizer storage boxes. The total length, width, and height of the applicator were 1,920, 620, and 520 millimeters. The skids of the applicator were positioned in the middle of

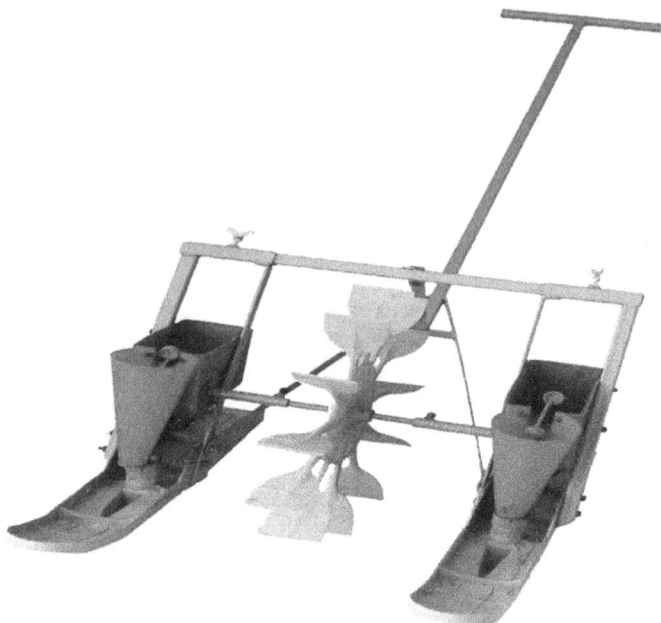

FIGURE 11.1   USG applicator

two rows of rice plants. The USG applicator was then advanced through the process by hand. The metering devices and cage wheel rotated as a result of this. It carries USG into the pockets and delivers it to the furrow openers as the metering devices rotate (Hoque et al. 2016). The USG is applied by a fertilizer applicator with a 20 cm row spacing, a 40 cm row width, and a 5–6 cm depth. Figure 11.1 shows a picture of a USG applicator.

STUDY AREAS AND ANALYTICAL TECHNIQUES

Quantitative and qualitative methods of data collection were used to gather the information required. Three major rice growing districts (administrative units) in Bangladesh were selected for face-to-face interview. The information was gathered by using a pretested interview schedule. The data was collected from winter-season rice growers. One upazila (sub-district) was selected randomly from each district with the highest concentration of rice fields. Three villages were selected from each upazila based on recommendations made by the agricultural extension office. Sixty (60) farmers were randomly selected from each upazila to be interviewed. Half of the 180 rice farmers who participated in the survey reported using a USG applicator.

We used both descriptive and inferential statistics to examine the data. The factors influencing USG applicator adoption in the study areas were identified using the Probit model. In this study, USG applicator adoption served as the dependent variable. The central premise is that farmers will choose to use USG applicators because doing so will provide them with more benefits than not doing so would. The farmers who don't adopt may be unwilling to do so because of their socio-economic constraints. Therefore, the independent variables included farmers' ages, levels of education and training, frequency of extension contacts, and ownership of farmland. The likelihood of adoption can be determined by employing the Equation (11.1):

$$\text{Prob}\left(\text{adoption of USG applicator} = 1\right) = \frac{\exp\left(bX_i\right)}{1+\exp\left(bX\right)} \quad (11.1)$$

Where *Prob* represents the probability, X_i are independent variables and b represents parameters to be estimated.

The likelihood of non-adoption can be written as follows:

$$\text{Prob}\left(\text{non}-\text{adoption of USG applicator} = 0\right) = \frac{1}{1+\exp\left(bX_i\right)} \quad (11.2)$$

The empirical Binary Probit model for adoption of USG application can be as follows:

$$Y_i = b_0 + \sum_{i=1}^{6} b_i X_i + u_i \quad (11.3)$$

Where Y_i represents the dependent variable (adoption status), X_i are independent variables, b represents parameters to be estimated, and u_i represents the error term.

To find out more about USG applicators' experiences on farms, we supplemented our quantitative field survey with qualitative focus group discussions (FGDs).

EMPIRICAL FINDINGS

SOCIO-ECONOMIC PROFILE OF THE FARMERS

Farmers' propensity to adopt new technologies is often influenced by their demographic characteristics and circumstances. Therefore, we evaluate the demographic differences between USG applicator users and non-users with regard to age, level of education, extension contact, and land ownership. According to the results, farmers' ages in both groups averaged out to be about 45 years. The education levels of users were higher than those of non-users. Ninety-one percent of USG applicator farmers had some level of education, while about 80% of non-users did. However, 41% of those who used USG applicators had only completed elementary school. This is why stressing the importance of education again is so crucial. Based on average farm size, it appears that both user groups are comprised of subsistence farmers. Both users and non-users owned roughly the same amount of land, at 0.51 and 0.53 ha on average. Agricultural extension workers have more regular contact with people who use USG applicators, which may encourage them to use the applicator.

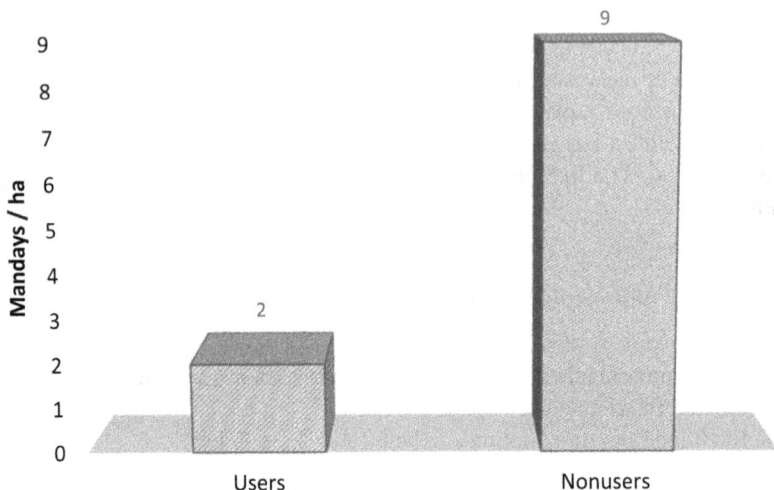

FIGURE 11.2   Comparative labor use pattern for urea application

EFFECT OF USG APPLICATOR ON UREA APPLICATION AND HUMAN LABOR

Our research showed that farmers who use USG applicators require only 2 man-days/ha of labor for urea application, compared to 9 man-days/ha for those who don't (Figure 11.2). Farmers who used an applicator needed 161 kg of urea per hectare, while those who didn't needed 183 kg. There was a statistically significant gap between the two groups. Studies have shown that using an applicator can reduce the amount of granular urea needed from 247 kg per hectare to only 160 kg (Hoque 2008). Chemical fertilizers are an essential part of today's farming practices. Bangladesh needs about 6 million metric tons of chemical fertilizers every year, and imports will make up about 80% of that total. Due to several recent international issues such as the Ukraine and Russia wars, chemical fertilizer prices have become a new source of worry for developing nations like Bangladesh. While the annual demand for urea in Bangladesh is 2.65 million metric tons, domestic production is only 1 million metric tons (Daily Sun 2022). Therefore, there's a lot of scope for the USG applicator, which uses less labor and less urea, to be adopted across the country. USG applicator has the potential to save a substantial amount of money that can be put toward other development projects.

COMPARATIVE FINANCIAL BENEFITS

The financial analysis of two groups of farmers is shown in Table 11.1. The average rice yield was 6.45 t/ha for users of USG applicators and 6.27 t/ha for non-users. The users received 0.18 t/ha more yield than non-users, which is statistically significant at the 5% level. The results are consistent with those of Alam et al. (2011) and Anik et al. (2022), who also found that the use of USG applicators increases yield and return. Our research indicated that non-users applied more urea than users.

TABLE 11.1
Comparative Financial Benefits Between USG Users and Non-users

Items	Users	non-users
Rice Yield (t/ha)	6.45	6.27
Gross return (Tk/ha)	110,940	107,844
Total cost (TK./ha)	73,398	76,175
Net return (Tk/ha)	37,542	31,669
BCR	1.51	1.41
Benefits over non-users		
Additional yield from USG applicator (t/ha)	0.18**	
Additional return from USG applicator (Tk/ha)	5873***	
Urea saved (kg/ha)	22**	

Note: **, and *** indicate significance at 5% and 1% levels, respectively; Tk is Bangladeshi currency. 01 USD = Tk 105.

As a result, they have increased their labor force, which increases their production costs and decreases their urea use efficiency. Additionally, it appears that, on average, 22 kg of urea could be saved by employing a USG applicator as opposed to a manual application. Prior research demonstrated that the nitrogen requirement and rate of urea are extremely high, whereas the efficiency of prilled urea is extremely low (Akter et al. 2018). Applying USG is more effective than applying prilled urea (Sarma et al. 2021). This kind of small-scale mechanization helps farmers increase output, decrease the cost of cultivation, and turn a profit by replacing labor-intensive tasks with power-intensive tasks. Furthermore, the cost of USG is higher than that of prilled urea. Thus, shifting subsidies away from prilled urea and toward USG could be an efficient strategy for boosting the use of USG.

WHAT FACTORS INFLUENCE THE ADOPTION OF USG APPLICATORS?

The results of applying an econometric model (Binary Probit model) to identify the factors influencing the adoption of USG applicators are presented in Table 11.2. The adoption of USG applicators was associated with education, training, and contact with agricultural extension agents. According to the findings, additional training increases the likelihood of adoption by 10.6%. Similarly, increased communication with agricultural extension workers can increase the likelihood of adoption by approximately 62.6%. If farmers have access to additional training and extension services, they can adopt the USG applicator more quickly. Farmers' willingness to base decisions on quantified cost and profit data often increases after receiving training (Sarma et al. 2021). Previous studies also suggested that increasing farmers' access to knowledge through training increases the likelihood that they will adopt cutting-edge agricultural technologies (Dimkpa et al. 2020; Leake & Adam 2015). During FGDs, one of the farmers mentioned that "I have a cell phone but am unable

TABLE 11.2

Factors Affecting the Adoption of USG Applicator: Estimates of a Probit Model

Variable	Coefficients	Standard error	z-value	Marginal effect
Age (years)	−0.060*	0.032	−1.86	−0.023*
Education (years)	0.107***	0.042	2.56	0.042***
Experience (years)	0.030	0.033	0.91	0.012
Training (number)	0.269***	0.078	3.44	0.106***
Extension contact (number)	1.57***	0.241	6.51	0.626***
Own land (ha)	−0.002	0.001	−1.24	−0.0008
Constant	−3.37***	0.996	−3.39	–
Log-likelihood		−43.29		
LR chi2		161.55***		
Pseudo R^2		0.55		
Observations (n)		180		

Note: *, and *** indicate significance at 10%, and 1% levels, respectively.

to access the internet for farming information". Therefore, familiarizing farmers with modern communication technologies can contribute to their knowledge and adoption. The preference for the USG applicator decreases with age, suggesting that younger people are more likely to use it than their more senior counterparts.

WILLINGNESS TO ADOPT THE USG APPLICATOR

Approximately 51% of non-users were willing to adopt USG applicator practices in the future (Table 11.3). The majority of respondents (49%) stated the need for less fertilizer as the primary reason for the future adoption of USG applicator practices, followed by the need for less labor (42%) and higher yield potential (41%). In contrast, 49% of non-users stated they are unwilling to adopt USG applicator practices in the future due to their lack of technical knowledge regarding USG applicator machines. According to the results, there is still a lot of space to advance USG applicator adoption. The majority of the farmers cited barriers to adoption that can be overcome. Educating farmers on the benefits of USG applicators, for instance, would increase their technical knowledge. Home-based training and demonstrations can help achieve this goal. Training sessions can include not only primary farmers but also other household members, especially the farmer's spouse, to ensure that everyone in the household benefits from using USG applicators. The agricultural extension service has the potential to significantly contribute to farmers' education and empowerment. Spare parts of the USG applicator machine should be available

TABLE 11.3

Willingness to Adopt USG in the Future in the Study Areas

Items	Percentage
Willing to adopt	51
Not willing to adopt	49
Stated reasons for adoption	
Require less fertilizer	49
b. Require less labor	42
d. Easy weeding	35
e. Higher yield	41
Stated reasons for non-adoption	
a. Lack of technical knowledge	44
c. USG applicator needs to be driven on foot	22
d. Machinery parts are not available	20

in the local market. Distributors and small-scale manufacturers can reach customers anywhere in the country and offer support even after a purchase has been made. Therefore, they can be involved in supplying the machine and parts all over the country. Adoption of the USG applicator can be facilitated by the introduction of a community-based approach to its sale if farmers are unable to do it on their own.

Constraints Faced by USG Applicator Users

This research revealed that many barriers prevented farmers from continuing the use of the USG applicator. Eighty-four percent of respondents cited difficulties caused by USG applicators getting bogged down in muddy soil due to their weight, and 70% stated a lack of USG applicators as a barrier. About 50% of the users mentioned that inadequate training and lack of capital often prevent them from using the USG applicator. One farmer mentioned regarding this during FGDs "Although I have had some training in using the USG, it was not nearly enough. The government needs to provide us with more opportunities to learn". Farmers during FGDs also mentioned that "Clogged soil prevents us from using USG applicators in muddy conditions. The problem of clogging must be resolved if USG applicator adoption is to continue". Farm equipment can be customized by local small manufacturers to meet the needs of farmers. Therefore, research institutions can provide manufacturers with technical assistance by providing design and drawing, and offering advice and suggestions based on their technical expertise. In addition, extension agents and other relevant stakeholders should assist rice farmers in forming credit-relief groups to address their financial needs, and connect them with financial institutions offering low-interest loans. Regular training in the use of the USG applicator can ease the burden of applying USG for rice farmers.

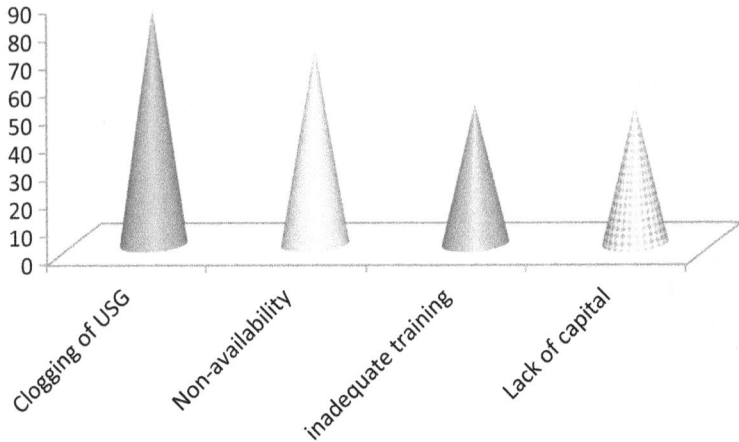

FIGURE 11.3   Percent of farmers facing different constraints

CONCLUSIONS

The survey results show that USG applicators help farmers achieve N-use efficiency. The USG applicator has emerged as an efficient and eco-friendly technology of N management. Adopters of USG applicators applied less amount of urea, and achieved higher yields, greater profits, and lower production costs compared to farmers who do not use USG. Nonuser plots required significantly more labor than USG applicator plots. The sample farmers who used the USG applicator noticed a decrease in urea use of 22 kg/ha and an increase in yield of 0.18 t/ha. This increase in yield resulted in an additional benefit from rice farming. Contact with extension workers and the total amount of training received are both important factors in determining whether or not people will adopt this technology. Fifty-one percent of the non-users were open to adopting the USG applicator practices in the future, while the rest of the non-users mentioned a lack of technical knowledge as the main reason they were not open to doing so. Adopting this technology extensively in rice -growing regions can lead to the attainment of self-sufficiency, the expansion of the input supplier's business, and substantial foreign currency savings. In this light, the following measures are to be considered for the future sustenance of the USG applicator, allowing for its greater expansion and dissemination.

- Government efforts should be increased to promote field demonstrations and farmer training on USG applicators for rice farming, as this method has been shown to be advantageous to crop production.
- Farmers should have access to USG applicators and spare components when needed.
- If we want to see widespread adoption of USG applicators, we need to improve communication between researchers, extension agents, and farmers.
- Implementation of a community-based sales strategy if farmers are unable to do so themselves can also facilitate adoption.

REFERENCES

Ahamed, M.S. (2012). *Improvement of Guti Urea Applicator.* Department of Farm Power and Machinery, Bangladesh Agricultural University, Mymensingh, Bangladesh.

Alam, M.S., Islam, S., & Islam, M.A. (2011). Farmers' efficiency enhancement through input management: The issue of USG application in modern rice. *Bangladesh Journal of Agricultural Research,* 36(1), 129–141.

Akter, R., Khan, M.M.H., & Rahman, M.Z. (2018). Profitability of urea super granule as a source of nitrogen on the yield and yield attributes of mustard. *International Journal of Biological Sciences,* 12(6), 220–224.

Anik, A., Begho, T., Sharna, S., Eory, V., & Rahman, M. (2022). Toward improving nitrogen use efficiency in rice production: The socio-economic, climatic and technological determinants of briquette urea adoption. *Renewable Agriculture and Food Systems,* 37(5), 417–428.

Chivenge, P., Sharma, S., Bunquin, M.A., & Hellin, J. (2021). Improving nitrogen use efficiency—A key for sustainable rice production systems. *Frontiers in Sustainable Food Systems,* 5, 737412.

Daily Sun. (2022). Fertilizer import and challenge for food security in Bangladesh. https://www.daily-sun.com/post/643340/Fertilizer-Import-and-Challenge-for-Food-Security-in-Bangladesh#:~:text=According%20to%20Bangladesh%20Fertilizer%20Association%202022%2C%20Bangladesh%20requires%202.65%20million,of%20DAP%20fertilizer%20every%20year.

Dimkpa, C.O., Fugice, J., Singh, U., & Lewis, T.D. (2020). Development of fertilizers for enhanced nitrogen use efficiency-trends and perspectives. *Science of the Total Environment,* 731, 139113.

Emran, S., Krupnik, T.J., Kumar, V., Ali, M.Y., & Pittelkow, C.M. (2019). Agronomic, economic, and environmental performance of nitrogen rates and source in Bangladesh's coastal rice agroecosystems. *Field Crops Research,* 241, 107567.

FAO. (2018). *FAOSTAT.* Food and Agriculture Organization of the United Nations (FAO), Rome.

Huda, A., Gaihre, Y.K., Islam, M.R., Singh, U., Islam, M.R., Sanabria, J., Satter, M.A., Afroz, H., Halder, A., & Jahiruddin, M. (2016). Floodwater ammonium, nitrogen use efficiency and rice yields with fertilizer deep placement and alternate wetting and drying under triple rice cropping systems. *Nutrient Cycling in Agroecosystems,* 104(1), 53–66.

Hoque, M.A., Karim, M.R., Miah, M.S., Rahman, M.A., & Rahman, M.M. (2016). Field performance of BARI urea super granule applicator. *Bangladesh Journal of Agricultural Research,* 41(1), 103–113.

Hoque, M.A., Wohab, M.A., Hossain, M.A., Saha, K.K., & Hassan, M.S. (2013). Improvement and evaluation of Bari USG applicator. *Agricultural Engineering International: CIGR Journal,* 15(2), 87–94.

Hoque, N. (2008). Saving of taka 250 crores and increasing yield by 20% by using Goti Urea (Bangla). Paper presented at the national workshop on Urea Super Granule Technology. BARC, Dhaka, Bangladesh, 25 June, 2008.

Iqbal, S.H. (2009). *Improvement of the Existing USG Fertilizer Applicator.* Dept. of Agricultural Engineering, Bangladesh Sheikh Mujibur Rahman Agricultural University, Gazipur, Bangladesh.

Ladha, J.K., Pathak, H., Krupnik, T.J., Six, J., & Kessel, C. (2005). Efficiency of fertilizer nitrogen in cereal production: Retrospect and prospects. *Advances in Agronomy,* 87, 85–156.

Leake, G., & Adam, B. (2015). Factors determining allocation of land for improved wheat variety by smallholder farmers of northern Ethiopia. *Journal of Development and Agricultural Economics,* 7(3), 105–112.

Mahajan, M., Gupta, P. K., Singh, A., Vaish, B., Singh, P., Kothari, R., & Singh, R. P. (2022). A comprehensive study on aquatic chemistry, health risk and remediation techniques of cadmium in groundwater. *Science of The Total Environment, 818*, 151784. https://doi.org/10.1016/j.scitotenv.2021.151784.

Mauget, S. A., Himanshu, S. K., Goebel, T. S., Ale, S., Lascano, R. J., & Gitz III, D. C. (2021). Soil and soil organic carbon effects on simulated Southern High Plains dryland Cotton production. *Soil and Tillage Research, 212*, 105040.

Miah, M.A.M., Gaihre, Y.K., Hunter, G., Sing, U., & Hossain, S.A. (2016). Fertilizer deep placement increases rice production: Evidence from farmers' fields in Bangladesh. *Agronomy Journal, 108*(2), 805–812.

Sarma, P.K. (2021). Adoption and impact of super granulated urea (guti urea) technology on farm productivity in Bangladesh: A Heckman two-stage model approach. *Environmental Challenges, 5*, 100228.

Singh, R. P., Mahajan, M., Gandhi, K., Gupta, P. K., Singh, A., Singh, P., ... & Kidwai, M. K. (2023). A holistic review on trend, occurrence, factors affecting pesticide concentration, and ecological risk assessment. *Environmental Monitoring and Assessment, 195*(4), 451. https://doi.org/10.1007/s10661-023-11005-2.

Tilman, D., Balzer, C., Hill, J., & Befort, B.L. (2011). Global food demand and the sustainable intensification of agriculture. *Proceedings of the National Academy of Sciences of the United States of America, 108*(50), 20260–20264.

USDA. (2018). Grain: World markets and trade. Foreign Agricultural Service, United States Department of Agriculture. https://apps.fas.usda.gov/psdonline/circulars/ grain.pdf.

Wohab, M.A., Islam, M.S., Hoque, M.A., Hossain, M.A., & Ahmed, S. (2009). Design and development of a urea super granule applicator for puddled rice field. *Journal of Agricultural Engineering, 37*, 57–62.

Yatoo, A.M., Ali, M.N., Zaheen, Z., Baba, Z.A., Ali, S., Rasool, S., Sheikh, T.A., Sillanpää, M., Gupta, P.K., Hamid, B. and Hamid, B., (2022). Assessment of pesticide toxicity on earthworms using multiple biomarkers: a review. *Environmental Chemistry Letters, 20*(4), 2573–2596. https://doi.org/10.1007/s10311-022-01386-0.

Yousaf, M., Li, X., Zhang, Z., Ren, T., Cong, R., Ata-Ul-Karim, S.T., Fahad, S., Shah, A.N., & Lu, J. (2016). Nitrogen fertilizer management for enhancing crop productivity and nitrogen use efficiency in a rice-oilseed rape rotation system in China. *Frontiers in Plant Sciences, 7*, 1496.

12 AI as Improved Agri-Tech Approach for Better Nutrients and Proper Irrigation Water Management
A Comparative Study

Lodsna Borkotoky and Hemanta Hazarika

INTRODUCTION

Building up ways to integrate food production into agricultural processes globally is of utmost importance. Focusing on technology development and employing various measures to improve food production in agriculture is immensely important to meet the needs of the growing population. The apparent shortfall in traditional practices of farming needs greater attention as it has, for some reason, cut down productivity and failed to meet the increasing demands of food production.

AI has emerged as a salient feature in meeting the needs of farmers within a short time. AI technology has successfully contributed to cultivating healthier crops, monitoring soil qualities, analyzing data for farmers, and other management activities in the agriculture sector. AI enables farmers to choose the optimum seed based on different climatic scenarios. AI-powered solutions will help farmers produce more with fewer resources, increase crop quality, and hasten product time to reach the market. AI application helps farmers choose the optimal time to sow their seeds and the desired nutrients required to increase the quality of the soil. The AI-powered system provides information regarding the health of the crops and the nutrients that need to be given to enhance yield quality and quantity. Employing AI-based farming techniques could be an efficient solution for the farmers to enhance food productivity. In this chapter, we will focus on the basic concepts of AI in the field of agriculture to help the academician as well as the farmer.

Agriculture is the main occupation of most people worldwide and with increasing populations there is more pressure on farmers to obtain a greater amount of crops

DOI: 10.1201/9781003441175-12

within a shorter amount of time. Traditional methods are not enough to handle the demand and a modern technique to increase production with reduced waste has to be adopted. In such cases, technology has redefined farming and has affected the agriculture industry in many ways. The challenge is to increase global food production by 25–70% by 2050 for 9.7[1] billion people. This corresponds to a growth rate of 2.4% each year; however, the actual growth rate is only 1.3%.[2] This could be achieved through proper investigation and smart production techniques. In recent years, there has been a growing interest in sustainable crop management. It is very important to use our resources efficiently to increase productivity and to meet the challenges in the agricultural field. Regarding this, AI technology has become an excellent modern tool to overcome all the challenges in traditional farming thereby becoming an integral part of future agricultural techniques. AI technology has found recognition in both crops as well as livestock farming.[3] AI techniques help farmers in various ways, such as predicting weather forecasts, monitoring crop disease and health, smart irrigation, crop readiness identification, yield prediction, weed/pest management, and precision farming. Due to its efficiency in farming, it has been spreading globally among farmers to make agriculture a profitable industry. In this chapter, we are mainly focusing on wastewater management using smart irrigation methods and improving nutrients in food production with AI-based technologies.

LIMITATIONS OF CONVENTIONAL FARMING

The challenges faced by farmers in traditional farming processes include soil management to harvesting and storage. A few problems incurred during the farming cycle involve:

o Weather: Frequent changes like rise and fall of temperature, rainfall, and humidity play an important role in making proper decisions for harvesting, sowing seeds and soil maintenance. Also, the quality and quantity of crops produced are affected by pollution which results in diminishing crop production.[4]

o Nutrient control: Proper use of nutrients when and where necessary in the soil plays a significant role in efficient farming. The personal monitoring of the farmers, in conventional methods, for the nutrient content in the soil either becomes insufficient or irregular.

o Rising global population: Conventional farming fails to meet the population demand because of a lack of manpower, skills, and cost factors. Increasing crop production will require the involvement of more and more trained manpower which will result in a very high cost.[1]

o Lack of sustainable irrigation systems: The daily task of monitoring whether the land is underwatered or overwatered appears to be quite difficult for farmers. The unsuitable water content in soil affects crop production by degrading its quality.[5]

Traditional agriculture analyzes and summarizes the crop growing factors through visual inspection or from individual experience. However, AI monitors the agricultural production process, such as the pH of soil, temperature, and humidity, in real-time through distributed sensors. The use of such sensing technologies generates a new era of smart agriculture with highly efficient data acquisition and analysis.[6] Those sensing technologies are applied with different working mechanisms like electrochemical,[7] resistive,[8] capacitive,[9] optical,[10] and thermal,[11] which possess high precision, high sensitivity, and diversified functions.

It is quite impracticable to meet all the limitations in traditional farming and nontechnical measures like proper training and support of the workforce. It is, therefore, beneficial to come up with the right technology to evaluate the problems and get instant help to overcome these difficulties.

MODERN TECHNOLOGIES IN AGRICULTURAL FIELDS

Agriculture is a vital element in the worldwide economy and has become a concern to all researchers. In this context, continuous development and application of new technologies are necessary to avoid laborious practices and maintain sustainability, food security, and competitiveness. Various technologies such as the development of microneedle technology for food and crop health,[12] geoinformation systems,[13] remote sensing,[14] electronics in plants,[15] etc., have been developed in this field. Smyth et al. presented the importance of adopting new agricultural food biotechnologies. This will be particularly important for improving food security, crop nutrient availability, and sustainability. Their article presents some of the global costs of not adopting genetically modified crops and genome edited breeding.[16] In 2022, Nikšić and others[17] presented an overview of the achievements in artificial cultivation of fruit bodies of a medicinally important mushroom species ignoring traditional farming practices. Overall artificial intelligence along with smart agricultural technologies can provide good contributions to agriculture.

Artificial intelligence is an interdisciplinary field that replicates human intelligence in robots, including learning and problem-solving. AI technology has now been recognized in agricultural productivity to address various problems in terms of yield and quality. This has led to the production of machines including manufacturing robots, smart assistants, media monitoring, image processing, sensors, automated financial investing, and many more. Adopting modern technologies such as artificial intelligence can help overcome the limitations of crop production to a great extent. First, it can identify the type and quantity of infected plants and provide measures to treat them. Image recognition technology easily helps to detect infected plants the correct measures to save them. AI uses a large database of images stored to diagnose sick plants. Agricultural production loss due to pests and diseases affecting crops is a matter of great concern worldwide.[18] As per reports, plant diseases result in 220 billion dollars of crop losses globally, where almost 30% of global crop production loss is due to pathogens such as bacteria, viruses, and fungi.[19] Therefore pesticides are usually required to overcome such food production losses despite their harmful effects on human health and the environment. Adequate use of pesticides is required to avoid unnecessary health and environmental risks, and for agricultural purposes continuous monitoring of the crops can be achieved only through automated technologies. In such cases, AI techniques could be applied such that such things can be easily tracked and supplied with only the minimum amount of pesticides required. Second, the labor shortage to meet the demands of the drastic rise in population has been greatly overcome by agricultural robots. These robots assist in providing more food production and therefore fulfilling market demands to reduce cost and maximize profit. AI-powered robots can be used to harvest crops and also in storing and packaging. A study on the application of agricultural robots found them to be more environmentally feasible compared to conventional systems.[20] The study was based on system analysis and an individual economic feasibility study showing the application of robotic weeding in high-value crops, crop scouting in cereals, and grass cutting on golf courses. Again, the Internet of Things (IoT) helps gather information regarding the upcoming bad weather conditions warning the farmers beforehand about the temperature, humidity, wind speed, and air pressure, thus providing them a chance to shield their crops from severe weather. AI weather forecasting is useful for farmers to have information on the weather, and they can plan accordingly for the type of crop to grow, seeds to sow, and when to harvest the crop. This would help in unwanted loss of crops due to bad weather and hence result in good profits. Keeping this in view, farmers have started implementing AI techniques and this uplift in agricultural processes will definitely result in fulfilling large population demands without any failure.

IMPORTANCE OF ARTIFICIAL INTELLIGENCE

Artificial intelligence is rapidly growing with many applications starting from health care, education, and in the workplace to industry. With the development of modern technologies, AI has been gradually applied in the field of agriculture. It is difficult to obtain a large amount of data in the field of plant science and agriculture. In this regard, AI plays a crucial role in meeting the needs of farmers with limited data within a short period of time. It helps in the agricultural sector by monitoring soil qualities, cultivating healthier crops, and analyzing data for the farmers (Figure 12.1). It also helps the farmers to choose the correct species of seed with changes in climatic conditions. The salient feature of AI technology is to reduce the time for the production of food with limited resources and low cost along with an increase in crop quality. Thus, AI has become a great tool for farmers to meet their requirements in an efficient way. A few of the examples are as follows. Steine et al.[21] adopted artificial intelligence-based mobile soil analysis for real-time, on-the-spot analysis of soil at low cost. The group reported a mobile chemical analysis system based on colorimetric paper sensors that operate under tropical field conditions. Their study consists of a data set of approximately 800 images of a colorimetric paper-based chemical analysis device captured with a custom mobile application in outdoor conditions. They have also included a data set comprising three csv files collecting the color information values and analysis results (pH values) as determined by the mobile application models from those paper-based device images. These files were captured in the field (pre-processed), before being processed in a post-field test and a set corresponding to the measurements on a compound sample combining the nine soil samples collected per hectare. In 2023, Aasim et al. adopted artificial

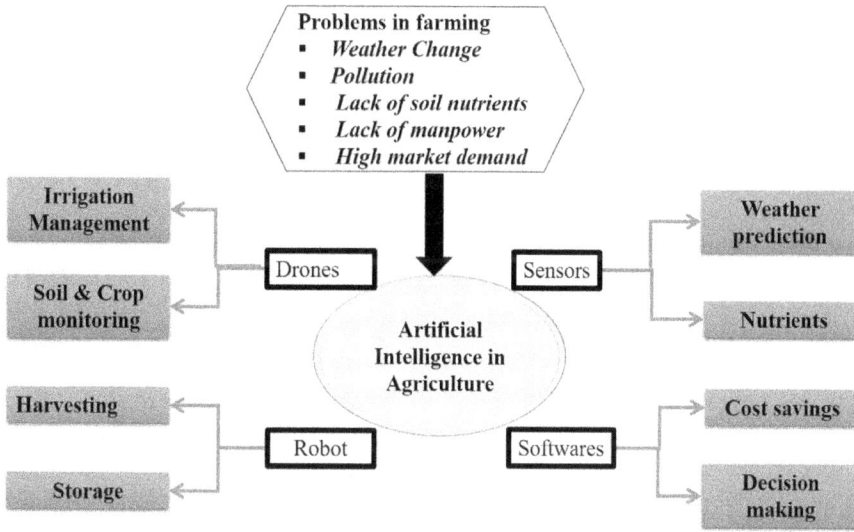

FIGURE 12.1 AI as a modern technique in the agricultural field

Neural Network modeling for deciphering the *in vitro*-induced salt stress tolerance in chickpeas.[22] As salt stress is a matter of concern, the group carried out the *in vitro* investigation of desi chickpea by continuous exposure of the seeds to a NaCl-containing medium. Different germination indices and growth indices of roots and shoots were recorded. The *in vitro* growth parameters and salt tolerance index values validated and predicted by the multilayer perceptron (MLP) model revealed a negative impact of elevated salinity levels on both root and shoot. A design of genetic algorithm-based and deep Neural Network technologies for the detection of tomato leaf disease shows the benefit of new digital technologies and artificial intelligence (AI).[23] It is important for early disease diagnosis of plants to prevent production loss, and for this the farmers need reliable automated tools which can easily detect the diseases. Moussafir et al. developed an AI-based hybrid model for tomato disease detection based on image data collection. They evaluated the performance of seven different architectures including VGG16, ResNet50, EfficientNetB0, EfficientNetB1, EfficientNetB2, EfficientNetB3, and EfficientNetB4. Among these, two models were selected to develop a hybrid model for plant disease identification. The model was applied to the plant village dataset, which contains nine classes of tomato diseases. The proposed model achieves an accuracy of 0.981.

Artificial intelligence-based robots can be adopted as an alternative to manual force in fields for harvesting, pesticide spraying, and maintaining water management systems in agriculture.[24] Various tasks, such as seed planting and pouring water, can be achieved with robots and no manpower is required as the robot is programmed and controlled. The system is based on forecast algorithms, particularly in agricultural water areas, which can detect and forecast precipitation patterns and climatic changes as well. Such robots contain inbuilt soil humidity content and rainfall forecast algorithms. Various tasks like digging, sowing seeds, checking the moisture content of the ground using a moisture sensor, and watering plants can be done with no manual labor. In addition, nutrients vital for plants can be supplied by fertilizers, and pesticides kill insects that harm crops. This multipurpose robot boosts crop production which in turn reduces the efforts of farmers and thereby increases the efficiency of sowing, fertilizers, and pesticide spraying. Sohail et al.[25] showed the inapplicability of conventional agrochemical-based agricultural practices and described the inclusion of artificial intelligence-based technologies for better efficiency and environmental sustainability. They have shown the hazardous effects of the use of agrochemicals on human health and the environment and urged for the need for robotics-assisted, organic agricultural-biotechnology based environment-friendly healthy food options. The practice of using artificial intelligence in organic farming will provide a route to efficient and precision agriculture. Sohail et al. further provided the inability to meet the demands of the population (According to FAO's recent report in 2020 ((FAO, 2020) FAO and UNICEF)), out of which two billion people (25.9%) experienced hunger or did not have regular access to nutritious food.

PURPOSE OF AI TO IMPROVE NUTRIENTS IN FOOD PRODUCTION

Introducing AI technology in agriculture has now become a topic of concern because of its proven contribution toward efficient farming. It is now quite clear that implementing AI will not only give a real-time analysis for sowing seeds, but will help to detect poor nutrition on farms. Improper use of nutrients may develop environmental pollution risks by releasing toxic heavy metals, antibiotics, and pathogens.[26] For instance, manure used in agriculture is generally obtained from animal farms. A study carried out on pig slurries from 36 pig farms in south-eastern Spain, for salt content (electrical conductivity, chloride and sodium), organic load, micronutrients (Fe, Cu, Mn and Zn), and heavy metals (Cd, Co, Cr, Ni and Pb) showed the need for sustainable use of animal manures as crossing the maximum input of the desired nutrients results in significant salinization risk and accumulation of unwanted metals in soils.[27] Therefore, proper monitoring of soil contents and supply of adequate nutrients is highly desirable. However, this task appears to be non-viable for the farmers using conventional techniques. This is where artificial intelligence could be a boon to agriculture. Use of AI sensors can be used to detect the proper nutrients required for crop production and hence an unusual waste of chemicals used for spraying on crops can be avoided thus giving cost-efficient farming. Khalil et al., in 2015, described the importance of soil properties for the healthy growth of plants.[28] The plant growth depends largely on the soil's chemical properties especially the content of the soil which in turn improves plant growth by controlling several factors like water penetration, surface sealing, and erosion. The pH of the soil is an important indication of nutrients in soil which indicates the presence of phosphorus, manganese, and calcium in the soil obtained by decomposition of organic matter. In agriculture, it is very important to gather the information for the desired fertilizer administration for a successful farming. AI can assist farmers in determining the absent nutrient in soil for crop production and thereby lower the risk of crop failure. AI-drone technology helps farmers virtually examine every crop and look for irregular crop degradation, collect data, analyze the cause, and provide the best possible nutrients. The quality of soil is degrading day by day with the increasing rate of deforestation, this in turn degrades the soil quality, and this makes it hard to determine. Nutrient loss should be balanced to prevent unwanted loss of crop production by practicing certain preventive courses of action like determining immediately the unhealthy plant and using adequate fertilizers and manure at the right time. This could be done successfully with the help of AI sensors in a short time. The practice of including AI sprayers in farming can widely reduce the unnecessary use of chemicals to be used on fields, and hence improve the quality of crops thereby saving money. "Plantix" by PEAT is an AI-based crop advisory app for soil quality including disease detection, pest control, nutrient deficiencies, and increased yield. This has been used by 15 million farmers to improve the crop quality. A few more examples of AI startups in agriculture are as follows.

1. Fasal: This app is an Indian start-up in agriculture which is used by nearly 9 billion people. It uses AI-enabled sensors to provide real-time data and insights to farmers. The app provides information about weather forecasts, and helps in disease and pest control, irrigation and finance management.
2. Blue River technology: this technology is a California-based start-up which started in the year 2011. The machines utilize machine learning, computer vision, and robotics in the field of agriculture.
3. Farm Bot: This bot is an open-source CNC precision farming which was started in 2011 to encourage growing crops by anyone.

AI IN IRRIGATION WATER MANAGEMENT

The optimal use of water during the farming cycle is an important factor in obtaining good success in terms of yield and quantity. In most of the developing countries in the world, agriculture has been considered the primary source of earnings. It has been observed that the agricultural sector consumes the largest water resources. However, a major portion of the water is drained off, with only a small portion left for efficient crop growth. To overcome this unnecessary wastage, a modern technology that can effectively provide all the solutions is the need of the hour. With the growth of modern scientific techniques, AI has provided us with all the necessary requirements for wastewater management in the field of agriculture. AI technology is set up in such a way that it can monitor the water content in crops, and compare the availability of water in the soil. It helps to determine the water requirements per crop. Various groups have reported different AI-based methods for water management in the agricultural field. Arif and their group[29] described ANN-based models to predict the content of moisture in soil. This method works by adopting a Dynamic ANN model that was used to estimate soil moisture with the inputs of reference evapotranspiration (ETo) and precipitation. Hinnell,[30] and co-workers also represented a method which is based on a Neural Network irrigation system. Moreover, Goap and their group[31] and Nawandar and Satpute[32] have described an IoT-based smart irrigation management system. The mechanism of this method is based on a smart algorithm. It considers sensed data along with the parameters of weather forecasts. The overall system has been developed on a pilot scale. The sensor node data are wirelessly collected over the cloud using web services and a web-based information visualization, which provides real-time information insights based on the analysis of sensor data and weather forecast data. The system has a provision for closed-loop control of the water supply to achieve a fully independent irrigation scheme. Choudhary et al. have represented an Automatic Irrigation System that is based on AI.[4] This method can independently irrigate fields via soil moisture data. This method basically consists of two components

1. Machine learning to predict the amount of rainfall and the soil moisture content.
2. Economic hardware implementation using the Internet of Things.

The mechanism in this method is based on the prediction of algorithms that can be used to identify historical weather data, patterns of rainfall, and changes in the climate. After that, a smart system is created which can irrigate crop fields only when it is required depending on the weather and real-time soil moisture conditions. It was found to be 80% accurate, thereby providing an efficient modern tool for the farmers.

CONCLUSION

To address the challenges in agriculture, intelligent, data-driven autonomous farm operations using digital technologies are highly crucial. AI adaptation to minimize unwanted water supply and save water and also to increase production efficiency using smart irrigation in standard environmental conditions is important. Changing from traditional farming to AI-based agriculture will enable us to use our resources more efficiently, which will help both farmers and the economy as a whole. This could be achieved by encouraging farmers to be more resourceful and environmentally sustainable and contributing new strategies to meet the rising demand for goods and services. To date, the gradual acquisition of experience with the implementation of drones for crop protection could be observed to a greater extent. However, to meet the strong demands, AI crop technology should be utilized in all aspects of farming. Current research takes up the present scenario of high population growth and finds ways to develop conceptual models and frameworks, which could aid in resource utilization, reducing labor requirements, monitoring crops, pests, and diseases, and tracking irrigation, harvest, and distribution. AI technology has come up to handle the future aspects of agriculture and thereby became an integral part of farming.

REFERENCES

1. Tayade, R.; Yoon, J.; Lay, L.; Khan, A. L.; Yoon, Y.; Kim, Y. Utilization of spectral indices for high-throughput phenotyping. *Plants.* 2022, *11*(13), 1712.

2. Ray, D. K.; Ramankutty, N.; Mueller, N. D.; West, P. C.; Foley, J. A. Recent patterns of crop yield growth and stagnation. *Nat. Commun.* 2012, *3*, 1293.

3. Hayden, M. A.; Barim, M. S.; Weaver, D. L.; Elliott, K. C.; Flynn, M. A.; Lincoln, J. M. Occupational safety and health with technological developments in livestock farms: A literature review. *Int. J. Environ. Res. Public Health* 2022, *19*(24), 16440.

4. Ahmed, M.; Hayat, R.; Ahmad, M.; Hassan, M. U.; Ahmed, M. S. K.; Hassan, F. U.; Rehman, M. H. U.; Shaheen, F. A.; Raza, M. A.; Ahmad, S. Impact of climate change on dryland agricultural systems: A review of current status, potentials, and further work need. *International Journal of Plant Production.* 2022, *16*, 341–363. (b) Schmidhuber, J.; Tubiello, F. N. Global food security under climate change. *PNAS* **2007**, *104*(50), 19703–19708.

5. Choudhary, S.; Gaurav, V.; Singh,A.; Agarwal, S. Autonomous crop irrigation system using artificial intelligence. *Int. J. Eng. Adv. Technology* 2019, *8*(5s), 46–51.

6. Dai, S.; Li, X.; Jiang, C.; Ping, J.; Ying, Y. Triboelectric nanogenerators for smart agriculture. *InfoMat.* 2023, *5*(2), 12391.

7. Schroeder, V.; Savagatrup, S.; He, M.; Lin, S.; Swager, T. M. Carbon nanotube chemical sensors. *Chem. Rev.* 2019, *119*(1), 599–663.

8. Narimani, K.; Nayeri, F. D.; Kolahdouz, M.; Ebrahimi, P. Fabrication, modeling and simulation of high sensitivity capacitive humidity sensors based on ZnO nanorods. *Sens. Actuat. B: Chem.* 2016, *224*, 338–343.

9. Tsouti, V.; Boutopoulos, C.; Zergioti, I.; Chatzandroulis, S. Capacitive microsystems for biological sensing. *Biosens. Bioelectron.* 2011, *27*(1), 1–11.

10. Perumal, J.; Wang, Y.; Attia, A. B. E.; Dinish, U. S.; Olivo, M. Towards a point-of-care SERS sensor for biomedical and agri-food analysis applications: A review of recent advancements. *Nanoscale* 2021, *13*(2), 553–580.

11. Liu, W.; Wang, X.; Song, Y.; Cao, R.; Wang, L.; Yan, Z.; Shan, G. Self-powered forest fire alarm system based on impedance matching effect between triboelectric nanogenerator and thermosensitive sensor. *Nano Energy* 2020, *73*, 104843.

12. Rad, Z. F. Microneedle technologies for food and crop health: Recent advances and future perspectives. *Adv. Eng. Mater.* 2023, *25*(4), 2201194.

13. Lekka, C.; Petropoulos, G. P.; Triantakonstantis, D.; Detsikas, S. E.; Chalkias, C. Exploring the spatial patterns of soil salinity and organic carbon in agricultural areas of Lesvos Island, Greece, using geoinformation technologies. *Environ. Monit. Assess.* 2023, *195*(3), 391.

14. Garajeh, M. K.; Salmani, B.; Naghadehi, S. Z.; Goodarzi, H. V.; Khasraei, A. An integrated approach of remote sensing and geospatial analysis for modeling and predicting the impacts of climate change on food security. *Sci. Rep.* 2023, *13*(1), 1057.

15. Stavrinidou, E.; Gabrielsson, R.; Gomez, E.; Crispin, X.; Nilsson, O.; Simon, D. T.; Berggren, M. Electronic plants. *Sci. Adv.* 2015, *1*(10), 1501136.

16. Paarlberg, R.; Smyth, S. J. The cost of not adopting new agricultural food biotechnologies. *Trends Biotechnol.* 2023, *41*(3), 304–306.

17. Nikšić, M.; Podgornik, B. B.; Berovic, M. Farming of medicinal mushrooms. *Adv. Biochem. Eng./Biotechnology.* https://doi.org/10.1007/10_2021_201.

18. Savary, S.; Ficke, A.; Aubertot, J.-N.; Hollier, C. Crop losses due to diseases and their implications for global food production losses and food security. *Food Sec.* 2012, *4*(4), 519–537.

19. Sarkozi, A. New standards to curb the global spread of plant pests and diseases. https://www.fao.org/news/story/en/item/1187738/icode/.

20. Pedersen, S. M.; Fountas, S.; Have, H.; Blackmore, B. S. Agricultural robots—System analysis and economic feasibility. *Precis. Agric.* 2006, *7*(4), 295–308.

21. Silva, A. F. D.; Ohta, R. L.; Azpiroz, J. T.; Fereira, M. E.; Marçal, D. V.; Botelho, A.; Coppola, T.; Oliveira, A. F. M. D.; Bettarello, M.; Schneider, L.; Vilaça, R.; Abdool, N.; Malanga, P. A.; Junior, V.; Furlaneti, W.; Steiner, M. Artificial intelligence enables mobile soil analysis for sustainable agriculture. *Materials Cloud Archive* 2022. 10.24435/materialscloud:vt-4t.

22. Aasim, M.; Akin, F.; Ali, S. A.; Taskin, M. B.; Colak, M. S.; Khawar, K. M. Artificial neural network modeling for deciphering the in vitro induced salt stress tolerance in chickpea (*Cicer arietinum* L). *Physiol. Mol. Biol. Plants* 2023, *29*(2), 289–304.

23. Moussafir, M.; Chaibi, H.; Saadane, R.; Chehri, A.; Rharras, A. E.; Jeon, G. Design of efficient techniques for tomato leaf disease detection using genetic algorithm-based and deep neural networks. *Plant Soil* 2022, *479*(1–2), 251–266.

24. Navya, P.; Sudha, D. Artificial intelligence-based robot for harvesting, pesticide spraying and maintaining water management system in agriculture using IoT. *AIP Conf. Proc.* 2023, *2523*, 020025.

25. Husaini, A. M.; Sohail, M. Robotics-assisted, organic agricultural-biotechnology based environment-friendly healthy food option: Beyond the binary of GM versus Organic crops. *J. Biotechnol.* 2023, *361*, 41–48.

26. Köninger, J.; Lugato, E.; Panagos, P.; Kochupillai, M.; Orgiazzi, A.; Briones, M. J. I. Manure management and soil biodiversity: Towards more sustainable food systems in the EU. *Agric. Syst.* 2021, *194*, 103251.

27. Moral, R.; Perez-Murcia, M. D.; Perez-Espinosa, A.; Moreno-Caselles, J.; Paredes, C.; Rufete, B. Salinity, organic content, micronutrients and heavy metals in pig slurries from South-Eastern Spain. *Waste Manag.* 2008, *28*(2), 367–371.

28. Khalil, H. P. S. A.; Hossain, M. S.; Rosamah, E.; Azli, N. A.; Saddon, N.; Davoudpoura, Y.; Islam, M. N.; Dungani, R. The role of soil properties and it's interaction towards quality plant fiber: A review. *Renew. Sustain. Energy Rev.* 2015, *43*, 1006–1015.

29. Arif, C.; Mizoguchi, M.; Mizoguchi, M.; Doi, R. Estimation of soil moisture in paddy field using artificial neural networks. *Int. J. Adv. Res. Artif. Intell* 2013, *1*(1), 17–21.

30. Hinnell, A. C. et al. Neuro-drip: Estimation of subsurface wetting patterns for drip irrigation using neural networks. *Irrig. Sci.* 2010, *28*(6), 535–544.

31. Goap, A.; Sharma, D.; Shukla, A. K.; Rama Krishna, C. An IoT based smart irrigation management system using machine learning and open source technologies. *Comput. Electron. Agric.* 2018, *155*, 41–49.

32. Nawandar, N. K.; Satpute, V. R. IoT based low cost and intelligent module for smart irrigation system. *Comput. Electron. Agric.* 2019, *162*, 979–990.

13 Plasma Technology
An Emerging Tool for Sustainable Agriculture

*Bhavna Nigam, Mangesh M. Vedpathak,
and Indra Jeet Chaudhary*

INTRODUCTION

In the current scenario, agriculture faces continuous global problems such as population growth, environmental pollution, loss of agricultural land, and climate change.
Atmospheric Pollution poses a major threat to agriculture worldwide (Gupta, 2020a,b;
Mahajan et al., 2022; Gupta et al., 2020; Gupta et al., 2022; Yatoo et al., 2022; Singh
et al., 2023) and causes negative effects on the growth and productivity of crops (Sun
et al., 2017; Rathore and Chaudhary, 2019, 2021; Pandya et al., 2022). The Food and
Agriculture Organization (FAO) reported that the estimated 20–45% maize, 5–50%
wheat, and 20–30% rice yield reductions are expected, respectively, by 2100 under
the current rates of climate change (Arora, 2019). Deforestation and pollution load
are the main contributors to climate change. Climate change has caused changes in
the distribution of plant infections, pests, and changes in the virulence of pathogens,
and the emergence of new illnesses (Bebber et al., 2013; Velásquez et al., 2018). This
alteration also changes the crop cultivation patterns and reduces the crop quality
and quantity (Pathak et al., 2018). Various studies by researchers have reported that
environmental pollution caused a negative impact on plants, especially agricultural
plants. Despite that researchers also reported higher yield loss due to salinity, ozone,
SPM, etc. (Sett, 2017; Chaudhary and Rathore, 2020, 2021; Nigam et al., 2022).

Therefore, for controlling abiotic and biotic stress on plants, various exogenous
protectants have been applied and also seen good results (Chaudhary and Singh,
2020; Chaudhary and Rathore, 2020; Nigam et al., 2022). Different methods and
tools have been used to adjust to the shifting agricultural conditions. Optimal land
use and management, changing patterns of food demand, and decreased food loss
and waste are frequently proposed as adaptation techniques (Anderson et al., 2019).
The issues connected to the environment can be successfully solved through technological agricultural advancements. Crop plants with improved yields and stress
tolerance are regularly created using genetic engineering and breeding-based technologies (Anderson et al., 2019). However, the complex processes involving numerous genes that govern crop yield and tolerance make genetically modified crops more

DOI: 10.1201/9781003441175-13

difficult to enhance. Another difficulty to the widespread use of genetic techniques is safety concerns.

Plasma is one of the considered technologies for advanced oxidation processes (Ekezie et al., 2017; Fan and Song, 2020). Although large-scale uses of plasma technology are still pricey, it has advantages over traditional therapies based on synthetic chemicals. The plasma technique has beneficial synergistic effects on seed germination and seedling vigour without leaving behind any artificial chemical residues. Typically, a plasma treatment apparatus consists of electrodes for generating plasma, a treatment chamber to house the electrodes, and electricity to provide current to the electrodes. Reactive oxygen species (ROS, such as superoxides, singlet oxygens, atomic oxygen, ozone, hydrogen peroxide, and hydroxyl radicals), reactive nitrogen species (RNS, such as nitric oxide, nitrogen dioxide, nitrate, nitrite, and peroxynitrite), and ultraviolet (UV) photons are primarily generated from the plasma discharge when a high electric discharge is applied to air or an aqueous solution in a chamber (Song et al., 2020). According to separate studies (Jisha et al., 2013; Arajo et al., 2016, Antoniou et al., 2016, Thomas and Puthur, 2017), ROS, RNS, and UV have been used to scarify seeds (seed coat softening process), inactivate seed-borne pathogens, and strengthen antioxidant defence systems in crop plants. Recent reports on the synergistic benefits of plasma treatment have been examined by Randeniya and de Groot (2015), Ito et al. (2018), and Adhikari et al. (2020) for a variety of crops.

However, the plasma technology used in different agricultural field for production of crops, but their role against stress will be useful tool for agricultural sustainability. Therefore this study will be helpful for people who don't realise how to used plasma technology for control of stress on plants. This chapter defines the effects of various plasma treatments on plant responses in terms of seed germination rate, inactivation of pathogens on seeds, physiological processes, and plant stress tolerance. From these points, recent data on plasma applications and their mechanisms against a variety of stresses were examined. Thus, this chapter suggests that plasma technology may be useful for reviving seeds and establishing seedlings in challenging environments.

PLASMA TECHNOLOGY

A partially or completely ionised gas called plasma can catch fire under low atmospheric pressure and is made up of charged species like electrons and positive and negative ions, neutral species like atomic and/or molecular radicals and non-radicals, electric fields, and photons, as well as neutral species like neutral species. Early in the 1960s, research on the effects of glow discharge on cotton, wheat, alfalfa, red clover, sweet clover, beans, and other types of grass seeds examined one of the earliest uses of plasma for treating seeds. It was demonstrated that plasma therapy affects bean hard seed content, moisture adsorption, and seed germination (Priatama et al., 2022). Since then, research into plasma treatment of seeds has increased thanks to the use of various plasma devices, which enable in-depth investigations into the physical, chemical, and biological mechanisms of plasma that can be sparked by the analysis of plasma components (Ranieri et al., 2021). The resultant change in the seed surface properties (e.g. wettability, chemical composition, adhesion) improved crop growth and yields. In the past decades, low-temperature plasma at atmospheric pressure has opened up a new research field in biology and medicine (Yousfi et al., 2014; Laroussi, 2020). Plasma treatment of seeds has been divided into two methods: direct and indirect based on the contact of the plasma with the samples. For plasma treatment, plasma sources, such as dielectric barrier discharge (DBD) (Gómez-Ramírez et al., 2017; Rahman et al., 2018; Bafoil et al., 2019), radio frequency (RF) plasma (Nakano et al., 2016; Volkov et al., 2019; Saberi et al., 2020), and atmospheric-pressure plasma jet (APPJ) (Zhou et al., 2016; Liu et al., 2023), have been used. The treatment is performed by controlling operating parameters, such as electrode structure, power source (voltage, frequency, and waveform), discharge gas (air, Ar, He, etc.), and other conditions (gas flow, gas pressure, gas temperature, etc.). Now these day's plasma technology is used for agricultural sustainability, such as seed quality, growth, biomass, and yield of agricultural crops (Ahmad et al., 2022). Plasma technology is also used in the field of biotic and abiotic stress tolerance (Gierczik et al., 2020).

PLASMA TECHNOLOGY AND IT'S USED IN AGRICULTURAL SUSTAINABILITY

Plasma helps in growth enhancement, seed sterilisation, and soil remediation and boosts plant growth and yield (Priatama et al., 2022). Plasma technology has recently received more attention in the agriculture sector (fertilisers, fungicides, plant growth regulators, etc.). Therefore, this technology can be called sustainable agricultural technology. Plasma science technology has the potential to make a significant contribution not only to reducing food loss/waste and improving food safety but also to improving food material production. "Plasmas in agriculture and the food cycle" is a newly emerging interdisciplinary field of plasma science and applications, and its popularity has recently continued to increase due to its high potential impact on our society. Plasmas can be used in a wide range of agricultural and food cycles from farm to table, including cultivation, storage, food safety, waste treatment, and so on. Plasma has been utilised for both constructive (seed germination, plant growth improvement, food material functionalisation, etc) and destructive functions (sterilisation, disinfection, etc) (Figure 13.1).

Role of Plasma on Seed

A seed typically consists of the cuticle, epidermis, hypodermis, and parenchyma as its four layers. Differences between species and even cultivars are determined by variations in the microstructures and chemistry of these layers. The seed coat functions similarly to a water modulator by regulating water entry so that it can be absorbed gradually by the cotyledons to reduce or prevent imbibition damage (Waskow et al., 2021). It has been hypothesised that removing the lipid layer improves access to water, which is necessary to initiate germination (Sehrawat et al., 2017). For agricultural reasons, plasma treatments are now being evaluated as a seed-processing technology where seeds are treated prior to pre-sowing to improve the seeds' capacity

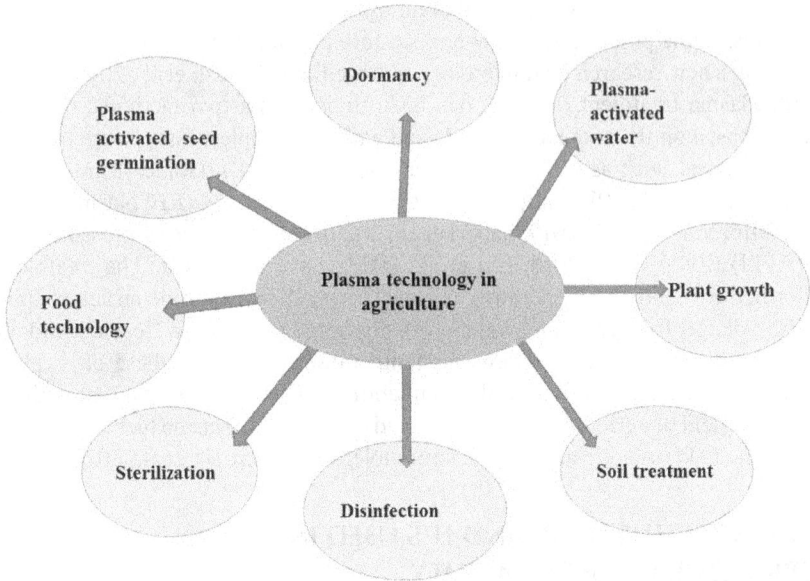

FIGURE 13.1   Plasma technology used in agriculture

for rapid nutrient uptake, rapid embryo growth, rapid water uptake, and increased germination (Waskow et al., 2021). The plasma therapy on seeds serves a variety of purposes, such as the inactivation of microorganisms, loosening of cell walls, quick absorption of nutrients, quick growth of the embryo, quick absorption of water, and the elimination of diseases transmitted by seeds (Figure 13.2).

All seeds have generally evolved to include everything they need to develop into plantlets once the environmental conditions are assessed as favourable, despite their great diversity. The seed coat (testa), which can differ across species and cultivars or depend on the plants being fertile or clones unable to create the next generation, protects the living tissues of seeds. The embryo is in a partially dried, quiescent state inside the dry mature seed, ready to germinate upon the addition of water, or in other words, imbibition. Following germination, which is the transformation of an inactive seed into an active one that grows, ruptures the seed coat, and develops from a seedling into a plantlet which is typically still frail and especially vulnerable to external stresses and finally into a more stress-resistant autotrophic plant it is then supplied with stored foods through the endosperm.

An ionised gas called plasma can be ignited at atmospheric or low pressures. Operating factors like voltage, frequency, humidity, flow rate, and gas mixing affect the plasma's composition. Electric fields can ionise gases like argon, oxygen, nitrogen, helium, and/or air to create electrons, ions, UV radiation, thermal radiation, and reactive species. Due to the fact that all of the components of plasmas may be found in nature and quickly recombine, plasma treatment may increase seed survival without leaving behind any hazardous residues. Furthermore, plasma components only allow for surface functionalisation due to their approximately 10 nm deep penetration

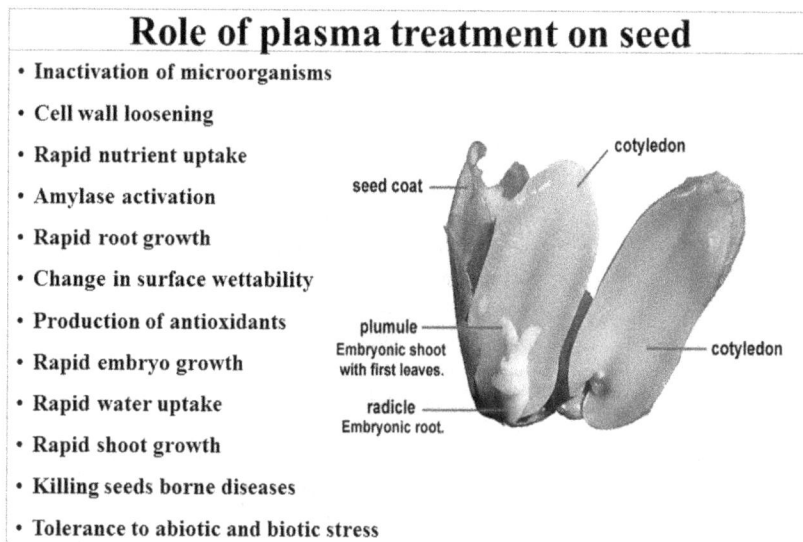

Role of plasma treatment on seed

- Inactivation of microorganisms
- Cell wall loosening
- Rapid nutrient uptake
- Amylase activation
- Rapid root growth
- Change in surface wettability
- Production of antioxidants
- Rapid embryo growth
- Rapid water uptake
- Rapid shoot growth
- Killing seeds borne diseases
- Tolerance to abiotic and biotic stress

FIGURE 13.2   The role of plasma technology on seeds

(Guo et al., 2018). Importantly, plasma therapies are thought to require little upkeep and use little energy (Randeniya and de Groot, 2015). As a result, numerous research has been conducted using plasma treatments of agronomically interesting plants, including quinoa, basil, tomato, wheat, radish, soybean, mung bean, rice, Ajwain and Umbu, as well as seeds considered significant for the landscape, like Norway spruce (Pauzaite et al., 2018).

Role of Plasma Technology on Seed Scarification and Germination

Seed scarification is conventionally done by acid treatment, heating, freeze-thawing, and a mechanical scarifier. Plasma technology, as a particle bombardment or an oxidising agent or both, can also be used for seed scarification at pre-sowing. The thickness and permeability of the seed coat control seed germination. As a result, seeds with thick coats cannot quickly absorb water unless they are scarified (Song et al., 2020). Legumes in particular develop thick, impenetrable seed coverings (like those of wild soybeans). According to a study (Zhou et al., 2010), an impermeable seed coat is present in soybeans when the seed coat/embryo weight ratio is higher than 0.1. As a result, seed scarification may frequently be necessary for seed vigour and seedling germination. According to the treatment circumstances of the plasma sources, such as radiofrequency (RF) discharge, DBD, and other types of plasma sources, Table 13.1 summarises the scarification effects of plasma treatments on seeds.

Depending on the crop species, feed gas, power, and exposure time, RF discharge is frequently employed to test plasma effects on seed invigoration of crops in a temperature-controlled chamber. For plasma creation fuelled by an RF discharge, air and helium are frequently utilised as feed gases. Wheat, maize, and mung bean

TABLE 13.1

Plasma Technology and Its Role in Seed Invigoration and Seed Germination

S. No.	Plasma	Treatment stages, medium and effects	References
	Radio frequency discharge	Dry seed treated with gas medium: Increased seed germination, higher invigoration percentage of wheat, soybean, and common bean, maize, pea, and mug bean.	Bormashenko et al. (2012); Filatova et al. (2013,2014)
	Emulated plasma-activated water	Increased seed quality by employing high germination index, seedling growth rate and fresh and dry weight of seedlings.	Ahmad et al. (2022)
	Inductive helium plasma discharge	Plasma-treated soybean seeds absorbed 14.03% water in comparison to the control which led to quick and higher germination percentage.	Ling et al. (2014)
	Dielectric barrier discharge	Dry seed and wet seed treated with gas medium: Increased seed germination, higher invigoration percentage of wheat, pea, and spinach.	Li et al. (2017); Stolárik et al. (2015); Park et al. (2018)
	Corona discharge	Dry seed treated with gas medium: Increased seed germination, higher invigoration percentage of rice.	Khamsen et al. (2016)
	Arc discharge	Dry seed treated with gas medium: Increased seed germination, higher invigoration percentage of spinach.	Shao et al. (2013)
	Glow discharge	Dry seed treated with gas medium: Increased seed germination, higher invigoration percentage of rice.	Chen et al. (2016)
	Microwave discharge	Dry seed treated with gas medium: Increased seed germination, higher invigoration percentage and growth of wheat.	Šerá et al. (2010)
	Radio frequency	Air plasma at a radio frequency of 13.56 MHz for six minutes with fixed 50 watts coupled to power and 0.7 mbar pressure increased wheat seed invigoration by water uptake through seed at 6 and 12 hours after soaking and also improved seed germination rate, time, speed and germination percentage (%).	Sharma et al. (2020)

seeds were given an extra boost by RF discharge in the air (Bormashenko et al., 2012; Filatova et al., 2013; Filatova et al., 2014; Sadhu et al., 2017). A discharge in the air in the instance of the common bean shortened the time needed to reach 50% germination, but the final percentage of germination was unaffected (Bormashenko et al., 2015). Additionally, excessive 20-minute exposure to an air-based discharge with a 100 W power intensity reduced the ultimate germination of wheat seeds (Filatova et al., 2013). When generated with a power intensity of 80 to 100 W for an exposure length of 0.25 min, RF discharge in helium increased seed invigoration with a 9% and 7% increase in the germination of soybeans and oilseed rapes, respectively, compared with that of the untreated control (Li et al., 2014; Li et al., 2015). However, Li et al. (2014) found no discernible impact on soybean seed germination when exposed to a helium-based RF discharge with a power level of less than 60 W. Additionally, when produced from hydrazine (N_2H_4), RF discharge occasionally had no visible impact on amount and time for maize to germinate (Volin et al., 2000).

When created under ideal circumstances, DBD has been shown to be effective for seed germination and seedling vigour. Under laboratory conditions, DBD in air induced quicker germination, improving germination rates and promoting early seedling growth in wheat, barley, and peas (Dobrin et al., 2015; Stolárik et al., 2015; Li et al., 2017; Park et al., 2018). When compared to the untreated control, a DBD with a power intensity of 1.5 to 2.7 W and an exposure time of 7 to 15 min increased wheat seed invigoration by up to 47% (Dobrin et al., 2015; Li et al., 2017). DBD in air increased seed invigoration in peas and barley by 31% and 51%, respectively, after exposure of 2 and 0.2 min at higher power intensities (370 to 400 W) compared to the untreated control (Stolárik et al., 2015; Park et al., 2018). Additionally, barley and spinach's early seedling growth and seed germination were improved by DBD's nitrogen content (Ji et al., 2016; Park et al., 2018).

Although their maximum values vary depending on the treatment conditions of the plasma sources, other discharge plasmas, such as corona, arc, glow, and microwave discharges, have also demonstrated stimulating effects on seed invigoration of some crops (erá et al., 2010; Shao et al., 2013; Jiang et al., 2014a; Chen et al., 2016; Khamsen et al., 2016; Roy et al., 2018). These findings imply that plasma treatment can improve seedlings' capacity to absorb water, improving crop vigour. Plasma treatment's favourable impact on seed invigoration is primarily related to its power (W) and exposure time (min). For instance, seed invigoration of crops can be improved by plasma treatment when generated with a power of 100 W and an exposure time of less than ten minutes. Based on the crop type, these plasma treatment parameters need to be further tuned.

The reproductive organs of plants with full potency, or the ability to grow into whole plants, are seeds. For a plant to survive, to reproduce, and to maintain its offspring, it needs seeds. To protect themselves from unfavourable environmental circumstances, seeds enter a period of dormancy at the time of dissemination. Even when the conditions are favourable, many dormant seeds do not germinate. Therefore, to maximise germination, the dormant state must be broken. In addition to environmental influences, the hard seed coat, the presence of inhibitors, the seed maturation time, the underdeveloped embryo, the seed coat's permeability to oxygen and water, and hormonal imbalances all have an impact on the latent phase and seed germination. Abscisic acid (ABA), a phytohormone, is in charge of preserving dormancy, whereas gibberellic acid (GA) is in charge of breaking the dormant state. In response to physical circumstances, seeds produce these phytohormones, which then activate the signalling pathways and enzymes that encourage the breakdown of seed reserves and start germination. To overcome dormancy, numerous seed treatment techniques have been used during the various phases of germination, which involve multiple physiological processes. Scarification, stratification, and chemical treatments are a few common seed treatment or priming techniques used to stimulate germination in dormant seeds (Adhikari et al., 2020). A novel technique to improve seed germination uses cold (non-thermal) air or low-pressure plasma (Tables 13.1).

The first instance of plasma being applied to seeds was in a US patent by Krapivina et al. (1994), where soybean seedlings were exposed to cold atmospheric-pressure plasma for 5 to 300 seconds while being exposed to a mixture of inorganic gases (atmospheric air, oxygen, and nitrogen) (Krapivina et al., 1994). Vegetables (tomato, radish, coriander, green peas, and sunflower) and crops (rapeseed, cotton, maize, oat, wheat, mustard, soybean, legumes, and honey clover) can now be treated using a variety of plasma sources, including DBD jet plasma, microwave discharge, radiofrequency (RF) discharge, and gliding discharge (Table 13.1). In most research, plasma treatment improved seed germination, while some investigations (Zhou et al., 2011; Mihai et al., 2014; Lindsay et al., 2014; Ahn et al., 2019; Judée et al., 2018; Kang et al., 2019) showed no changes in the germination percentage. Plasma increased the rate of germination, the percentage of germination overall plant growth and development.

Role of Plasma Technology on Inactivation of Seed-Borne Diseases

Plasma treatment prevents plant diseases by inactivating pathogens (Table 13.2). According to various studies, power (W) and exposure time (min) are among the most important factors of seed-borne pathogen inactivation (Table 13.2). During crop growing seasons, Plant diseases are caused by seed-borne pathogens, which reduce seed germination and seedling establishment in crops; however, RF discharge can inactivate seed-borne bacteria and fungi. *Alternaria* and *Fusarium* species cause a wide range of economically significant diseases in a wide range of crops, along with legumes, cereals, and vegetables (Thomma, 2003; Khan et al., 2006). In a vacuum chamber, an eight-minute RF plasma treatment with a power intensity of 77 W decreased fungal infection by the *Alternaria* and *Fusarium* species by up to 71% and 99% in wheat and maize, respectively (Filatova et al., 2014). An RF discharge inactivated the bacterial pathogen *Ralstonia solanacearum* after 0.25 minutes of exposure at a similar power intensity of 80 W in a vacuum (Jiang et al., 2014b). On young tomato plants, *Ralstonia solanacearum* causes bacterial wilt symptoms (Eljounaidi et al., 2016).

Other plasma sources, such as DBD and arc discharge, have also been found to have inactivation effects on a broad range of filamentous fungi, resulting in higher crop survival. Under environmentally controlled conditions, plasma treatments with DBD and arc discharge effectively inactivated *Gibberella fujikuroi* (synonym,

TABLE 13.2
Plasma Technology for Inactivation of Seed-Borne Diseases

S. No.	Plasma	Plants, Treatment stages, medium and effects	References
	Radio frequency discharge	Dry seed of wheat and maize using gaseous medium: Inactivation of fungal (Alternaria spp. and Fusarium spp.)	Filatova et al. (2014)
	Low-pressure plasma	Low-pressure plasma treatment in the inactivation of the seed-borne plant pathogenic bacterium, Xanthomonas campestris, inoculated on cruciferous seeds, was evaluated. The highest inactivation effect was observed when the treatment voltage and argon gas flow rate were 5.5 kV and 0.5 L/min, respectively.	Nishioka et al. (2016)
	Dielectric barrier discharge	Dry seed treated with gaseous medium: Tomato: bacterial inactivation (*Ralstonia solanacearum*), rice: fungal inactivation (*Gibberella fujikuroi*), sweet basil: fungal inactivation (*Alternaria* spp., *Aspergillus* spp., and *Penicillium* spp.) and wheat, chickpea, barley, oat, rye, lentil, maize: fungal inactivation (*Aspergillus* spp. and *Penicillium* spp.).	Jiang et al. (2014b); Jo et al. (2014); Ambrico et al. (2017); Zahoranová et al. (2016); Selcuk et al. (2008)
	Ozone and arc discharge plasma	Dry seed of rice using gaseous medium: fungal inactivation (*Fusarium fujikuroi*). About 80% of rice seeds were disinfected by treatment with arc plasma discharged at 12 Hz for 30 min.	Kang et al. (2015)
	Emulated plasma-activated water	Emulated plasma treatment on pepper seeds killed fungal pathogens as it destroyed the cell wall that may be caused by the generation of ROS and the contribution of its physical properties (low pH and high EC) in the emulated plasma condition.	Ahmad et al. (2022)

Fusarium fujikuroi) growth and thus lowered the fungal infection of rice by up to 51%. (Jo et al., 2014; Kang et al., 2015). Similarly, when generated at a power intensity of 6.5 W for five minutes, DBD in air effectively prevented fungal infection of sweet basil in an in vitro seed culture from naturally established fungi such as *Alternaria, Aspergillus,* and *Penicillium* species by approximately 44% (Ambrico et al., 2017). After four minutes of in vitro exposure to wheat seeds, DBD at a high-power intensity of 400 W completely inhibited the growth of fungal pathogens such as *Fusarium nivale, Fusarium culmorum, Aspergillus flavus,* and *Trichothecium roseum* (Zahoranová et al., 2016). DBD treatment effectively reduced fungal contamination by *Aspergillus* and *Penicillium* species even in the case of a seed mixture (Selcuk et al., 2008). However, if no seed treatment is used, fungal pathogens such as *Fusarium* and *Aspergillus* species as well as *Trichotheciumroseum* are capable of inhibiting plant growth and development in legume and cereal crops, resulting in severe crop loss (Palencia et al., 2010; Scherm et al., 2013). Furthermore, the seed-borne *Aspergillus, Penicillium* and *Fusarium* species produce a large number of mycotoxins that are potentially toxic to both human and animal health (Amaike and Keller, 2011; Ráduly et al., 2020).

Plasma in seed application can be utilised to prevent plant diseases in a wide variety of crops, which including legumes, vegetables and cereals. Plasma application may necessitate a high power and long exposure to effectively inactivate seed-borne pathogens throughout seed germination (Table 13.2). A plasma exposure with a power of 100 W and an exposure time of ten minutes, for example, almost inactivates seed-borne pathogens in wheat. Other factors must be considered when determining the standardised values for the plasma treatment's power (W) and exposure time (min), because different parameters for pathogen inactivation can occur depending on the environmental conditions (Table 13.2).

ROLE OF PLASMA TECHNOLOGY ON PLANT GROWTH, BIOMASS, AND YIELD

The life cycle of plants is divided into four distinct phases: juvenile, vegetative, reproductive, and seed formation. Vegetative growth is an important phase in which plants perform photosynthesis, increase their biomass, synthesise the reserve food, and prepare for reproduction. It is also a very sensitive stage because growth is influenced by environmental factors (i.e. heat, drought, pathogen, alkalinity, UV rays) and biological stimuli. Overall crop productivity depends on the vegetative growth stage, and, therefore, the regulation of vegetative growth is critical for plant development and survival (Huijser and Schmid, 2011). Plasma can regulate the vegetative growth phase of plants, and plasma-seed treatment has long-term effects on early vegetative growth (Adhikari, 2020). Plasma technology in seed increases the germination rate and enhances the antioxidative properties of plants. Therefore, the enhancing defensive mechanisms in plants provide tolerance against environmental stresses. Biotic and abiotic stress tolerant plants ultimately increased plant growth and biomass. Stress especially abiotic such as salt, ozone, particulate matter and drought causes overproduction of reactive oxygen species and finally causes water loss in plant cells (Chaudhary and Rathore, 2018; Rathore and Chaudhary, 2019, 2021). Higher water loss in plant cells degrades the membrane lipid and membrane damage. Therefore, the application of plasma technology such as emulated plasma-activated water improved the quality of seed and enhanced seed germination, seedling germination index, growth, and biomass (Ahmad et al., 2022). The various plasma technologies and their role in the improvement of growth and biomass of agricultural crops are given in Table 13.3.

TABLE 13.3
Plasma Technology for Plant Growth and Biomass

S. No.	Plasma	Effects of plasma on plants growth, biomass and yield	References
	Emulated plasma-activated water	Emulated plasma-activated water improved the quality of pepper seed and enhanced seed germination, seedling germination index, growth, and biomass.	Ahmad et al. (2022)
	Low-pressure plasma	Low-pressure plasma-treated at the reproduction stage on oilseed rape plant increased in pod numbers (13.8%) and grain weight (8.2%).	Adhikari et al. (2020)
	Plasma Jet and Dielectric Barrier Discharge	Positive effect of the plasma treatment was found for biomass such as average length, dry weights of roots and sprouts, and other of wheat cultivars.	Velichko et al. (2019)
	Low-pressure radio frequency plasma	Improved plant health status and crop yield of some important agricultural plants such as maize, narrow-leaved lupine and winter wheat. Yield increase was seen in cultivar winter wheat grain yield (2.3%), maize (1.7%), and narrow-leaved lupine (26.8%) as compared to control plants.	Filatova et al. (2020)

Enhancement of Antioxidant Defence Systems

Plants have effective antioxidant systems that scavenge intracellular ROS and protect cells from oxidative damage. In plant cells, the antioxidant defence system consists of two components viz. non-enzymatic (polyphenols and quercetin) and enzymatic (SOD, CAT and POD) activities (Gill and Tuteja, 2010; Agati et al., 2012). When cells are exposed to excessive ROS, the antioxidant enzymes viz. SOD, CAT, and POD work together, despite working in different subcellular compartments (Mittler, 2002; Mittler et al., 2004; Sharma et al., 2012). SODs catalyse the dismutation (or partitioning) of the superoxide (O_2^-) radical into hydrogen peroxide (H_2O_2) and ordinary molecular oxygen in almost all cellular compartments (O_2). Finally, CAT in peroxisomes and PODs in the cytosol catalyse the reduction of hydrogen peroxide (H_2O_2) to water (H_2O). Non-enzymatic antioxidants, when combined with these enzymes, could provide cells with high-efficiency machinery for detoxifying active oxygen molecular species. The enhancement of these antioxidant systems after plasma treatment has been reported recently in a few crops (Table 13.4).

Plasma applications with short exposure (min) and low power (W) efficiently improve cellular antioxidant systems (Table 13.4). Under environmentally controlled conditions, an radio frequency discharge with a power intensity of 80 to 100 W for a

TABLE 13.4

Plasma Technology in Antioxidant Defence Systems of Plants

S. No.	Plasma	Treatment stages, medium and effects	References
	Radio frequency discharge	Dry seeds of Tomato and Oilseed rape treated with a gaseous medium increased the activity of antioxidant enzymes such as POD, CAT, and SOD.	Jiang et al. (2014b); Li et al. (2015)
	Dielectric barrier discharge	Dry seeds of Wheat, Maize, Spinach, Coriander and Soybean treated with a gaseous medium increased the activity of antioxidant enzyme POD, CAT, and SOD, and antioxidant content (total phenolic Compounds).	Li et al. (2017); Henselová et al. (2012); Zhang et al. (2017); Ji et al. (2015, 2016)
	Arc discharge	The dry seed of tomato treated with a gaseous medium increased the increased POD activity.	Yin et al. (2005)
	Microwave discharge	Increased the activity of antioxidants (some phenolic Compounds) wheat and rice seed treated with gaseous medium.	Šerá et al. (2010);Chen et al. (2016)
	Low-pressure radio frequency plasma	Plasma treatment enhanced the accumulation of non-enzymatic antioxidants and proline in plants.	Filatova et al. (2020)

short exposure time of 0.25 min quickly increased the activity of POD in seedlings of tomato and the CAT and SOD activities of oil rapeseed seedlings in comparison to the untreated control seedlings (Jiang et al., 2014b; Li et al., 2015). Similarly, after DBD exposure to their seeds, growing plants of soybeans, maize, and wheat had higher activities in the antioxidant enzymes viz, POD, CAT, or SOD (Henselová et al., 2012; Guo et al., 2017; Iranbakhsh et al., 2017; Li et al., 2017; Zhang et al., 2017). DBD in air, increased the SOD and POD activities of wheat seedlings after up to four minutes of exposure at a power intensity of 1.5 W (Guo et al., 2017; Li et al., 2017). Wheat seedlings directly exposed to DBD for one minute had higher activity of POD compared with untreated control (Iranbakhsh et al., 2018). In the case of arc discharge, a brief exposure increased POD activity in tomato seedlings (Yin et al., 2005).

Under laboratory conditions, dielectric barrier discharge has shown positive effects on the total phenolic compound content, though the effects vary depending on the type of feed gas. Ji et al. (2016) found that a nitrogen-based DBD application with an increasing duration of up to five minutes increased total phenolic compounds in seedlings of spinach, whereas an air-based DBD application with the same exposure time showed a reduction in total phenolic compounds. A similar study using micro found that coriander seedlings had a higher content of total phenolic compounds after a nitrogen-based application lasting up to one minute (Ji et al., 2015). In contrast, a discharge in air enhanced the total phenolic compounds in the growing wheat and rice plants when their seeds were exposed for a longer period of time (Šerá et al., 2010; Chen et al., 2016). These findings suggest that plasma applications produced with a low power and short exposure time (viz. 10 W and ten minutes) can boost cellular antioxidant systems (Table 13.4). When compared to the untreated control, antioxidant enzyme activities are enhanced by 8% to 100%. More experimental evidence is required to support the enhancement of the cellular antioxidant systems after plasma treatment. More research is needed to endorse the improvement of the cellular antioxidant systems after plasma application.

PLASMA TECHNOLOGY ITS ROLE IN PLANT STRESS TOLERANCE

Plasma technology has the potential to crop stress tolerance under environmentally controlled conditions. Application of plasma on agricultural plants enhanced antioxidative properties, photosynthetic activity, metabolite, and growth of plants. For minimisation of environmental stress on plants various plasma treatments are applied day by day and improved growth and yield of agricultural crops. In previous studies, seeds treated with radio frequency discharge plasma, generated a power intensity of 40 to 100 W and an exposure time of 0.25 min, which improved crop tolerance to drought stress in oilseed rape (Li et al., 2015) and alfalfa (Feng et al., 2018). According to Li et al. (2015), plasma treatment improved the seeds' ability to absorb water, raised the activity of antioxidant enzymes, and boosted the build-up of soluble sugars and proteins as osmolytes while under stress. A related investigation on the treatment of RF discharge revealed plant disease resistance to the tomato bacterial wilt-causing pathogen *Ralstonia solanacearum*. In tomato leaves inoculated with *R. solanacearum*, Jiang et al. (2014b) found that plasma treatment caused a quick rise in the H_2O_2 content, which in turn boosted the activities of defence-related and antioxidant enzymes. In the case of DBD plasma, the optimum exposure also increased plant resistance to two seed-borne illnesses of rice seedlings (bakanae disease and bacterial seedling blight) by activating defence-related responses via $H_2O_2^-$ mediated signalling (Ochi et al., 2017). Wheat seeds treated with DBD had higher antioxidant and defence-related enzyme activity, which raises the possibility that plasma therapy may help crops tolerate stress (Iranbakhsh et al., 2017). Guo et al. (2017) showed that DBD treatment improved wheat seedlings' ability to absorb water as a result of surface alteration. Through the regulation of several biological processes, this change reduced the oxidative damage brought on by dryness (e.g. hormone-mediated signalling, drought-tolerant-related gene expression, antioxidant enzyme activation, and osmolyte accumulation). Interestingly, direct application of DBD to wheat seedlings reduced oxidative damage brought on by salt by upregulating stress-related genes and turning on defence- and antioxidant-related enzymes (Iranbakhsh et al., 2018). By raising the concentration of flavonoid glycosides, Bußler et al. (2015) found that repeated DBD treatment led to plant adaptation to various oxidative stimuli (such as ROS, RNS, and UV) in peas in the early growth stage.

TABLE 13.5

Role of Plasma Technology on Plants Stress Tolerance

S. No.	Plasma	Effects on plant stress tolerance	References
	Radio frequency discharge	Crops: oilseed rape. Stage and medium: dry seed and gas. Improved water absorption, activity of enzyme CAT, SOD, cell membrane stability, osmotic adjustment ability, and tolerance to drought stress.	Li et al. (2015)
	Low-pressure radio frequency plasma	Plasma treatment enhanced the resistance of crop plants by promoting the accumulation of non-enzymatic antioxidants and proline in plants.	Filatova et al. (2020)
	Plasma-activated water	Plasma-activated water improved the stress tolerance of barley plants. H_2O_2 and NO are involved in the activation of protective mechanisms against salt stress. Damage can be mitigated by PAW-hardening through its effect on GSH, Chl, and Car, which are involved in the control of ROS.	Gierczik et al. (2020)
	Radio frequency discharge	In tomato leaves, plasma treatment caused a quick rise in the H_2O_2 content and boosted the activities of plant defence system and antioxidant enzymes.	Jiang et al. (2014b)
	Dielectric barrier discharge	Optimum exposure increased plant resistance to two seed-borne illnesses of rice seedlings (bakanae disease and bacterial seedling blight) by activating defence-related responses via H_2O_2-mediated signalling.	Ochi et al., 2017
	Dielectric barrier discharge	Plasma-treated wheat seeds enhanced antioxidant and defence-related enzyme activity, which raises tolerance.	(Iranbakhsh et al., 2017). Guo et al. (2017)

Plasma therapies can therefore improve crop tolerance prior to stressful situations. By modifying the seed surface environment (such as scarification and pathogen inactivation) and physiological processes (such as enhanced antioxidant systems and activated defence responses) in seeds under stressful conditions like drought, salinity, and pathogen infection, the optimal plasma exposure can improve seed germination and seedling growth (Table 13.5).

THE MECHANISMS OF PLASMA IN PLANTS UNDER STRESS

The plasma-induced stress tolerance may be the outcome of different interactions between the plasma and, respectively, the seed surface, seed-borne pathogens, and cellular homeostasis (e.g. ROS, RNS, and UV). On the basis of a study of the literature on plasma-induced plant responses, it is possible to offer plausible mechanisms for crop stress tolerance even though the effects of plasma on plants are still not fully understood. Before being exposed to environmental stressors, the ROS, RNS, and UV of plasma can alter the physical and chemical properties of the seed surface, making seeds more hydrophilic and permeable to water. Studies using scanning electron microscopy have revealed that the cellulose in plasma-treated seeds is partially degrading and that fractures are developing on their surface (Ji et al., 2016; Zhou et al., 2016; Li et al., 2017; Guo et al., 2018). Furthermore, studies using optical emission spectra and Fourier transform infrared spectroscopy have demonstrated the development of oxygen- and nitrogen-containing groups at the surface of treated materials (Filatova et al., 2013; Guo et al., 2017; Wang et al., 2017). Last but not least, plasma-treated seeds contain fissures in the seed coat and are hydrophilic, both of which are advantageous for water absorption before being exposed to drought stress (Li et al., 2015; Guo et al., 2017). By controlling endogenous hormones and hydrolytic enzymes during imbibition and sequentially providing nutrients to an active growing embryo, the improvement of water absorption can therefore result in faster germination and earlier seedling vigour (Chen et al., 2016; Ji et al., 2016; Guo et al., 2017; Sadhu et al., 2017). Furthermore, under stressful circumstances, plasma-induced ROS, RNS, and UV can improve seed germination and seedling establishment as follows: 1) lessen pathogenic microbial infection in developing or germinating plants (Zahoranová et al., 2016; Ono et al., 2017; Lee et al., 2019); 2) control ROS homeostasis by controlling plant antioxidant machinery (Chen et al., 2016; Ji et al., 2016; Guo et al., 2017; Li et al., 2017; Zhang et al., 2017); and (Iranbakhsh et al., 2017; Iranbakhsh et al., 2018). Numerous studies have demonstrated that exogenous ROS, RNS, or UV radiation applied to seeds has similar effects that increase their ability

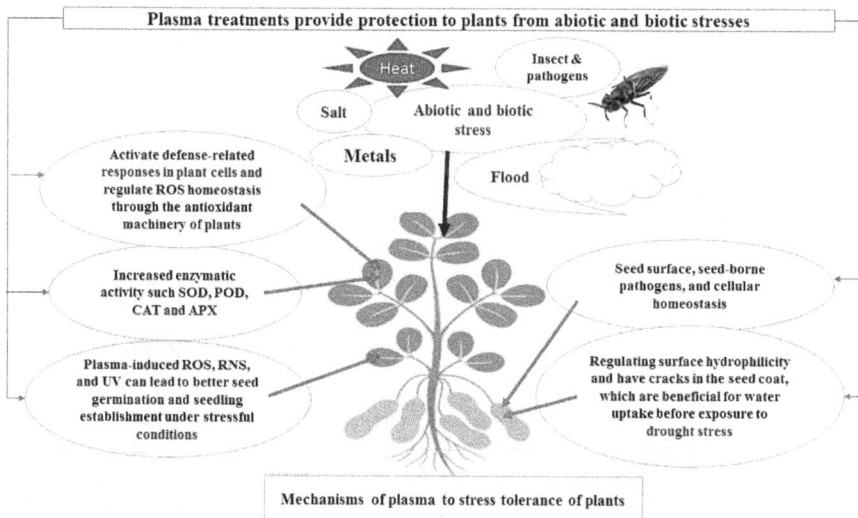

Plasma treatments provide protection to plants from abiotic and biotic stresses

Heat

Insect & pathogens

Salt

Abiotic and biotic stress

Metals

Activate defense-related responses in plant cells and regulate ROS homeostasis through the antioxidant machinery of plants

Flood

Increased enzymatic activity such SOD, POD, CAT and APX

Seed surface, seed-borne pathogens, and cellular homeostasis

Plasma-induced ROS, RNS, and UV can lead to better seed germination and seedling establishment under stressful conditions

Regulating surface hydrophilicity and have cracks in the seed coat, which are beneficial for water uptake before exposure to drought stress

Mechanisms of plasma to stress tolerance of plants

FIGURE 13.3 The mechanism of plasma in plants under stress

to withstand stress before stress occurrences (Antoniou et al., 2016; Thomas and Puthur, 2017). Together, plasma-induced ROS, RNS, and UV could increase crop tolerance to stress by influencing surface hydrophilicity, microbial infection, and many cellular processes (Figure 13.3).

FUTURE PROSPECTS

Under stressful conditions, seed invigoration and seedling establishment are extremely variable depending on a variety of elements, which includes biological viz. crop species, cultivar, and growth and developmental phase and environmental elements including timing, duration, and intensity of the exposure to the stressor). There is little to no information available for developing crop stress management practices utilising plasma technology. To our information, there are hardly any studies on crop tolerance to disease (Jiang et al., 2014b; Ochi et al., 2017), salinity (Iranbakhsh et al., 2018), drought (Li et al., 2015; Guo et al., 2017; Feng et al., 2018), and oxidative stress (Bußler et al., 2015; Iranbakhsh et al., 2017) using plasma technology under laboratory conditions. As a result, more research is needed to improve the effectiveness of plasma treatment on crop tolerance to a wide range of stressors. Flooding, for example, is extremely damaging to upland crops (Arduini et al., 2019; Arduini et al., 2020). As a result, as demonstrated by a similar seed treatment using a magnetic field, plasma treatment could be a useful tool in alleviating flooding stress (Balakhnina et al., 2015). Future research should look into plasma-induced stress tolerance in the laboratory, greenhouse, and field. The growth of a crop is influenced not only by environmental or biological elements, but also by culture (viz, irrigation, tillage, and fertilisation) and plasma operating elements (viz. plasma type, exposure time, and power). Additional research is needed to identify the physiological, biochemical, and molecular mechanisms of stress tolerance in seeds or plants treated with plasma. More information about the embryo (which is a young plant) within the seed is needed to better understand the mechanisms of plant responses after plasma exposure. More specifically, a better understanding of epigenetic changes and longlasting plasma effects on the entire plant life cycle without mutations is required. We may eventually be able to learn enough about plasma-seed priming to make it a viable seed-processing technology. Plasma priming treatments will hopefully be another useful agricultural technology.

CONCLUSION

Plasma applications in agriculture are just getting started on a global scale, and it is one of the current hotspots in agro-research. Environmental stress such as heat, cold, drought, salinity, and nutrient stress are the main contributors to agricultural loss. For the minimisation and control of abiotic and biotic stress on plants, various technologies are used in the agricultural field and also various abatements are applied such as nutrient management technology, nanotechnology, and biofertification. Plasma technology is one of the important applications for minimising the environmental impacts on plants and also controlling the paste attack on seed and whole plant parts. Nowadays plasma technologies are used for seed treatments, control of pathogen attacks, and enhancement of defensive mechanisms in plants such as enzymatic and non-enzymatic activities. These are the basic tools of plant growth and development. Therefore the positive impacts of plasma on seed germination, seedling growth, pathogen deactivation, tolerance and productivity of crops may mitigate the food loss. Despite that to understand and usefulness of plasma treatments on plants under various stress-prone areas and varieties of crops, more research is required.

REFERENCES

Adhikari, B., Adhikari, M., and Park, G. 2020. The effects of plasma on plant growth, development, and sustainability. *Appl. Sci.* 10(17), 6045. doi: 10.3390/app10176045

Agati, G., Azzarello, E., Pollastri, S., and Tattani, M. 2012. Flavonoids as antioxidants in plants: Location and functional significance. *Plant Sci.* 196, 67–76. doi: 10.1016/j.plantsci.2012.07.014

Ahmad, A., Sripong, K., Uthairatanakij, A., Photchanachai, S., Pankasemsuk, T., and Jitareerat, P. 2022. Decontamination of seed borne disease in pepper (Capsicum annuum L.) seed and the enhancement of seed quality by the emulated plasma technology. *Sci. Hortic.* 291, 110568. doi: 10.1016/j.scienta.2021.110568

Ahn, C., Gill, J. and Ruzic, D.N., 2019. Growth of plasma-treated corn seeds under realistic conditions. *Scientific reports*, 9(1), p.4355.

Amaike, S., and Keller, N. P. 2011. Aspergillus flavus. *Annu. Rev. Phytopathol.* 49, 107–133. doi: 10.1146/annurev-phyto-072910-095221

Ambrico, P. F., Šimek, M., Morano, M., De Miccolis Angelini, R. M., Minafra, A., Trotti, P., et al. 2017. Reduction of microbial contamination and improvement of germination of sweet basil (Ocimum basilicum L.) seeds via surface dielectric barrier discharge. *J. Phys. D: Appl. Phys.* 50(30), 305401. doi: 10.1088/1361-6463/aa77c8

Anderson, R., Bayer, P. E., and Edwards, D. 2019. Climate change and the need for agricultural adaptation. *Curr. Opin. Plant Biol.* 13, 1–6.

Antoniou, C., Savvides, A., Christou, A., and Fotopoulos, V. 2016. Unravelling chemical priming machinery in plants: The role of reactive oxygen-nitrogensulfurspecies in abiotic stress tolerance enhancement. *Curr. Opin. Plant Biol.* 33, 101–107. doi: 10.1016/j.pbi.2016.06.020

Araújo, S. S., Paparella, S., Dondi, D., Bentivoglio, A., Carbonera, D., and Balestrazzi, A. 2016. Physical methods for seed invigoration: Advantages and challenges in seed technology. *Front. Plant Sci.* 7, 646. doi: 10.3389/fpls.2016.00646

Arduini, I., Kokubun, M., Shao, G., and Licausi, F. (Ed.) 2020 *Crop Response to Waterlogging*. Lausanne: Frontiers Media SA. doi: 10.3389/978-2-88963-366-1

Arduini, I., Baldanzi, M., and Pampana, S. 2019. Reduced growth and nitrogen uptake during waterlogging at tillering permanently affect yield components in late sown oats. *Front. Plant Sci.* 10, 1087. doi: 10.3389/fpls.2019.01087

Arora, N. K. 2019. Impact of climate change on agriculture production and its sustainable solutions. *Environ. Sustain.* 2(2), 95–96.

Bebber, D.P., Ramotowski, M.A. and Gurr, S.J., 2013. Crop pests and pathogens move polewards in a warming world. *Nature climate change*, 3(11), pp.985–988.

Bafoil, M., Le Ru, A., Merbahi, N., Eichwald, O., Dunand, C., and Yousfi, M. 2019. New insights of low-temperature plasma effects on germination of three genotypes of Arabidopsis thaliana seeds under osmotic and saline stresses. *Sci. Rep.* 9(1), 8649.

Balakhnina, T., Bulak, P., Nosalewicz, M., Pietruszewski, S., and Włodarczyk, T. 2015. The influence of wheat Triticum aestivum L. seed pre-sowing treatment with magnetic fields on germination, seedling growth, and antioxidant potential under optimal soil watering and flooding. *Acta Physiol. Plant.* 37(3), 59. doi: 10.1007/s11738-015-1802-2

Bormashenko, E., Grynyov, R., Bormashenko, Y., and Drori, E. 2012. Cold radiofrequency plasma treatment modifies wettability and germination speed of plant seeds. *Sci. Rep.* 2, 741–748. doi: 10.1038/srep00741

Bormashenko, E., Shapira, Y., Grynyov, R., Bormashenko, Y., Drori, E., and Drori, E. 2015. Interaction of cold radiofrequency plasma with seeds of beans (Phaseolusvulgaris). *J. Exp. Bot.* 66(13), 4013–4021. doi: 10.1093/jxb/erv206

Laroussi, M. 2020. Cold plasma in medicine and healthcare: The new frontier in low temperature plasma applications. *Front. Phys.* 8, 74. doi: 10.3389/fphy.2020.00074

Bußler, S., Herppich, W. B., Neugart, S., Schreiner, M., Ehlbeck, J., Rohn, S., et al.2015. Impact of cold atmospheric pressure plasma on physiology and flavonol glycoside profile of peas (Pisum sativum 'Salamanca'). *Food Res. Int.* 76, 132–141. doi: 10.1016/j.foodres.2015.03.045

Chaudhary, I.J., and Rathore, D. 2018. Suspended particulate matter deposition and its impact on urban trees. *Atmospheric Pollution Research* 9; 1072–1082.

Chaudhary, I. J., and Rathore, D. 2020. Relative effectiveness of ethylene diurea, phenyl urea, ascorbic acid and urea in preventing groundnut (Arachis hypogaea L) crop from ground level ozone. *Environ. Technol. Innov.* 19, 100963.

Chaudhary, I. J., and Rathore, D. 2021. Assessment of ozone toxicity on cotton (Gossypium hirsutum L.) cultivars: Its defensive system and intraspecific sensitivity. *Plant Physiol. Biochem.* 166, 912–927. doi: 10.1016/j.plaphy.2021.06.054

Chaudhary, I. J., and Singh, V. 2020. Titanium dioxide nanoparticles and its impact on growth, biomass and yield of agricultural crops under environmental stress: A review. *Res. J. Nanosci. Nanotechnol.* doi: 10.3923/rjnn.2019

Chen, H. H., Chang, H. C., Chen, Y. K., Hung, C. L., Lin, S. Y., and Chen, Y. S. 2016. An improved process for high nutrition of germinated brown rice production: Low-pressure plasma. *Food Chem.* 191, 120–127. doi: 10.1016/j.foodchem.2015.01.083

Dobrin, D., Magureanu, M., Mandache, N. B., and Ionita, M. D. 2015. The effect of non thermal plasma treatment on wheat germination and early growth. *Innov. Food Sci. Emerg. Technol.* 29, 255–260. doi: 10.1016/j.ifset.2015.02.006

Ekezie, F. G. C., Sun, D. W., and Cheng, J. H. 2017. A review on recent advances in cold plasma technology for the food industry: Current applications and future trends. *Trends Food Sci. Technol.* 69, 46–58. doi: 10.1016/j.tifs.2017.08.007

Eljounaidi, K., Lee, S. K., and Bae, H. 2016. Bacterial endophytes as potential biocontrol agents of vascular wilt diseases – Review and future prospects. *Biol. Control* 103, 62–68. doi: 10.1016/j.biocontrol.2016.07.013

Fan, X., and Song, Y. 2020. Advanced oxidation process as a postharvest decontamination technology to improve microbial safety of fresh produce. *J. Agric. Food Chem..* doi: 10.1021/acs.jafc.0c01381

Feng, J., Wang, D., Shao, C., Zhang, L., and Tang, X. 2018. Effects of cold plasma treatment on a seed growth under simulated drought stress. *Plasma Sci. Technol.* 20(3), 035505. doi: 10.1088/2058-6272/aa9b27

Filatova, I., Lyushkevich, V., Goncharik, S., Zhukovsky, A., Krupenko, N., and Kalatskaja, J. 2020. The effect of low-pressure plasma treatment of seeds on the plant resistance to pathogens and crop yields. *J. Phys. D: Appl. Phys.* 53(24), 244001

Filatova, I. I., Azharonok, V. V., Goncharik, S. V., Lushkevich, V. A., Zhukovsky, A. G., and Gadzhieva, G. I. 2014. Effect of RF plasma treatment on the germination and phytosanitary state of seeds. *J. Appl. Spectrosc.* 81(2), 250–256. doi: 10.1007/s10812-014-9918-5

Filatova, I. I., Azharonok, V., Lushkevich, V., Zhukovsky, A., Gadzhieva, G., Spaisić, K. et al. 2013. "Plasma seeds treatment as a promising technique for seed germination improvement." In *Proceedings of the 31st International Conference on Phenomena in Ionized Gases* (Granada, Spain: ICPIG) (International Conference on Phenomena in Ionized Gases)). pp. 4–7.

Gierczik, K., Vukušić, T., Kovács, L., Székely, A., Szalai, G., Milošević, S., Kocsy, G., Kutasi, K., and Galiba, G. 2020. Plasma-activated water to improve the stress tolerance of barley. *Plasma Process Polym.* 17(3), e1900123.

Gill, S. S., and Tuteja, N. 2010. Reactive oxygen species and antioxidant machinery in abiotic stress tolerance in crop plants. *Plant Physiol. Biochem.* 48(12), 909–930. doi: 10.1016/j.plaphy.2010.08.016

Gómez-Ramírez, A., López-Santos, C., Cantos, M., García, J. L., Molina, R., Cotrino, J.; Espinós, J.P., and González-Elipe, A. R. 2017. Surface chemistry and germination improvement of Quinoa seeds subjected to plasma activation. *Sci. Rep.* 7(1), 5924.

Guo, Q., Meng, Y., Qu, G., Wang, T., Yang, F., Liang, D., et al. 2018. Improvement of wheat seed vitality by dielectric barrier discharge plasma treatment. *Bioelectromagnetics* 39(2), 120–131. doi: 10.1002/bem.22088

Guo, Q., Wang, Y., Zhang, H., Qu, G., Wang, T., Sun, Q., et al. 2017. Alleviation of adverse effects of drought stress on wheat seed germination using atmospheric dielectric barrier discharge plasma treatment. *Sci. Rep.* 7(1), 16680. doi: 10.1038/s41598-017-16944-8

Gupta, P. K. (2020a). Pollution load on Indian soil-water systems and associated health hazards: a review. *Journal of Environmental Engineering, 146*(5), 03120004. https://doi.org/10.1061/(ASCE)EE.1943-7870.0001693.

Gupta, P. K. (2020b). Fate, transport, and bioremediation of biodiesel and blended biodiesel in subsurface environment: a review. *Journal of Environmental Engineering, 146*(1), 03119001. https://doi.org/10.1061/(ASCE)EE.1943-7870.0001619.

Gupta, P. K., Gharedaghloo, B., Lynch, M., Cheng, J., Strack, M., Charles, T. C., & Price, J. S. (2020). Dynamics of microbial populations and diversity in NAPL contaminated peat soil under varying water table conditions. *Environmental Research, 191*, 110167. https://doi.org/10.1016/j.envres.2020.110167.

Gupta, P. K., Mustapha, H. I., Singh, B., & Sharma, Y. C. (2022). Bioremediation of petroleum contaminated soil-water resources using neat biodiesel: A review. *Sustainable Energy Technologies and Assessments, 53*, 102703. https://doi.org/10.1016/j.seta.2022.102703.

Henselová, M., Slováková, Ľ., Martinka, M., and Zahoranová, A. 2012. Growth, anatomy and enzyme activity changes in maize roots induced by treatment of seeds with low-temperature plasma. *Biologia* 67(3), 490–497. doi: 10.2478/s11756-012-0046-5

Huijser, P., and Schmid, M. 2011. The control of developmental phase transitions in plants. *Development.* 138(19), 4117–4129.

Iranbakhsh, A., Ardebili, N. O., Ardebili, Z. O., Shafaati, M., and Ghoranneviss, M. 2018. Nonthermal plasma induced expression of heat shock factor A4A and improved wheat (Triticum aestivum L.) growth and resistance against salt stress. *Plasma Chem. Plasma Process.* 38(1), 29–44. doi: 10.1007/s11090-017-9861-3

Iranbakhsh, A., Ghoranneviss, M., Ardebili, Z. O., Ardebili, N. O., Tackallou, S. H., and Nikmaram, H. 2017. Nonthermal plasma modified growth and physiology in Triticum aestivum via generated signaling molecules and UV radiation. *Biol. Plant.* 61, 702–708. doi: 10.1007/s10535-016-0699-y

Ito, M., Oh, J. S., Ohta, T., Shiratani, M., and Hori, M. 2018. Current status and future projects of agricultural applications using atmospheric-pressure plasmatechnologies. *Plasma Process Polym.* 15, e1700073. doi: 10.1002/ppap.201700073

Ji, S. H., Choi, K. H., Pengkit, A., Im, J. S., Kim, J. S., Kim, Y. H., et al. 2016. Effects of high voltage nanosecond pulsed plasma and micro DBD plasma onseed germination, growth development and physiological activities in spinach. *Arch. Biochem. Biophys.* 605, 117–128. doi: 10.1016/j.abb.2016.02.028

Ji, S. H., Kim, T., Panngom, K., Hong, Y. J., Pengkit, A., Park, D. H., et al. 2015. Assessment of the effects of nitrogen plasma and plasma-generated nitric oxideon early development of coriandum sativum. *Plasma Process Polym.* 12(10), 1164–1173. doi: 10.1002/ppap.201500021

Jiang, J., He, X., Li, L., Li, J., Shao, H., Xu, Q., et al. 2014a. Effect of cold plasma treatment on seed germination and growth of wheat. *Plasma Sci. Technol.* 16(1), 54–58. doi: 10.1088/1009-0630/16/1/12

Jiang, J., Lu, Y., Li, J., Li, L., He, X., Shao, H., et al. 2014b. Effect of seed treatment by cold plasma on the resistance of tomato to Ralstonia solanacearum (bacterial wilt). *PLOS ONE* 9(5), e97753. doi: 10.1371/journal.pone.0097753

Jisha, K. C., Vijayakumari, K., and Puthur, J. T. 2013. Seed priming for abiotic stress tolerance: An overview. *Acta Physiol. Plant.* 35(5), 1381–1396. doi: 10.1007/ s11738-012-1186-5

Jo, Y. K., Cho, J., Tsai, T. C., Staack, D., Kang, M. H., Roh, J. H., et al. 2014. A nonthermal plasma-seed treatment method for management of a seed-borne fungal pathogen on rice seed. *Crop Sci.* 54(2), 796–803. doi: 10.2135/cropsci2013. 05.0331

Judée, F., Simon, S., Bailly, C., and Dufour, T. 2018. Plasma-activation of tap water using DBD for agronomy applications: Identification and quantification of long lifetime chemical species and production/consumption mechanisms. *Water Res.* 133, 47–59.

Kang, M. H., Pengkit, A., Choi, K., Jeon, S. S., Choi, H. W., Shin, D. B., et al. 2015. Differential inactivation of Fungal spores in water and on seeds by ozone and arc discharge plasma. *PLOS ONE* 10(9), e0139263. doi: 10.1371/journal.pone.0139263

Kang, M. H., Jeon, S. S., Shin, S. M., Veerana, M., Ji, S.-H., Uhm, H.-S., Choi, E.H., Shin, J.H., and Park, G. 2019. Dynamics of nitric oxide level in liquids treated with microwave plasma-generated gas and their effects on spinach development. *Sci. Rep.* 9(1), 1011.

Khamsen, N., Onwimol, D., Teerakawanich, N., Dechanupaprittha, S.,Kanokbannakorn, W., Hongesombut, K., et al. 2016. Rice (Oryza sativa L.)seed sterilization and germination enhancement via atmospheric hybridnonthermal discharge plasma. *ACS Appl. Mater. Interfaces* 8(30), 19268–19275. doi: 10.1021/acsami.6b04555

Khan, M. R., Fischer, S., Egan, D., and Doohan, F. M. 2006. Biological control of Fusarium seedling blight disease of wheat and barley. *Phytopathology* 96(4), 386–394. doi: 10.1094/PHYTO-96-0386

Krapivina, S. A., Filippov, A. K., Levitskaya, T. N., and Bakhvalov, A. 1994. Gas plasma treatment of plant seeds. U.S. Patent No. US5281315A, 25 January 1994

Yousfi, M., Merbahi, N., Pathak, A., and Eichwald, O. 2014. Low-temperature plasmas at atmospheric pressure: Toward new pharmaceutical treatments in medicine. *Fundam. Clin. Pharmacol.* Apr 28(2), 123–135. doi: 10.1111/fcp.12018. Epub 2013 Feb 25. PMID: 23432667

Lee, E. J., Khan, M. S. I., Shim, J., and Kim, Y. J. 2019. Roles of oxides of nitrogen on quality enhancement of soybean sprout during hydroponic production using plasma discharged water recycling technology. *Sci. Rep.* 8(1), 16872. doi: 10.1038/s41598-018-35385-5

Li, L., Jiang, J., Li, J., Shen, M., He, X., Shao, H., et al. 2014. Effects of cold plasma treatment on seed germination and seedling growth of soybean. *Sci. Rep.* 4, 5859–5865. doi: 10.1038/srep05859

Li, L., Li, J., Shen, M., Zhang, C., and Dong, Y. 2015. Cold plasma treatment enhances oilseed rape seed germination under drought stress. *Sci. Rep.* 5, 13033. doi: 10.1038/srep13033

Li, Y., Wang, T., Meng, Y., Qu, G., Sun, Q., Liang, D., et al. 2017. Air atmospheric dielectric barrier discharge plasma induced germination and growth enhancement of wheat seed. *Plasma Chem. Plasma Process.* 37(6), 1621–1634.doi: 10.1007/s11090-017-9835-5

Liu, Y., Sun, Y., Wang, Y., Zhao, Y., Duan, M., Wang, H., Dai, R., Liu, Y., Li, X. and Jia, F., 2023. Inactivation mechanisms of atmospheric pressure plasma jet on Bacillus cereus spores and its application on low-water activity foods. *Food Research International*, 169, p.112867.

Lindsay, A., Byrns, B., King, W., Andhvarapou, A., Fields, J., Knappe, D.; Fonteno W., and Shannon, S. 2014. Fertilization of radishes, tomatoes, and marigolds using a large-volume atmospheric glow discharge. *Plasma Chem. Plasma Process.* 34(6), 1271–1290.

Ling, L., Jiafeng, J., Jiangang, L. et al. 2014. Effects of cold plasma treatment on seed germination and seedling growth of soybean. *Sci. Rep.* 4, 5859. doi: 10.1038/srep05859

Mahajan, M., Gupta, P. K., Singh, A., Vaish, B., Singh, P., Kothari, R., & Singh, R. P. (2022). A comprehensive study on aquatic chemistry, health risk and remediation techniques of cadmium in groundwater. *Science of The Total Environment*, *818*, 151784. https://doi.org/10.1016/j.scitotenv.2021.151784.

Mihai, A. L., Dobrin, D., Magureanu, M., and Popa, M. E. 2014. Positive effect of non-thermal plasma treatment on radish seed. *Rom. Rep. Phys.* 66, 1110–1117.

Mittler, R. 2002. Oxidative stress, antioxidants, and stress tolerance. *Trends Plant Sci.* 7(9), 405–410. doi: 10.1016/S1360-1385(02)02312-9

Mittler, R., Vanderauwera, S., Gollery, M., and Van Breusegem, F. 2004. Reactive oxygen gene network of plants. *Trends Plant Sci.* 9(10), 490–498. doi: 10.1016/ j.tplants.2004.08.009

Nakano, R. T., Aijima, K. R., and Hayashi, N. 2016. Effect of oxygen plasma irradiation on gene expression in plant seeds induced by active oxygen species. *Plasma Med.* 6(3–4), 303–313.

Nigam, B., Dubey, R. S., and Rathore, D. 2022. Protective role of exogenously supplied salicylic acid and PGPB (Stenotrophomonas sp.) on spinach and soybean cultivars grown under salt stress. *Sci. Hortic.* 293(5), 110654. doi: 10.1016/j.scienta.2021.110654

Nishioka, T., Takai, Y., Mishima, T., Kawaradani, M., Tanimoto, H., Okada, K., Misawa, T., and Kusakari, S. 2016. Low-pressure plasma application for the inactivation of the seed-borne pathogen Xanthomonas campestris. *Biocontrol Sci.* 21(1), 37–43.

Ochi, A., Konishi, H., Ando, S., Sato, K., Yokoyama, K., Tsushima, S., et al. 2017. Management of bakanae and bacterial seedling blight diseases in nurseries by irradiating rice seeds with atmospheric plasma. *Plant Pathol.* 66(1), 67–76. doi: 10.1111/ppa.12555

Ono, R., Uchida, S., Hayashi, N., Kosaka, R., and Soeda, Y. 2017. Inactivation of bacteria on plant seed surface by low-pressure RF plasma using a vibrating stirring device. *Vacuum* 136, 214–220. doi: 10.1016/j.vacuum.2016.07.017

Palencia, E. R., Hinton, D. M., and Bacon, C. W. 2010. The black Aspergillus species of maize and peanuts and their potential for mycotoxin production. *Toxins* 2(4), 399–416. doi: 10.3390/toxins2040399

Pandya, S., Gadekallu, T. R., Maddikunta, P. K. R., and Sharma, R. 2022. A study of the impacts of air Pollution on the agricultural community and Yield crops (Indian context). *Sustainability* 14(20), 13098.doi: 10.3390/su142013098

Park, Y., Oh, K. S., Oh, J., Seok, D. C., Kim, S. B., Yoo, S. J., et al. 2018. The biological effects of surface dielectric barrier discharge on seed germination and plant growth with barley. *Plasma Process Polym.* 15, e1600056. doi: 10.1002/ppap.201600056

Pathak, T., Maskey, M. L., Dahlberg, J. A., Kearns, F., Bali, K. M., and Zaccaria, D. 2018. Climate change trends and impacts on California agriculture: A detailed review. *Agronomy* 8(3), 25.

Pauzaite, G., Malakauskiene, A., Nauciene, Z., Zukiene, R., Filatova, I., Lyushkevich, V., et al. 2018. Changes in Norway spruce germination and growth induced by pre-sowing seed treatment with cold plasma and electromagnetic field: Short-term versus long-term effects. *Plasma Proc. Polym.* 15(2), 1700068. doi: 10.1002/ppap.201700068

Priatama, R. A., Pervitasari, A. N., Park, S., Park, S. J., and Lee, Y. K. 2022. Current advancements in the molecular mechanism of plasma treatment for seed germination and plant growth. *Int. J. Mol. Sci.* 23(9), 4609. doi: 10.3390/ijms23094609

Rahman, M. M., Sajib, S. A., Rahi, M. S., Tahura, S., Roy, N. C., Parvez, S.; Reza, M.A., Talukder, M.R., and Kabir, A.H. 2018. Mechanisms and signaling associated with LPDBD plasma mediated growth improvement in wheat. *Sci. Rep.* 8(1), 10498.

Randeniya, L. K., and de Groot, G. J. J. B. 2015. Non-Thermal plasma treatment of agricultural seeds for stimulation of germination, removal of surface contamination and other benefits: A review. *Plasma Proc. Polym.* 12(7), 608–623. doi: 10.1002/ppap.201500042

Ranieri, P., Sponsel, N., Kizer, J., Rojas-Pierce, M., Hernández, R., Gatiboni, L., and Stapelmann, K. 2021. Plasma agriculture: Review from the perspective of the plant and its ecosystem. *Plasma Processes Polym..* doi: 10.1002/ppap.202000162

Rathore, D., and Chaudhary, I. J. 2019. Ozone risk assessment of castor (Ricinus communis L.) cultivars using open top chamber and ethylenediurea (EDU). *Environ. Pollut.* 244, 257–269.

. Rathore, D., and Chaudhary, I. J. 2021. Effects of tropospheric ozone on groundnut (Arachis hypogea L.) cultivars: Role of plant age and antioxidative potential. *Atmos. Pollut. Res.* 12(3), 334–348. ISSN: 1309-1042. doi: 10.1016/j.apr.2021.01.005

Ráduly, Z., Szabó, L., Madar, A., Pócsi, I., and Csernoch, L. 2020. Toxicological and medical aspects of Aspergillus-derived mycotoxins entering the feed and food chain. *Front. Microbiol.* 10, 2908. doi: 10.3389/fmicb.2019.02908. PMID: 31998250. PMCID: PMC6962185

Roy, N. C., Hasan, M. M., Talukder, M. R., Hossain, M. D., and Chowdhury, A. N. 2018. Prospective applications of low frequency glow discharge plasmas on enhanced germination, growth and yield of wheat. *Plasma Chem. Plasma Process.* 38(1), 13–28. doi: 10.1007/s11090-017-9855-1

Saberi, M., Modarres-Sanavy, S. A. M., Zare, R., and Ghomi, H. 2020. Improvement of photosynthesis and photosynthetic productivity of winter wheat by cold plasma treatment under haze condition. *J. Agric. Sci. Technol.* 21, 1889–1904.

Sadhu, S., Thirumdas, R., Deshmukh, R. R., and Annapure, U. S. 2017. Influence of cold plasma on the enzymatic activity in germinating mung beans (Vignaradiate). *LWT Food Sci. Technol.* 78, 97–104. doi: 10.1016/j.lwt.2016.12.026

Scherm, B., Balmas, V., Spanu, F., Pani, G., Delogu, G., Pasquali, M., et al. 2013. Fusarium culmorum: Causal agent of foot and root rot and head blight onwheat. *Mol. Plant Pathol.* 14(4), 323–341. doi: 10.1111/mpp.12011

Sehrawat, R., Thakur, A. K., Vikram, A., Vaid, A., and Rane, R. 2017. Effect of cold plasma treatment on physiological quality of okra seed. *J. Hill Agric.* 8(1), 66–71. doi: 10.5958/2230-7338.2017.00010.6

Selcuk, M., Oksuz, L., and Basaran, P. 2008. Decontamination of grains and legumes infected with Aspergillus spp. and Penicillum spp. by cold plasma treatment. *Biores. Technol.* 99, 5104–5109. doi: 10.1016/j.biortech.2007.09.076

Šerá, B., Špatenka, P., Šerý, M., Vrchotova, N., and Hruskova, I. 2010. Influence of plasma treatment on wheat and oat germination and early growth. *IEEE Trans. Plasma Sci.* 38(10), 2963–2967.

Sett, R. 2017. Responses in plants exposed to dust pollution. *Hortic. Int. J.* 1(2), 53–56.

Shao, C. Y., Wang, D., Tang, X., Zhao, L., and Li, Y. 2013. Stimulating effects of magnetized arc plasma of different intensities on the germination of old spinach seeds. *Math. Comput. Modell.* 58(3–4), 814–818. doi: 10.1016/j.mcm.2012.12.022

Sharma, P., Jha, A. B., Dubey, R. S., and Pessarakli, M. 2012. Reactive oxygen species, oxidative damage, and antioxidative defense mechanism in plants under stressful conditions. *J. Bot.*, 217037. doi: 10.1155/2012/217037

Sharma, R., Pandey, S. T., Verma, O., Srivastava, R. C., and Guru, S. K. 2020. Physiological seedling vigour parameters of wheat as influenced by different seed invigoration techniques. *Int. J. Chem. Stud.* 8(1), 1549–1552.

Singh, R. P., Mahajan, M., Gandhi, K., Gupta, P. K., Singh, A., Singh, P., ... & Kidwai, M. K. (2023). A holistic review on trend, occurrence, factors affecting pesticide concentration, and ecological risk assessment. *Environmental Monitoring and Assessment*, 195(4), 451. https://doi.org/10.1007/s10661-023-11005-2.

Song, J. S., Lee, M. J., Ra, J. E., Lee, K. S., Eom, S., Ham, H. M., et al. 2020. Growth and bioactive phytochemicals in barley (Hordeum vulgare L.) sprouts affected by atmospheric pressure plasma during seed germination. *J. Phys. D: Appl. Phys.* 53(31), 314002. doi: 10.1088/1361-6463/ab810d

Stolárik, T., Henselová, M., Martinka, M., Novák, O., Zahoranová, A., and Cernák, M. 2015. Effect of low-temperature plasma on the structure of seeds, growth, and metabolism of endogenous phytohormones in pea (Pisum sativum L.). *Plasma Chem. Plasma Process.* 35(4), 659–676. doi: 10.1007/s11090-015-9627-8

Sun, F., DAI, Y., and Yu, X. 2017. Air pollution, food production and food security: A review from the perspective of food system. *J. Integr. Agric.* 16(12), 2945–2962. doi: 10.1016/S2095-3119 (17)61814-8

Thomas, T. T. D., and Puthur, J. T. 2017. UV radiation priming: A means of amplifying the inherent potential for abiotic stress tolerance in crop plants. *Environ. Exp. Bot.* 138, 57–66. doi: 10.1016/j.envexpbot.2017.03.003

Thomma, P. H. J. 2003. Alternaria spp.: from general saprophyte to specific parasite. *Mol. Plant Pathol.* 4, 225–236. doi: 10.1046/j.1364-3703.2003.00173.xl

Velásquez, A. C., Castroverde, C. D. M., and He, S. Y. 2018. Plant–pathogen warfare under changing climate conditions. *Curr. Biol.* 28(10), R619–R634.

Velichko, I., Gordeev, I., Shelemin, A. et al. 2019. Plasma jet and dielectric barrier discharge treatment of wheat seeds. *Plasma Chem. Plasma Process.* 39(4), 913–928. doi: 10.1007/s11090-019-09991-8

Volin, J. C., Ferencz, S. D., Raymond, A. Y., and Park, S. M. T. 2000. Modification of seed germination performance through cold plasma chemistry technology. *Crop Sci.* 40(6), 1706–1718. doi: 10.2135/cropsci2000.4061706x

Volkov, A. G., Hairston, J. S., Patel, D., Gott, R. P., and Xu, K. G. 2019. Cold plasma poration and corrugation of pumpkin seed coats. *Bioelectrochemistry* 128, 175–185.

Wang, X. Q., Zhou, R. W., de Groot, G., Bazaka, K., Murphy, A. B., and Ostrikov, K. K. 2017. Spectral characteristics of cotton seeds treated by a dielectric barrier discharge plasma. *Sci. Rep.* 7(1), 5601. doi: 10.1038/s41598-017-04963-4

Waskow, A., Howling, A., and Furno, I. 2021. Mechanisms of plasma-seed treatments as a potential seed processing technology. *Front. Phys.* 9, 617345. doi: 10.3389/fphy.2021.617345

Yatoo, A.M., Ali, M.N., Zaheen, Z., Baba, Z.A., Ali, S., Rasool, S., Sheikh, T.A., Sillanpää, M., Gupta, P.K., Hamid, B. and Hamid, B., (2022). Assessment of pesticide toxicity on earthworms using multiple biomarkers: a review. *Environmental Chemistry Letters*, 20(4), 2573–2596. https://doi.org/10.1007/s10311-022-01386-0.

Yin, M., Huang, M., Ma, B., and Ma, T. 2005. Stimulating effects of seed treatment by magnetized plasma on tomato growth and yield. *Plasma Sci. Technol.* 7(6), 3143–3147. doi: 10.1088/1009-0630/7/6/017

Zahoranová, A., Henselova´, M., Hudecova´, D., Kalinˇaˊkova´, B., Kovaˊcˇik, D., Medvecka, V., et al. 2016. Effect of cold atmospheric pressure plasma on the wheat seedlings vigor and on the inactivation of microorganisms on the seeds surface. *Plasma Chem. Plasma Process.* 36, 397–414. doi: 10.1007/s11090-015-9684-z

Zhang, J. J., Jo, J. O., Huynh, D. L., Mongre, R. K., Ghosh, M., Singh, A. K., et al. 2017. Growth-inducing effects of argon plasma on soybean sprouts via the regulation of demethylation levels of energy metabolism-related genes. *Sci. Rep.* 7, 41917. doi: 10.1038/srep41917

Zhou, R., Zhou, R., Zhang, X., Zhuang, J., Yang, S., Bazaka, K., et al. 2016. Effects of atmospheric-pressure N2, He, air, and O2 micro plasmas on mung bean seed germination and seedling growth. *Sci. Rep.* 6, 32603. doi: 10.1038/srep32603

Zhou, S., Sekizaki, H., Yang, Z., Sawa, S., and Pan, J. 2010. Phenolics in the seed coat of wild soybean (Glycine soja) and their significance for seed hardness and seed germination. *J. Agric. Food Chem.* 58(20), 10972–10978. doi: 10.1021/jf102694k

Zhou, Z., Huang, Y., Yang, S., and Chen, W. 2011. Introduction of a new atmospheric pressure plasma device and application on tomato seeds. *Agric. Sci.* 2(1), 23–27.

14 Application of Metaheuristic Optimizations for Unconfined Aquifer Parameter Estimation to Improve the Irrigation Water Management

Sharad Patel and Uttam Puri Goswami

INTRODUCTION

Aquifer parameters such as hydraulic conductivity, specific yield, and storativity can provide valuable information on the aquifer's characteristics and help to develop effective strategies for irrigation water management. The estimation of these parameters can help to improve irrigation water management by optimizing well placement, reducing water waste, promoting groundwater recharge, and protecting groundwater quality (Dillon et al., 2006; Mahajan et al., 2022; Yatoo et al., 2022; Singh et al., 2023). With accurate estimates of these parameters, farmers can optimize the use of groundwater resources and maximize crop yields (Guzman et al., 1999). In general, an unconfined aquifer is a dependable source of groundwater for irrigation, which can enhance crop quality and productivity (Raghunath, 1987). This is crucial in regions with scarce or erratic rainfall as well as those with few or no surface water resources. To improve the management of such aquifers, it is necessary to accurately estimate them. For example, by understanding the hydraulic conductivity of an aquifer, farmers can determine the rate at which water can be extracted from the aquifer and ensure that water is pumped at a sustainable rate, preventing the depletion of the aquifer. Similarly, the specific yield and storativity of the aquifer can help farmers determine how much water is available for irrigation and how long the water will last during a drought or low rainfall period. In addition to these benefits,

DOI: 10.1201/9781003441175-14

aquifer parameter estimation can also be used to develop groundwater models that can simulate the behavior of the aquifer under different conditions (Blin et al., 2022). These models can be used to predict how changes in pumping rates or weather patterns will affect groundwater resources and help farmers make informed decisions about water management. The pumping test and graphical matching techniques of parameter estimation are based on closed-form solutions and are typically limited to homogeneous and isotropic aquifer domains. Therefore, an alternative mathematical approach, inverse groundwater modeling, to estimate aquifer parameters can be considered a valuable tool for improving irrigation water management in the field of agriculture (Romero et al., 2012).

Inverse groundwater modeling (IGM) using a simulation–optimization (SO) approach is a method for estimating the hydraulic properties of an aquifer from observed hydraulic data (Gupta et al. 2023). This approach uses a numerical groundwater flow model to simulate the hydraulic behavior of the aquifer, and then optimizing the model parameters to match observed hydraulic head data. A SO approach typically follows three steps, i.e., hydraulic modeling, parameter estimation, and model validation. In the first step, a numerical groundwater flow model is developed to simulate the hydraulic behavior of the aquifer. This model requires the hydrogeological properties of the aquifer, such as hydraulic conductivity, specific yield, and porosity, and the boundary conditions, such as recharge and discharge rates. The finite difference method (FDM) based MODFLOW (McDonald & Harbaugh, 2003) is a popular example of such a model. In the second step, model parameters are optimized to match observed hydraulic data. This is typically done using an optimization algorithm, which adjusts the model parameters to minimize the difference between the simulated and observed hydraulic data. In the final step, optimized model parameters are validated by comparing the simulated hydraulic behavior with new hydraulic data. This is an important step to ensure that the model accurately represents the hydraulic behavior of the aquifer. Among the different steps of the SO approach, the second step related to optimization plays a critical role. Optimization models use mathematical algorithms to search for the optimal combination of parameter values that result in the best match between the observed and simulated hydraulic data. It is highly significant due to its ability to solve complex optimization problems involving multiple variables and constraints. These models can incorporate a wide range of optimization techniques, including genetic algorithms (GA), simulated annealing (SA), particle swarm optimization (PSO), differential evolution (DE), and others, to search for the best solution. By using an optimization model, the simulation–optimization approach can estimate the hydraulic properties of an aquifer more accurately and efficiently than traditional gradient-based optimization methods. This is because optimization models can assimilate multiple sources of data, such as hydraulic data from monitoring wells, geologic data, and topographic data, to inform the estimation of hydraulic parameters. Furthermore, metaheuristic optimization models allow for the incorporation of uncertainty and variability in the input data, which is an essential feature of groundwater modeling. Uncertainty and variability can arise from various sources, including measurement errors, modeling assumptions, and natural variability in the aquifer properties. By incorporating uncertainty and variability, optimization models can provide more realistic and reliable estimates of the hydraulic properties of the aquifer.

Metaheuristic optimizations are powerful tools that can be used for optimizing complex systems. Metaheuristic optimizations provide a way to improve the parameter estimation process by exploring a large search space and finding the optimal parameter values that minimize a given objective function. Certain methods of this class that have been successfully applied to the inverse groundwater problems are: GA (Harrouni et al., 1996; Lakshmi Prasad & Rastogi, 2001), simulated annealing (Zheng & Wang, 1996), PSO (Ch & Mathur, 2012), ant colony optimization (ACO; Abbaspour et al., 2001), among others. The global stochastic population-based optimization due to its strong ability to handle discrete problems replaced conventional numerical optimization methods in the SO approach. Among different notably used stochastic search methods, DE and PSO are prominently used in inverse groundwater problems. DE has a problem of slothful convergence characteristic as it continues to search the solution space based on its crossover characteristic while the PSO restricts the search based on particle velocity bounds, therefore, it may suffer from premature convergence. As a proposed hybridization, DE and PSO are combined together in a series manner and each population that is processed through DE will pass on to PSO for further improvement. The idea to combine these well-tested metaheuristics resulted in a hybrid version of DE and PSO optimization, i.e., DE-PSO. Therefore as per the objective of this chapter a comparative performance of different popularly used metaheuristic optimizations (i.e., DE, PSO, and DE-PSO) are presented by its application on a synthetic problem. This study will be highly useful to researchers and practitioners in the field of groundwater hydrology to select a better alternative as an optimization for obtaining the precise values of aquifer parameters.

METHODOLOGY

The main objective of this study is to explore the ability of different selected optimization algorithms to solve the inverse problem when applied to a synthetic two-dimensional unconfined aquifer. In the proposed approach the real field continuous parameters are assumed to be distributed throughout the domain using parameterization.

GROUNDWATER SIMULATION

According to Willis and Yeh (1987), the groundwater flow regulating equation for an unconfined aquifer under transient conditions, with variability such as anisotropy, non-homogeneity, and areal recharge involving pumping or draft or both, is as follows:

$$\frac{\partial}{\partial x}\left(k_x h \frac{\partial h}{\partial x}\right) + \frac{\partial}{\partial y}\left(k_y h_y \frac{\partial h}{\partial y}\right) = S\left(\frac{\partial h}{\partial t}\right) \pm Q_w\left(x - x_p, y - y_p\right) + R \qquad (14.1)$$

The initial condition can be as follows:

$$h(x, y, 0) = h_0(x, y) \qquad x, y \in \Omega \tag{14.2}$$

The boundary condition can be the constant groundwater head (Dirichlet boundary) and boundary flux (Neumann boundary), which are described as follows:

$$h(x, y, t) = h_1(x, y, t) \qquad x, y \in \partial\Omega_1 \tag{14.3}$$

$$kh\left(\frac{\partial h}{\partial n}\right) \text{ or } \left[k_x h\left(\frac{\partial h}{\partial x}\right)\right] + \left[k_y h\left(\frac{\partial h}{\partial y}\right)\right] \cdot q_2(x, y, t) \qquad x, y \in \partial\Omega_2 \tag{14.4}$$

where $h(x,y,t)$ is the groundwater head (m) at time t; k, hydraulic conductivity (m/day); k_x and k_y are hydraulic conductivities along principal axes (m/day); and S_y, specific yield (dimensionless); Q_w, source or sink term (m/day); (x_p, y_p), coordinate for the well location (m); δ, Dirac delta function with the property that if $x = x_p$ and $y = y_p$ then $\delta = 1$ else $\delta = 0$; R, areal recharge (m/day); t, time (day); h_0, initial known groundwater head distribution (m); h_1, known groundwater head values at the boundary (m); q_2, known boundary flux (m³/day/m); (l_x, l_y), direction cosine of the outward normal at certain node on Neumann boundary (dimensionless); Ω the computational domain; $\partial\Omega$, the boundary $\partial\Omega_1 U \partial\Omega_2 = \partial\Omega$ of computational domain; $(\partial /\partial n)$ the normal derivative.

In this study, Galerkin's Finite Element Method (FEM) based simulation model code is developed using MATLAB® programming. Using this method, the entire aquifer domain is discretized by different triangular elements, which are formed by joining distributed discrete nodes throughout the aquifer. Here the solution is obtained for the given aquifer geometry, initial and boundary conditions, flux across the boundary, rate of recharge, and rate of pumping. However, for approximation of temporal derivative implicit is employed and results in solution of head (h) as:

$$\left\{[G] + \frac{1}{\Delta t}[P]\right\}\{h_I^{t+\Delta t}\} = \frac{1}{\Delta t}[P]\{h_I^t\} + \{F\} \tag{14.5}$$

where $[G]$ is conductance matrix containing transmissivity terms; $[P]$, storage matrix containing storativity terms; Δt the time-step size; $\{F\}$ represents flux vector; and $\{h\}$ the unknown head vector.

OPTIMIZATION ALGORITHMS

The metaheuristic-based global optimization models like DE and PSO are general approximate algorithms which are inspired by the theory of evolution, and social behavior of colony-based species (birds, ants, fish, termites, bees) or other less complex optimization methods and are capable of solving any kind of optimization problem. Therefore the DE, PSO, and DE-PSO are adopted as optimization in this study in their standard forms.

DE Optimization Algorithm

The DE, a type of evolutionary algorithm (EA) was introduced by Storn and Price (1997) which can be considered as an advanced version of GA and more suitable for real-valued function optimization (Price et al., 2005). DE follows the same operators to search optima from candidate solution as GA viz. initialization, mutation, crossover, and selection but with a more effective and rigorous search procedure. It approaches optima by introducing perturbation information from the distance between the vectors from the population itself. The relevant description of different DE operators is elaborated as follows:

Initialize the population $P_i^G = p_{1,i}^G, p_{2,i}^G, p_{3,i}^G \ldots\ldots p_{D,i}^G$

where $i = 1, 2\ldots M$ and $G = 0, 1 \ldots G_{max}$. M and G_{max} represent the population size and maximum no. of iterations respectively.

For $I = 1$ to M (Population size) do

Select $r_1, r_2, r_3 \in \mathbb{N}$ (Natural numbers) and $r_1 \neq r_2 \neq r_3$

For $j = 1$ to D (dimension of the problem) do

　Select $j_{rand} \in D$

　If $\left(rand() \prec Cr \; or \; j = j_{rand} \right)$

// where rand () represents a uniformly distributed random number between 0 and 1

　　$V_{i,G} = P_{r_3,G} + F(P_{r_1,G} - P_{r_2,G})$ where F($\in$ 0,1) is a constant scaling factor

　End if

　If $E\left(U_{i,G} \right) \leq E\left(P_{i,G} \right)$ then $P_{i,G+1} = U_{i,G}$

　Else $P_{i,G+1} = P_{i,G}$

　End if

End for

End for

where $V_{j,i}^G$ represents the velocity of the particle in position at j dimensional space at current G^{th} generations; $V_{j,i}^{G+1}$ represents the velocity of particle i in position at j^{th} dimensional space after $G + 1^{st}$ generation.

PSO Algorithm

PSO is an iterative stochastic optimization method envisaged by Eberhart and Kennedy (1995) to get global optima. It belongs to the category of swarm intelligence which imitates social models like swarm theory, bird flocking and fish schooling to simulate its search method to reach the best possible solution. In this method, the possible candidate solution is known as particles which are distributed throughout the solution space (collectively these particles are known as a swarm) and each one is associated with a randomized position and velocity. These individual particles are able to update their previous location by perturbation provided by the combined effect of their own individual best cognitive experience (*pbest*) and collectively best social interaction (*gbest*) of a swarm. The working process of PSO is elaborated as follows:

Initialize the particles $P_i^G = p_{1,i}^G, p_{2,i}^G, p_{3,i}^G \ldots\ldots p_{D,i}^G$ and their associated velocity where $i = 1, 2 \ldots M$ and $G = 0, 1 \ldots G_{max}$. M and G_{max} represent the population size and maximum no. of iterations respectively.

For $G = 1$ to G_{max} (Maximum number of generations) do
For $i = 1$ to M (population size) do

  $Ep = E(p)$

  If $Ep \prec E(pbest)$ then $pbest = p$
  End if
End for

  $gbest = best\ in\ p$

For each particle $p_{1,i}^G$, to $p_{D,i}^G$ do

$$V_{j,i}^{G+1} = \omega \times V_{j,i}^G + C_1 \times rand_1 \times \left(pbest_{j,i}^G - TP_{j,i}^G \right) + C_2 \times rand_2 \times \left(gbest_i^G - P_{j,i}^G \right)$$

$$P_{j,i}^{G+1} = P_{j,i}^G + V_{j,i}^{G+1}$$

End for
End for

The DE-PSO Algorithm

In the hybrid version, the DE-PSO algorithm follows a similar hierarchy as a standard DE up to the crossover operation. In the selection operation if the DE phase is unable to produce an offspring with better fitness compared to the current location then the proposed algorithm enters the PSO phase. It generates a new one after the completion of each iteration of DE, which is again compared with the fitness of its current location. If it provides an improved location then it is selected otherwise current location will continue as the next generation. This entire process is repeated until the required stopping criterion is attained. The inclusion of the PSO phase creates a new location and provides perturbation to the existing population. It helps to diversify the DE population and eventually helps to provide more choices to get higher fitness or to reach greater precision. The working of DE-PSO in the following steps:

Initialize the population $P_i^G = p_{1,i}^G, p_{2,i}^G, p_{3,i}^G \ldots\ldots p_{D,i}^G$

where $i = 1, 2 \ldots M$ and $G = 0, 1 \ldots G_{max}$. M and G_{max} represent the population size and maximum no. of iterations respectively.

For $i = 1$ to M (Population size) do

Select $r_1, r_2, r_3 \in \mathbb{N}$ and $r_1 \neq r_2 \neq r_3$

For $j = 1$ to D (dimension of the problem) do

Select $j_{rand} \in D$

If $\left(rand() \prec Cr \text{ or } j = j_{rand} \right)$

// where rand () represents a uniformly distributed random number between 0 and 1

$V_{i,G} = P_{r_1,G} + F(P_{r_2,G} - P_{r_3,G})$ where F($\in 0,1$) is a constant scaling factor

End if

If $E\left(U_{i,G} \right) \leq E\left(P_{i,G} \right)$ then $P_{i,G+1} = U_{i,G}$

Else

PSO activated to generate a new particle (TP_i^{G+1}) as

$$TP_{j,i}^{G+1} = TP_{j,i}^G + V_{j,i}^{G+1}$$

where $V_{j,i}^{G+1} = \omega \times V_{j,i}^G + C_1 \times rand_1 \times \left(pbest_{j,i}^G - TP_{j,i}^G \right) + C_2 \times rand_2 \times \left(gbest_i^G - P_{j,i}^G \right)$

If $E\left(TP_i^{G+1} \right) \leq E\left(P_i^G \right)$ then $P_{i,G+1} = TP_i^{G+1}$

Else $P_{i,G+1} = P_{i,G}$

End if

End if

End for

End for

where $v_{j,i}^G$ represents the velocity of particle in position at j dimensional space at current G^{th} generations; $pbest_{j,i}^G$ represents the personal best performance of an individual particle till G^{th} generations; $gbest_i^G$ represents the best performance of whole swarm till G^{th} generations; $v_{j,i}^{G+1}$ represents the velocity of particle i in position at j^{th} dimensional space after $G + 1^{st}$ generation; ω is the inertia weight, which gives weight to the velocity of a particle at its previous position; C_1 and C_2 are the acceleration constants; $rand_1$ and $rand_2$ represent the random number between 0 and 1. $TP_{j,i}^G$ represents the current location of an individual particle I by PSO in G^{th} generations.

Objective Function

In this study, the FEM simulation model is coupled with DE, PSO, and DE-PSO to develop three different SO models. Optimization models provide the variable input to the simulation model by their individual internal mechanisms in terms of aquifer parameters. Subsequently, these parameters are utilized to evaluate the objective function. Since the main objective of the study is to minimize the fitting error between the observed and simulated head, therefore the parameter corresponding to the lowest objective function value will be the optimal representative aquifer parameters. This objective function used here is defined by the sum of squared difference (SSD) which is represented as (Lakshmi Prasad & Rastogi, 2001):

$$Min \ E(P) = \gamma_{l,t} \sum_{l=1}^{L} \sum_{t=t_0}^{t_t} \left[h_{l,t}^{obs} - h_{l,t}^{sim}(P) \right]^2 \qquad (14.6)$$

Subjected to:

$$P_i^{lb} \leq P_i \leq P_i^{ub} \qquad (14.7)$$

where E represents an objective function to be minimized; $h_{l,t}^{sim}$ is calculated groundwater head at observation well l at time t with parameter (P) as input [m]; $h_{l,t}^{obs}$ is observed groundwater head at observation well l at time t; P_i is aquifer parameter at zone i; L is the total number of observation wells; t_0 and t_t are beginning and ending time of observations [day]; lb and ub are the superscripts representing the lower and upper bounds on the parameters and $\gamma_{l,t} \in [0,1]$ is the weighing coefficient whose value is chosen according to confidence on measured groundwater head at a certain observation well location.

VERIFICATION OF DE, PSO, AND DE-PSO ON TWO CLASSICAL BENCHMARK PROBLEMS

To verify the applicability of proposed DE, PSO, and DE-PSO optimizations, two classical unconstrained benchmark problems (i.e., sphere and Ackley) with known minima are considered. Mathematical expressions of these functions are presented in Table 14.1 (Pant et al., 2011). For both cases, the size of the solution space is kept [−10, 10]. The maximum number of function evaluations is set to 300 for all ten-dimensional test functions using DE, PSO, and DE-PSO optimizations. The population size for both functions is kept uniform i.e., 20. Values of all five control parameters used in DE, PSO, and DE-PSO are presented in Table 14.2. These values are problem specific which are decided after numerous model runs.

After feeding the required input control and strategy parameters the performance of all three optimization models. In both problems, all three optimization models attained functional convergence around 100 generations. The fitness value obtained after termination is presented in Table 14.3. In each case, the estimated fitness value

TABLE 14.1

Expression of Two Different Benchmark Problems with Their Known Fitness Values

Sr. No.	Benchmark function	Expression of objective function	Known solution
1	Sphere	$$f_{sphere} = \sum_{i=1}^{n} x_i^2$$	0
2	Ackley	$$f_{Ackley} = 20 + e - 20\exp\left(-0.2\sqrt{\frac{1}{D}\sum_{i=1}^{D} x_i^2}\right) - \exp\left(\frac{1}{D}\sqrt{\sum_{i=1}^{D}\cos(2\pi x_i)}\right)$$	0

TABLE 14.2

Summary of the Best Control Parameters Used in DE, PSO and DE-PSO for Two Benchmark Problems

No.	Control parameter	Suggested range	Tuned value	Optimization	
				DE	**DE-PSO**
1	Population size (N)		25		
2	Mutation weighing factor (F)	0.3–0.5	0.4	DE	DE-PSO
3	Crossover rate (Cr)	0.7–1.0	0.7		
4	Inertia weight (ω)	0.8–0.3	Linearly varying from 0.8 to 0.3	PSO	
5	Acceleration constants (C1 = C2)	1.9	1.9		

was found very near to the known minima, hence all three algorithms can be considered as optimization models for the solution of inverse groundwater problem in the upcoming application section. Among different heuristic methods, DE-PSO is the most efficient optimization with the lowest objective function value and it is able to explore the solution more profoundly. This model runs on benchmark problems and also proves the superiority of the hybrid model over individual optimization. PSO is the worst-performing algorithm this may be due to the lack of diversity in the population after certain generations.

TABLE 14.3
Solution Obtained Through DE, PSO and DE-PSO for Two Benchmark Problems for Two Benchmark Problems

	Benchmark problems			Estimated objective function value		
Function	Known objective function value	Dimension	Size of solution space	DE	PSO	DE-PSO
Sphere	0	10	$x \epsilon [-10,10]$	0.1578	0.0027	2.80E-08
Ackley	0	10	$x \epsilon [-10,10]$	0.0692	1.6679	0.02489

RESULTS AND DISCUSSION

PROBLEM DESCRIPTION: SYNTHETIC UNCONFINED IRREGULAR DOMAIN PROBLEM

For testing the applicability of different developed SO models on unconfined problems a synthetic aquifer problem by Sharief et al. (2008) is considered (Figure 14.1). This selected domain is surrounded by two impervious boundaries situated in the north and south direction while two constant head boundaries are also considered

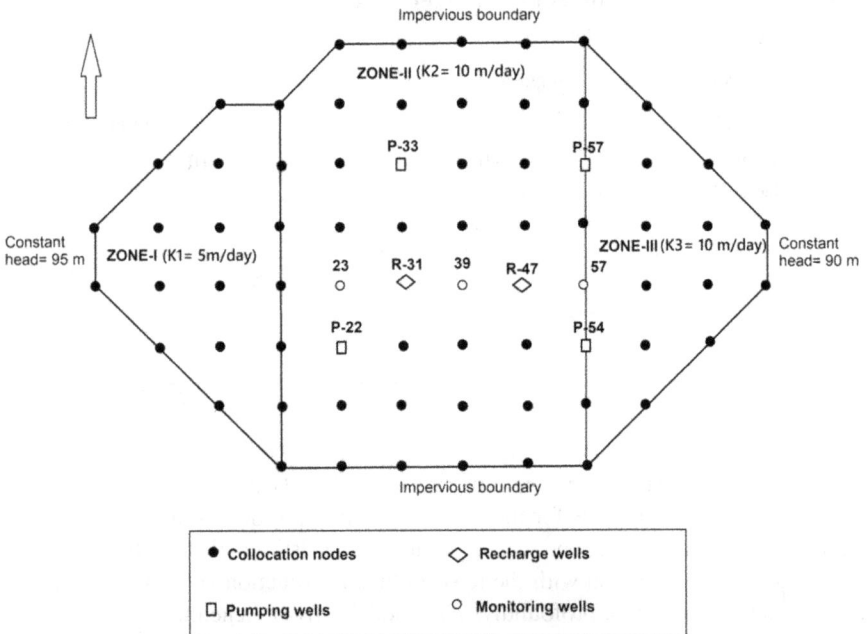

FIGURE 14.1 Synthetic unconfined aquifer domain showing zonation pattern, boundary conditions with pumping and recharge activities

in the east (90 m) and west direction (95 m). Four pumping wells (P_{22} = 500 m³/day, P_{33} = 800 m³/day, P_{54} = 600 m³/day, P_{57} = 1000 m³/day) with two recharge wells (R_{31} = 500 m³/day, R_{47} = 800 m³/day) are situated in the domain to represent the dynamic activities of synthetic problem. Three observation wells are located at nodes number 23, 39 and 57 to monitor the variability in the groundwater head. A uniform value 0.2, is assumed as a specific yield of considered domain.

MODEL INPUT

In this case study, three zonal hydraulic conductivity values are considered to be unknown for testing of all the developed models. The assumed values of hydraulic conductivity for the entire aquifer vary between 1 to 50 m/day. Using the known data like specific yield, boundary conditions, pumping activities and zonation pattern, the specified SO model tries to minimize the error between observed and simulated head to estimate the zonal hydraulic conductivity values using an indirect approach. For evaluating the objective function, estimated head values are obtained by FEM based simulator using 72 nodes and 116 triangular elements. Using both simulators, each iteration of the SO model is performed for 100 days with one day as a time-step size. Control parameters associated with DE, PSO, and DE-PSO based three SO models are calibrated after performing numerous model runs. The problem-specific tuned values of these control parameters are presented in Table 14.4.

MODEL OUTCOME AND ITS ANALYSIS

The above obtained tuned values of control or strategy parameters are fed to DE, PSO, and DE-PSO based SO models and model runs are performed up to 200 generations. The variation of the objective function with each generation is presented in terms of a convergence graph as depicted in Figure 14.2 (and Table 14.5). DE-PSO

TABLE 14.4
Summary of the Best Suited Values of Control Parameters Used in the Developed DE, PSO and DE-PSO Models for Synthetic Unconfined Aquifer Problem

No.	Control parameter	Suggested range	Tuned value	Optimization	
1	Population size (N)		10		
2	Mutation weighing factor (F)	0.3–0.5	0.3	DE	DE-PSO
3	Crossover rate (Cr)	0.8–1	0.6		
4	Inertia weight (ω)	0.8–0.3	Linearly varying from 0.8 to 0.3	PSO	
5	Acceleration constants (C1 = C2)	1.5–2	1.9		

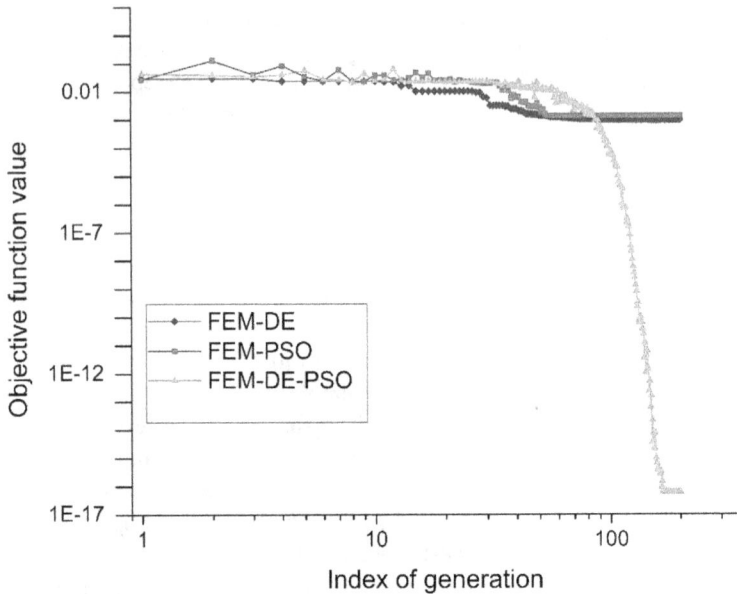

FIGURE 14.2   The performance of different SO models is presented based on the objective function convergence graph for the synthetic unconfined aquifer problem

is the best-performing algorithm compared to its other counterparts with the lowest objective function value. It takes nearly 120 generations to get steady global convergence of all three hydraulic conductivity values. After objective functional evaluation, the time consumption to perform one iteration of all the developed models is presented in Table 14.5. In terms of computational time the DE-PSO optimization performed poorly because the model has to shift from the DE phase to the PSO phase in each iteration with the expectation of further improvement in solution.

Due to the stochastic nature of all the developed models each model run is performed ten times and its mean value is taken as representative zonal hydraulic conductivity value as shown in Figure 14.3 and showed greater agreement with true value in all developed eight models (Table 14.6).

TABLE 14.5

Time Required to Perform One Iteration Using Eight Different SO Models for Synthetic Unconfined Aquifer Problem

Sr. no.	SO model	Time required for one generation (sec.)	Lowest value of objective
1	FEM-DE	2.1696	0.000989358
2	FEM-PSO	4.3211	0.001374345
3	FEM- DE-PSO	90.222	5.94171E-17

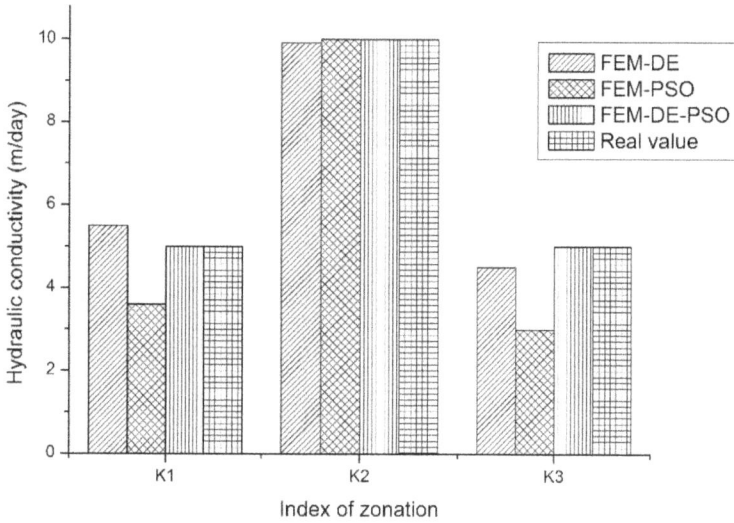

FIGURE 14.3 Average values of each parameter after eight times model run by eight different methods and their comparison with the known value

TABLE 14.6
Average Values of Each Parameter After Ten Times Model Run by Eight Different Methods

Sr. no.	SO models	K_1	K_2	K_3
1	FEM-DE	5.041	9.919	5.078
2	FEM-PSO	5.000	10.000	5.000
4	FEM-CMA-ES	5.000	10.000	5.000
Real value		5.000	10.000	5.000

CONCLUSIONS

In above above-discussed synthetic unconfined problem, all the developed models ended up with nearly accurate values of aquifer parameters. Although some of the individual heuristics-based SO models are not able to explore the solution space profoundly their estimated parameter values showed unanimity with true value. There are two possible reasons behind it: (1) there is clearly a large head variation after the simulation run to perform dynamic activities therefore a specific SO does not need to explore the solution space with higher accuracy. (2) An unconfined synthetic problem is associated with a small number of dimensions therefore all the parameters attain convergence with higher objective function value. On the other hand, DE-PSO produces a diverse population (aquifer parameters) using the perturbation of the PSO phase to DE, which empowers the proposed algorithm to explore the solution space profoundly and reach a desired minimum value of the objective function. The combination of both algorithms is able to estimate the groundwater parameters in an unconfined aquifer which is a daunting task indeed.

REFERENCES

Abbaspour, K. C., Schulin, R., & van Genuchten, M. T. (2001). Estimating unsaturated soil hydraulic parameters using ant colony optimization. *Advances in Water Resources*, 24(8), 827–841. https://doi.org/10.1016/S0309-1708(01)00018-5

Blin, N., Hausner, M., Leray, S., Lowry, C., & Suárez, F. (2022). Potential Impacts of climate change on an aquifer in the arid Altiplano, northern Chile: The case of the protected wetlands of the Salar del Huasco basin. *Journal of Hydrology: Regional Studies*, 39, 100996.

Ch, S., & Mathur, S. (2012). Particle swarm optimization trained neural network for aquifer parameter estimation. *KSCE Journal of Civil Engineering*, 16(3), 298–307. https://doi.org/10.1007/s12205-012-1452-5

Dillon, P., Pavelic, P., Toze, S., Rinck-Pfeiffer, S., Martin, R., Knapton, A., & Pidsley, D. (2006). Role of aquifer storage in water reuse. *Desalination*, 188(1–3), 123–134.

Eberhart, R., & Kennedy, J. (1995). A new optimizer using particle swarm theory. MHS'95. In *Proceedings of the Sixth International Symposium on Micro Machine and Human Science*, 39–43. https://doi.org/10.1109/MHS.1995.494215

Gupta, P. K., Gharedaghloo, B., & Price, J. S. (2023). Multiphase flow behavior of diesel in bog, fen, and swamp peats. *Journal of Contaminant Hydrology*, 255, 104162.

Guzman, S. M., Paz, J. O., Tagert, M. L. M., & Mercer, A. E. (2019). Evaluation of seasonally classified inputs for the prediction of daily groundwater levels: NARX networks vs support vector machines. *Environmental Modeling & Assessment*, 24(2), 223–234.

Harrouni, K. E., Ouazar, D., Walters, G. A., & Cheng, A. H.-D. (1996). Groundwater optimization and parameter estimation by genetic algorithm and dual reciprocity boundary element method. *Engineering Analysis with Boundary Elements*, 18(4), 287–296. https://doi.org/10.1016/S0955-7997(96)00037-9

Lakshmi Prasad, K., & Rastogi, A. K. (2001). Estimating net aquifer recharge and zonal hydraulic conductivity values for Mahi Right Bank Canal project area, India by genetic algorithm. *Journal of Hydrology*, 243(3–4), 149–161. https://doi.org/10.1016/S0022-1694(00)00364-4

Mahajan, M., Gupta, P. K., Singh, A., Vaish, B., Singh, P., Kothari, R., & Singh, R. P. (2022). A comprehensive study on aquatic chemistry, health risk and remediation techniques of cadmium in groundwater. *Science of The Total Environment*, 818, 151784. https://doi.org/10.1016/j.scitotenv.2021.151784.

McDonald, M. G., & Harbaugh, A. W. (2003). The history of Modflow. *Ground Water*, 41(2), 280.

Pant, M., Thangaraj, R. and Abraham, A. (2011). DE-PSO: a new hybrid meta-heuristic for solving global optimization problems. *New Mathematics and Natural Computation*, 7(03), pp.363–381.

Price, K., Storn, R. M., & Lampinen, J. A. (2005). *Differential Evolution*. Springer-Verlag. https://doi.org/10.1007/3-540-31306-0

Raghunath, H. M. (1987). *Ground Water: Hydrogeology, Ground Water Survey and Pumping Tests, Rural Water Supply and Irrigation Systems*. New Age International.

Romero, R., Muriel, J.L., García, I. and de la Peña, D.M., (2012). Research on automatic irrigation control: State of the art and recent results. *Agricultural water management*, 114, pp.59–66.

Sharief, S. M. V., Eldho, T. I., & Rastogi, A. K. (2008). Optimal pumping policy for aqui-fer decontamination by pump and treat method using genetic algorithm. *ISH Journal of Hydraulic Engineering*, 5010(14), 1–17. https://doi.org/10.1080/09715010.2008 .10514901

Singh, R. P., Mahajan, M., Gandhi, K., Gupta, P. K., Singh, A., Singh, P. ... & Kidwai, M. K. (2023). A holistic review on trend, occurrence, factors affecting pesticide concen-tration, and ecological risk assessment. *Environmental Monitoring and Assessment*, *195*(4), 451. https://doi.org/10.1007/s10661-023-11005-2.

Storn, R., & Price, K. (1997). Differential evolution – A simple and efficient heuristic for global optimization over continuous spaces. *Journal of Global Optimization*, 11(4), 341–359. https://doi.org/10.1023/A:1008202821328

Willis, R. and Yeh, W.W., (1987). *Groundwater systems planning and management.*

Yatoo, A.M., Ali, M.N., Zaheen, Z., Baba, Z.A., Ali, S., Rasool, S., Sheikh, T.A., Sillanpää, M., Gupta, P.K., Hamid, B. and Hamid, B., (2022). Assessment of pesticide toxicity on earthworms using multiple biomarkers: a review. *Environmental Chemistry Letters*, *20*(4), 2573–2596. https://doi.org/10.1007/s10311-022-01386-0.

Zheng, C., & Wang, P. (1996). Parameter structure identification using tabu search and simu-lated annealing. *Advances in Water Resources*, 19(4), 215–224. https://doi.org/10.1016 /0309-1708(96)00047-4

15 Application of Remote Sensing, GIS, and AI Techniques in the Agricultural Sector

Mridu Kulwant and Divya Patel

INTRODUCTION

The monitoring of the surface of the planet's features can be considerably aided by remote sensing records since they provide rapid, synoptic, economical, and repeatable information (Balsamo et al., 2018). The primary uses of remote sensing (RS) in agriculture include plotting and estimating crop area, monitoring vegetation vigour and drought stress, estimating biomass and yield, and assessing crop phenological development (Atzberger, 2013). "Remote sensing involves gathering data about objects or events without touching them. Remote sensing gives tools and concepts to detect objects and phenomena. Remote sensing involves developing and applying analytical methods to get meaningful data" (Chauhan et al., 2018). Remote sensing is the process of getting information about the Earth's surface through data collected by a device located at a distance from the surface, most often satellites circling the Earth or aircraft. Geographic information systems (GIS) are computer-based systems that acquire, store, analyse, and display geographic data. These two methodologies are extensively used, often in tandem, to evaluate natural resources and monitor environmental changes (Manson et al., 2015). By recording the radiation energy that is reflected and released RS is the course of acquiring information, identifying, analysing, and tracking the physical features of a part without coming into direct physical touch with the thing being examined. Obtaining the reflected energy or radiation allows for this (Nath, 2021; Navalgund, 2002).

As it can complement more established agrometeorological data-gathering techniques, the acceptance of satellite remote sensing innovations as an essential component of agrometeorological data is expanding (Sivakumar and Hinsman, 2004; Himanshu et al., 2023a, 2023b). GIS is a vital piece of mechanised programme development that also provides knowledge through charts and diagrams and plays a key role in the attempt to eradicate world hunger. It does this by identifying areas of need and the root causes of food insecurity by utilising data from numerous

 DOI: 10.1201/9781003441175-15

instruments, including mobile devices (Gomiero, 2016). Based on multiple crop and soil features, geographical information and RS systems equipment have the ability to revolutionise the characterisation and detection of farming output, and therefore to expand agricultural productivity to feed the world's growing population. Even at modest or big agricultural land holdings, modern technologies like RS and GIS are extremely advantageous for identifying and dealing with a myriad of crop concerns (Balafoutis et al., 2020).

In order to accelerate the development of products, analyses, and forecasts that impact agricultural crop management decisions, irrigation planning, commodities trading, market prices, as well as ecosystem management and conservation, systems for recording agricultural weather and climate data are necessary(Stone and Meinke, 2006; Ale et al., 2022, 2023). The fundamental concept of the region of the electromagnetic (visible, infrared, and microwaves) is what drives spatial in nature) to measure the characteristics of the Earth. According to the targets' typical responses to different wavelengths. Although they might be discarded, they are beneficial for removing plants, bare topsoil, water, and additional similar occurrences, regions remain different (Ashraf et al., 2011). Information from satellites like LANDSAT, RADARSAT, TerraSAR-X, SRTM, EOS, ERS, Sentinel, and numerous other satellites is used to make a number of significant choices (Balzter et al., 2015). Using interpolation, spatially referenced data in a GIS system enables the visualisation of the distribution of different parameters from one site to other methodologies and assesses how historical data has changed over time. In order to assist in keeping a sustainable food and water supply in accordance with the potential for the environment, technologies like GIS must be applied due to the increased pressure on land resources brought on by the expanding global population (Khouni et al., 2021).

This study aims to provide a comprehensive overview of the literature on the use of RS and GIS for the implicational appraisal of RS and GIS in agriculture to encourage future research and development in the field of agriculture. There are countless GIS and artificial intelligence based applications that could be used both nationally and locally. For instance, agricultural planners may utilise geographic data for the goal of determining both dimensions and placement of ecologically suitable regions by combining information on soils, elevation, and rainfall to identify the optimal zones for a cash crop (Sivakumar, 2003).

New tools like GIS can manage massive amounts such as conventional digital maps, databases, models, etc., of data. An environmental inspection spacecraft is a man-made Earth satellite that collects information a sort of satellite upon that Earth system is indeed a meteorological satellite, the atmospheric observatory that performs climatic measurements (Sivakumar and Hinsman, 2004). GIS are made to collect, store, integrate, analyse, and present data from a geographical perspective. It is a potent collection of implements for gathering, packing, and recuperating facts. To be successful besides profitable, a farm must balance its inputs and outputs. You may utilise GIS, which integrates location figures through quantitative and qualitative specific features, it is possible to use maps and charts to visualise, analyse, as well as report information (Sood et al., 2015).

Khanal (2020) and his colleagues have observed the literature produced between the years 2000 and 2019 that concentrated on the use of remote sensing productive agriculture methods, such as field preparation, planting, and in-season applications, and gathering in order to advance scientific knowledge regarding decision-supporting capabilities of RS technologies during various construction periods. Most RS research has centred on monitoring crop health in-season and soil moisture, not crop grain quality, subsurface drainage, or soil compaction. In conclusion, RS technologies can make it possible for site-specific management choices to be made at various phases of crop development to maximise crop output while also taking environmental quality, profitability, and sustainability into consideration (Khanal et al., 2020). Sensors record object data to detect them. Most sensors are active or passive. Object detection systems should be real-time, affordable, and compact. The right sensor gives our system these traits (Abbasi et al., 2017). Figure 15.1 and Figure 15.2 show different types of sensors, which can be categorised further.

Active Remote Sensing

Definition of Active remote sensing:

"Remote sensing techniques that generate their own electromagnetic radiation to light the object. A flash camera is another example of active remote sensing" (Kairu, 1982).

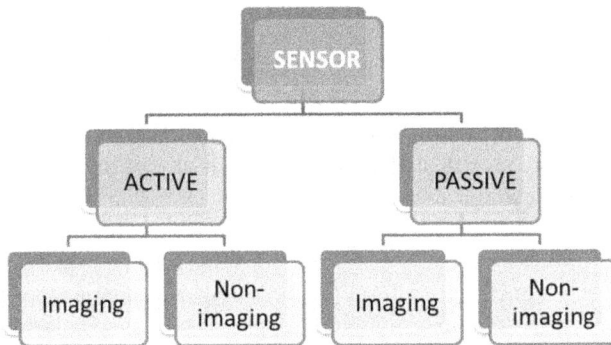

FIGURE 15.1   Displaying the kinds of sensors depending on lighting

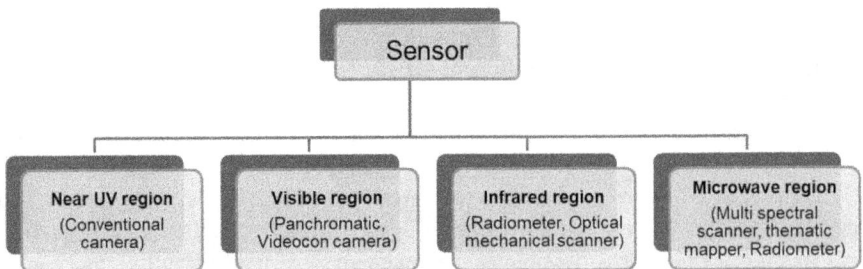

FIGURE 15.2   Sensor types are represented by spectral areas

Direct 3D evaluation of the topography and structure of forest canopies is now possible thanks to a new generation of active, slightly raised sensor technologies, such as interferometric synthetic aperture radar (IFSAR) and aerial laser scanning (LIDAR). The features of the individual treetops that make up the above-ground foliage forest may be measured using high-resolution LIDAR. Additionally, key forest structure factors like dominant branch diameter, understory, elevation, and output have a strong correlation with metrics based on LIDAR height distribution. Multispectral sonar and LIDAR (SAR) remain examples of active sensors (Andersen et al., 2006; Lugari, 2014).

PASSIVE REMOTE SENSING

Definition of Passive remote sensing:

"Passive remote sensing is the detection of energy that is spontaneously reflected or emitted from a target" (Kairu, 1982). The sun is the planet's primary source of energy, and remote sensing is basically dependent on it as well. For VIS-NIR wavelengths, the sun energy is reflected; for thermal infrared wavelengths, it is absorbed and then reemitted. This means that when all reflected or reemitted solar energy is available during the sun's illumination period on Earth's surface, passive sensors are able to detect energy. Landsat series, IRS series, SPOT series, IKONOS, Quickbird, and other passive sensors are examples (GV et al., 2016; Gupta, 2018).

AERIAL PHOTOGRAPHY

Since the crucial days of RS, aerial photography has generally been recognised as an indispensable factor for witnessing, assessing, describing, and passing judgement about our surroundings. There are two types of conventional remote sensing: from satellites and aircraft (Matese et al., 2015).

SATELLITE IMAGERY

Satellite imagery comes from the transmission of information via radiation through the atmosphere from an item to a receiver (observer). The contact of the particle emission with the subject of attention transmits the necessary evidence about the thing's nature (e.g. reflection coefficient, emittance, roughness). Active radars include Radar, LIDAR, PSInSAR, SAR, SRT, even Squee SAR. Passive remote sensing occurs when a sensor notices the sun's reflection. Passive remote sensing techniques include FLIR, geodetic survey, and aerial photography (Danklmayer et al., 2009). Analysers, SWIR, stereo aerial photographs, FLIR, geographic coordinate survey, hyperspectral imaging, long-wave infrared, hyperspectral tomography, and near-infrared (NIR) surveys are forms of oblique airborne and terrestrial visible spectrum and computed tomography imaging (Bitelli et al., 2017).

Table 15.1 Different Satellite Data with Their Resolutions and Wavelength (USGS)

Landsat 1–5 Multispectral Scanner (MSS)

Landsat1–3	Landsat4–5	Wavelength (micrometres)	Resolution (metres)
Band 4 – Green	Band 1 – Green	0.5–0.6	60
Band 5 – Red	Band 2 – Red	0.6–0.7	60
Band 6 – Near-Infrared (NIR)	Band 3 – Near-Infrared (NIR)	0.7–0.8	60
Band 7 – Near-Infrared (NIR)	Band 4 – Near-Infrared (NIR)	0.8–1.1	60

Landsat 4–5 Thematic Mapper (TM)

Landsat 4–5	Wavelenth (micrometres)	Resolution (metres)
Band 1 – Blue	0.45–0.52	30
Band 2 – Green	0.52–0.60	30
Band 3 – Red	0.63–0.69	30
Band 4 – Near-Infrared (NIR)	0.76–0.90	30
Band 5 – Near-Infrared (NIR)	1.55–1.75	30
Band 6 – Thermal	10.40–12.50	120 (30)
Band 7 – Shortwave Infrared (SWIR)	2.08–2.35	30

STANDARDS OF REMOTE SENSING

OPERATIONAL UTILITIES OF REMOTE SENSING

Our understanding of a variety of environmental sectors is improved by the development of inverse retrieval techniques, the ongoing technical advancement of a satellite platform's sensors, as well as accessibility of various restrictions defining the atmosphere-ocean arrangement. Tiny RS from spacecraft permits the study of various both space and time scales of interactions between it ocean and the atmosphere extending beyond air-sea interactions through planetary- and watershed exchanges (Loisel et al., 2009). The application of RS to tricky proof of identity, procedure design, policy implementation, control, and assessment is illustrated with examples. While at first RS was largely used to identify environmental issues plus enact policies, attention has more recently broadened to include applications for controlling and evaluating policies (De Jong et al., 2007; De Leeuw et al., 2010).

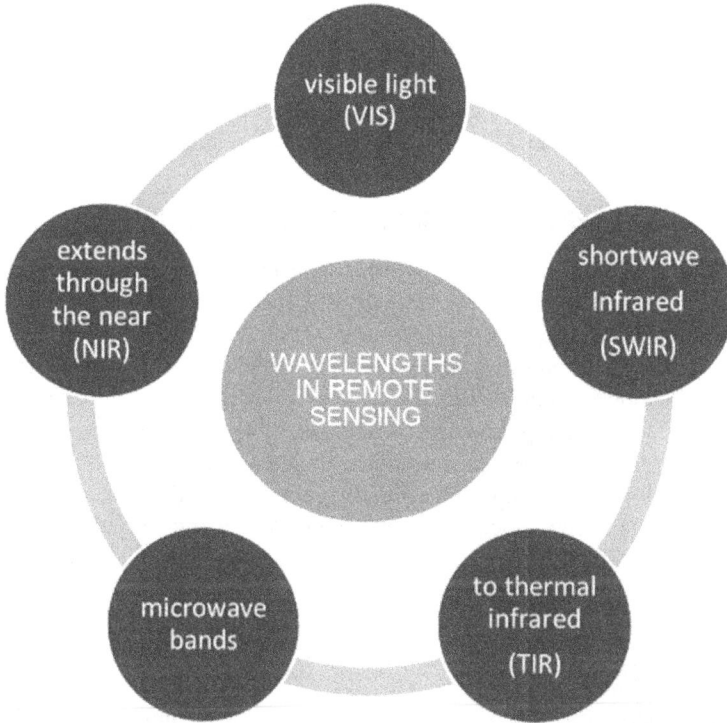

FIGURE 15.3   The most useful wavelengths in remote sensing (Wójtowicz et al., 2016)

Physical Utilities of Remote Sensing

The first method of RS used aerial photography in the visible province of the electromagnetic range, but technical advancements have made it possible to collect data at additional wavelengths such as microwave, near-infrared, and thermal infrared (Figure 15.3). Multispectral or hyperspectral data are sets of information gathered across a variety of wavelength bands (Kairu, 1982). This combines contextual classification with segmentation, which is a key step in the strategy. The technique of segmentation separates an image into homogenous pixel groups (segments), which are then organised into classes depending on their absorption spectra, geometric, topographical, as well as other features all through the semantic classification stage (Veljanovski et al., 2011).

REMOTE SENSING IN AGRICULTURE

The foundation of the Indian economy is agriculture, which is also crucial for providing food security. India ranks among the top producers of pulses, jute, fruits, vegetables, rice, wheat, sugarcane, and cotton in the world. In addition, it is a major producer of plantation crops, fish, poultry, milk, and animals (Singh et al., 2022).

Harvest documentation, crop situation monitoring, pest and disease detection, vegetative indices, canopy transpiration, and crop stress are all areas where RS and GIS are used in cultivation (Sishodia et al., 2020). Primarily in industrialised nations, there is a tendency to encourage more energy inputs through heavy machinery and higher chemical and fertiliser applications through the industry.

APPLICATIONS OF REMOTE SENSING IN AGRICULTURE

Despite, the negative social and environmental effects of these techniques, soil destruction, soil salinity, loss of topsoil fertility, subsoil compression, and contamination of the soil and water, they have typically been successful in meeting the nutrition and fibre requirements of a swift expanding populace (Naik et al., 2022). Stress response detection, vegetative indices, canopy transpiration, crop stress, cropping structural analysis, yield prediction, erosion inventory, sand dune assessment, soil texture and hydraulic properties, soil drainage, soil surface hardness, and precision farming techniques are all available. In comparison to traditional approaches, the GIS allows for relatively quick data capture and processing (Repullo-Ruibérriz de Torres et al., 2018). Many applications have been suggested by Marek Wójtowicz et al. (2016).

VEGETATION MONITORING

It is straightforward to identify the stages of crop development and the relationships between them thanks to conservational restoration elements and RS indicators (such as NDVI). Remote sensing-based plant disease detection and recognition aid farmers in making effective management choices that will maintain crop production (Messina and Modica, 2022). Felix Owusu Anyimah et al. have done a case study on cocoa homesteads Using GIS and RS to detect the stress area. According to the findings, change in the research area's rainfall patterns is specifically responsible for the apparent rise in greenery over a period of several months. Many raindrops were reported throughout the showery season (March–August), which caused water to be retained. This increased soil moisture in the area, which in turn increased vegetation (NDVI) (Anyimah et al., 2021).

> The way vegetation reflects sunlight will reveal information about the object's reflection coefficient, spectrum variation, and consequently, the nature of the object (green trees, etc.).
>
> **(Ribeiro da Luz and Crowley, 2007)**

CROP CONDITION ASSESSMENT

Crop condition is also assessed using a Normalised Difference Vegetation Index (NDVI) based technique that compares seasonal dynamics and spatial variability over years (Zhang et al., 2014). Traditional uses of agrometeorology, such as crop production forecasts, are becoming more and more significant, as are used for the environment and human security (Hoogenboom, 2000). Horticulturalists are progressively gathering in order to improve rural livelihoods, efficacy, and/or plant quality with cutting-edge, highly technical and scientific estimations. Smart agriculture and innovative agricultural technologies combine with biology; in an organised farm management system, computers and electronics communicate with one another autonomously. With integrated pest control and/or biocontrol agents, cultivators can minimise plantation inputs (pesticides and nutrients) without boosting output (Ennouri and Kallel, 2019).

WATER AND NUTRIENTS

In agrometeorology, we use all the information available on the land to describe a particular situation, including the infrastructure (highways, railroads, power, or satellite communications), availability of water, soil composition, woodlands and grasslands, meteorological conditions, geology, population, land-use, and administrative borders are all elements of geospatial data can be analysed (O'Donoghue et al., 2021).

Information on the size and intensity of raindrops is revealed by microwave radiation that is broadcast from a radar system and scattered from a rain cloud in the back direction to a receiver (Cafiero et. al.). Local point surveys, airborne RS, and spatially satellite pictures of spatiotemporal nutrient distributions were utilised to generate realistic environmental patterns. In this context, ecosystem state descriptors such as total bottom temperature, Secchi disc depth, or sediments (or turbidity) seem to be some elements of transparency, as well as it give priority to whole nitrogen, whole phosphorus, chlorophyll-a concentration, coloured dissolved organic matter (dissolved OC or TOC), and excessive algal outbreaks (including cyanobacterial botulism neurotoxin, microcystin density) (Chang et al., 2015).

CONTROLLING WEEDS

Advanced vision-guided robotics that can be adopted for site-specific weed management (SSWM) are transgenic herbicide-resistant crops, weed control and spraying robots, decision support systems, and pattern recognition modelling (Rosle et al., 2021). The assessment of climate risk, which corresponds to the chance that specific meteorological phenomena could occur and harm crops or infrastructure, is one of the most significant agroclimatology applications (Martínez et al., 2012). Drones, Artificial Intelligence (AI), and a number of instruments, such as hyperspectral, multispectral, and RGB sensors, are all combined to increase the possibilities of managing weed issues more effectively. The majority of the significant or minor problems brought on by a vegetation invasion may be resolved by integrating remote sensing machinery into numerous cultivated chores. It spans many fields and disciplines, with techniques including spectroscopy, optics, computer, cinematography, communications satellites, semiconductors, and communications (Mohd Roslim et al., 2021).

PESTS AND ILLNESSES

The spectroscopic nature of live creatures is crucial to RS assertions. Insect pests and plant diseases can now be detected, predicted, and managed via the practice of RS in a wide series of modern orchards and agronomic settings (Buja et al., 2021). All of these programmes were established and created with the ultimate goal of reducing the environmental destruction caused by synthetic pesticides and enhancing the quality of data used in decision-making related to insect pests. Insect pest management and weed detection are two key areas where airborne RS have shown great potential and utility. In addition, RS utilising satellite data has shown to be a descriptive research design to observe and anticipate the expansion of insect pest species. Multispectral techniques, which rely on gradual shifts in canopy colour to detect mite infestation, have also been put into use to aid producers in this regard. Reduce pest damage and management expenses with the help of RS's accurate and timely prediction of selected insect pests (Abd El-Ghany et al., 2020).

PRECISION AGRICULTURE

Exacting farming methods increasingly more academics, technicians, and even farm owners are turning to RS as an integral part of precision farming. Precision farming utilises data collected through sensors installed in farm equipment to lessen input costs, boost management capabilities, and ensure efficient use. The pinnacle of smart agriculture innovation is floating interest technology (VRT). On-board electronics using sensors attached to mobile farm machinery use GPS coordinates to propose inputs and regulate the machines' use of those recommendations. Accuracy farming's main advantage is that it enables farmers to collect crop data with both the precision and timeliness necessary for effective management. Undoubtedly, RS is a useful technique for gathering this data. The utilisation of multispectral RS for site-specific nitrogen fertiliser control was described in *International Agrophysics* (26:103–108). Pictures of a 23-hectare corn field near Iran were taken by the Improved Spaceborne Thermal Emission and Reflection Radiometer (Aster). For identifying patterns in the climate, like other uses of RS technology, meteorological satellites are crucial in predicting the weather. Satellites used mostly for climatology collect data on things like air pressure, humidity, and weather patterns. Changes in overall canopy temperature may point to spots with and without sufficient field water supplies. Canopy air temperature difference (CATD) might be an indication of yield moisture stress and canopy temperature variability (CTV) is utilised in water control. Choices in the agricultural sector have increasingly relied on drought assessments informed by RS information. Drought assessment and monitoring at the provincial level with NDVI derived from NOAA-AVHRR information provides prompt preventative and remedial action to be taken (Shanmugapriya et al., 2019; Steven and Clark, 2013).

AGRONOMY

This developing and expanding field should be advantageous to the agronomic community, which includes farmers, land managers, other scientists, policymakers, and the general public. It covers site-specific farming (SSF), or managing a farm in accordance with the varying soil and microclimate conditions that exist in most fields. The use of GIS and related landscape technologies in many aspects of agronomy will be illustrated (Petersen et al., 1995). The GIS, in conjunction with further companion machinery, like RS, GPS, AI, computation arrangements, and figures analytics, has performed a crucial part in crop observation in putting the best administration practices into place in an effort to increase crop productivity (Parmita and Siva, 2022).

CROP IDENTIFICATION

Since the 1960s, when the United States first began its civilian RS initiative, recognising crops as well as estimating their areas was a primary goal. A few of the earliest research efforts comprised controlled trials, containing Crop Identification Technology Assessment for Remote Sensing (CITARS) and the Large Area Crop Identification Investigation (LACIE). Indian researchers first used aerial photographs captured in multi-band and colour infrared (CIR) throughout 1974–1975 to investigate the feasibility of RS for agricultural use. As part of the Indian Geographic Information Satellite-Utilisation Programme (IRSUP), scientists planned field experiments utilising a multi-band radiometer to learn more about crop signatures for the purposes of i) agricultural manufacture predicting, ii) drought revealing, and iii) harvest yield demonstrating. As a consequence of several works, the approach and implementation of projects on a huge level have been made more practical. Agriculture acquired a much-needed boost with the inauguration of Indian IRS-1A, IB, and 1C (Remote Sensing Satellites) equipped with longitudinal image self-scanning sensors (LISS I, II, and III). The sensors aboard IRS-1C were really a one-of-a-kind combination: i) a Wide-Angle Sensor (WiFS) with 188 m spatial and temporal resolution, different red and infrared light spectral bands, an 810 km swath, and a five-day repetition cycle; (ii) a Longitudinal Imaging Auto Scanning Sensor (LISS-III) with 23.5 m spatial and temporal resolution with the green, red, and near IR region and 70.5. Farming RS has been given a fresh start following RISAT's launch since the Indian satellite now possesses the potential of providing knowledge in all circumstances (Steven and Clark, 2013)

STRESSED PLANTS

Phosphorus' significance in controlling plant development function has been extensively researched. An inorganic phosphorus (Pi) molecule's function throughout lowering abiotic stressors, including heat, acidity, salinity, saltiness, and droughts, is not well understood. There is a relationship between plant development and low/no-phosphorus (P) stresses as well as famine conditions when one or both of these stresses are present (Bechtaoui et al., 2021). Based on measured soil moisture data, this subsystem determines the present plant water deficit (also known as water stress) for each plot's specific plant phonological stage, while the weather prediction is used to determine the predicted plant water deficit for the next days. The shortfall is required in order to determine the amount of plant water required to make up the current deficit during irrigation the next day. The physical knowledge of the soil–plant–water conditions is lacking (Trout et al., 2015).

YIELD ESTIMATION AND YIELD MAPS

Yield manufacture includes the yield of crops with crop acreage; since the former's evaluation is very difficult due to the elevation level of variability involved. Harvest yield information is a crucial component of the production estimation process. Every plant genotype seems to have a production potential that can be approximated in one optimal experimental context. However, in the real domain, factors including land, climate, and agribusiness techniques, as well as planting period, irrigation, and nourishment, have an impact on crop yield (Momtaz and Shameem, 2016). Additionally, biotic stressors like disease and pests have an impact on crop productivity. While 12-monthly fluctuation in a brief period of time is primarily explained by weather variability, the variability over a period of 10 to 20 years is primarily explained by farming practices and innovative varieties (Huseynova et al., 2014). The primary factors affecting agricultural productivity over longer time periods are soil improvement degradation or weather changes. Yield assessment is a more difficult process due to the interdependence of all these elements. Understanding the variability in the aforementioned criteria and describing how they relate to ultimate crop production are thus two ways to forecast the yield. Because it delivers a rapid, comprehensive, extensive, and objective evaluation of multiple crop characteristics, satellite-based remote sensing is a suitable alternative for crop condition and yield evaluation and prediction (Maximillian et al., 2019). Initially in the CAPE project, depending on divisional field observations via DES, time-series-based trends as well as arima models were built and multiplied by RS-derived area estimations to calculate production. For a range of conditions, crops, and regions, further agrometeorology models, spectral copies, combinations testing, and application of these designs (Hamel and Ismael, 2022).

The data is kept in a conventional ASCII setup. The graphic with both harvest route maps was prepared as row data and transferred from a floppy disc (RDS system) onto a personal computer (Neményi et al., 2003).

SOIL ANALYSIS PLANT BREEDING

The usage of RS is crucial during agronomical research because of their extraordinary sensitivity to fluctuations in changes toward climate, soil, as well as other physico-chemical factors. Whenever agricultural production systems are being kept under observation having regard towards the biological life cycle for crops, clear seasonal trends are seen. All of these components have a good deal of spatial and temporal flexibility. Those very same innovations are utilised in agriculture for a variety of purposes, including estimating crop farmlands, crop growth monitoring, soil water estimation, soil fertility evaluation, crop stress recognition, disease, and insect infestation detection, monitoring for drought and flood conditions, soil-water quality , requiring, weather forecasting, as well as precision agriculture (Gupta, 2020a,b; Mahajan et al. 2022; Gupta et al. 2020; Gupta et al. 2022; Yatoo et al. 2022; Singh et al. 2023). These practices help to maintain the resilience of agricultural systems to foster a country's economic growth (Kumar et al., 2022; Palanisamy et al., 2019; Weiss et al., 2019).

LAND COVER MAPPING

Bibliometric and network analysis were used to assess the present state of knowledge in many fields of artificial intelligence and sustainable agriculture. Additionally, the goal of this bibliometric study was to serve as an inspiration for future uses of AI in supportable cultivation by compiling key developments in the area. To obtain comprehensive data for precise results, the mapping of land cover was done based on seasonal and annual data (Bhagat et al., 2022).

CROP WATCHING

Since crop information is constantly needed during the growing season and there is a range of sensors with varying spatial, spectral, and temporal resolutions available, regular monitoring is practical at different scales (Jafarbiglu and Pourreza, 2022). A novel idea was developed as a result of the realisation of the value of information from numerous sources, such as weather, economics, and field surveys, for a robust approach for multiple crop forecasts: FASAL stands for Forecasting Agricultural Output Using Space, Agro-Meteorology, and Land Based Observations (Parihar and Oza, 2006a; Parihar and Oza, 2006b).

SOIL AND WATER MONITORING

The study uses low-resolution (MODIS) and intermediate-resolution (Landsat-8) satellite imagery to measure harvest water shortage within the subject area's canal command areas (CCAs) in close to real-time. Utilising a time series of the NDVI, a yield arrangement and cropping outline were developed using Landsat imagery. The MODIS NDVI product was used to construct a yearly situation NDVI series based on imagery from the previous 13 years and monitor the current state of crop health (Yousaf et al., 2021). Remote sensing was used in an automated integrated submission for the Water Resources and Agriculture Spatial Indicators System (WRASIS). Facts were created to track and evaluate various agricultural development projects. Satellite image time-series information analysis is the foundation of this system (Agriculture Moderate Resolution Imaging Spectra-radiometer MODIS NDVI – Rain-Fall Estimate RFE 2.0) (Zahran et al., 2022).

MONITORING AND PREDICTING CLIMATE CHANGES

In addition to other RS claims, weather forecasting is one area where meteorological satellites are important. The purpose of meteorological satellites is to gauge the atmosphere's temperature, wind, moisture, and low clouds. Varying overhead conditions may be an indication of where there is adequate and inadequate water in a certain field. It is possible to use canopy air temperature differential as a sign of crop water stress and canopy temperature variability as a tool for managing irrigation. Agriculture is a subject where drought assessment is important and remote sensing data has been utilised to make management decisions. In order to combat drought, timely preventative and corrective steps can be taken thanks to the district-level monitoring and evaluation of something like drought utilising NDVI produced from NOAA-AVHRR information (Shanmugapriya et al., 2019).

Pest/Weed Detection

For measuring agricultural stress brought on by biotic and abiotic variables, remote sensing has become a key technique. To establish plans to prevent their expansion and implement efficient control measures, it is necessary to perfect remote sensing approaches for identifying insect breeding areas. With the assumption that these variations are connected, generally categorised, and construed, the RS tactic to correlate variations in colour variation to chlorosis, yellowing of foliage, with vegetation decrease during a certain time period, insect defoliation assessment and monitoring has really been applied. RS has been used for a variety of purposes, such as locating as well as mapping defoliation, identifying patterning disruptions, including giving information into insect control decision provision arrangements. To distinguish between healthy and unhealthy vegetation cover, various vegetation indices were examined on Landsat imagery taken both before and after defoliation. It was found that MODIS data offer a crucial tool for determining vegetation indices at the plot scale and defoliation caused by insects. Plants with insect infestations and diseases can be easily and affordably identified using remote sensing technologies. RS performances are used by insect pests to differentiate between people with oats and people with insect damage. . They proposed using remote sensing to assess the canopy properties and spectral reflectance variations between damage caused by infected oat crop canopies including insect infestations and disease-related damage (Shanmugapriya et al., 2019).

The Landsat 5 TM image can be utilised to precisely detect and quantify illness used to site-specific Wheat Streak Mosaic illness control in wheat yield, according to a report by Mirik and his co-workers (Mirik et al., 2013). Wheat containing fungal growth might well be detected utilising high-resolution hyperspectral data from RS according to some research (Franke and Menz, 2007). The evaluation and management of agricultural activities largely depend on remote sensing, coupled with other cutting-edge instruments like GPS and GIS. Such techniques are employed in agriculture for a variety of purposes, which would include estimating crop acreage, crop growth monitoring, moisture estimation, soil fertility evaluation, plant stress detection, disease and pest infestation detection, observing flood and drought conditions, requiring, weather prediction, and precision cultivation, which help to maintain the viability of the agricultural systems and boost the country's economic growth (Shanmugapriya et al., 2019).

CROP PHENOLOGICAL DATA COLLECTION

Harvest growth phases are often seen from the ground, which takes time and has little geographic variation. Land surface phenology (LSP) cultivated evolution periods have been mapped using RS Vegetation Index (VI) period statistics, mostly preceding the growing season. Harvest phenology during the mounting season has gone up dramatically which has been mapped in almost real-time using data from RS with great temporal and geographical resolution. To calculate yield development stages, curve-based techniques blend the most recent observations with period sequences VIs and crop growth phases from previous times. Short-term predictions are possible using curve-based methods. With the use of increasing or diminishing momentum and VI criteria, fashion-based techniques identify time-series information patterns either upward or downward. Only recent observations are used in trend-based techniques. In mapping agricultural growth phases in a timely manner, combining trend and regression techniques show promise. Throughout the growing season, the incidence and convenience of cloud clarifications affect how accurately crop phenology is detected. It looks optimistic that current satellite datasets, like harmonised Landsat and Sentinel-2 (HLS), would be able to monitor agricultural phenology across a vast area during the season. It is possible to use operational applications in the near future (Gao and Zhang, 2021).

PRACTICAL APPLICATIONS OF AI IN AGRICULTURE

AI-based technology has strengthened and increased the productivity of businesses that are based in agriculture. AI technologies may be able to assist farmers in dealing with issues like changing weather patterns and crop-depleting weed and pest infestations. Applications like automatic machine changes for illness or pest diagnostics and disease or weather predictions are currently being used using AI (Thakur and Singh, 2021). Several studies have demonstrated that high-tech technologies, such as AI, the Internet of Things (IoT), and robotics, can be used for virtually all steps in agriculture, from seed or crop selection to marketing, farm maintenance and monitoring, yield estimation, pathogens and weed control, harvesting, and storing, and can almost always increase the efficiency of these practices (Bhagat et al., 2022; Mathushika et al., 2022).

AI technology is used to acquire data from field sensors on soil moisture, weather, fertiliser quantities, irrigation systems, soil composition, and temperature. This aids in raising agricultural yield, which ultimately results in increased farm income (Nigam et al., 2020). Remote sensing technology may now be expanded and used in more places to identify and treat plants, weeds, pests, and diseases thanks to advances in computers, artificial intelligence, and machine learning. It offers a unique chance to create clever planting strategies for precise fertilising. Solutions based on artificial intelligence can help farmers increase product quality, decrease waste, and assure quicker market access. This section will explore several AI technologies for enhancing crops and their yield, issues that farmers run into while applying AI, Future potential of AI in farming start-ups (Raina and Sharma, 2022).

There are several aspects where AI helps in improving applications of GIS in the field of agriculture. Some are listed and discussed below.

CROP/SOIL MONITORING

Crop monitoring, harvesting, storage, and distribution are all parts of crop management, which starts with the sowing of the seed. This is a succinct summary of the activities that raise agricultural product output and growth. Crop output will undoubtedly increase with a thorough awareness of the various crop classes and their scheduling requirements as well as preferred soil types. In order to maximise profitability and safeguard the environment, precision crop management (PCM), a type of agricultural management, targets crop and soil inputs in accordance with field needs. PCM has been hampered by a lack of time, and broadly disseminated information on crop and soil conditions. Farmers must use a number of crop management strategies to address a water shortage caused by the soil, the climate, or inadequate irrigation. It is preferable to use flexible decision-based crop management systems. When deciding between cropping options, drought timing, intensity, and predictability are crucial factors. The crop prediction approach is used to predict the right crop by identifying several soil characteristics and indicators associated with the atmosphere. Depth, temperature, precipitation, humidity, soil type, pH, nitrogen, phosphate, potassium, organic carbon, calcium, magnesium, sulphur, manganese, copper, and iron are a few examples of these variables. Demeter is a speed-rowing device that is computer-controlled and equipped with two cameras and a GPS. It can plan out harvesting activities for a whole field, complete them by cutting crop rows, rotating to cut succeeding rows, moving throughout the field, and seeing unforeseen obstructions (Eli-Chukwu and Ogwugwam, 2019)

INSECT AND PLANT DISEASE DETECTION

Several terms, including chemical classes, functional groups, modes of action, and toxicity, are used to categorise pesticides. Because pesticides use chemicals to control pests and manage weeds, they can be damaging to non-target plants, fish, birds, beneficial insects, soil, water, crops, and non-target animals (Tudi et al., 2021). Farmers must have access to the latest technologies and practices in order to get the most yield possible from their crops. A wide range of industries are using artificial intelligence. Artificial intelligence can be a huge help in combating crop illnesses because of its capacity to recognise issues, identify the proper causes of them, and design effective treatments (Et.al, 2021).

LIVESTOCK MONITORING

In order to enhance animal welfare and health and make a profit, smart farming uses AI technology. Data collection, processing, assessment, and analysis in the areas of detecting animal behaviour, monitoring diseases, estimating growth, and monitoring the environment at the experimental farm accounted for the majority of scientific studies on animal farming supported by sensors and AI models, with the pig (37.95%), cattle (37.44%), and poultry (16.92%) farm animal species receiving the significant attention (Bao and Xie, 2022). Global food security is being hampered by certain unanticipated shocks. Artificial intelligence is one of the contemporary techniques being employed in many stages of the food system, and numerous interventions have been made to improve food security. The use of AI throughout the total food production ecosystem is assessed in this study, covering the cultivation of crops and livestock, their harvesting and slaughter, post-harvesting management, food processing, distribution, and consumption, as well as the management of food waste (Kutyauripo et al., 2023).

SMART SPRAYING

An advanced sensing system for sprayers used on tree crops is being developed and tested. It makes use of artificial intelligence and sensor fusion, including LIDAR, machine vision, and GPS. The inexpensive prototype was tested in citrus groves. Based on tree size and leaf density, it could identify trees and spray accordingly. Compared to conventional spraying, it lowered the volume of the spray by 28% (Partel et al., 2021). AI can make a substantial contribution to improving agronomic practices and achieving the goal of increasing the productivity of alternative arable cropping systems when combined with precision agriculture (PA) and other cutting-edge technology. Farmers may increase yields, maintain their crops, and have a far more dependable source of food by using AI techniques and machine teaching (Naresh et al., 2020).

AUTO-WEEDING

Through the development of intelligent systems that can monitor, manage, and visualise a range of agricultural processes in real-time and with intelligence comparable to that of human professionals, innovations in these digital technologies have led to revolutionary breakthroughs in agriculture. Building smart agricultural instruments, irrigation systems, weed and pest control, fertiliser application, greenhouse culture, storage structures, plant protection drones, crop health monitoring, etc. may be achieved by the IoT and AI (Subeesh and Mehta, 2021). Still today, crop breeders must choose between trade-offs like increased yield and marketable beauty. Furthermore, even the utmost advanced AI cannot ensure the success of a new variety. However, as AI is incorporated into agriculture, some crop researchers envision an agricultural revolution that will be led by computer science (Beans, 2020).

AERIAL IMAGING

Although the fourth digital revolution in agriculture was made possible by technological advancements, there are still many obstacles to overcome, including a lack of arable land, declining water supplies, and climate change. In order to feed the world's population, it is clear that extraordinary efforts are required to create agricultural resilience. GIS has been instrumental in monitoring crops and implementing the best management practices to boost crop productivity. It works in conjunction with other partner technologies like RS, GPS, artificial intelligence, computational systems, and data analytics (Ghosh and Kumpatla, 2022). Drones, also referred to as unmanned aerial vehicles (UAVs), have made significant advancements in recent years. They have revolutionised agricultural operations by providing farmers with significant cost savings, improved operational effectiveness, and higher profitability. In relation to agricultural drones, important subjects include remote sensing, precision agriculture, deep learning, machine learning, and the Internet of Things (Rejeb et al., 2022).

PRODUCE SORTING

Agricultural robotics, predictive analytics, and soil and crop monitoring are the three main areas where AI is starting to show up. In this sense, farmers are utilising sensors and soil samples more frequently to collect data that farm management systems can use for more research and analysis. Through an overview of AI applications in the agriculture industry, this article makes a contribution to the area. It begins with an introduction to AI and a survey of all AI techniques applied in agriculture, including machine learning, the Internet of Things, expert systems, image processing, and computer vision (Elbaşı et al., 2023).

AGRICULTURE'S AI FUTURE: FARMERS AS AI ENGINEERS

To absorb satellite data at the highest spatial and spectral resolutions, sophisticated weather prediction models are required. To enhance the accuracy of weather forecasts, it places new demands on the precision and spectral resolution of soundings. Fishery fleets currently use satellite data for operational purposes. The main driving force behind oceanic motions is wind and the accompanying surface stress. Forecasting ocean circulation requires information on a precise wind field. Monitoring of climate change and variability is becoming more and more dependent on wind data from space. On all spatial scales, the troposphere's chemical composition is shifting. The temperature and chemical balance of the Earth/atmosphere system may be impacted by an increase in trace gases with extended atmospheric residence periods (Fu et al., 2020).

ADVANTAGES AND DISADVANTAGES OF REMOTE SENSING AND GIS APPLICATIONS IN THE AGRICULTURE FIELD

RS offers the advantage of allowing for repeated data collection without crop damage, which can be used to acquire valuable data for precision agricultural applications (Praseartkul et al., 2022; Bawa et al., 2023). The identification and characterisation of agricultural production using RS technologies based on the biophysical characteristics of crops and/or soils could revolutionise how the agricultural output is described (Bégué et al., 2018). The majority of these responses are challenging to accurately and precisely measure visually, however remote sensing applications may offer a more accurate technique to measure stress than visual interpretation (Pause et al., 2016). Precision agriculture is intellectualised through a systematic approach in order to reorganise the entire agricultural system in the direction of low-input, high-efficiency, sustainable agriculture (Yang et al., 2021).

Superior spatial resolution, very simple photography and film processing, affordable equipment, and the provision of a large amount of information are some of its advantages. However, because the film emulsion uses current technology, the sensitivity range is restricted to the visible and near-infrared regions (0.4 to 1 m) (Jangir et al., 2022). The multispectral scanners on the satellites scan the Earth line by line in a variety of distinct light quality ranges (spectrum bands) in the visible and thermal spectra (0.3 to 14.0 m). Due to physical limitations, RS offers the advantage of enabling data and information to be obtained from far locations. In addition to helping decision-makers with disaster management strategies and application schedules, RS has advantages with its ability to monitor disasters and environmental occurrences while saving time and labour (Topçu and Güvel, 2021). RS and GIS have the potential to determine the geographical distribution of environmentally challenged areas on cocoa plantations on a larger scale, which could aid in their early management and mitigation. This can boost cocoa output and provide environmental protection. Numerous uses of remote sensing in agricultural production, particularly cocoa growing, have proven to be beneficial (Rojas-Briceño et al., 2022). The benefits are numerous and crucial, particularly for the quick cross-sector exchanges and the creation of clear and synthesised information for decision-makers. The most significant informational contribution to GIS comes from remote sensing, which offers fundamental informational layers with the best time and space resolutions (Hargreaves and Watmough, 2021).

Even though the majority of satellite RS applications require high spatial resolution, these applications often include huge area studies. However, there is a significant amount of potential for its application in large-scale investigations now that high-resolution sensors have recently developed (Anyimah et al., 2021).

CONCLUSION AND FUTURE PROSPECT

In the past few decades, the use of RS and GIS along with AI in agriculture has grown quickly, and the number of applications and importance of RS and GIS have grown even more in recent years. This is due to the sustainable integration of remote sensing, geographic information systems, and artificial intelligence through key partner technologies for assessing agriculture, soil, crop production estimation, crop disease monitoring, and management, as well as other environmental phenomena like managing natural disasters and climate change. Every stage of the agricultural value chain uses RS, GIS and Artificial Intelligence, as we have discussed in this chapter. In addition to the historical, current, and widespread uses in land suitability/use planning and management of water, soil, biotic, and abiotic stresses, the development of digital agricultural equipment and technologies has increased the use of Remote Sensing and artificial intelligence in high-fidelity crop monitoring, yield prediction, precision farming, and supply chain management for both primary produce and biomass application towards energy production and economic growth of the nation. The fact that artificial intelligence, RS and GIS can collect and analyse data in real-time and give a lot of information about it has created even more importance for providing the location/spatial intelligence that farms need to be more productive and generate more revenue. With their current and future applications, as well as their existing and newer partner technologies, RS and AI are key to making agriculture more productive in a sustainable way. To use the best crop information to help any country's economy, we need to build an information system at the state or district level or taluka based on our current understanding of different crops and agricultural activity.

REFERENCES

Abbasi T, Shams V, Azari B, Shamshirdar F, Baltes J, Sadeghnejad S. Inter-humanoid robot interaction with emphasis on detection: A comparison study – Addendum. *The Knowledge Engineering Review* 2017; 32.

Abd El-Ghany NM, Abd El-Aziz SE, Marei SS. A review: Application of remote sensing as a promising strategy for insect pests and diseases management. *Environmental Science and Pollution Research International* 2020; 27(27): 33503–33515.

Ale, S., Su, Q., Singh, J., Himanshu, S., Fan, Y., Stoker, B., ... & Wall, J. (2022). A Mobile App for Cotton Irrigation Management. *Resource Magazine*, 29(4), 6–8.

Ale, S., Su, Q., Singh, J., Himanshu, S., Fan, Y., Stoker, B., ... & Wall, J. (2023). Development and Evaluation of a Decision Support Mobile Application for Cotton Irrigation Management. *Smart Agricultural Technology*, 5, 100270.

Andersen H-E, Reutebuch S, McGaughey R. *Active Remote Sensing*, pp. 43–66, 2006.

Anyimah FO, Osei Jnr EM, Nyamekye C. Detection of stress areas in cocoa farms using GIS and remote sensing: A case study of Offinso Municipal & Offinso North district, Ghana. *Environmental Challenges* 2021; 4: 100087.

Ashraf M, Maah M, Yusoff I. *Introduction to Remote Sensing of Biomass*, 2011.

Atzberger C. Advances in remote sensing of agriculture: Context description, existing operational monitoring systems and major information needs. *Remote Sensing* 2013; 5(2): 949–981.

Balafoutis AT, Evert FKV, Fountas S. Smart farming technology trends: Economic and environmental effects, labor impact, and adoption readiness. *Agronomy* 2020; 10(5): 743.

Balsamo G, Agusti-Panareda A, Albergel C, Arduini G, Beljaars A, Bidlot J, et al. Satellite and in situ observations for advancing global earth surface modelling: A review. *Remote Sensing* 2018; 10(12): 2038.

Balzter H, Cole B, Thiel C, Schmullius C. Mapping CORINE land cover from Sentinel-1A SAR and SRTM digital elevation model data using random forests. *Remote Sensing* 2015; 7(11): 14876–14898.

Bao J, Xie Q. Artificial intelligence in animal farming: A systematic literature review. *Journal of Cleaner Production* 2022; 331: 129956.

Bawa A, Samanta S, Himanshu SK, Singh J, Kim J, Zhang T., ... Ale, S. (2023). A support vector machine and image processing based approach for counting open cotton bolls and estimating lint yield from UAV imagery. *Smart Agricultural Technology* 2023; 3: 100140.

Beans C. Inner Workings: Crop researchers harness artificial intelligence to breed crops for the changing climate. *Proceedings of the National Academy of Sciences of the United States of America* 2020; 117(44): 27066–27069.

Bechtaoui N, Rabiu MK, Raklami A, Oufdou K, Hafidi M, Jemo M. Phosphate-dependent regulation of growth and stresses management in plants. *Frontiers in Plant Science* 2021: 2357.

Bégué A, Arvor D, Bellon B, Betbeder J, De Abelleyra D, Ferraz PDR, et al. Remote sensing and cropping practices: A review. *Remote Sensing* 2018; 10(2): 99.

Bhagat PR, Naz F, Magda R. Artificial intelligence solutions enabling sustainable agriculture: A bibliometric analysis. *PLoS One* 2022; 17(6): e0268989.

Bitelli G, Blanos R, Conte P, Mandanici E, Paganini P, Pietrapertosa C. *Hyperspectral Data classification to Support the Radiometric Correction of Thermal Imagery*, 2017.

Buja I, Sabella E, Monteduro AG, Chiriacò MS, De Bellis L, Luvisi A, Maruccio G. Advances in plant disease detection and monitoring: From traditional assays to in-field diagnostics. *Sensors (Basel)* 2021; 21(6).

Cafiero G, Cammarano D, Das H, De Simone L, D'Urso G, Sehgal VMV, et al. *Remote Sensing and GIS Applications in Agrometeorology*.

Chang N-B, Imen S, Vannah B. Remote sensing for monitoring surface water quality status and ecosystem state in relation to the nutrient cycle: A 40-year perspective. *Critical Reviews in Environmental Science and Technology* 2015; 45(2).

Chauhan AS, Sharma R, Kumar A, Malik K, Dagar H. *Applications of Remote Sensing in Agriculture*, 2018, pp. 141–146.

Danklmayer A, Doring BJ, Schwerdt M, Chandra M. Assessment of atmospheric propagation effects in SAR images. *IEEE Transactions on Geoscience and Remote Sensing* 2009; 47(10): 3507–3518.

De Jong S, Meer F, Clevers JGPW. *Basics of Remote Sensing*, 2007, pp. 1–15.

De Leeuw J, Georgiadou PY, Kerle N, Gier A, Yoshio I, Ferwerda J, et al. The function of remote sensing in support of environmental policy. *Remote Sensing* 2010; 2(7): 1731–1750.

Elbaşı E, Mostafa N, AlArnaout Z, Zreikat A, Cina E, Shdefat A, et al. Artificial intelligence technology in the agricultural sector A systematic literature review. *IEEE Access* 2023; 11: 171–202.

Eli-Chukwu N, Ogwugwam E. Applications of artificial intelligence in agriculture: A review. *Engineering, Technology and Applied Science Research* 2019; 9(4): 4377–4383.

Ennouri K, Kallel A. Remote sensing: An advanced technique for crop condition assessment. *Mathematical Problems in Engineering* 2019; 2019: 1–8.

Et.al GSS. *Application of Artificial Intelligence in Detection of Diseases in Plants: A Survey*, 2021.

Franke J, Menz G. Multi-temporal wheat disease detection by multi-spectral remote sensing. *Precision Agriculture* 2007; 8(3): 161–172.

Fu W, Ma J, Chen P, Chen F. Remote sensing satellites for digital Earth. In: Guo H, Goodchild MF, Annoni A, editors. *Manual of Digital Earth*. Springer, Singapore, Singapore, pp. 55–123.

Gao F, Zhang X. Mapping crop phenology in near real-time using satellite remote sensing: Challenges and opportunities. *Journal of Remote Sensing* 2021.

Ghosh P, Kumpatla S. *GIS Applications in Agriculture*, 2022.

Gomiero T. Soil degradation, land scarcity and food security: Reviewing a complex. *Challenge. Sustainability* 2016; 8: 281.

Gupta S. *Active and Passive Remote Sensing*, 2018.

Gupta, P. K. (2020a). Pollution load on Indian soil-water systems and associated health hazards: a review. *Journal of Environmental Engineering*, 146(5), 03120004. https://doi.org/10.1061/(ASCE)EE.1943-7870.0001693.

Gupta, P. K. (2020b). Fate, transport, and bioremediation of biodiesel and blended biodiesel in subsurface environment: a review. *Journal of Environmental Engineering*, 146(1), 03119001. https://doi.org/10.1061/(ASCE)EE.1943-7870.0001619.

Gupta, P. K., Gharedaghloo, B., Lynch, M., Cheng, J., Strack, M., Charles, T. C., & Price, J. S. (2020). Dynamics of microbial populations and diversity in NAPL contaminated peat soil under varying water table conditions. *Environmental Research, 191,* 110167. https:// doi.org/10.1016/j.envres.2020.110167.

Gupta, P. K., Mustapha, H. I., Singh, B., & Sharma, Y. C. (2022). Bioremediation of petro-leum contaminated soil-water resources using neat biodiesel: A review. *Sustainable Energy Technologies and Assessments, 53,* 102703. https://doi.org/10.1016/j.seta.2022 .102703.

Hamel A, Ismael B. *Time Series Forecasting Using ARIMA Model,* 2022.

Hargreaves PK, Watmough GR. Satellite Earth observation to support sustainable rural devel-opment. *International Journal of Applied Earth Observation and Geoinformation* 2021; 103: 102466.

Himanshu, S. K., Pandey, A., Madolli, M. J., Palmate, S. S., Kumar, A., Patidar, N., & Yadav, B. (2023a). An ensemble hydrologic modeling system for runoff and evapotranspira-tion evaluation over an agricultural watershed. *Journal of the Indian Society of Remote Sensing, 51*(1), 177–196.

Himanshu, S. K., Pandey, A., Karki, K., Pandey, R. P., Palmate, S. S., & Datta, A. (2023b). Assessing the Applicability of Variable Infiltration Capacity (VIC) Model using Remote Sensing Products for the Analysis of Water Balance: Case Study of the Tons River Basin, India. *Journal of the Indian Society of Remote Sensing,* 51, 2323–2341.

Hoogenboom G. Contribution of agrometeorology to the simulation of crop production and its applications. *Agricultural and Forest Meteorology* 2000; 103(1–2): 137–157.

Huseynova I, Sultanova N, Mammadov A, Suleymanov S, Aliyev J. *Biotic Stress and Crop Improvement,* 2014, pp. 91–120.

Jafarbiglu H, Pourreza A. A comprehensive review of remote sensing platforms, sensors, and applications in nut crops. *Computers and Electronics in Agriculture* 2022; 197: 106844.

Jangir N, Suman S, Kumar N, Saxena S. *Remote Sensing Technology and Its Applications in Agriculture*2022, p. 100.

Kairu E. An introduction to remote sensing. *GeoJournal* 1982; 6(3): 251–260.

Khanal S, KC K, Fulton JP, Shearer S, Ozkan E. Remote sensing in agriculture—Accomplishments, Limitations, and Opportunities. *Remote Sensing* 2020; 12: 3783.

Khouni I, Louhichi G, Ghrabi A. Use of GIS based Inverse Distance Weighted interpola-tion to assess surface water quality: Case of Wadi El Bey, Tunisia. *Environmental Technology and Innovation* 2021; 24: 101892.

Kumar M, Pimprikar A, Afonso M. *A Study of the Application of Remote Sensing in Agriculture and Plant Sciences,* 2022, pp. 91–102.

Kutyauripo I, Rushambwa M, Chiwazi L. Artificial intelligence applications in the agrifood sectors. *Journal of Agriculture and Food Research* 2023; 11: 100502.

Loisel H, Jamet C, Riedi J. *Journal of Remote Sensing* 2009; 187.

Lugari A. *Active and Passive Remote Sensing Techniques and Artificial Neural Networks in Support of Buildings Seismic Vulnerability Assessment,* 2014.

Mahajan, M., Gupta, P. K., Singh, A., Vaish, B., Singh, P., Kothari, R., & Singh, R. P. (2022). A comprehensive study on aquatic chemistry, health risk and remediation techniques of cadmium in groundwater. *Science of The Total Environment, 818,* 151784. https://doi .org/10.1016/j.scitotenv.2021.151784.

Manson S, Bonsal D, Kernik M, Lambin E. *Geographic Information Systems and Remote Sensing. International Encyclopedia of the Social & Behavioral Science,* 2015.

Martínez R, Hemming D, Malone L, Bermudez N, Cockfield G, Diongue A, et al. Improving climate risk management at local level – Techniques, case studies, good practices and guidelines for World Meteorological Organization members. In: Nerija B, editor. *Risk Management.* IntechOpen, Rijeka, 2012.

Matese A, Toscano P, Di Gennaro SF, Genesio L, Vaccari FP, Primicerio J, et al. Intercomparison of UAV, aircraft and satellite remote sensing platforms for precision viticulture. *Remote Sensing* 2015; 7(3): 2971–2990.

Mathushika J, Vinushayini R, Gomes C. *Smart Farming Using Artificial Intelligence, the Internet of Things, and Robotics: A Comprehensive Review*, 2022, pp. 1–19.

Maximillian J, Brusseau ML, Glenn EP, Matthias AD. Chapter 25. Pollution and environmental perturbations in the global system. In: Brusseau ML, Pepper IL, Gerba CP, editors. *Environmental and Pollution Science* (Third Edition). Academic Press, 2019, pp. 457–476.

Messina G, Modica G. The role of remote sensing in olive growing farm management: A research outlook from 2000 to the present in the framework of precision agriculture applications. *Remote Sensing* 2022; 14(23): 5951.

Mirik M, Ansley RJ, Price JA, Workneh F, Rush CM. *Remote Monitoring of Wheat Streak Mosaic Progression Using Sub-Pixel Classification of Landsat 5 TM Imagery For Site Specific Disease Management in Winter Wheat*, 2013.

Mohd Roslim MH, Juraimi A, Che'Ya N, Sulaiman N, Manaf M, Ramli Z, Motmainna M. Using remote sensing and an unmanned aerial system for weed management in agricultural crops: A review. *Agronomy* 2021; 11(9): 1809.

Momtaz S, Shameem MIM. Household assets and capabilities. In: Momtaz S, Shameem MIM, editors. *Experiencing Climate Change in Bangladesh*. Academic Press, Boston, 2016, pp. 69–86.

Naik P, Singamaneni A, Saritha D, Gokul AP. Application of remote sensing and geographical information system in agriculture 2022; 13: 0976–0997.

Naresh R, Chandra M, Vivek, SS, GRC, Chaitanya J, et al. *The Prospect of Artificial Intelligence (AI) in Precision Agriculture for Farming Systems Productivity in Sub-Tropical India: A Review*, 2020.

Nath V. *Remote Sensing and Its Applications*, 2021.

Navalgund R. Remote sensing. *Resonance* 2002; 6: 51–60.

Neményi M, Mesterházi PÁ, Pecze Z, Stépán Z. The role of GIS and GPS in precision farming. *Computers and Electronics in Agriculture* 2003; 40(1–3): 45–55.

Nigam S, Begam S, Naha S, Devi S, Chaurasia H, Kumar D, et al. *Role of Artificial Intelligence (AI) and Internet of Things (IoT) in Mitigating Climate Change-1*, 2020, pp. 465–472.

O'Donoghue C, Buckley C, Chyzheuskaya A, Green S, Howley P, Hynes S, et al. The spatial impact of rural economic change on river water quality. *Land Use Policy* 2021; 103: 105322.

Palanisamy S, Selvaraj R, Ramesh T, Ponnusamy J. Applications of remote sensing in agriculture - A review. *International Journal of Current Microbiology and Applied Sciences* 2019; 8(1): 2270–2283.

Parihar J, FASAL, Oza M. *An Integrated Approach for Crop Assessment and Production Forecasting*. Vol. 6411, 2006a.

Parihar JS, Oza MP. FASAL: An integrated approach for crop assessment and production forecasting. *Agriculture and Hydrology Applications of Remote Sensing*. SPIE 2006b; 6411: 641101.

Parmita G, Siva PK. GIS applications in agriculture. In: Yuanzhi Z, Qiuming C, editors. *Geographic Information Systems and Applications in Coastal Studies*. IntechOpen, Rijeka, 2022.

Partel V, Costa L, Ampatzidis Y. Smart tree crop sprayer utilizing sensor fusion and artificial intelligence. *Computers and Electronics in Agriculture* 2021; 191: 106556.

Pause M, Schweitzer C, Rosenthal M, Keuck V, Bumberger J, Dietrich P, et al. In situ/remote sensing integration to assess forest health—A review. *Remote Sensing* 2016; 8(6): 471.

Petersen GW, Bell JC, McSweeney K, Nielsen GA, Robert PC. Geographic information systems in agronomy. In: Sparks DL, editor. *Advances in Agronomy*. Vol. 55. Academic Press, 1995, pp. 67-111.

Praseartkul P, Taota K, Pipatsitee P, Tisarum R, Sakulleerungroj K, Sotesaritkul T.,... Cha-um, S. Unmanned aerial vehicle-based vegetation monitoring of aboveground and belowground traits of the turmeric plant (Curcuma longa L.). *International Journal of Environmental Science and Technology* 2022: 1–14.

Raina M, Sharma DA. *Intelligent Farming With Artificial Intelligence*, 2022.

Rejeb A, Abdollahi A, Rejeb K, Treiblmaier H. Drones in agriculture: A review and bibliometric analysis. *Computers and Electronics in Agriculture* 2022; 198: 107017.

Repullo-Ruibérriz de Torres MA, Ordóñez-Fernández R, Giráldez JV, Márquez-García J, Laguna A, Carbonell-Bojollo R. Efficiency of four different seeded plants and native vegetation as cover crops in the control of soil and carbon losses by water erosion in olive orchards. *Land Degradation and Development* 2018; 29(8): 2278–2290.

Ribeiro da Luz B, Crowley JK. Spectral reflectance and emissivity features of broad leaf plants: Prospects for remote sensing in the thermal infrared (8.0–14.0 μm). *Remote Sensing of Environment* 2007; 109(4): 393–405.

Rojas-Briceño NB, García L, Cotrina-Sánchez A, Goñas M, Salas López R, Silva López JO, et al. Land suitability for cocoa cultivation in Peru: AHP and MaxEnt modeling in a GIS environment. *Agronomy* 2022; 12(12): 2930.

Rosle R, Che'Ya NN, Ang Y, Rahmat F, Wayayok A, Berahim Z, et al. Weed detection in rice fields using remote sensing technique: A review. *Applied Sciences* 2021; 11(22): 10701.

Shanmugapriya P, Rathika S, Ramesh T, Janaki P. Applications of remote sensing in agriculture-A Review. *International Journal of Current Microbiology and Applied Sciences* 2019; 8(1): 2270–2283.

Singh K, Nair Singh B, Chand P, Sharma SK, Bhati AS. *Sustaining Vegetables Production through Farmers' Own Seed: An Integrated Approach*, 2022, pp. 285–306.

Singh, R. P., Mahajan, M., Gandhi, K., Gupta, P. K., Singh, A., Singh, P., ... & Kidwai, M. K. (2023). A holistic review on trend, occurrence, factors affecting pesticide concentration, and ecological risk assessment. *Environmental Monitoring and Assessment*, 195(4), 451. https://doi.org/10.1007/s10661-023-11005-2.

Sishodia RP, Ray RL, Singh SK. Applications of remote sensing in precision agriculture: A review. *Remote Sensing* 2020; 12(19): 3136.

Sivakumar M. *Satellite Remote Sensing and GIS Applications in Agricultural Meteorology*. World Meteorological Organisation, 2003.

Sivakumar M, Hinsman D. *Satellite Remote Sensing and GIS Applications in Agricultural Meteo-Rology and WMO Satellite Activities*, 2004.

Sood K, Singh S, Rana R, Rana A, Kalia V, Kaushal A. *Application of GIS in Precision Agriculture*, 2015.

Steven M, Clark JA. *Applications of Remote Sensing in Agriculture*. Elsevier, 2013.

Stone RC, Meinke H. Weather climate, and farmers: An overview. *Meteorological Applications* 2006; 13(S1): 7–20.

Subeesh A, Mehta CR. Automation and digitization of agriculture using artificial intelligence and internet of things. *Artificial Intelligence in Agriculture* 2021; 5: 278–291.

Thakur R, Singh B. *Importance of Artificial Intelligence in Agriculture*, 2021.

Topçu E, Güvel Ş. *Drought Assessment by Using Geographic Information Systems and Remote Sensing*, 2021.

Trout T, Ahuja L, Ma L, McMaster GS, Nielsen DC, Andales A, et al. Quantifying crop water stress factors from soil water measurements in a limited irrigation experiment. *Agricultural Systems* 2015; 137: 191–205.

Tudi M, Daniel Ruan H, Wang L, Lyu J, Sadler R, Connell D, et al. Agriculture development, pesticide application and its impact on the environment. *International Journal of Environmental Research and Public Health* 2021; 18(3).

USGS. *What Are the Band Designations for the Landsat Satellites?*

Veljanovski T, Kanjir U, Oštir K. Object-based image analysis of remote sensing data. *Geodetski vestnik* 2011; 55(4): 641–664.

Weiss M, Jacob F, Duveiller G. Remote sensing for agricultural applications: A meta-review. *Remote Sensing of Environment* 2019; 236.

Wójtowicz M, Wójtowicz A, Piekarczyk J. Application of remote sensing methods in agriculture 2016; 11: 31–50.

Yang X, Shu L, Chen J, Ferrag MA, Wu J, Nurellari E, et al. A survey on smart agriculture: Development modes, technologies, and security and privacy challenges. *IEEE/CAA Journal of Automatica Sinica* 2021; 8(2): 273–302.

Yatoo, A.M., Ali, M.N., Zaheen, Z., Baba, Z.A., Ali, S., Rasool, S., Sheikh, T.A., Sillanpää, M., Gupta, P.K., Hamid, B. and Hamid, B., (2022). Assessment of pesticide toxicity on earthworms using multiple biomarkers: a review. *Environmental Chemistry Letters*, 20(4), 2573–2596. https://doi.org/10.1007/s10311-022-01386-0.

Yousaf W, Awan WK, Kamran M, Ahmad SR, Bodla HU, Riaz M, et al. A paradigm of GIS and remote sensing for crop water deficit assessment in near real time to improve irrigation distribution plan. *Agricultural Water Management* 2021; 243: 106443.

Zahran SAE-S, Saeed RA-H, Elazizy IM. Remote sensing based water resources and agriculture spatial indicators system. *The Egyptian Journal of Remote Sensing and Space Science* 2022; 25(2): 515–527.

Zhang M, Wu B, Yu M, Zou W, Zheng Y. Crop condition assessment with adjusted NDVI using the uncropped arable land ratio. *Remote Sensing* 2014; 6(6): 5774–5794.

16 Flood Modeling Using the AHP Method in a GIS Environment of the Iril River Catchment, Manipur, India

Sandhip Khundrakpam and
Thiyam Tamphasana Devi

INTRODUCTION

Flood is a primary natural disaster (Khan et al., 2011) that has been occurring in all parts of the world. As the flood event increases, it is necessary to conduct assessment studies to mitigate the causes of the flood (Jonkman et al., 2012). There are various forms of flood, including flash floods, urban floods, coastal floods, and land ponding floods, which have distinct impacts and disturbances to the environment and ecosystem. Flooding can occur as a result of extreme precipitation, quick snowfall, or storm surge from a tropical cyclone or tsunami in coastal locations (Gupta et al., 2019). Dam collapse, snowmelt, deforestation, climate change, emissions of greenhouse gases, etc., are also other factors causing floods. A flood is an unwanted event which impacts living things on Earth in different directions and is a threat to its survival (Drdácký, 2010). Floods caused by heavy rains triggered overflows in rivers, lakes, and ponds, as well as caused drainage system failure. Health systems, facilities, and services were disrupted; leaving people without access to healthcare is the most disturbing effect of flood. During monsoon season, flooding is a major natural hazard in any region, destroying crops and people's dwellings (Manipur Science and Technology Council, MASTEC, 2000; Madolli et al., 2022). When agricultural activities (crops as well as livestock) are affected by floods, there is a loss of livelihood to several people and subsequently this affects the overall development of the region in different dimensions and thus its overall critical infrastructure (a network of different facilities to maintain a normal daily life which includes transportation network, utilities and buildings, etc.) is disturbed (Deshmukh et al., 2011). Hence, it is very much imperative to understand flood scenarios before any

DOI: 10.1201/9781003441175-16

planning is done in the region. Thus, prediction of flood and its mapping can be achieved using satellite data and GIS (Geographical Information System) techniques (Himanshu et al., 2023a, 2023b). Several researchers have used GIS techniques to quantify and map flood-affected zones. An approach to the Analytical Hierarchy Process (AHP) (Saaty, 1980) is a GIS-based model (Saaty, 2008) for flood modeling which is frequently adopted by researchers (Ouma and Tateishi, 2014; Gigovic et al., 2017; Khaleghi and Mahmoodi, 2017; Das, 2018). A combined approach of AHP with GIS techniques and satellite data can be the best agri-tech approach in the water management of agricultural activities if the weights assigned are suitably taken considering all the possible factors. Deshmukh et al. (2011) conducted an assessment of a real flood event that occurred in 2008 in the United States using a decision-making methodology (Severity Assessment Tool). They report the severity of the impact of floods in different cities of the United States on critical infrastructure such as transportation networks (highways, railways, port ways, connecting bridges and tunnels), utility facilities for food, medicine, and water supplies, and on the building of different uses. They develop a relationship between the serviceability (activities) and functionality (infrastructure) for the post-recovery flood. Based on the developed relationship between serviceability and functionality, they suggest few support systems to manage the problems created by flood. Such support systems were developed (framed) based on (i) alternate infrastructure available (such as for commuting and shopping) and (ii) absence of infrastructure alternates available (such as for supply of electricity, water, gas and telecommunications &postal services, etc.). Flood hazard maps provide vital and useful information that assists in disaster management and mitigation by taking care of people and authorities in identifying the risk posed by natural catastrophes. Application of this GIS-AHP in flood risk mapping in Indian catchments was also conducted by different authors (Sinha et al., 2008; Ajin et al., 2013; Dash and Sar, 2020; Vignesh et al., 2021) and regionally by Pradyumna and Devi (2022). In all these studies, the disturbance caused by floods on the human ecosystem was reported so that prior information by the concerned authority could be gathered.

The major causes of flooding in Manipur Valley (the surrounding region of the present study catchment area) are heavy runoff and low infiltration in degraded watersheds in the higher reaches of rivers during rainy seasons. In this study region, MASTEC, which is a government autonomous body taking responsibility for conducting flood risk assessment, published a detailed report in 2000 about the scenario of flood (flood risk maps) in the state. It reports that due to insufficient drainage, flash floods occur practically every year during the rainy season. The present study region has a subtropical monsoon environment and the wet season is quite long (early May to mid-October every year) during September 1999, the Manipur Valley experienced flooding which became a customary event each year during this season. From August 24 until September 3, 1999, there was nonstop rain causing floods that primarily devastated the valley's southern reaches. It also reports that there are 7300 impacted households and 15,300 hectares of rice fields (MASTEC, 2000). Apart from such reports published by government organizations, which provide generalized information for a larger region, there is very little information available, particularly for a specific catchment region. As a result, this present study is being conducted so that this gap of non-availability or very little information can be filled, especially for the Iril River catchment, as it is an important region whose downstream mostly impacts the central city part of the state, thus its effect is highly significant on the overall well-being of the state.

Therefore, in this study, the prediction and mapping of flood-affected areas of the Iril River catchment in Manipur is conducted using the AHP method with GIS techniques to quantify the risk of flooding and its effect on the vegetation-covered land as well as built-up areas.

STUDY AREA

The study area (Iril catchment) is located in the districts of Senapati, Imphal east and west and Ukhrul of Manipur state (Northeastern part of India) shown in Figure 16.1.

The geographical area lies between latitude $24^040'$ N–$25^025'$ N and longitude $93^055'$ E–$94^020'$ E. The catchment area is estimated to be around 2985.5 km^2. The pour point to delineate the catchment area is taken at Lamboikhul Tiger Camp (on the riverbed under the Eereima Suspension Bridge) having latitude $24^055'58.38''$N and longitude $94^02'46.41''$E, which is marked in Figure 16.1. Vulnerable points of the Iril River basin are at Pukhao Iranpham, Nongthombam Rajen mapa of Phaknungmayai Leikai, Khoisam mapa to Yambem mapa of Moirangkampu, Thoudam mapa, Naharup, Saikhom mapa of Naharup, Major Ingkhol mapa, Keirao Makting,

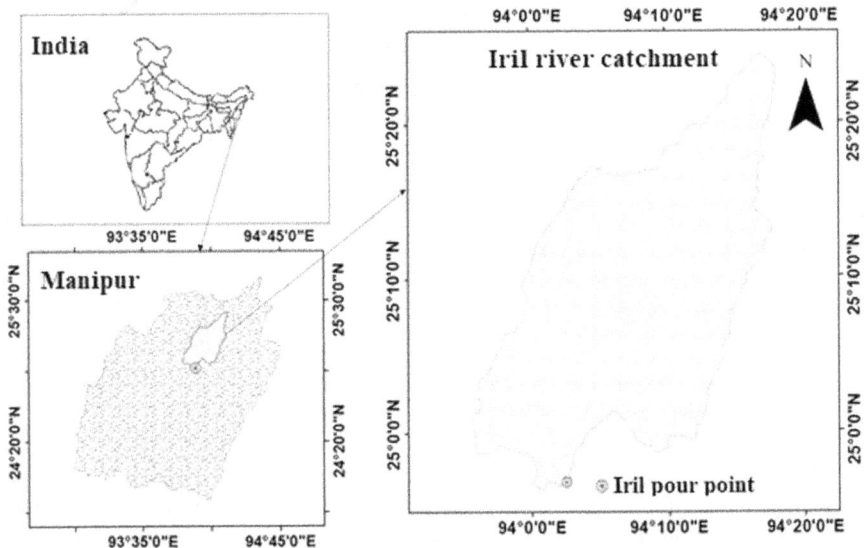

FIGURE 16.1   Study area (Iril River catchment)

Rahamat mapa, Keirao Makting, Konsam Momon mapa and Arapti Mamang Leikai (Disaster Management System of Manipur, DMSM, 2019). These places are frequently and severely affected by flooding most of the time due to heavy rain.

DATA USED AND METHOD

DATA USED

The required data are collected and shown in Table 16.1. The data that are collected are DEM (Digital Elevation Model), Landsat 8/9 (2017, 2019, 2021) from USGS (United States Geological Survey), Earth Explorer; rainfall (Directorate of Environment and Climate Change, Government of Manipur) and soil type (Food and Agricultural Organization, United Nation). The Landsat 8/9 data were taken for the month of October.

METHOD

Prediction and mapping of flood-affected zones in the study region using the GIS-AHP method includes collection of required data from different sources, generation of a database of flood mapping (selection of input parameters), and finally development of a flood risk map through the AHP approach and then create a hierarchical structure to provide preferable solutions for flood risk analysis as a result or outcome of the study. AHP is a decision-making approach that analyses judgment calls. Theoretically, the process of AHP includes Step I: Problem modeling, Step II: Determination of priorities, Step III: Deriving the overall relative weights, Step IV: Judgment and conclusion. And, in the GIS model, it is broadly classified as primary and secondary processing. As part of primary processing, the generation of thematic layers of the selected input parameters is performed. The selected criteria or input parameters (Gigovic et al., 2017) for the flood mapping are rainfall, slope,

TABLE 16.1
Input Parameters Use in the AHP Method

Sl. No.	Data	Source	Resolution	Extracted data
1.	DEM	USGS, Earth Explorer	Spatial (30 m)	Slope, Drainage density
2.	NDVI	USGS, Earth Explorer Landsat 8/9	Temporal (2015, 2017, 2019, 2021)	NDVI
3.	Rainfall	Directorate of Environment and Climate Change	Temporal (2015, 2017, 2019, 2021)	Rainfall
4.	TWI	USGS, Earth Explorer	Spatial (30 m)	TWI
5	Soil type	Food and Agricultural Organization (FAO)	Spatial (30 m)	Soil type of study Iril River catchment

soil type, *NDVI* (Normalized Difference Vegetation Index), *TWI* (Topographical Wetness Index), and drainage density (Ouma and Tateishi, 2014) based on the significant influence on causing floods. The theoretical concepts of each parameter will be briefly presented in the following section. A flowchart of the methodology adopted in this study is provided in Figure 16.2.

The data collected from different sources, which is shown in Table 16.1, is utilized and processed for the required format of the thematic layer using GIS tools (ArcGIS10.3®). In order to generate the secondary data from the primary data, the IDW (Inverse Distance Weighting) algorithm is used, which is within the GIS platform. Once the required input parameters are generated (secondary data or thematic layers), individual evaluation is made and fixed for the modeling process. Then, the weights of the parameters are calculated using the Saaty (1980) approach and

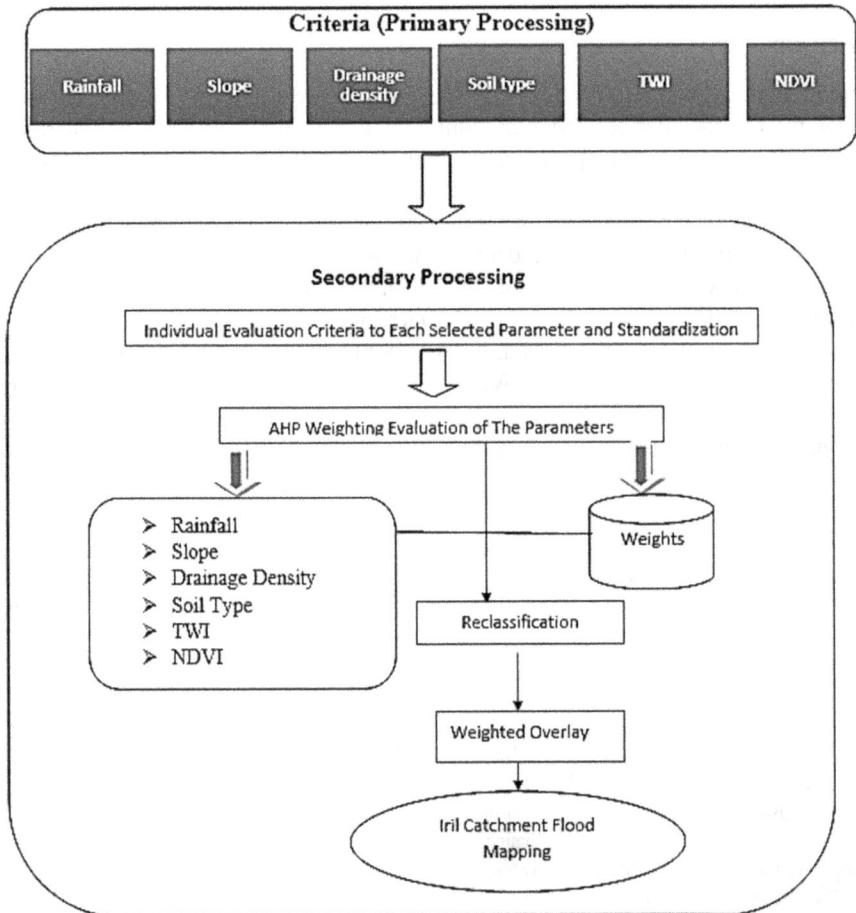

FIGURE 16.2   Flowchart of methodology

accordingly re-classification is done. Accurate weights are assigned to each input parameter to generate the final flood risk map (for the years 2015, 2017, 2019, and 2021) by the weighted overlay method.

INPUT PARAMETERS – THEORETICAL CONCEPT

There are six inputs considered for flood modeling in this study and the significance of the theoretical concept is provided in the following section.

SLOPE

The slope of a terrain is a critical aspect in the determination of its dependability. The direction and quantity of surface runoff or subsurface drainage that reaches an area are determined by the slope. The contribution of rainfall to stream flow is dominated by slope. It regulates the length of overland flow, infiltration, and subterranean flows are all examples of flow. Slope is presented in percentage and calculated as rise divided by run multiplied by 100.

DRAINAGE DENSITY

If the density is high, the catchment region will be more prone to degradation, ending in deposition on the deeper grounds. It is calculated as stream length divided by basin area. And its unit is km/km^2.

SOIL TYPE

Soil texture is an essential component and property of soils. Clay soils are far less transparent and hold water for a greater amount of time than sandy soils. It demonstrates that locations with clay soils are more prone to flooding. When measurements are unavailable, the feel and appearance of the soil can be used to infer soil moisture and serve as a barrier. It serves as the boundary between the land surface and atmosphere, and it is important in the division of rainfall into runoff and water table storage.

TWI

TWI (Nsangoua et al., 2021) indicates the amount of water that is accumulated in a specific area, which is given in index as high value gives high potential and vice versa. It ranges from –3 to 30. It is calculated as:

$$TWI = \ln\frac{U_{as}}{\tan\beta} \tag{16.1}$$

Where U_{as} is the area contributing to its upstream side, β is the slope gradient.

RAINFALL DISTRIBUTION

Heavy rainfall is a primary cause of flooding heavy rains, which prohibit natural watercourses from channeling surplus water, and they are the most prevalent cause of flooding. The amount of runoff generated by a site is related to the amount of rain received. Heavy rains raise the level of water in rivers and lakes. When the water level exceeds the riverbanks or dams, the water begins to overflow, resulting in river-based floods. Water overflows from all the water bodies.

NDVI

In *NDVI*, using reflected light in the visible (*VIS*) and near-infrared wavelengths (*NIR*), the *NDVI* analyses and assesses the presence of live greenery. Simply, *NDVI* is a gauge of the greenness of vegetation, as well as its richness and health. Thus, *NDVI* is calculated as:

$$NDVI = \frac{(NIR - VIS)}{(NIR + VIS)} \tag{16.2}$$

AHP WEIGHTS

The major purpose of multi-criteria decisions based on AHP is a ranking process. The effectiveness of available the quality of prioritization has a direct impact on resources. In most situations, the decision maker's primary judgment is used. In this study, experts and decision makers with their technical skills and know-how to solve the problems are considered. Several field surveys were conducted in and around the study area interacting with the people of local communities. With the understanding resulting from the community interaction and consultation with them,, the weights are assigned to the selected input parameters on a scale of 1–5 which is very low to very high classification. The standardized raster layers are weighted using an eigenvector to demonstrate the significance, considering each criterion in regard to the other elements in the flood risks resulting. Table 16.2 shows the outcomes of the pairwise comparison and criterion ranking.

From Table 16.3, the Consistency Index (*CI*) is found by the Matrix, $AX = \lambda_{max} X$. In which X is the priority vector (weight) and λ_{max} is calculated as 6.52, then, *CI* is calculated by Equation (16.3) as:

$$CI = (6.52 - n)/(n - 1) \tag{16.3}$$

Where, n = number of input parameters, which are six numbers for this study. Then, the Consistency Ratio (*CR*) is calculated as $CR = CI / RI = 0.08$. Random Index (*RI*) is obtained from Saaty's (1980) table and in the table, its value is given as 1.25 for six numbers of input parameters in the case of this present study. The calculated *CI*

TABLE 16.2
AHP Pairwise Comparison Matrix

Factors	Rainfall	Drainage	Slope	TWI	Soil	NDVI
Rainfall	1	3	5	3	3	9
drainage	0.333	1	3	0.333	0.333	3
Slope	0.2	0.333	1	0.5	0.5	3
TWI	0.333	3	2	1	0.333	5
soil	0.333	3	2	3	1	3
NDVI	0.11	0.333	0.333	0.2	0.333	1

TABLE 16.3
Calculation of Priority Vector

Factors	Rainfall	Drainage	Elevation	Slope	Soil	NDVI	Avg (X)
Rainfall	0.433	0.281	0.375	0.373	0.545	0.375	0.397
Drainage	0.144	0.094	0.225	0.041	0.061	0.125	0.115
Slope	0.086	0.031	0.075	0.062	0.091	0.125	0.078
TWI	0.144	0.281	0.15	0.124	0.061	0.208	0.161
Soil	0.144	0.281	0.15	0.373	0.182	0.125	0.209
NDVI	0.048	0.031	0.025	0.025	0.061	0.042	0.038

value is provided in Table 16.3. The calculated CR is substantially less than the lower limit of 0.1, indicating a less than the cut-off value.

The weight is given in Table 16.4 and reclassified into five classes based on the ranges of the criteria taken for flood mapping to identify the flood-affected area. Using the assigned ranks and weights, these thematic layers (input parameters) are overlaid by using the weighted overlay method and finally a flood risk map (flood-affected area) is generated for the years 2015, 2017, 2019, and 2021.

RESULT AND DISCUSSION

GENERATED INPUT PARAMETERS

A generated slope (%, which is rise divided by run multiplied by 100) and drainage density (km/km²) map is provided in Figure 16.3. Drainage density is high for the major streams (Iril River) and nearby to the river and its tributaries. For soil (Figure 16.4), it has been generated using FAO (Food and Agricultural Organization, United Nations) data which is given as clay loamy soil type for the study area and TWI is also shown in Figure 16.4 (b). It is observed that the entire study region is covered by clay loam soil which has the combined properties of low drainage capacity, moderate fertility and good water-holding potential.

TABLE 16.4
Assigned AHP Rankings and Weights for Flood Mapping

Criteria	Units	Year				Class	Rank	Weight
		2015	2017	2019	2021			
NDVI	Level	-1-0	-1-0	-1-0	-1-0	Very high	5	4%
Slope	%	0-6	0-6	0-6	0-6	Very high	5	8%
		6-16	6-16	6-16	6-16	High	4	
		16.01-40	16.01-40	16.01-40	16.01-40	Moderate	3	
		40.01-60	40.01-60	40.01-60	40.01-60	Low	2	
		60.1-183.8	60.1-183.8	60.1-183.8	60.1-183.8	Very low	1	
TWI	Level	-0.58-5	-0.58-5	-0.58-5	-0.58-5	Very high	1	12%
		5.1-9	5.1-9	5.1-9	5.1-9	High	2	
		9.1-11	9.1-11	9.1-11	9.1-11	Moderate	3	
		12-19	12-19	12-19	12-19	Low	4	
		20-21	20-21	20-21	20-21	Very low	5	
Drainage density	km/km^2	0-2	0-2	0-2	0-2	Very low	1	16%
		2.1-4	2.1-4	2.1-4	2.1-4	Low	2	
		4.1-4.5	4.1-4.5	4.1-4.5	4.1-4.5	Moderate	3	
Soil type	Level	Clay Loam	Clay Loam	Clay Loam	Clay Loam	Moderate	3	21%
Rainfall	mm	59.87-402.7	83.3-600	761-914	1040-1100	Very low	1	39%
		402.8-642.7	601-1000	915-1010	1100-1140	Low	2	
		642.8-833.7	1000-1400	1020-1090	1150-1180	Moderate	3	
		833.8-1030	1410-1700	1100-1160	1190-1210	High	4	
		1031-1309	1710-2230	1170-1240	1220-1250	Very high	5	

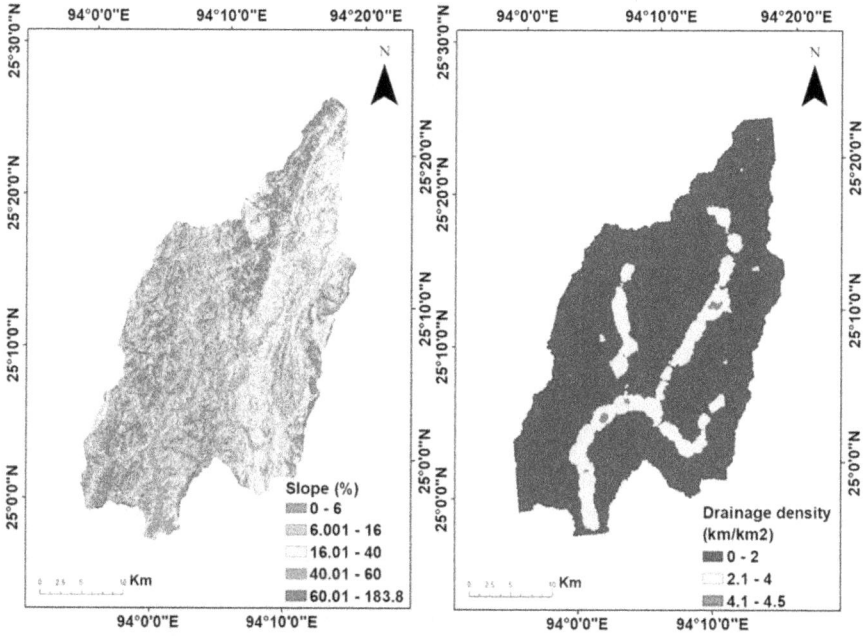

FIGURE 16.3 (a) Slope and (b) drainage density

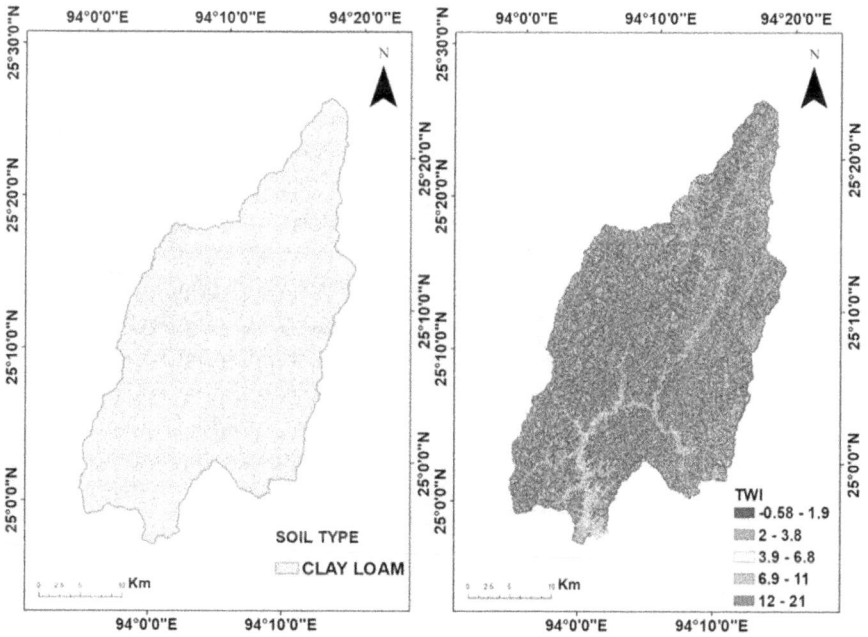

FIGURE 16.4 (a) Soil type and (b) *TWI*

By the IDW method, the rainfall ranges of the given study area are extracted and provided in Figure 16.5 for the years 2015, 2017, 2019, and 2021. High rainfall is concentrated mostly in the northern part of the study region in all the predicted years. The highest rainfall goes up to around 2230 mm in a year which was observed in the

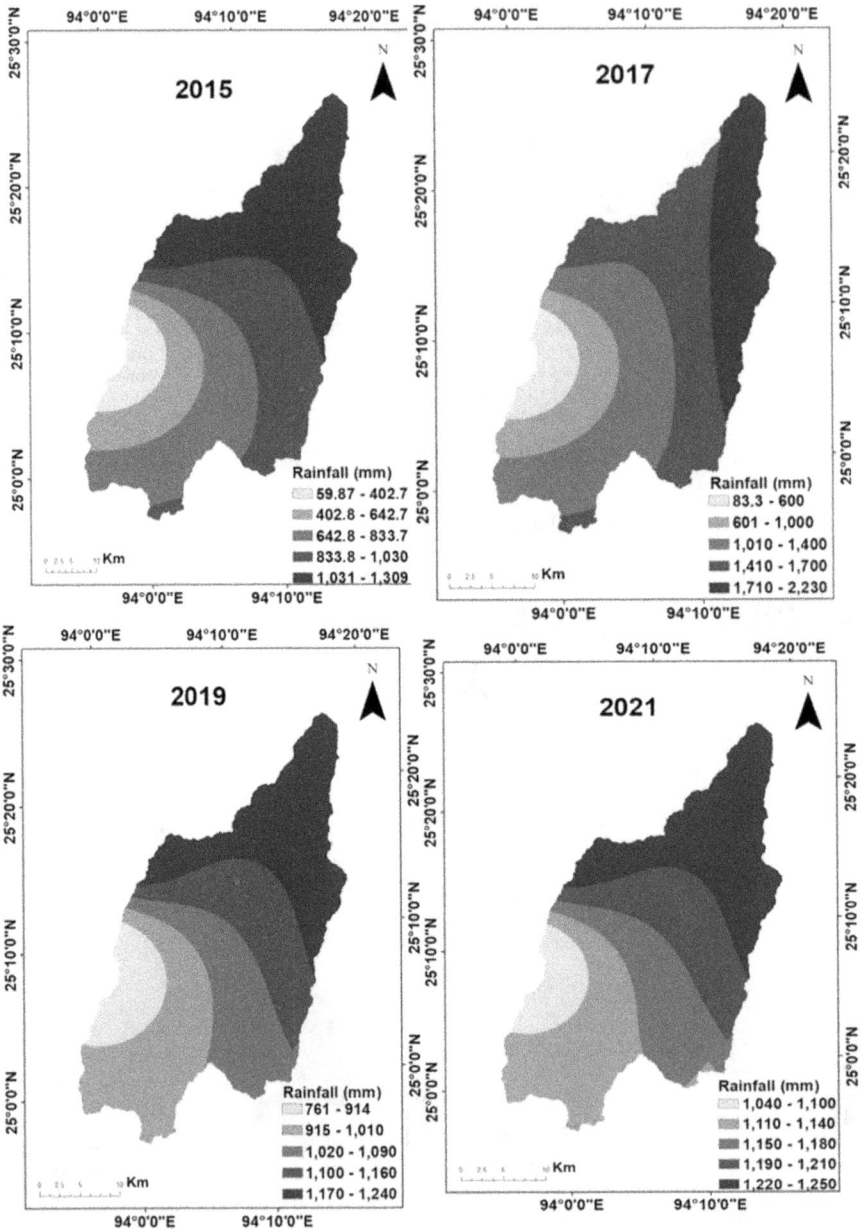

FIGURE 16.5   Rainfall for the years (a) 2015, (b) 2017, (c) 2019, and (d) 2021

TABLE 16.5

Flood Class Based on Predicted, Simulated and Observed Data

Year	Present study (High flood, km²)	Predicted discharge (SWAT model)[1] (m³/sec)	Observed discharge (IFCD)[2] (m³/sec)	Validation remarks		
				Present study (AHP model)	SWAT Model[1]	Ground data[2]
2013	–	907.981	922.863	–	No flood	No flood
2014	–	929.45	914.9501	–	No flood	No flood
2015	39.4	2452.87	2414.15	**Very high flood**	**Very high flood**	**Very high flood**
2017	38.1	–	–	High flood	–	–
2019	25.7	–	–	High flood	–	–
2021	26.6	–	–	High flood	–	–

[2]observed discharge (IFCD) or ground data, result collected from Irrigation and Flood Control Department (IFCD), Government of Manipur.

note: [1]predicted discharge (swat model), result taken from Pradyumna and Devi (2022)

year 2017 followed by 1309 mm in the year 2015 and lower rainfall observed in 2019 and 2021 and the lowest is observed in the year 2015 (60 mm per year) followed by in the year 2017 which is around 83 mm. So, there is a large uneven distribution of rainfall in these years which suggests uncertain climatic conditions. However, in the years 2019 and 2021, the minimum rainfall observed is around 761 mm to 1040 mm and the maximum is 1240 mm to 1250 mm respectively which indicates moderate climate conditions as per the web report of the Indian Meteorological Department (IMD, 2022). Table 16.5

Then, the *NDV* (range of –1 to 0) of these years was calculated and found that it has less vegetation in the month of October (Figure 16.6) which is the month of every year in this study.

The predicted flood risk map for the years 2015, 2017, 2019, and 2021 is shown in Figure 16.7 (a–d). Flood-affected zone is classified as very low, low, moderate, and high for each year. Table 16.6

From Figure 16.7, all the years (2015, 2017, 2019, and 2021) are affected by low to moderate floods and very small areas by high floods. Figure 16.8 shows the area (%) of predicted flood. But the high category flood is the most significant to cause damage to agricultural activities as well as to properties. Thus, it is effective to study further, selecting only the flood-affected areas classified as high-risk in the present study. Therefore, the flood-affected area in 2015 is observed to be the highest, at 39.2 km² (1.32%), followed by the year 2017 at 38.1 km² (1.28%) in the category of high flood as compared with other years, as shown in Figure 16.8.

Such a study of flood prediction and mapping of flood-affected areas in a catchment or region could provide early information to the concerned, such as farmers,

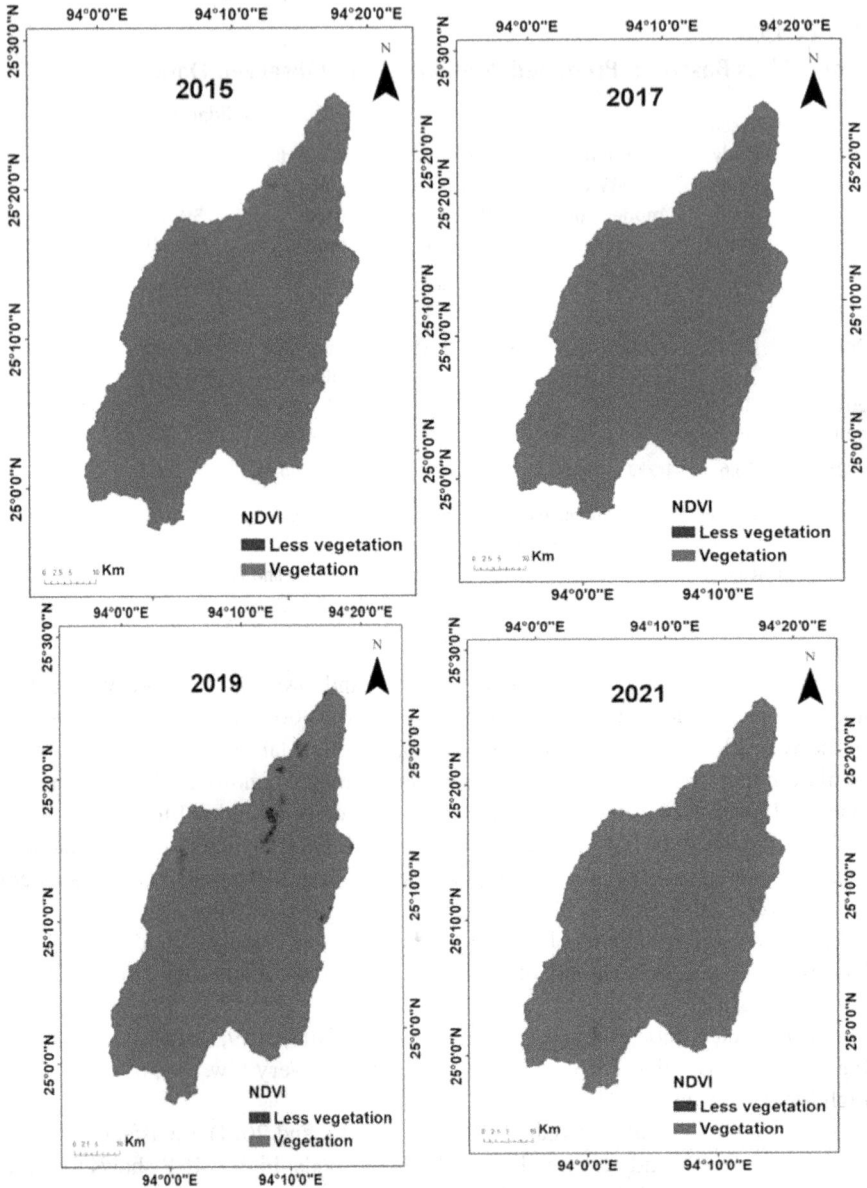

FIGURE 16.6   NDVI for the years (a) 2015, (b) 2017, (c) 2019, and (d) 2021

if we look for agricultural activities in the region. The agricultural activities such as water resource management in terms of irrigation systems or drainage systems in and around agricultural land can be enhanced by the similar results of the present study. In this study, the decision-making method AHP combined with the techniques of GIS and several satellite data sets is a kind of agri-tech tool or approach that

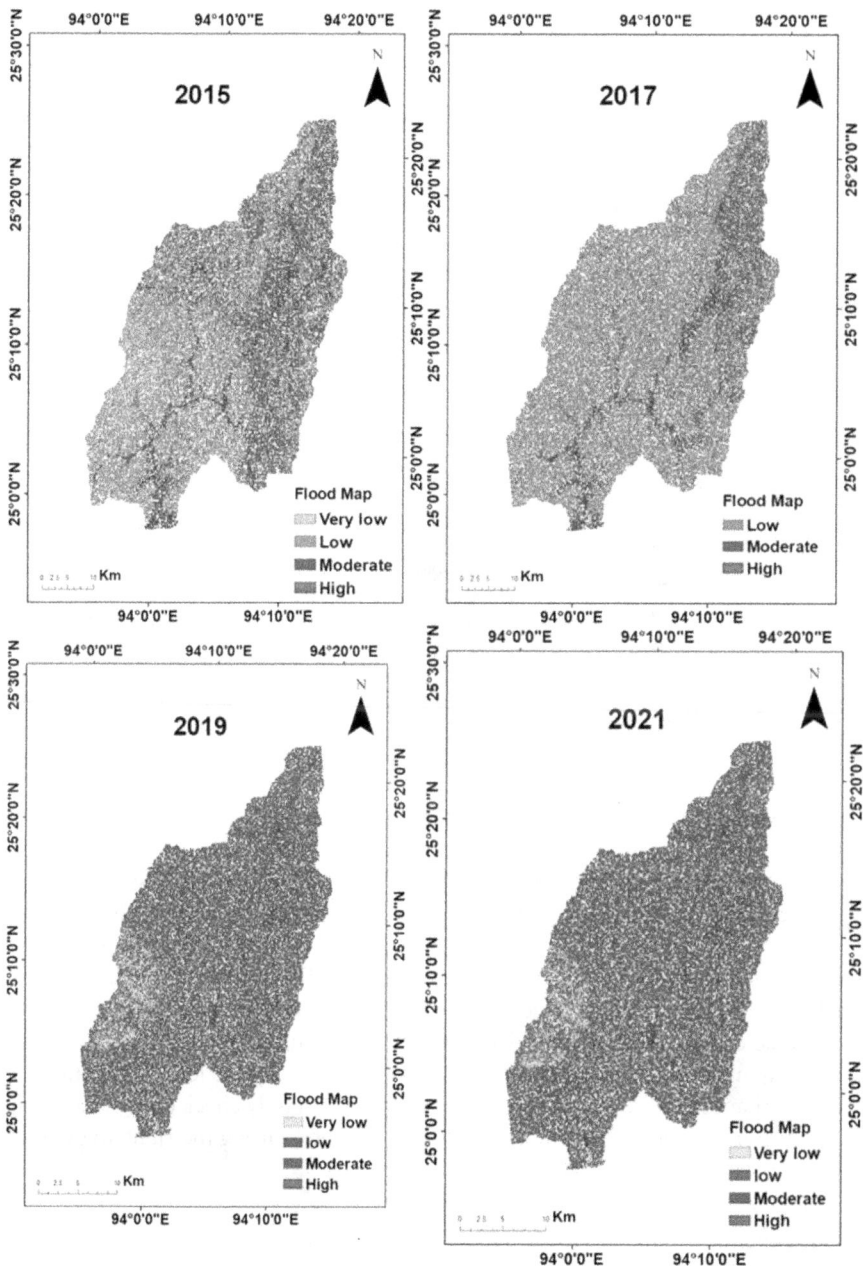

FIGURE 16.7 Flood risk map for the years (a) 2015 (b) 2017, (c) 2019, and (d) 2021. Table 16.5 shows the predicted flood area in km² and % for the years 2015, 2017, 2019, and 2021. The north-west side of the study region is more prone to flood as compared to the south-east side

TABLE 16.6
Predicted Flood-Affected Area

Flood class	Flood-affected area							
	(km²)				(%)			
	2015	2017	2019	2021	2015	2017	2019	2021
Very low	39.2	0.00	72.00	58.30	1.31	0.00	2.41	1.95
Low	1940.9	2257.80	2149.50	2136.90	65.01	75.63	72.00	71.58
Moderate	966	689.60	738.30	763.70	32.36	23.10	24.73	25.58
High	39.4	38.10	25.70	26.60	1.32	1.28	0.86	0.89

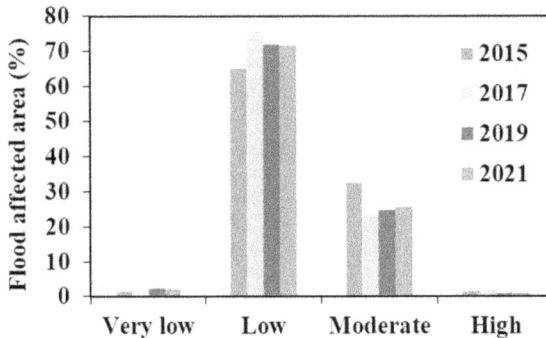

FIGURE 16.8   Predicted flood-affected area (%)

serves the purpose of achieving simulated flood information in any specific catchment area. This geospatial model has the potential to improve the results further if we could further accurately assign the weights; thus, its application is not limited and always has the scope for improvement. The AHP method is interrelated with machine learning methods, which have vast applications in the agricultural sector to enhance crop productivity and its nutrients. Thus, the scope of this present study can be expanded further to apply the AHP with other machine learning methods in genetic enhancement research of crop varieties that not only check crop productivity but also check nutrient benefits. It can also be applied to using the right amount and timing of fertilizer.

VALIDATION

The predicted (present study), simulated result (literature) and ground data (IFCD) were compared for the respective years (2015, 2017, 2019, and 2021). For the year 2015, the predicted area inundated by flood water is 39.4 km² with a simulated discharge of 2452.87 m³/sec which is very high compared with the other years in both the cases (predicted and simulated). The collected ground data is also the highest (2414.15 m³/sec) in the year 2015 as compared with other years. Thus, this very high flood classification for the year 2015 is compared with the simulated SWAT model (Pradyumna and Devi, 2022) result and ground data and it has been found that they match the same result (high flood class).

CONCLUSION

The predicted flood-affected was highest in the year 2015 followed by 2017, 2019, and 2021. Thus, it is concluded that throughout the year, this study region is affected by low to high flood intensity. In such a situation, it is not recommended for continuous cropping practice unless a proper irrigation system is provided. Therefore, it is suggested to provide a proper drainage system to drain out the flood water or divert the excess surface water to those areas where it is needed or develop of storage mechanism. Thus, a combined approach of AHP with GIS tools and satellite data can be used in water management of agricultural practices and can become one of the best agri-tech approaches in this field.

REFERENCES

Ajin, R.S., Krishnamurthy, R.R., Jayaprakash, M. and Vinod, P.G. Flood hazard assessment of Vamanapuram river basin, Kerala, India: An approach using remote sensing & GIS techniques. *Advances in Applied Science Research*, 4(3): 263–274, 2013.

Aldharab, H. S., Ali, S. A. and Ghareb, J. I. S. A. Analysis of basin geometry in ATAQ region, part of Shabwah Yemen: Using remote sensing and geographic information system techniques. *Bulletin of Pure and Applied Sciences*. (Geology), No.1, 38 F: 1–15, 2019.

Behera, P.K., Devi, T.T. Study on Impact of Urbanization by SWAT Model in Iril River, Northeast India. In: Jha, R., Singh, V.P., Singh, V., Roy, L.B., Thendiyath, R. (eds) *Hydrological Modeling. Water Science and Technology Library*, 109: 385–393, 2022.

Das, S. Geographic information system and AHP-based flood hazard zonation of Vaitarna basin, Maharashtra, India. *Arabian Journal of Geosciences*, 11(19): 1–13, 2018.

Dash, P. and Sar, J. Identification and validation of potential flood hazard areas using GIS-based multi-criteria analysis and satellite data-derived water index. *Journal of Flood Risk Management*, 13(3): 1–14, 2020.

Deshmukh, A., Oh, E.H. and Hastak, M. Impact of flood damaged critical infrastructure on communities and industries. *Built Environment Project and Asset Management*, 1(2): 156–175, 2011.

DMSM (Disaster Management System of Manipur). *District Disaster Management Plan, (2018–2019), report* volume II, 2019.

Drdácký, M.F. Impact of Floods on Heritage Structures. *Journal of Performance of Constructed Facilities*, 24(5): 430 – 431, 2010.

Gigovic, L., Pamučar, D., Bajić, Z. and Drobnjak, S. Application of GIS-interval rough AHP methodology for flood hazard mapping in urban areas. *Water*, 9(6): 360–-366, 2017.

Gupta, S., Gupta, A., Himanshu, S.K. and Singh, R. Analysis of the extreme rainfall events over upper catchment of Sabarmati River basin in Western India using extreme precipitation indices. In *Advances in Water Resources Engineering and Management: Select Proceedings of Trace 2018*, pp. 103–111, 2019. Springer, Singapore.

Himanshu, S.K., Pandey, A., Madolli, M.J., Palmate, S.S., Kumar, A., Patidar, N. and Yadav, B. An ensemble hydrologic modeling system for runoff and evapotranspiration evaluation over an agricultural watershed. *Journal of the Indian Society of Remote Sensing*, 51(1): 177–196, 2023a.

Himanshu, S. K., Pandey, A., Karki, K., Pandey, R. P., Palmate, S. S., & Datta, A. (2023b). Assessing the Applicability of Variable Infiltration Capacity (VIC) Model using Remote Sensing Products for the Analysis of Water Balance: Case Study of the Tons River Basin, India. *Journal of the Indian Society of Remote Sensing*, 51 (11), 2323–2341.

IMD (Indian Meteorological Department). *State-wise rainfall information*, Web Report, 2022.

Jonkman, S.N. and Dawson, R.J. Issues and challenges in flood risk management—Editorial for the special issue on flood risk management. *Water*, 4(4): 785–792, 2012.

Khaleghi, S. and Mahmoodi, M. Assessment of flood hazard zonation in a mountainous area based on GIS and analytical hierarchy process. *Carpathian Journal of Earth and Environmental Sciences*, 12(1): 311–322, 2017.

Khan, S.I., Hong, Y., Wang, J., Yilmaz, K.K., Gourley, J.J., Adler, R.F., Brakenridge, G.R., Policelli, F., Habib, S. and Irwin, D. Satellite remote sensing and hydrologic modeling for flood inundation mapping in Lake Victoria basin: Implications for hydrologic prediction in ungauged basins. *IEEE Transactions on Geoscience and Remote Sensing*, 2011(49): 85–95, 2011.

Kumar, S. and Kushwaha, S.P.S. Modelling soil erosion risk based on RUSLE-3D using GIS in a Shivalik sub-watershed. *Indian Academy of Sciences Journal of Earth System Science*, 122(2): 389–398, 2013.

Madolli, M.J., Himanshu, S.K., Patro, E.R. and De Michele, C. Past, present and future perspectives of seasonal prediction of Indian summer monsoon rainfall: A review. *Asia-Pacific Journal of Atmospheric Sciences*, 58(4): 591–615, 2022.

MASTEC (Manipur Science and Technology Council). *Flood Hazard Zonation in Manipur Valley*, final report, 2000.

Nsangou, D., Kpoumié, A., Mfonka, Z., Ngouh, A.N., Fossi, D.H., Jourdan, C., Mbele, H.Z., Mouncherou, O.F., Vandervaere, J.P. and Ngoupayou, J.R.N. Urban flood susceptibility modelling using AHP and GIS approach: Case of the Mfoundi watershed at Yaoundé in the South-Cameroon plateau. *Scientific African*, 15: e01043 2021. https://doi.org/10.1016/j.sciaf.2021.e01043

Ouma, O.Y. and Tateishi, R. Urban flood vulnerability and risk mapping using integrated multi-parametric AHP and GIS: Methodological overview and case study assessment. *Water*, 6(6): 1515–1545, 2014.

Saaty, T.L. Decision making with the analytic hierarchy process. International Journal of Services Sciences, 1(1): 83–98, 2008.

Saaty, T.L. *The Analytic Hierarchy Process*. Mcgraw Hill, New York, p. 70, 1980.

Sinha, R., Bapalu, G.V. Singh, L.K. and Rath, B. Flood risk analysis in the Kosi River basin, north Bihar using the multi-parametric approach of analytical hierarchy process (AHP). *Journal of the Indian Society of Remote Sensing*, 36(4): 335–349, 2008.

Vignesh, K.S., Anandakumar, I., Ranjan, R. and Borah, D. Flood vulnerability assessment using an integrated approach of multi-criteria decision-making model and geospatial techniques. *Modeling Earth Systems and Environment*, 7(2): 767–781, 2021.

17 Agricultural Drought Modelling Through Drought Indices in the Thoubal District, Manipur, India

Denish Okram and Thiyam Tamphasana Devi

INTRODUCTION

Agriculture plays a major role in the national economy in every nation. India is also a country that depends on agriculture. Today we have many advanced technologies, and agriculture is taking advantage of these different new technologies (Ale et al., 2023; Zobeidi et al., 2021; Das et al., 2023; Zhang et al., 2023), but still, we face many difficulties, among these droughts are one of the major challenges confronting farmers (Udmale et al., 2014). Drought is a slow recurring and non-predictable disaster that leads to serious impacts on livestock, humans, and the environment throughout the world. Drought is defined as a prolonged shortage of water supply due to a lack of significant precipitation, over-utilisation of water, and interrupted weather patterns which disturb the water cycle (Gupta et al., 2011). Many parts of the world are facing extreme drought conditions. It is a complicated natural phenomenon which has various severe impacts on humans and the environment (Khayyati and Aazami, 2016). Unlike many natural disasters, drought is barely noticeable and is difficult to recognise when a drought starts (Hollins and Dodson, 2013); the end of a drought can take days, months, or even more as the onset of drought is gradual (Maybank et al., 1995). Most droughts occur when usual weather patterns are disturbed which leads to drastic changes in the water cycle. Nowadays, satellite-based remote sensing techniques (West et al., 2019; Vreugdenhil et al., 2022) with geographical information system (GIS) tools (Sahana et al., 2021; Singh and Devi; Saha et al., 2022) use high spatial resolution and high temporal resolution for observing the Earth (Tanguy et al., 2023). The surface characteristics of land and atmosphere can be derived from remotely sensed images (Himanshu et al., 2023a, 2023b). Currently, researchers (Aadhar and Mishra, 2017; Peng et al., 2020; Prodhan et al., 2021; van Hateren et al., 2023) are demanding an increasing development of remote sensing data for drought monitoring.

DOI: 10.1201/9781003441175-17

Abood et al. (2018) conducted a study with the help of GIS and remote sensing data to analyse drought risk zones facing agriculture and meteorological drought in Maysan and Iraq. For the years 2013, 2015, and 2017 the drought risk area was forecasted using drought indices like the normalised difference vegetation index (NDVI) (Landsat 8), rainfall variation, and standard precipitation index (SPI) values (meteorological data). A combination weighted overlay method was used after preparing the thematic map. Subsequently, for the years 2013, 2015, and 2017 drought risks were identified in three cases: no drought, slight drought, and moderate drought. After analysing all the parameters, it was found that there is a shortage of rainfall and very little vegetation in the southern part of Maysan province (Iraq), therefore this was the most frequent area of drought. Dodamani et al. (2015) conducted a study on agricultural drought modelling using remote sensing. The drought-affected areas were identified using SPI and NDVI collected from MODIS data (MOD13Q1) and Sunspot number which indicates solar activity in the part of Krishna Basin for drought modelling. To forecast agricultural drought for the Kharif season, two multiple linear regression models were established considering the record of 13 years from 2000 to 2012. The variables operated for the first model were NDVI and SPI values, while the second model used Sunspot data in addition to the NDVI and SPI values. After analysing the two models, the second model had a higher significance in the assessment of solar activity in the existence of drought than the first model. In this study, the relationship between NDVI and SPI was used, and significant relationships of various time lags in the rainfall and scarcity zones were obtained. From the forecasted NDVI for major crops, the crop yield model was established and verified with the actual yield. Singh and Devi (2022) conducted a study on estimating drought-prone areas in the low-lying topography of India's north-eastern region (Imphal west district in Manipur state). In their study, it was observed in the year 2019 that a condition similar to drought with a terrible amount of available surface water arose which seriously affected agriculture and the livelihood of the region. To forecast drought-prone areas, two methods were used, i.e. analytic hierarchy process (AHP) and multi-influencing factor (MIF). The drought zone region was summarised considering several parameters (like rainfall, temperature, slope, infiltration, vegetation cover, density, and soil) as acute (22.82% by AHP and 39.42% by MIF), moderate (60.10% and 54.71%), critical (16.16% and 5.55%), and extreme (0.92% and 0.32%). After analysing all the data, it was recommended that the MIF approach is more precise displaying 43.71% of drought zones as acute and 51.32% as moderate drought-prone areas.

The aim of this study is to assess drought using satellite data and GIS techniques in Thoubal district, Manipur, and the main objective of this study is the identification and mapping of drought-prone areas using drought indices, i.e. SPI, temperature content index (TCI), and NDVI, for the nine conjugative years, i.e. 2013 to 2021.

STUDY AREA

Thoubal district (Figure 17.1) is one of Manipur's districts in north-eastern India. The district is located in the eastern Manipur Valley, where it makes up a larger

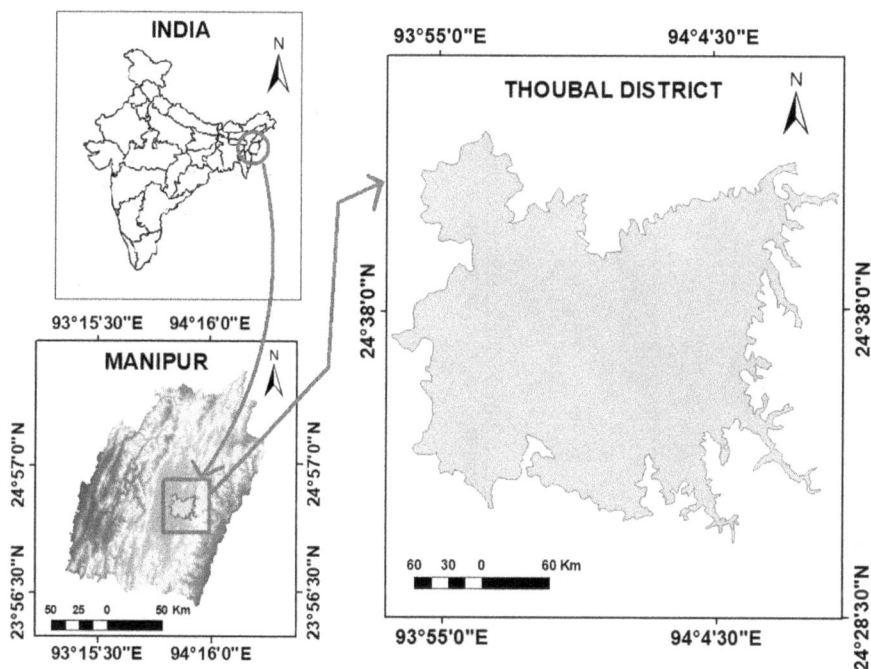

FIGURE 17.1   Location of the study area (Thoubal district)

portion of the state. The district covers an area of 324 km². It is located between 23° 45' and 24° 45' N latitude and 93° 45' and 94° 15' E longitude. The district is generally located at an altitude of 790 metres above mean sea level. The district hardly has any hillocks or hills with a low height. Of these, Punam Hill is located at a height of 1009 metres above sea level. Figure 17.2 shows the digital elevation model (DEM) and slope of the study area.

The major rivers in the study area are the Imphal River, which originates in the Senapati district, and the Thoubal River, which emerges from the hill ranges of the Ukhrul district and flows through the Thoubal district. It has a moderate climate that varies seasonally. The major source of income for the population of Manipur depends on agriculture and the activities related to it. And 70% of the population in the district is associated with farming, since the topography of Thoubal district facilitates irrigation significantly. Sugarcane, pineapple, and rice are the most cultivated crops. Animal husbandry and fishing also contribute to the economy of the district.

METHODOLOGY

The study is conducted using two different sources, i.e. metrological data and satellite data, and a schematic presentation of the methodology that has been followed is shown in Figure 17.3.

Metrological data has been collected for nine conjugative years, from 2013 to 2021. Annual rainfall at nine rain stations has been used to derive SPI. In our study

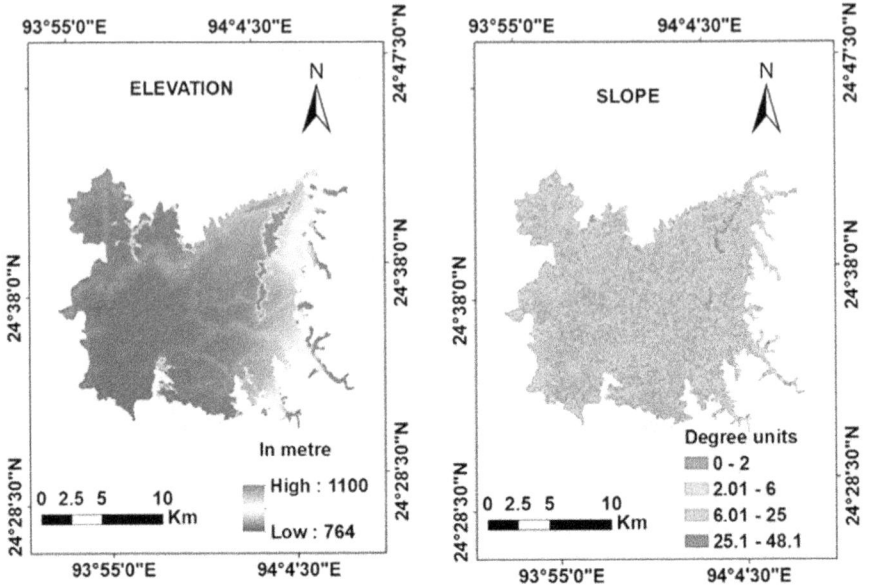

FIGURE 17.2 DEM (left) and slope map (right)

area, there is only one rainfall station, so we have collected eight more from another district as well. For every station, SPI has been calculated, and those calculated SPI are being interpolated with the help of inverse distance weighting (IDW) through a GIS tool (ArcGIS®). From the interpolated SPI map, our study area has been extracted. From the satellite data, again, two sources are used: (i) MODIS data and (ii) Landsat-8 data. MODIS data has been collected for nine years (2013–2021), and from those data, land surface temperature (LST) maximum and minimum have been derived, and using a formula given in Equation 17.3, TCI for the years 2013 to 2021 has been calculated. From the Landsat-8 data, NDVI is derived using a GIS tool for nine consecutive years, i.e. 2013–2021. All the indices have been reclassified, and using the weighted overlay tool, a drought-prone zone is predicted.

Drought Indices

SPI

The SPI is a drought monitoring tool that collects and analyses precipitation data over time. It has an intensity scale where the SPI positive values denote wet conditions and the SPI negative values show drought conditions. Its purpose is to standardise the rarity of the current drought. The formula for SPI is given in Equation (17.1) as:

$$SPI = \frac{(X_i - X_{i\ mean})}{\sigma} \tag{17.2}$$

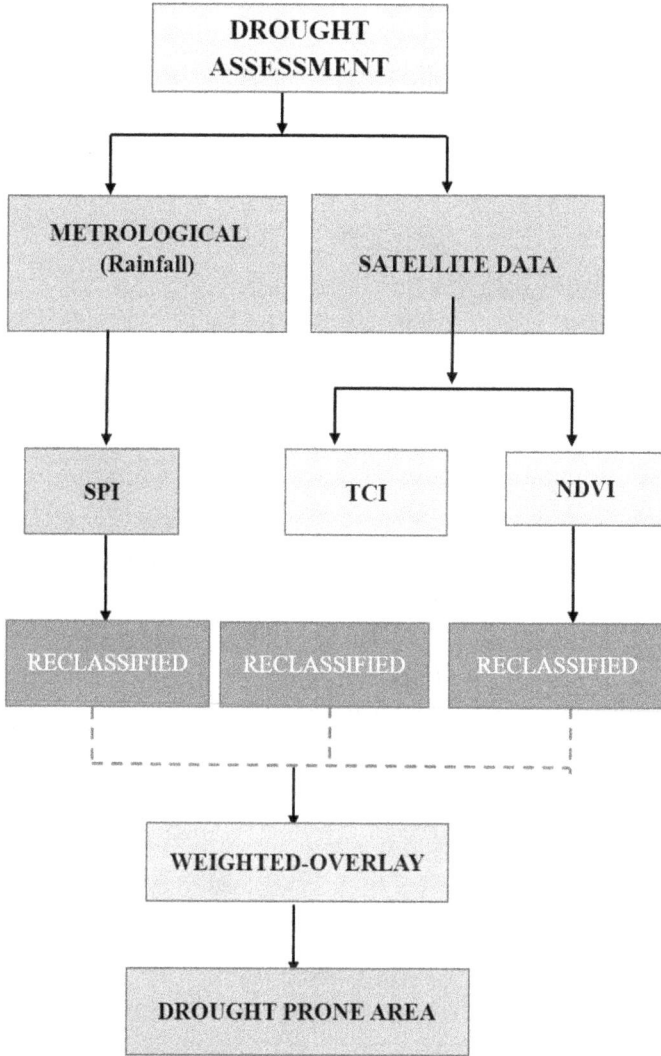

FIGURE 17.3 Flowchart of methodology

Where X_i = significant precipitation, $X_{i\,mean}$ = average precipitation, and σ = the standard deviation of the selected time. The range of SPI values is given in Table 17.1.

NDVI

The NDVI is a commonly used vegetation index to understand vegetation health. It computes the difference between visible and near-infrared to determine the density of green vegetation. High NDVI values show dense green vegetation, while the lower

TABLE 17.1

Meteorological Drought Classification Using SPI Values (McKee et al., 1993), NDVI (Aziz et al., 2018) and TCI (Bhuiyan, 2008)

SPI	Drought Category	NDVI	Drought Category	TCI	Drought Category
> 2	Extremely wet	≥ 6	Extremely wet	≥ 40	No drought
1.50_1.99	Very wet	0.4–0.6	Wet	< 40	Mild drought
1.00_1.49	Moderately wet	0.2–0.4	Moderate	< 30	Moderate drought
0.99_0	Mild wet	0–0.2	Dry	< 20	Severely drought
0_−0.99	Mild dry	< 0	Extremely dry	< 10	Extremely drought
−1.00_−1.49	Moderate dry	–	–	–	–
−1.5_−1.99	Severe dry	–	–	–	–
−2 and less	Extreme dry	–	–	–	–

values denote sparse vegetation like barren areas, snow, or sand. The formula used to compute NDVI is as shown in Equation (17.2):

$$NDVI = \frac{(NIR - Red)}{(NIR + Red)} \qquad (17.2)$$

Where NIR = near-infrared light and Red = visible red light. The range of NDVI value is given in Table 17.1.

TCI

The Temperature Conditions Index is an index used to estimate vegetation stress affected by temperature and stress caused by an extreme amount of wetness. They are determined based on the maximum and minimum values of temperature. The unfavourable conditions are denoted by high temperatures, while the favourable situation is shown by low temperatures. The expression of TCI is given in Equation (17.3):

$$TCI_j = \frac{(LST_{max} - LST_j)}{(LST_{max} - LST_{min})} * 100 \qquad (17.3)$$

Where LST is land surface temperature and values are based on the long-term record of remote sensing images during a specific time period. The range of TCI values is given in Table 17.1.

MATERIALS REQUIRED

The data required with spatial and temporal resolution for this study and their sources are given in Table 17.2.

TABLE 17.2
Data Source Used in the Study

Sl. No.	Data	Source	Resolution (Spatial/ Temporal)	Extracted Data Type
1.	STRM DEM	USGS, Earth Explorer	Spatial (30 m resolution)	Slope
3.	Landsat 7/8	USGS, Earth Explorer	Temporal (2013–2021)	NDVI
4	LST	MODIS	Temporal (2013–2021)	TCI
4.	Meteorology	DoECC, Manipur	Temporal (2013–2021)	Rainfall and Temperature
5.	District Map of Manipur	MARSAC, Manipur	Spatial (30 m resolution)	District Map of Thoubal

RESULTS AND DISCUSSION

In this section, understanding how rainfall deviations, vegetation density, and temperature are determined as well as how drought indices like SPI, NDVI, and TCI behave during the study period are analysed.

Drought Assessment

SPI Map

The annual SPI value for the year (2013–2021) was examined to demonstrate the spatial pattern during these years. Figure 17.4 illustrates the SPI for the Thoubal district. An IDW approach was used to interpolate the obtained SPI value in order to determine the drought zone. Figure 17.5 (a–i), there is no sign of drought in 2013 and 2017 as the SPI value of 2013 and 2017 lies between the range of 0 and 99, which

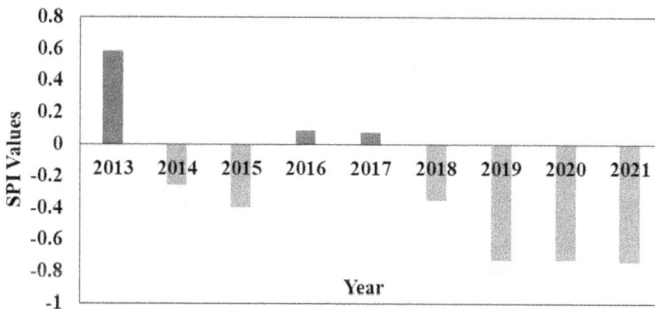

FIGURE 17.4 SPI values (2013–2021)

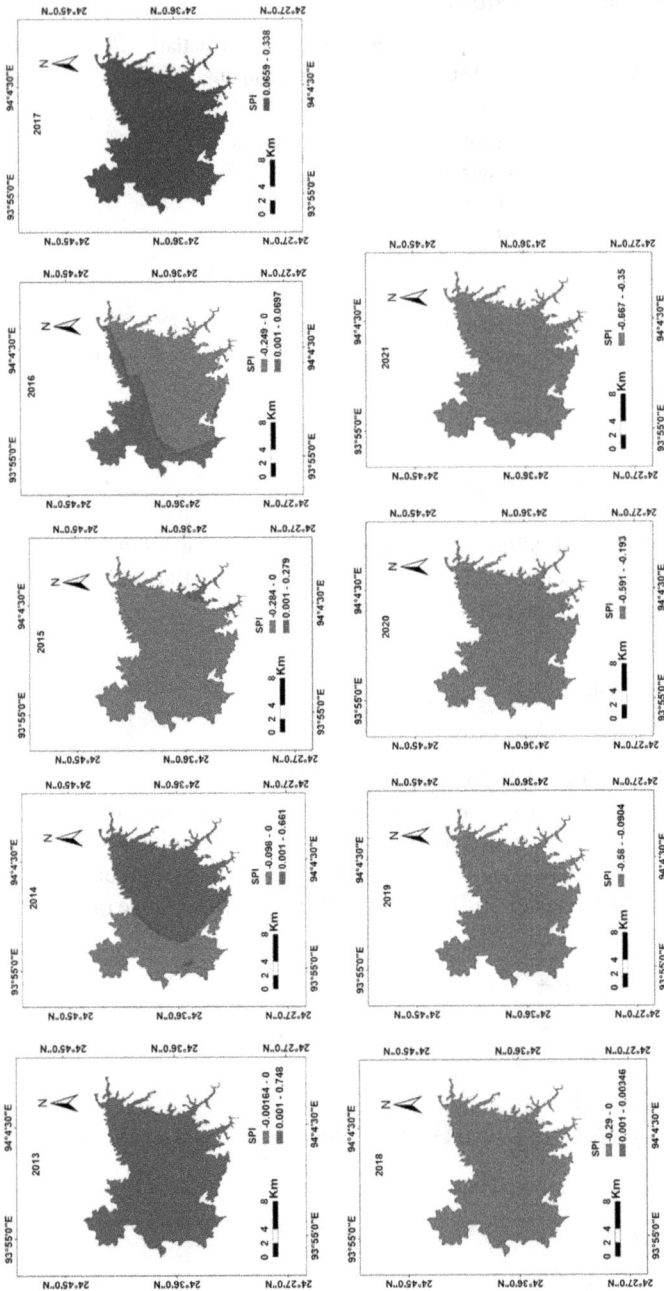

FIGURE 17.5 (a–i): SPI map (2013–2021)

is a sign of mild wetness as shown in Figure 17.5 (a and e). While in 2014 and 2016, some regions of the Thoubal district faced mild wetness and some regions faced mild drought, as shown in Figures 17.5(b) and (d). In the past four years, i.e. 2018, 2019, 2020, and 2021, the Thoubal district

After analysing the result of NDVI (Figure 17.6) from the Landsat-8 image for the month of November (2013–2021), it was found that in the years 2013, 2019, 2020, and 2021, the NDVI value is mostly in the range of 0–0.2, which indicates moderate condition as per the drought indices value provided in Table 17.1. While the years 2014, 2015, 2016, and 2018 are of mixed condition, only in 2017 can some difference be seen, i.e. in the range of 0.2–0.4 and greater than 0. It indicates that some parts are extremely dry and some areas are in moderate condition.

TCI

After analysing the results of TCI derived from MODIS data shown in Figure 17.7, it was found that the years 2013, 2014, 2015, 2017, 2018 and 2021 are mostly in the range of no drought as per Table 17.1. For 2019, some parts of the study area do not have data, and in 2018 and 2020, it was found that there is a mixed situation among no drought, mild drought, moderate drought, severe drought, and extreme drought.

Overlay Analysis

For a period of nine years (from 2013 to 2021), and through weighted overlay analysis combining SPI, NDVI, and TCI thematic layers, the final map for each year (Figure 17.8) is calculated. In the overlay analysis, SPI is given 50% weight, NDVI 40%, and TCI 10%, respectively (Aziz et al., 2018). It is observed that for the years 2013 and 2014, the district can be divided into two conditions: a near-normal condition and a mildly wet condition. In 2013, the western part of the district was in mildly wet condition, while in 2014, the eastern part of the district was in mildly wet condition. For the years 2015, 2018, 2020, and 2021, the majority of the region of the whole district is in near-normal conditions, but only some areas of the north-eastern part of the district in 2020 are in moderately dry condition. In 2016, the district was moderately dry, resulting in drought-like conditions, and in 2017, the whole district received sufficient precipitation and was mildly wet. For the year 2019, the records were not sufficiently available for the whole district, and only a few regions of the western side of the district were analysed, which indicated that the district was near normal and a small portion of the south-eastern side was moderately dry.

VALIDATION

The predicted drought-affected zone is compared with the ground data values, which have been taken from the Department of Agriculture, Government of Manipur, as shown in Figure 17.9. A survey has also been done in various parts of the study area with the local community, including locations such as Heirok Part-3 Ngarouthen, Heirok Part-2 Khunou, Langmeithet Mamang Leikai, Langmeithet Maning Leikai,

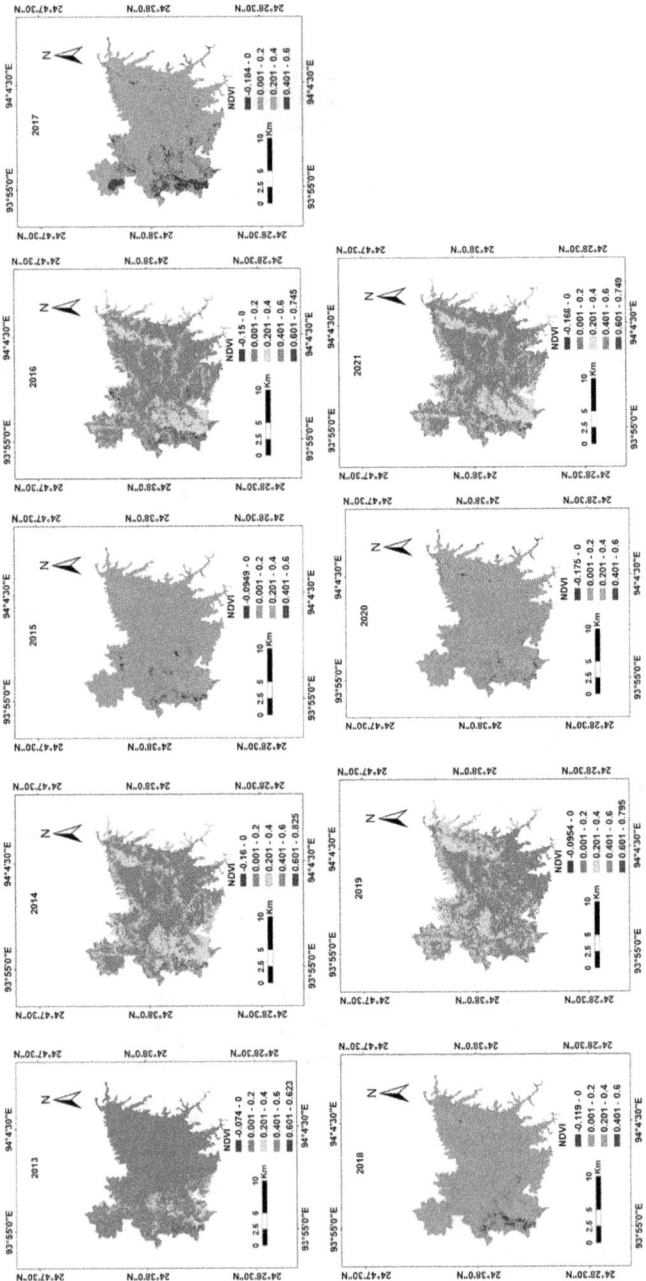

FIGURE 17.6   (a–i): NDVI map (2013–2021)

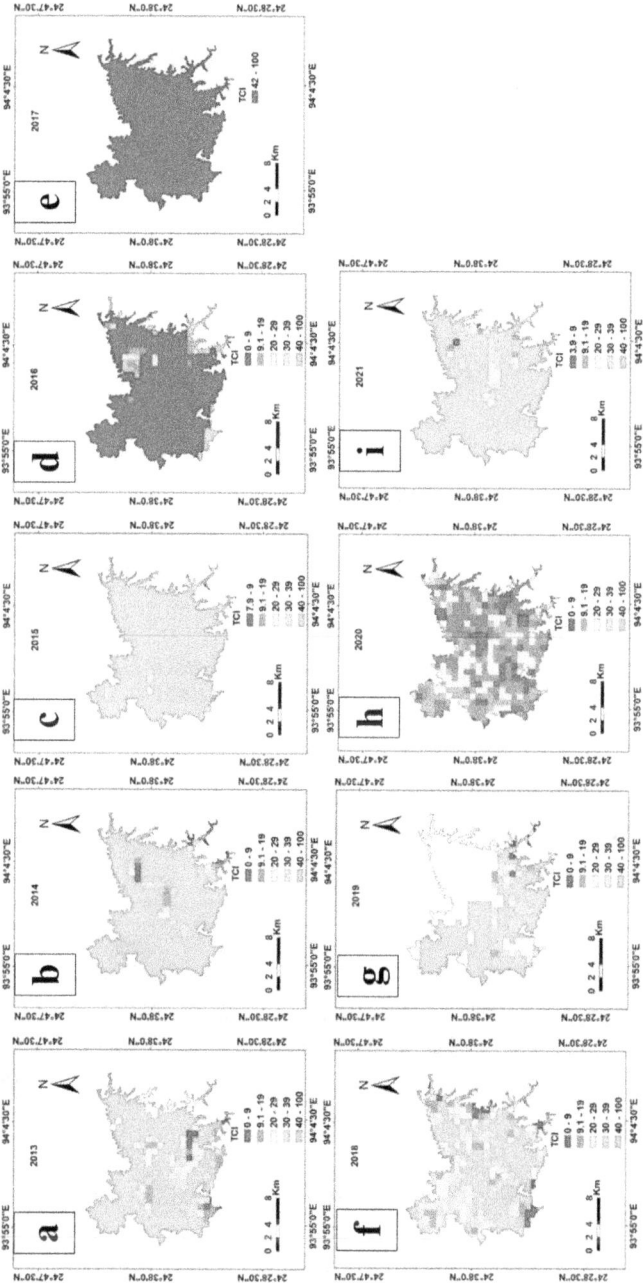

FIGURE 17.7 TCI map (2013–2021)

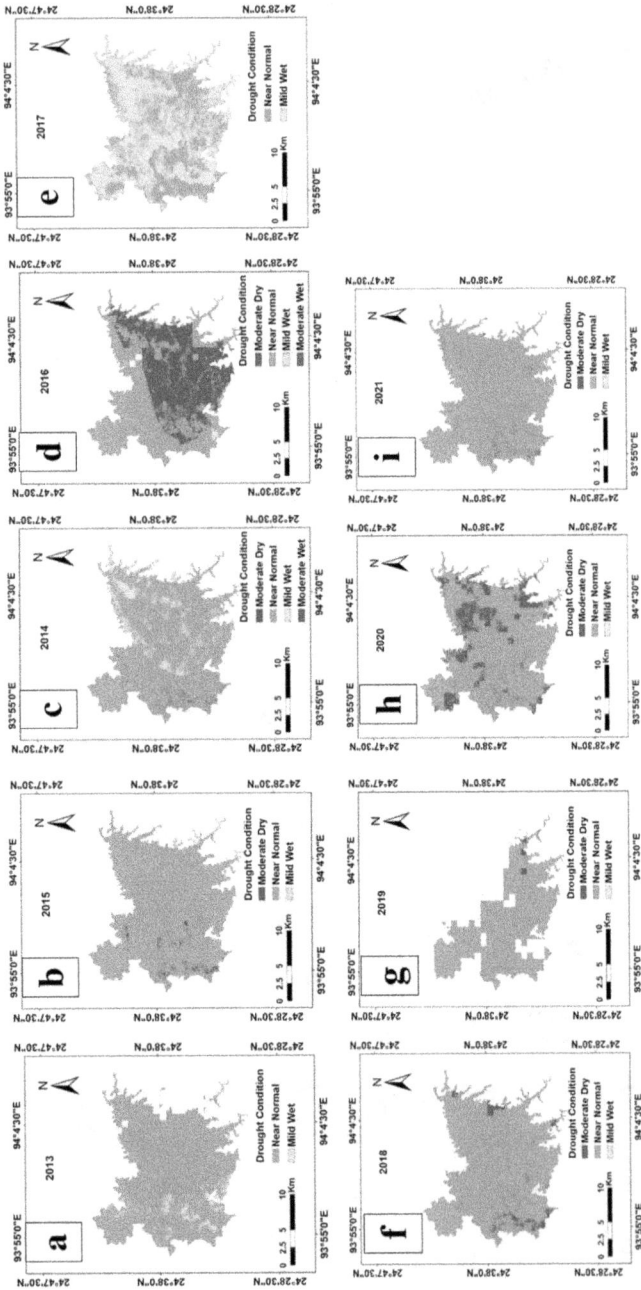

FIGURE 17.8 (a–i): Drought analysis map (2013–2021)

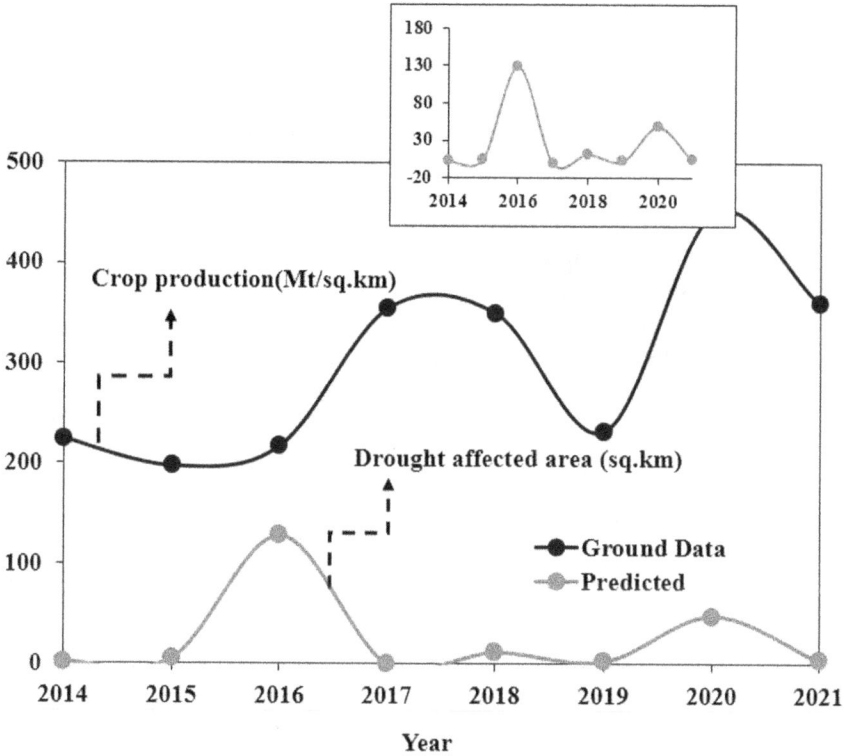

FIGURE 17.9   Crop production (ground data) in drought-affected areas (predicted result)

and Ukhongshang Laikol leirak. In the survey, the first author interacted with several people whose profession is agriculture and who are also familiar with agricultural practices dating back several years. They have been asked about the difficulties faced during the farming season, which may be due to a flood or drought-like situation, i.e. scarcity of available surface water (in the study region, the main source of water for agriculture is only surface water) or excess water during the growing season. After the survey was conducted, it was found that in the years 2013, 2014, and 2017, this study location faced a flood-like situation, while in the year 2016, a drought-like situation was witnessed in the study region. Such findings are also matched with the present study of the predicted drought-affected zone in a particular year, which is evident in Figure 17.9. In the year 2016, crop production was low as per collected ground data, and there is the highest area of affected drought as per prediction; thus, this graph follows the pattern of high crop production and low drought prediction in the years 2017, 2018, and 2021. But in the year 2019, crop production is low, and there is also a low prediction for drought. So, it may be due to other factors that crop production is low and not necessarily due to drought. In the year 2020, drought prediction is comparatively high, but crop production is also high, and this is the exception in the drought prediction result of this study.

CONCLUSION

Among the three indices (SPI, NDVI, and TCI), SPI and TCI give similar results, while the NDVI result deviates slightly in drought prediction. For the year 2017, no drought was predicted by the SPI and TCI indices, while the NDVI predicted moderate to extreme drought. And for the years 2016 and 2020, these three indices predict a similar result, i.e. a moderate drought. It is concluded that SPI and TCI give nearly similar results when compared with NDVI. For the study period (2013–2021), this study region faced mild to moderate drought, which affected crop production. From the validation of predicted results (drought-affected area) with crop yield, it is understood that when the crop yield is low, there are chances of drought, though severity may vary. Thus, this study demonstrates that the use of GIS tools with satellite data can predict the drought situation, and its assessment is useful for effective water resource management. The predicted drought situation of the study was not expected as other parts of the state faced frequent seasonal floods during the monsoon season and thus, this gives important information to the farmer about the unpredictable situation of rainfall variation. Lastly, this study could give preliminary information for proper planning of irrigation systems, such as how excess water from other parts can be diverted to these regions where there is a lack of available water for agricultural purposes.

REFERENCES

M. U. Aziz, M.S. Khan, M.N. Javed, H. Gao, M. Mansha, S.B. Farhan, I. Iqbal and S. Abdullah (2018). Assessment of drought conditions using HJ-1A/1B data: A case study of Potohar region, Pakistan, *Geomatics, Natural Hazards and Risk*, 9(1), pp. 1019–1036.

A.C. Das, S.A. Shahriar, M.A. Chowdhury, M.L. Hossain, S. Mahmud, M.K. Tusar, R. Ahmed and M.A. Salam (2023). Assessment of remote sensing-based indices for drought monitoring in the north-western region of Bangladesh, *Heliyon*, e10316, pp. 1–14.

A.K. Gupta, P. Tyagi and V.K. Sehgal (2011). Drought disaster challenges and mitigation in India: Strategic appraisal, *Current Science*, 100(12), pp. 1795–1806.

B.M. Dodamani, R. Anoop and D.R. Mahajan (2015). Agricultural drought modeling using remote sensing, *International Journal of Environmental Science and Development*, 6(5), pp. 326–331.

C. Bhuiyan, Desert vegetarian during droughts: Response and Sensitivity. Proceedings of the International Archives of the Photogrammetry, *Remote Sensing and Spatial Information Sciences*, Vol. XXXVII. Part B8, Beijing, 2008, pp.907–912.

F.A. Prodhan, J. Zhang, F. Yao, L. Shi, T.P.P. Sharma, D. Zhang, D. Cao, M. Zheng, N. Ahmed and H.P. Mohana (2021). Deep learning for monitoring agricultural drought in South Asia using remote sensing data, *Remote Sensing*, 13(9), p. 1715.

H. West, N. Quinn and M. Horswell (2019). Remote sensing for drought monitoring & impact assessment: Progress, past challenges and future opportunities, *Remote Sensing of Environment*, 232, p. 111291.

J. Peng, S. Dadson, F. Hirpa, E. Dyer, T. Lees, D.G. Miralles, S.M. Vicente-Serrano and C. Funk (2020). A Pan-African high-resolution drought index dataset, *Earth System Science Data*, 12(1), pp. 753–769.

M. Khayyati and M. Aazami (2016). Drought impact assessment on rural livelihood systems in Iran, *Ecological Indicators*, 69, pp. 850–858.

M. Tanguy, M. Eastman, E. Magee, L.J. Barker, T. Chitson, C. Ekkawatpanit, D. Goodwin, J. Hannaford, I. Holman, L. Pardthaisong, S. Parry, D.R. Vicario and S. Visessri (2023). Indicator-to-impact links to help improve agricultural drought preparedness in Thailand, EGUsphere [preprint], https://doi.org/10.5194/egusphere-2023-308.

M. Vreugdenhil, I. Greimeister-Pfeil, W. Preimesberger, S. Camici, W. Dorigo, M. Enenkel, R. van der Schalie, S. Steele-Dunne and W. Wagner (2022). Microwave remote sensing for agricultural drought monitoring: Recent developments and challenges, *Frontiers in Water*, 4, p. 1045451.

N.M. Singh and T.T. Devi (2022). Assessment and identification of drought prone zone in a low laying area by AHP and MIF method: A GIS based study, IOP Conference Series: Earth and Environmental Science, pp. 1–12.

P. Udmale, Y. Ichikawa, S. Manandhar, H. Ishidaira and A.S. Kiem (2014). Farmers' perception of drought impacts, local adaptation and administrative mitigation measures in Maharashtra State, India, *International Journal of Disaster Risk Reduction*, 10(A), pp. 250–269.

R.H. Abood and R.R. Mahmoud (2018). Drought assessment using gis and meteorological data in Maysan Province /Iraq, *International Journal of Civil Engineering and Technology (IJCIET)*, 9, pp. 516–524.

S. Aadhar and V. Mishra (2017). High-resolution near real-time drought monitoring in South Asia, Nature, *Scientific Data*, 4, p. 170145.

S. Ale, Q. Su, J. Singh, S. K. Himanshu, Y. Fan, B. Stoker... & J. Wall (2023). Development and Evaluation of a Decision Support Mobile Application for Cotton Irrigation Management. *Smart Agricultural Technology*, 5, 100270.

S. K. Himanshu, A. Pandey, K. Karki, R. P. Pandey, S. S. Palmate & A. Datta (2023). Assessing the Applicability of Variable Infiltration Capacity (VIC) Model using Remote Sensing Products for the Analysis of Water Balance: Case Study of the Tons River Basin, India. *Journal of the Indian Society of Remote Sensing*, 51, 2323–2341.

S. K. Himanshu, A. Pandey, M. J. Madolli, S. S. Palmate, A. Kumar, N. Patidar & B. Yadav (2023). An ensemble hydrologic modeling system for runoff and evapotranspiration evaluation over an agricultural watershed. *Journal of the Indian Society of Remote Sensing*, 51(1), 177–196.

S. Hollins and J. Dodson (2013). Drought. In: P.T. Bobrowsky (ed.) *Encyclopedia of Natural Hazards*. Encyclopedia of Earth Sciences Series. Springer, Dordrecht, pp. 189–197.

T. Zobeidi, M. Yazdanpanah, N. Komendantova, S. Sieber and K. Löhr (2021). Factors affecting smallholder farmers' technical and non-technical adaptation responses to drought in Iran, *Journal of Environmental Management*, 298, p. 113552.

T.B. McKee, N.J. Doesken, and J. Kleist, The relation of drought frequency and duration to time scales. Proceedings of the Eighth Conference on Applied Climatology. *American Meteorological Society Boston*, 179–184, 1993.

T.C., van Hateren, M. Chini, P. Matgen and A.J. Teuling (2023). High resolution soil moisture drought monitoring over Luxembourg, EGU General Assembly 2023, Vienna, Austria, 24–28 Apr 2023, EGU23-5839.

V. Sahana, A. Mondal and P. Sreekumar (2021). Drought vulnerability and risk assessment in India: Sensitivity analysis and comparison of aggregation techniques, *Journal of Environmental Management*, 299, p. 113689.

Y. Zhang, D. Xie, W. Tian, H. Zhao, S. Geng, H. Lu, G. Ma, J. Huang and K.T.C.L.K. Sian (2023). Construction of an integrated drought monitoring model based on deep learning algorithms, *Remote Sensing*, 15(3), p. 667.

18 Seed Priming
Potential Nutrient Management Tool for Improving Crop Productivity Under Abiotic Stress

Debesh Das, Hayat Ullah, Sushil Kumar Himanshu, and Avishek Datta

INTRODUCTION

Seed priming is the controlled hydration of seeds for inducing pre-germinative metabolic activities to promote seed germination, germination uniformity, and stand establishment. Simple, eco-friendly, and practical nutrient management practices can be adopted by farmers to improve germination and seedling vigor, growth, and finally crop yield under abiotic stress conditions (Raj and Raj, 2019). Nowadays, seed priming is a promising tool that enhances seed quality toward better germination and seedling growth, which is a major consideration as high-quality seeds are greatly needed in the agricultural market. Seed priming enhances the capability of resistance against abiotic stresses by improving physiological and biochemical regulation (Rajjou et al., 2012). Plants raised from primed seeds show vigorous seedling growth and tolerance against stress, which might be due to smooth reserve mobilization, enzyme activation, embryo enlargement, energy metabolism, and DNA and RNA synthesis (Jisha et al., 2013). Besides, seed priming stimulates various signaling cascades by building up signaling proteins toward faster and more efficient defense responses in plants (Schwember and Bradford, 2010). Priming triggers the repair of various metabolic damages and DNA replication resulting in enhanced vigor, uniform emergence, and seedling growth under field conditions (Farooq et al., 2008; Hussain et al., 2016a). Seed priming is a potential alternative that safeguards rapid and uniform seed germination by lowering imbibition time, activating enzymes, repairing DNA, and regulating hydration levels (Cheng et al., 2017; Marthandan et al., 2020). There are several types of seed priming, such as hydropriming, osmopriming, hormonal priming, nutripriming, biopriming, and

DOI: 10.1201/9781003441175-18

nano-priming, based on priming agents which are used in experimentation and even farmers' practice as well (Farooq et al., 2006a, 2006b, 2006c; Farooq et al., 2013). The physiological, biochemical, and subcellular basis of seed priming depicts the hidden fact that seed priming is a rapid, uniform germination process followed by vigorous seedling growth and subsequent development to cope with adverse environmental conditions. There is much literature available regarding individual seed priming material on specific environmental conditions, but the interaction of different seed priming agents on complex stress conditions has been less studied. Therefore, the attention was centered on the potential of commonly usable seed priming materials on growth, physiological, biochemical response, and crop yield under multiple abiotic stresses.

PHYSIOLOGY OF SEED PRIMING

Seed priming is a physiological process of controlled seed hydration to enhance sufficient pre-germinative metabolic processes that ensure dormancy breakdown and efficient water uptake. Seed imbibition is the initial phase where the seed uptakes water rapidly as the water potential of the seed content is lower than pure water. The imbibition rate is much higher for hydropriming due to the higher water potential gradient between pure water and seed contents. Imbibition brings about the activation phase or lag phase (Stage II) in which different metabolic events take place such as de novo synthesis of protein and mitochondria, activation of enzymes, antioxidative system, and DNA repair toward prompt and uniform seed germination. Figure 18.1

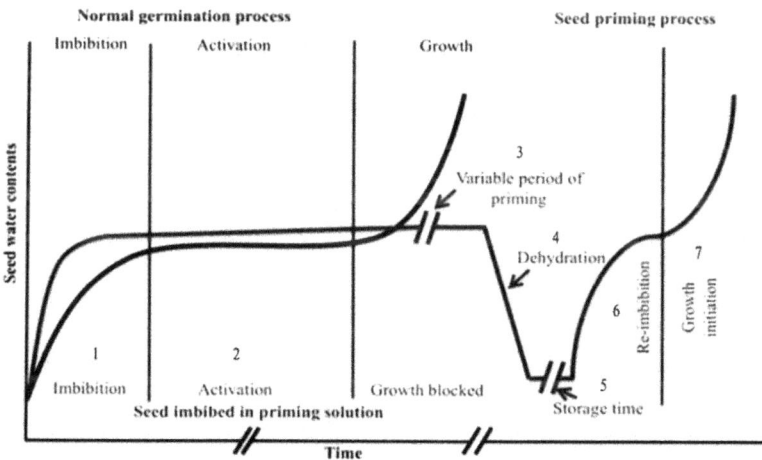

FIGURE 18.1 Diagrammatic representation of sequential events for seed priming followed by germination in non-primed and primed seeds. The figure is partly adapted from Rajjou et al. (2012). Sequential steps ① = Seed imbibition in a specific solvent; ② = Activation of enzyme and metabolism; ③ = Variable period of priming; ④ = Dehydration; ⑤ = Certain period storage of primed seed; ⑥ = Re-imbibition; ⑦ = Rapid cell division and elongation for vigorous seed germination

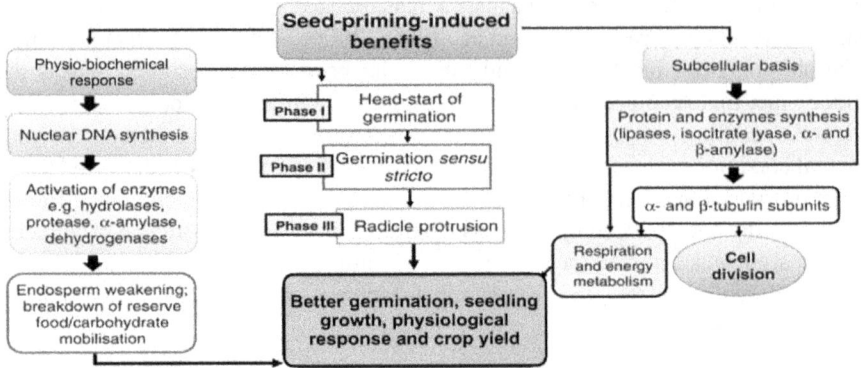

FIGURE 18.2   Physiological, biochemical, and cellular basis of seed priming toward better crop production (Farooq et al., 2019)

Moreover, seed priming induces different antioxidant enzymes such as peroxidase (POD), catalase (CAT), and superoxide dismutase (SOD) which are used to maintain a balance between generation and utilization of ROS under stress (Wojtyla et al., 2016; Farooq et al., 2017). Basically, a series of metabolic and biochemical repairing processes occur in the embryo during priming in which primed seeds undergo phase I (hydration) and phase II (lag phase) without entering the initial growth phase (phase III), rather than dehydration and storage. Seed dehydration after priming (drying-back) is essential to allow positive effects of seed priming without quality loss (Ibrahim, 2016). The longevity of primed seeds depends on the dehydration and storage conditions in which slow dehydration increases the subsequent longevity of primed seeds more than non-primed seeds. During the storage period, different latent defense proteins accumulate that assist stress response and tolerance capacity against abiotic stresses (Taiz and Zeiger, 2010; Jisha et al., 2013; Borges et al., 2014). After a certain period of resting stage, seeds are induced for rehydration during which primed seeds allow seeds to restart again from phase I (rehydration), phase II (lag phase), and phase III of germination, such as radicle protrusion under favorable environmental conditions (Farooq et al., 2019). During the storage period, seed reserves are converted into a form that nourishes the embryo during germination for vigorous seedling. Sometimes, seeds are exposed to a stressful environment during priming when defense responses are activated by signaling molecules. Plants grown from primed seeds under abiotic stresses (such as drought, heat, cold, heavy metals, and salinity) exhibited better performance in comparison to non-primed plants by strengthening molecular mechanisms for improving tolerance. Figure 18.2

ABIOTIC STRESSES FOR CROP PRODUCTION

DROUGHT STRESS

Drought is the foremost provocation to meet global food demand for an ever-growing population under adverse climatic conditions. Agricultural production is highly dependent on climatic variability in efficient water management is of paramount consideration to meet global food security (Gosling and Arnell, 2016; Ale et al., 2021). Water is the most important natural resource and input for food production, which is becoming increasingly scarce around the world, necessitating water conservation for sustainable crop production (Neupane et al., 2021). Irrigation water-use efficiency is of prime consideration along with getting maximum crop yield regardless of environmental conditions (Himanshu et al., 2021, 2023). However, sustainable crop production is seriously threatened by multiple challenges in which irrigation water deficit is the prime limiting ecological factor that jeopardizes crop productivity (Ullah et al., 2018a, 2018b; Das et al., 2021a). For example, the drought-driven yield loss is projected to increase by 20% of global rice production in the near future (Horie et al., 2005; Huang et al., 2017), whereas global freshwater availability for irrigation is decreasing due to climate change and over-extraction of groundwater, which will threaten regional and global food security (Ullah et al., 2019; Wu et al., 2017). It has been reported that consumptive water use (precipitation and irrigation) for food production is increasing by 0.7% annually and different regions of the world have been suffering severe drought with significant yield loss (IPCC, 2013). Figure 18.3

Furthermore, progressive freshwater scarcity is deteriorating annual grain production by 7–8% endangering regional and global food security in suboptimal climatic conditions (Maclean et al., 2013). Drought stress regulates leaf relative water content (LRWC) and leaf greenness, carbon fixation, and nutrient uptake toward reduced carbohydrate synthesis resulting in significant yield loss (Lesk et al., 2016; Zhao et al., 2020; Das et al., 2021c).

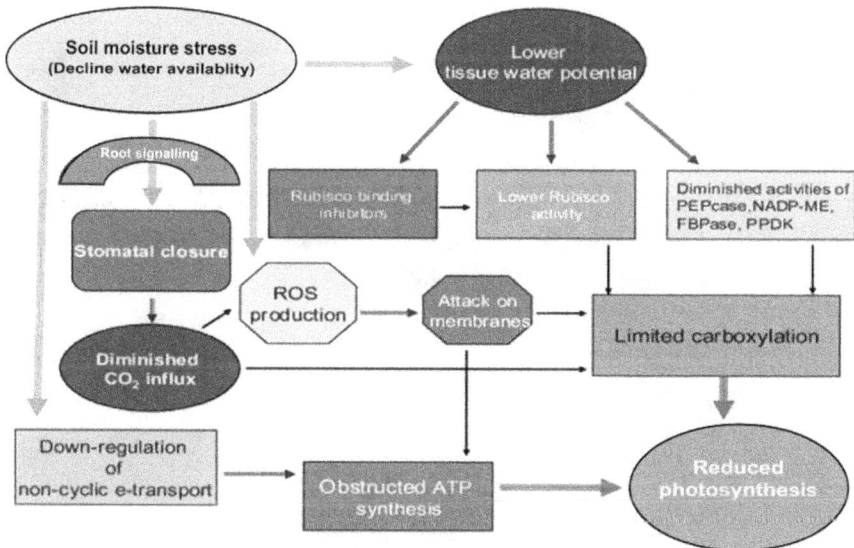

FIGURE 18.3 Drought-induced declined plant photosynthetic pathway (Farooq et al., 2009)

SALT STRESS

Soil salinity is the dominant environmental constraint that interrupts normal physiological and biochemical processes to significantly diminish potential crop yield (Farooq et al., 2010; Das et al., 2017). Excessive salt concentration in soil declines soil water potential that decreases soil moisture availability for plants and accumulates diverse salts in soil providing osmotic and oxidative stress. High concentration of salt (sodium chloride) leads to excessive ion accumulation in the cytosol and causes ionic imbalance (Ali et al., 2021). Saline areas are rising at an alarming rate and account for approximately 800 million hectares worldwide, which encompasses more than 6% of the world's total land area. Saline soil possesses an exchangeable salt content of 15% and an electrical conductivity (EC) of the saturation extract (ECe) in the rhizosphere is greater than 4 dS/m (about 40 mM NaCl) at which growth and developmental phenomena, even yield of the glycophytes, are decreasing (Jamil et al., 2013). It has been estimated that worldwide 20% of total cultivated and 33% of irrigated lands are affected by the high degree of soil salinity. It has been projected that more than 50% of arable land will be salinized by 2050, which would drastically reduce agricultural production globally (Mostofa et al., 2015). Salinity adversely affects crop production by impairing physiological and biochemical mechanisms associated with growth and developmental processes (Gupta and Huang, 2014). High concentrations of soluble salts reduce carbon assimilation, transpiration rate, stomatal conductance, and leaf enlargement, as well as leaf senescence during the vegetative stage (Hussain et al., 2017). The complex interactions among morphological, physiological, and biochemical processes that control plant growth, development, and biomass production are inhibited by salt stress (Shrivastava and Kumar, 2015). Figure 18.4

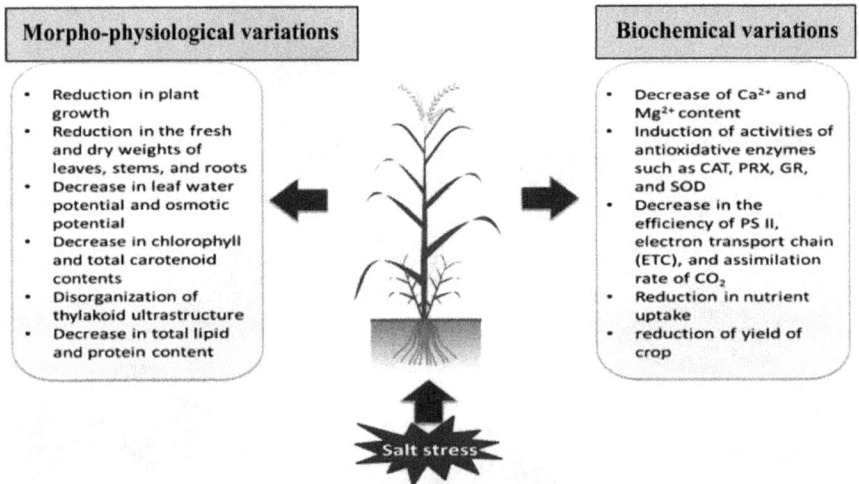

Morpho-physiological variations

- Reduction in plant growth
- Reduction in the fresh and dry weights of leaves, stems, and roots
- Decrease in leaf water potential and osmotic potential
- Decrease in chlorophyll and total carotenoid contents
- Disorganization of thylakoid ultrastructure
- Decrease in total lipid and protein content

Biochemical variations

- Decrease of Ca^{2+} and Mg^{2+} content
- Induction of activities of antioxidative enzymes such as CAT, PRX, GR, and SOD
- Decrease in the efficiency of PS II, electron transport chain (ETC), and assimilation rate of CO_2
- Reduction in nutrient uptake
- reduction of yield of crop

Salt stress

FIGURE 18.4 Morpho-physiological and biochemical responses of plants under salt stress (Yadav and Atri, 2020)

Moreover, the osmotic potential of soil solution at the root zone reduces due to high salt concentration that decreases the turgor pressure of root cells (Roy et al., 2017; Al-Taey et al., 2018). This sudden change in turgor pressure inhibits the cell division and elongation processes, resulting in an inhibition of plant height under saline conditions. Relative chlorophyll content or leaf greenness (SPAD value) is an important indicator of stress tolerance and leaf senescence that can be altered in response to salt stress. Actually, a salt stress-mediated decrease in leaf greenness might be attributable to the degradation of chlorophyll, which strongly influences photochemical reaction for carbon fixation. Salt stress reduces leaf relative water content and leaf greenness through chlorosis that inhibits net photosynthetic rate, phloem translocation, and finally yield (et al., 2023). Moreover, excess accumulation of salt deteriorates carbohydrate metabolism, cell division in floral organs toward limiting fertility, and filled-grain percentage for cereals (Hussain et al., 2017; Najafabadi and Ehsanzadeh, 2017). Spikelet sterility under salt stress is a common phenomenon that reduces pollen viability or stigmatic receptivity impairing pollination and fertilization processes with reduced carbon fixation (Fu et al., 2011; Gonzalez et al., 2019). Remobilization of photoassimilate from leaf to grain and translocation of photosynthates from the source to the sink directly affect the grain-filling process in cereals (Rang et al., 2011; Das et al., 2021b).

HEAT STRESS

Extreme temperature coupled with inconsistent precipitation is the most recurrent form of abiotic stress that causes the danger of heat stress to successful crop production. Among the other environmental factors, high temperature is a devastating ecological factor that is becoming more frequent and severe in many regions of the world (Hatfield and Prueger, 2015). Heat stress is the rising of temperature beyond a threshold level for a certain period to induce irreversible damage to the plant metabolic system. Heat stress is directly associated with plant developmental processes including seed germination, photosynthesis, respiration, transpiration, phytohormone synthesis, and seed setting (Hasanuzzaman et al., 2012). Elevated temperature increases the fluidity and leakiness of root cell membranes, which reduces root growth, permeability, water, and nutrient uptake (Fahad et al., 2015a, 2015b). Plants accumulate specialized compounds named heat shock proteins (HSPs) in response to heat stress that protect cells against oxidative damage. Plants have evolved a plethora of mechanisms to cope with heat stress, which generate reactive oxygen species (ROS) in cellular compartments that can be scavenged through elevated levels of antioxidants and osmoregulation (Dietz et al., 2016). Accumulation of volatile organic compounds is another potential mechanism to mitigate the effects of oxidative stress under high temperatures (Fahad et al., 2016). Different secondary metabolites such as isoprene and monoterpenes have often been shown to increase heat resistance in plants by stabilizing protein in cell membranes and thylakoid membranes. It was observed that heat stress significantly declines on assimilate partitioning due to abnormal behavior of the source–sink relationship. Figure 18.5

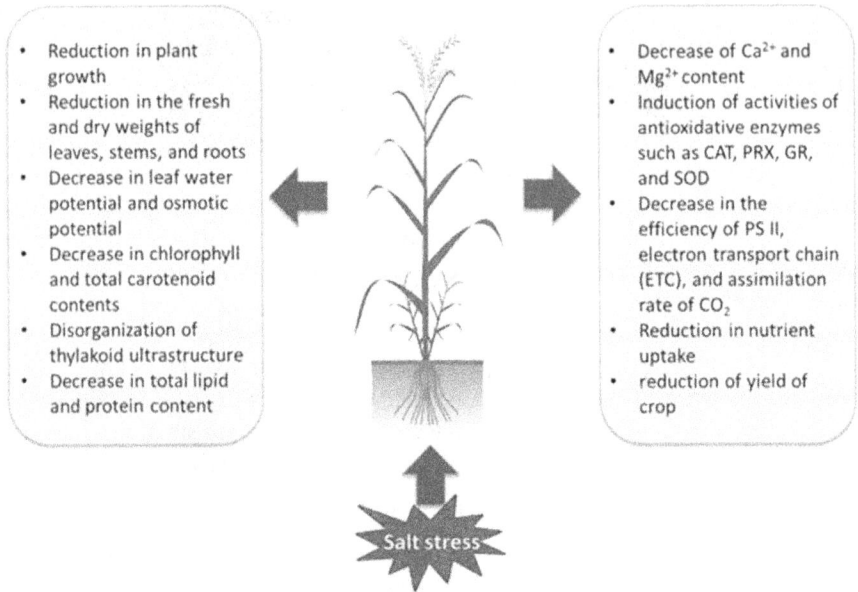

- Reduction in plant growth
- Reduction in the fresh and dry weights of leaves, stems, and roots
- Decrease in leaf water potential and osmotic potential
- Decrease in chlorophyll and total carotenoid contents
- Disorganization of thylakoid ultrastructure
- Decrease in total lipid and protein content

- Decrease of Ca^{2+} and Mg^{2+} content
- Induction of activities of antioxidative enzymes such as CAT, PRX, GR, and SOD
- Decrease in the efficiency of PS II, electron transport chain (ETC), and assimilation rate of CO_2
- Reduction in nutrient uptake
- reduction of yield of crop

Salt stress

FIGURE 18.5 Responses of plants to heat stress (Kumari et al., 2020; dos Santos et al., 2022)

HEAVY METAL STRESS

Heavy metals (HM) are metals or metalloids that have higher relative density and atomic mass over 20 (excluding the alkali metals); specific gravity of more than 5 and atomic density > 6 gcm^{-3} (except for As, B, and Se) (Park et al., 2011; Rascio and Navari-Izzo, 2011). Heavy metals are naturally occurring elements in the Earth's crust that comprise essential (Cu, Fe, Ni, and Zn) and nonessential metals (Cd, Hg, Ag, Pt, and Pb) for plants (Mahajan et al., 2022; Yatoo et al., 2022; Singh et al., 2023). Rhizospheres' heavy metal contamination is a hidden culprit that limits nutrient uptake, availability, uptake, and biomass production (dos Santos et al., 2022). Basically, heavy metals enter into plants primarily via the root and are stored in cell walls or intercellular space, or in vacuoles in plant tissue that impair the physiological and metabolic reaction directly or indirectly (Cai et al., 2012). Heavy metal concentration, especially Cu, Cd, Cr, Mn, Ni, and Zn, in soil solution, affects seed imbibition which has a substantial deleterious effect on the remobilization of endosperm (Chigbo and Batty, 2013; Kalai et al., 2014). Heavy metal stress stimulates a deficiency of Fe or Mg, which is a central element of chlorophyll molecules, resulting in chlorosis toward decreased mesophyll thickness and stomatal density (Tang et al., 2013). Reduction in photosynthetic function might be due to the alteration of supramolecular conformation of the light-harvesting pigment-protein complex that affects adversely on the photosynthetic electron transport system (Marques and Nascimento, 2013; Molins et al., 2013). Moreover, toxic levels of heavy metals

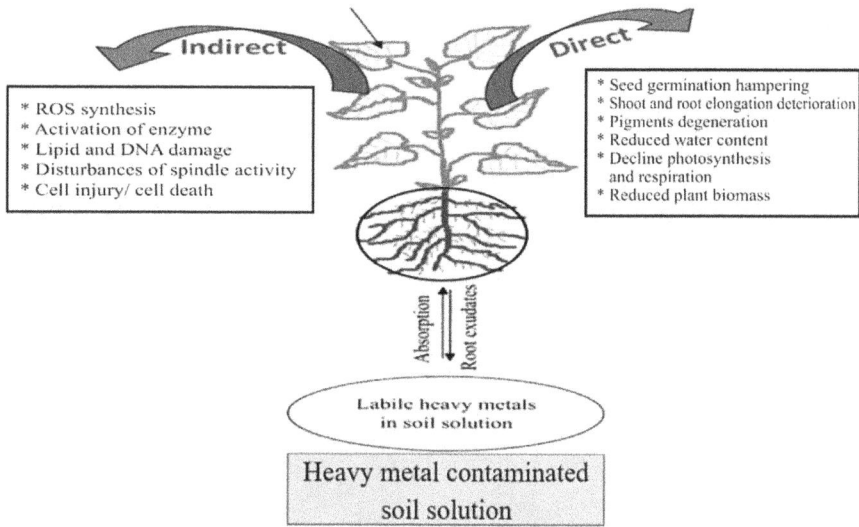

FIGURE 18.6 Heavy-metal-induced direct and indirect effects on plant growth and developmental process (Shahid et al., 2014)

produce severe toxicity symptoms in plants including low biomass, chlorosis, photosynthesis, nutrient assimilation, and senescence, which ultimately causes plant death. Therefore, heavy metals are toxic for plants, causing phytotoxicity followed by chlorosis, reduced plant growth, and yield. Figure 18.6

SEED PRIMING TO ALLEVIATE ABIOTIC STRESSES

DROUGHT STRESS

Drought is a devastating abiotic factor that induces oxidative damage by impairing most physiological processes, including net photosynthetic rate, respiration, and electrolyte leakage in plants (Abdelaal et al., 2020; Das et al., 2022). Seed treatment with different exogenous protectants has been reported to alleviate the harmful effects of drought stress on crop plants. Silicon (Si) is the second most important constituent of the Earth's crust (about 32% by weight) and is considered a beneficial element for crop nutrition (Neeru et al., 2019). Si is the most potential exogenous protectant for alleviating deleterious effects of soil moisture stress in both agronomic and horticultural crops (Sirisuntornlak et al., 2019; Alam et al., 2020). Seed priming with potassium nitrate (KNO_3) has been reported for enhancing seed germination and vigorous seedlings in rice, wheat, maize, cotton, melon, Chinese cabbage, and tomato under stressed conditions by improving hydraulic conductivity, enhancing cell wall integrity, promoting light energy distribution in chloroplasts, enhancing antioxidant defense capacity (Chen et al., 2011; Liu et al., 2015a; Cao et al., 2017; Cao et al., 2020). It was found that seed treatment with proline combined with KNO_3 boosted the membrane stability index and osmotic balance of salt-stressed muskmelon. At 0, 15, and 30 kPa, seed priming with *Trichoderma* increased grain yield of rice by 52%, water productivity by 70%, and water productivity by 66%, respectively, implying that seed priming with *Trichoderma* is recommended where frequent irrigation is difficult to practice (Das et al., 2021b). Seed priming with salicylic acid (300 ppm SA) in cantaloupe resulted in 48% and 59% higher fruit yield and water productivity, respectively, compared with non-primed control at 50% FC (Alam et al., 2022). It was revealed that melatonin seed priming mitigates the reduction in chlorophyll content, and maintains chloroplast integrity, cell expansion, and stomatal traits under deficit irrigation conditions (Khan et al., 2019). Similarly, tomato plants supplemented with 0.1 mM melatonin also showed enhancement of root vigor and detoxification of cellular ROS ((Liu et al., 2015b). The summary presented in Table 18.1 encompasses the impact of seed priming-induced nutrient management on the growth, physiological, and biochemical responses, as well as the yield of various crops under water-deficit conditions Table 18.1.

SALT STRESS

Soil salinity adversely impacts crop growth and yield by upsetting seed germination, seedling growth, photosynthetic electron transport chain, and photorespiration but seed priming improves seed germination and seedling growth under saline conditions (Ibrahim, 2016). Osmolytes are neutral molecules and compatible solutes such as proline, mannitol, trehalose, sucrose, fructans, glycine betaine, and polyamines that play an important role in the stabilization of proteins and membrane damage against abiotic stresses without disrupting plant metabolism (Khan et al., 2015; Lutts et al., 2016). In both stress and normal conditions, proline acts as a proteogenic amino acid that accumulates in plants to restrain the osmotic effects by scavenging free radicals (Kavi Kishor et al., 2015). Seed priming with Si has been shown to increase plant tolerance against soil salinity. The Salt-Overly Sensitive (SOS) pathway has the potential role to enhance salt tolerance in plants in which Na^+ is excluded from the cytosol. SOS1 is a vital component of the SOS pathway where the Na^+/H^+ antiporter maintains ion homeostasis (Gupta et al., 2021). By increasing the expression of oxidative stress genes and antioxidant enzyme activity (such as APX, SOD, and catalase), Si improves salinity tolerance in both Si-accumulators (e.g., rice, wheat, and barley) and non-accumulators (e.g., tomato) for minimizing oxidative damage (Anjum et al., 2015; Thorne et al., 2020).

In faba bean, salicylic acid induces a high selectivity of K^+/Na^+ ratio and downregulates ion leakage and lipid peroxidation that has a substantial impact on the accumulation of osmolytes, carotenoids, and antioxidant enzyme activity (CAT, POD, APX, and GR) (Azooz, 2009). At a salinity level of 6 dS m^{-1}, seed priming with SA increased root biomass, leaf relative water content, free proline content, and cob yield, by 43%, 5%, 7%, and 45%, respectively (Islam et al., 2022). The effect of seed priming with SA is dose-dependent but seed priming with 100 µM SA had a favorable impact on carbon fixation capacity, transpiration rates, chlorophyll concentration, stomatal conductance, and ion homeostasis. (Boukari et al., 2019). Chitosan is a polysaccharide obtained during seafood processing as waste material that accelerates reserve mobilization, activation of enzymes, and DNA and RNA

TABLE 18.1

Growth, Physiological, Biochemical Response, and Yield of Different Crops Influenced by Seed Priming Under Drought Stress

Crop	Priming treatments	Growing condition	Physiological and biochemical interventions	References
Rice	Seed priming with Zn (0.5 M, 24 h)	Field condition	Increased grain yield by 31% and grain nutrient by 27%.	Farooq et al. (2018)
	Osmopriming (Se 50 μM) and hormonal priming (Salicylic acid 100 mg/L)	Chilling stress under greenhouse conditions	Upregulate starch metabolism, respiration; reduced lipid peroxidation; improved antioxidative defense system.	Hussain et al. (2016b)
	Phosphorus priming (200 mM KH_2PO_4)	Phosphorus-deficient soil under greenhouse	Promoting germination, seedling growth, and leaf nutrient content with enhanced P uptake.	Pame et al. (2015)
	Osmopriming with KCl (0.15, 0.30, and 0.45 M)	Moisture stress under greenhouse	Improved carbohydrate mobilization; reduced lipid peroxidation; enhanced antioxidative system.	Ella et al. (2012)
	Hydropriming and osmopriming ($CaCl_2$)	Soil moisture stress under greenhouse	Enhanced stand establishment, accumulation of phenols, flavonoids.	Hussain et al. (2017)
	Sodium nitroprusside (100 mg/L)	Drought stress under greenhouse	Boosts up antioxidant enzymes, cell membrane stability, photosynthesis, and leaf water status by synthesizing compatible solutes.	Farooq et al. (2009b)
	Nutripriming (B)	Alternate wetting and drying (AWD) condition	Improved plant water relationships, chlorophyll content, grain nutrient.	Rehman et al. (2014)
Wheat	Osmopriming with $CaCl_2$ (1.5%)	Drought stress under field condition	Enhanced leaf area and tissue water status with higher accumulation of osmolyte.	Tabassum et al. (2018b)
	Osmopriming with zinc sulfate (0.5 M)	Field condition	Improved gain yield and grain mineral content by 21 and 55%, respectively.	Ali et al. (2020)
	Seed priming with *Pseudomonas* sp. (Mn 12 M + Zn 0.5 M, 12 h)	Field condition	Grain yield increased by 18–27%.	Rehman et al. (2018)

(Continued)

TABLE 18.1 (CONTINUED)
Growth, Physiological, Biochemical Response, and Yield of Different Crops Influenced by Seed Priming Under Drought Stress

Crop	Priming treatments	Growing condition	Physiological and biochemical interventions	References
	Hydropriming	Field experiment, zero tillage	Improved stand establishment, grain yield, and profitability.	Mustafa et al. (2018)
Maize	Nutripriming (4 mM Zn, $ZnSO_4\,H_2O$ + 2.5 mM Mn, $MnSO_4$ for 24 h)	Growth chamber	Performed yield advantage of about 15%.	Imran et al. (2015)
	Osmopriming (0.5% Zn)	Field experiment	Enhanced yield 10%.	Rasool et al. (2019)
	Osmopriming (0.01% Mn)	Field experiment	Improved grain nutrient content by about 32%.	Rasool et al. (2019)
	Nutripriming (B + Mn)	Semi-arid conditions	Early emergence; higher grain yield and protein content.	Rasool et al. (2019)
Barley	Osmopriming ($CaCl_2$) and biopriming (*Enterobacter* sp. FD-17)	Drought stress under field experiment	Improved leaf area, chlorophyll content, phenolics accumulation, antioxidant activity, and grain yield.	127Tabassum et al. (2018a)
Soybean	Priming with KNO_3 and KH_2PO_4	Drought stress under control condition	Increased pod and grain numbers per plant by about 300%.	Ghassemi-Golezani et al. (2011)
Chickpea	Osmopriming ($CaCl_2$)	Chilling stress	Enhanced stand establishment, water relationships, photosynthesis, sugar metabolism, and antioxidant enzyme activities.	Farooq et al. (2017)
Quinoa	Osmopriming (H_2O_2 80 mM)	Drought stress under control condition	Maintained turgor pressure; enhanced photosynthetic rate, antioxidant activity, membrane stability, and osmotic adjustment.	Iqbal et al. (2018)
Tomato	Seed priming and soil application of Si	Soil moisture stress in polyhouse	Better fruit yield, quality, and water productivity at moderate stress.	Chakma et al. (2021)

synthesis during osmotic priming (Hameed et al., 2014). Melatonin is a well-known bio-stimulant for improving seed germination and seedling growth under salt-stress conditions because it protects plant cells through free radical scavenging. Melatonin scavenges free radicals, enhances antioxidant enzymes, promotes the efficiency of the mitochondrial transport chain, and deteriorates the generation of free radicals (Janas and Posmyk, 2013). Moreover, nanoparticles coated with polyacrylic acid (PNC) improved salt tolerance by maintaining ROS homeostasis and scavenging ROS in rapeseed under saline conditions implying the clue that PNC has mimicked the catalase-like activity to scavenge the over-accumulated ROS (Zhou et al., 2021). Similarly, Manganese oxide nanoparticles (Mn_3O_4 NPs) possess diverse antioxidant enzyme-mimicking properties that contribute directly to ROS scavenging for enhancing cucumber salt tolerance by increasing antioxidative defense (Lu et al., 2020). Seed treatment with beta-amino butyric acid (BABA) significantly reduced MDA content in the seedlings and resulted in enhanced activity of nitrate reductase enzyme and antioxidant enzymes (Jisha et al., 2016). Growth, physiological, and biochemical response, and yield of different crops influenced by seed priming-induced nutrient management under salt stress are summarized in Table 18.2. Figure 18.7

Heat Stress

Temperature is a key environmental factor influencing crop growth and productivity, however, extreme temperature or heat stress causes detrimental effects on crop growth and development. The application of plant growth regulators or secondary metabolites as a seed treatment is highly beneficial to enhance heat stress tolerance in crop plants through modulating metabolic processes in plants including leaf chlorophyll content and antioxidant activity (Lohani et al., 2020). Furthermore, thiourea is an important stress-relieving compound that scavenges hydroxyl or superoxide radicals. Thiourea priming improved seedling growth and biomass production under heat stress by upregulating plant physiological attributes (Waraich et al., 2021). It has been observed that thiourea seed priming improved tolerance capacity in plants under high temperatures through an accumulation of secondary metabolites, antioxidative defense, and maintenance of plant water status (Ahmad et al., 2022). Brassinosteroids (BRs) are a group of growth regulators that modulate different biological processes associated with stress management. Seed priming along with a foliar spray of BRs significantly improved growth and physiological and biochemical traits with an enhanced antioxidant defense under heat stress.

Seed treatment with salicylic acid increased the chlorophyll content, soluble protein, proline, and grain yield content in leaves by 18%, 21%, 40%, and 19%, respectively, under heat stress depicting that seed priming led to stimulating tolerance in local wheat genotypes (Kousar et al., 2018). Seed osmopriming with salicylic acid and moringa leaf extract improved seedling vigor, relative water content, soluble solids, total phenolics content, and antioxidant content in wheat under heat stress (Mahboob et al., 2018). Upon exposure to heat stress, male sterility is a very common phenomenon in which pollen viability is disrupted and the stigmatic surface becomes dry to restrict fertilization. Qi et al. (2018) reported that melatonin (20 µM) seed priming

TABLE 18.2

Effects of Seed Priming on Growth, Physiological, Biochemical Response, and Yield of Different Crops under Salt Stress

Crop	Priming treatments	Growing condition	Physiological and biochemical interventions	References
Rice	Osmopriming with $CaCl_2$	Salt stress under greenhouse	Improved seed germination, seedling growth, and chlorophyll content.	Afzal et al. (2013)
	Nutripriming (B)	Salt stress with alternate wetting and drying conditions	Enhanced water relationships, grain yield, and grain quality.	Rehman et al. (2014)
	Spermidine 5 mM	Salt stress	Upregulation of several stress-responsive genes and transcription factors with expression of membrane Na^+ efflux pumps.	Paul et al. (2017)
Wheat	Silicon priming (Na_2SiO_3) 30 mM	Salt stress	Shoot Ca^{2+} and K^+ content increased.	Azeem et al. (2015)
	Seed priming with benzyl aminopurine	Salt stress under field condition	Accumulation of phenolics and total sugars with enhanced total protein and enzyme activity.	Bajwa et al. (2018)
	Ascorbic acid priming	Salt stress	Proteins associated with metabolism, energy, disease, defense, and storage showed increased abundance.	Fercha et al. (2014)
	Salicylic acid, 0.5 mM	Salt stress	Improved total phenols, flavonoids, carotenoids, phenylalanine ammonia-lyase, and ascorbic acid oxidase activity.	Yucel and Heybet (2016)
	Osmopriming with $CaCl_2$ (1.5%)	Salt stress under control condition	Osmotic adjustment; better antioxidant defense system; reduced LPO; improved water relationships, osmolyte accumulation, and yield.	Tabassum et al. (2017)
	Osmopriming with Moringa leaf extract	Salt stress under control condition	Enhanced net photosynthetic rate and osmotic adjustment for improving grain yield (18.5%).	Yasmeen et al. (2013)

(Continued)

TABLE 18.2 (CONTINUED)

Effects of Seed Priming on Growth, Physiological, Biochemical Response, and Yield of Different Crops under Salt Stress

Crop	Priming treatments	Growing condition	Physiological and biochemical interventions	References
Maize	Melatonin 0.4, 0.8 and 1.6 mM	150 mM salt stress	Improved germination energy, seedling growth, seedling vigor, and relative water content.	Jiang et al. (2016)
Cotton	Melatonin 25 µM	100 mM salt stress	Elevate photosynthetic efficiency and scavenge ROS.	Zhang et al. (2021)
Sunflower	KNO$_3$ 24 hours	Field experiment under saline condition	Better osmotic regulation.	Bajehbaj (2010)
Mustard	Methyl jasmonate	25 µmol L^{-1}	Indolic glycosylates, glucobrassicin, anthocyanins, and chlorogenic acid derivatives increased.	Hussain et al. (2017)
Mungbean	Gamma-amino butyric acid (GABA), 1 mM 6 hours	Salt stress	Promotes accumulation of total carbohydrates, total protein, proline, nitrate reductase, superoxide dismutase, and guaiacol peroxidase.	Jisha and Puthur (2016)
Bell pepper	Glycine betaine 10 mM	Saline soil	Higher accumulation of proline and decline in MDA levels.	Roychoudhury and Banerjee (2016)
Pea	Biopriming with *Typha angustifolia*	Salt stress under control condition	Better membrane integrity and photosynthetic pigments synthesis with promoting proline synthesis.	Ghezal et al. (2016)
Lettuce	Gibberellic acid (GA$_3$) 4.5 mM, 5 hours	Salt condition	ABA biosynthesis is suppressed which triggers seed germination.	Ella et al. (2012)
Cucumber	NaCl solution 100 mM, 36 hours	Field experiment under the salt condition	Increased proline, soluble carbohydrates, and antioxidants.	Farhoudi et al. (2011)

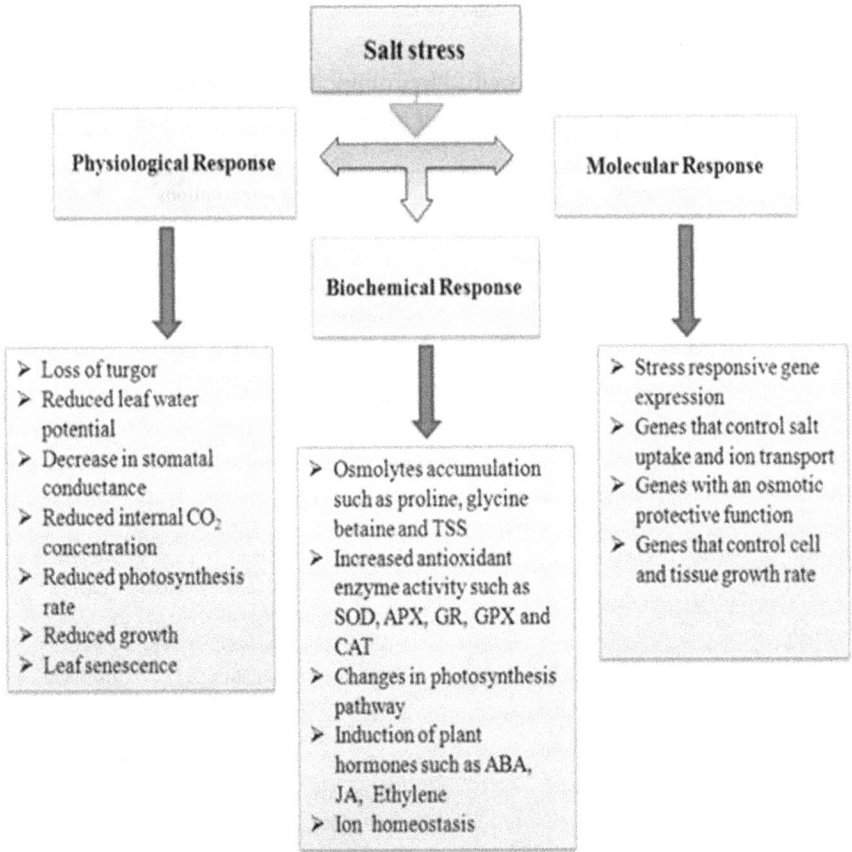

FIGURE 18.7 Physicochemical and molecular basis of salt stress tolerance in plants (De Oliveira et al., 2013)

along with soil incorporation minimizes the impact of heat stress on physicochemical and biological processes in tomatoes. Seed priming stimulates the accumulation of soluble protein, secondary metabolites, and enzymatic antioxidants to eliminate high-temperature-induced oxidative stress (Khan et al., 2019). Seed priming boosts the synthesis and activity of heat shock proteins and chaperones to prevent protein denaturation in rice during heat-stress conditions (Chakraborty and Dwivedi, 2021). Tamindžić et al. (2023) investigated that germination rate shoot length, root length, relative water content, biomass accumulation, and nutrient uptake were significantly higher in primed plants in comparison to non-primed plants under heat stress.

HEAVY METAL STRESS

HM toxicity is a vital environmental constraint that limits crop productivity while seed priming methods have been considered a unique approach to get rid of that stress by enhancing seed germination, seedling vigor, photosynthesis, biomass accumulation, and crop yield. Phytohormones regulate heavy metal absorption from soil and have significant roles in biochemical signaling, and defense pathways in plants (Bücker-Neto et al., 2017). Moreover, phytohormone priming is carried out to enhance crop productivity under metal stress. Plant growth performance under HM stress can be promoted by the exogenous application of phytohormones under HMs stress (Sytar et al., 2019). Hormonal priming such as auxin, gibberellin, cytokinin, abscisic acid, and ethylene improved seed germination in pigeon peas under cadmium (Cd) stress. Morphological, physiological, biochemical, and metabolic regulation of rice were significantly improved by salicylic acid priming under Lead (Pb) stress. In maize, pretreatment of salicylic acid combines with sodium hydrosulfide promotes glycinebetaine and nitric oxide contents with Pb stress (Kohli et al., 2018). Similarly, magneto-priming is the pretreatment of seed using a magnetic field before sowing that plays an important role in morpho-physiological, and biochemical traits. Priming with a static magnetic field increases germination capacity, germination speed, plant height, leaf area, efficiency of PS II, net photosynthetic rate, and yield of soybean plants under salt and UV-B stress (Kataria et al., 2020). It was reported that magneto-priming mitigates the adverse effect of cadmium stress by reducing the level of malondialdehyde, H_2O_2, and promoting nitric oxide content (Anand et al., 2019). Furthermore, melatonin application combined with Si reduced arsenic (As) and Cd uptake more efficiently than individual silicon. Nano-priming (NPs) with titanium dioxide (TiO_2) alleviates the Cd toxicity in plants by enhancing growth and developmental phenomena toward higher biomass formation. Si nano-priming protects maize seedlings against arsenic (As) toxicity by improving the synthesis of antioxidative enzymes like ascorbic acid peroxidase (APX), superoxide dismutase (SOD), glutathione reductase (GR) dehydroascorbate reductase (DHAR), that limits the accumulation of As and ROS (Tripathi et al., 2016). Furthermore, these NPs stimulate the upregulation of aquaporin genes in germinating seeds for increasing the growth, yield contributing attributes and yield of wheat plants under cadmium-contaminated soil (Mahakham et al., 2017).

CONCLUSIONS AND FUTURE PERSPECTIVE

Climate change-induced abiotic stresses are a major environmental hazard that will hamper sustainable agricultural production to meet global food security. Over the last decade, seed priming as a part of nutrient management has become a promising stress management strategy in modern agriculture that protects crop plants against various abiotic stresses such as drought, salinity, heat, and heavy metal stress through improving physiological and biochemical interventions. It is highly imperative to enhance the tolerance capacity of crop plants against catastrophic abiotic stresses by a paramount shifting of traditional cultivation practices along with the introduction of a new dimension of nutrient management such as seed priming for a better version of crop growth and yield. Significant advancements have been made in physiological, biochemical, and molecular mechanisms regarding seed priming-induced enhancement of irrigation water productivity and nutrient dynamics in an agroecosystem. Seed priming-mediated nutrient management is closely associated with accumulation in hyperactive signaling proteins that could amplify signal trans-duction for triggering defense responses. Priming-induced cross-talk among various signaling agents impart stress tolerance efficiency in crop plants under adverse envi-ronmental conditions. Finally, it was inferred that different seed priming materials turn on long-lasting stress memory by regulating cellular metabolic and biochemical responses to improve growth, yield, and tolerance capacity in plants against abiotic stresses.

REFERENCES

Abdelaal KA, Attia KA, Alamery SF, El-Afry MM, Ghazy AI, Tantawy DS, AlDoss AA, El-Shawy ESE, Abu-Elsaoud AM, Hafez YM. 2020. Exogenous application of proline and salicylic acid can mitigate the injurious impacts of drought stress on barley plants associated with physiological and histological characters. *Sustainability* 12(5): 1736.

Adhikary D, Das D, Ali MY, Ullah H, Datta A. 2023. Growth, grain yield, and water produc-tivity of traditional rice landraces from coastal Bangladesh, as affected by salt stress. *J Crop Improv* 37(1): 60–73.

Afzal I, Basra SMA, Cheema MA, Farooq M, Jafar MZ, Shahid M, Yasmeen A. 2013. Seed priming: A shotgun approach for alleviation of salt stress in wheat. *Int J Agric Biol* 15: 1199–1203.

Ahmad M, Waraich EA, Hussain S, Ayyub CM, Ahmad Z, Zulfiquar U. 2022. Improving heat stress tolerance in *Camelina sativa* and *Brassica napus* through thiourea seed priming. *J Plant Growth Regul* 41(7): 2886–2902.

Alam A, Hariyanto B, Ullah H, Salin KR, Datta A. 2020. Effects of silicon on growth, yield and fruit quality of cantaloupe under drought stress. *Silicon.* https://doi.org/10.1007/s12633-020-00673-1.

Alam A, Ullah H, Thuenprom N, Tisarum R, Cha-um S, Datta A. 2022. Seed priming with salicylic acid enhances growth, physiological traits, fruit yield, and quality parameters of cantaloupe under water-deficit stress. *S Afr J Bot* 150: 1–12.

Ale S, Himanshu SK, Mauget SA, Hudson D, Goebel TS, Liu B, Baumhardt, RL, Bordovsky, JP, Brauer, DK, Lascano, RJ, Gitz III, DC. 2021. Simulated dryland cotton yield response to selected scenario factors associated with soil health. *Front Sustain Food Syst* 4: 617509.

Ali N, Khan MN, Ashraf MS, Ijaz S, Saeed-ur-Rehman H, Abdullah M, et al. 2020. Influence of different organic manures and their combinations on productivity and quality of bread wheat. *J Soil Sci Plant Nutr* 20(4): 1949–1960.

Al-Taey DKA, Alazawi SSM, Al-Shareefi MJH, Al-Tawaha A. 2018. Effect of saline water, NPK and organic fertilizers on soil properties and growth, antioxidant enzymes in leaves and yield of lettuce (*Lactuca Sativa* Var. Parris Island). *Res Crops* 19: 441–449.

Anand A, Kumari A, Thakur M, Koul A. 2019. Hydrogen peroxide signaling integrates with phytohormones during the germination of magneto-primed tomato seeds. *Sci Rep* 9(1): 8814.

Anjum SA, Tanveer M, Hussain S, Bao M, Wang L, Khan I. et al. 2015. Cadmium toxicity in Maize (*Zea mays* L.): Consequences on antioxidative systems, reactive oxygen species and cadmium accumulation. *Environ Sci Pollut Res Int* 22(21): 17022–17030.

Azeem M, Iqbal N, Kausar S, Javed MT, Akram MS, Sajid MA. 2015. Efficacy of silicon priming and fertigation to modulate seedling's vigor and ion homeostasis of wheat (*Triticum aestivum* L.) under saline environment. *Environ Sci Pollut Res Int* 22(18): 14367–14371.

Azooz MM. 2009. Salt stress mitigation by seed priming with salicylic acid in two faba bean genotypes differing in salt tolerance. *Int J Agric Biol* 11: 343–350.

Bajehbaj AA. 2010. The effects of NaCl priming on salt tolerance in sunflower germination and seedling grown under salinity conditions. *Afr J Biotech* 9: 1764–1770.

Bajwa AA, Farooq M, Nawaz A. 2018. Seed priming with sorghum extracts and benzyl aminopurine improves the tolerance against salt stress in wheat (*Triticum aestivum* L.). *Physiol Mol Biol Plants* 24(2): 239–249.

Borges AA, Jiménez-Arias D, Expósito-Rodríguez M, Sandalio LM, Pérez JA. 2014. Priming crops against biotic and abiotic stresses: MSB as a tool for studying mechanisms. *Front Plant Sci* 5: 642.

Boukari N, Jelali N, Renaud JB, Youssef RB, Abdelly C, Hannoufa A. 2019. Salicylic acid seed priming improves tolerance to salinity, iron deficiency and their combined effect in two ecotypes of alfalfa. *Environ Exp Bot* 167: 103820.

Bücker-Neto L, Paiva ALS, Machado RD, Arenhart RA, Margis-Pinheiro M. 2017. Interactions between plant hormones and HMs responses. *Genet Mol Biol* 40: 373–386.

Cai L, Xu Z, Ren M, Guo Q, Hu X, Hu G, Wan H, Peng P. 2012. Source identification of eight hazardous heavy metals in agricultural soils of Huizhou, Guangdong Province, China. *Ecotoxicol Environ Saf* 78: 2–8.

Cao BL, Ma Q, Xu K. 2020. Silicon restrains drought-induced ROS accumulation by promoting energy dissipation in leaves of tomato. *Protoplasma* 257(2): 537–547.

Cao BL, Wang L, Gao S, Xia J, Xu K. 2017. Silicon-mediated changes in radial hydraulic conductivity and cell wall stability are involved in silicon-induced drought resistance in tomato. *Protoplasma* 254(6): 2295–2304.

Chakma R, Saekong P, Biswas A, Ullah H, Datta A. 2021. Growth, fruit yield, quality, and water productivity of grape tomato as affected by seed priming and soil application of silicon under drought stress. *Agric Water Manag* 256: 107055.

Chakraborty P, Dwivedi P. 2021. Seed priming and its role in mitigating heat stress responses in crop plants. *J Soil Sci Plant Nutr* 21(2): 1718–1734.

Chen W, Yao X, Cai K, Chen J. 2011. Silicon alleviates drought stress of rice plants by improving plant water status, photosynthesis and mineral nutrient absorption. *Biol Trace Elem Res* 142(1): 67–76.

Cheng J, Wang L, Zeng P, He Y, Zhou R, Zhang H, Wang Z. 2017. Identification of genes involved in rice seed priming in the early imbibition stage. *Plant Biol* 19(1): 61–69.

Chigbo C, Batty L. 2013. Effect of combined pollution of chromium and benzo (a) pyrene on seed growth of *Lolium perenne*. *Chemosphere* 90(2): 164–169.

Das D, Basar NU, Ullah H, Salin KR, Datta A. 2021a. Interactive effect of silicon and mycorrhizal inoculation on growth, yield and water productivity of rice under water-deficit stress. *J Plant Nutr* 44(18): 2756–2769.

Das D, Basar NU, Ullah H, Attia A, Salin KR, Datta A. 2021b. Growth, yield and water productivity of rice as influenced by seed priming under alternate wetting and drying irrigation. *Arch Agron Soil Sci* 68(11): 1515–1529.

Das D, Ullah H, Tisarum R, Cha-um S, Datta A. 2021c. Morpho-physiological responses of tropical rice to potassium and silicon Fertilization under water-deficit stress. *J Soil Sci Plant Nutr* https://doi.org/10.1007/s42729-021-00712-9.

Das D, Ullah H, Himanshu SK, Tisarum R, Cha-um S, Datta A. 2022. Arbuscular mycorrhizal fungi inoculation and phosphorus application improve growth, physiological traits, and grain yield of rice under alternate wetting and drying irrigation. *J Plant Physiol* 278: 153829.

Das D, Ali MA, Sarkar TA, Ali MY. 2017. Germination and seedling growth of indigenous aman rice under salt stress. *J Bangladesh Agric Univ* 15(2): 182–187.

De Oliveira BDO, Nara LMA, Eneas G-F. 2013. Comparison between the water and salt stress effects on plant growth and development. *Resp Organ Water Stress* 4: 67–94.

Dietz K-J, Turkan I, Krieger-Liszkay A. 2016. Redox- and reactive oxygen species dependent signaling into and out of the photosynthesizing chloroplast. *Plant Physiol* 171(3): 1541–1550.

dos Santos TB, Ribas AF, de Souza SGH, Budzinski IGF, Domingues DS. 2022. Physiological responses to drought, salinity, and heat stress in plants: A review. *Stresses* 2(1): 113–135.

Ella ES, Dionisio–Sese ML, Ismail AM. 2012. Seed pre-treatment in rice reduces damage, enhances carbohydrate mobilization and improves emergence and seedling establishment under flooded conditions. *AoB Plants* 2011. plr007.

Fahad S, Hussain S, Matloob A, Khan FA, Khaliq A, Saud S, et al. 2015a. Phytohormones and plant responses to salinity stress: A review. *Plant Growth Regul* 75(2): 391–404.

Fahad S, Hussain S, Saud S, Tanveer M, Bajwa AA, Hassan S. 2015b. A biochar application protects rice pollen from high-temperature stress. *Plant Physiol Biochem* 96: 281–287.

Fahad S, Hussain S, Saud S, Khan F, Hassan S, Amanullah et al. 2016. Exogenously applied plant growth regulators affect heat-stressed rice pollens. *J Agron Crop Sci* 202(2): 139–150.

Farhoudi R, Saeedipour S, Mohammadreza D. 2011. The effect of NaCl seed priming on salt tolerance, antioxidant enzyme activity, proline and carbohydrate accumulation of muskmelon (Cucumis melo L.) under saline condition. *Afr J Agric Res* 6: 1363–1370.

Farooq M, Basra SMA, Hafeez K. 2006a. Seed invigoration by osmohardening in coarse and fine rice. *Seed Sci Technol* 34(1): 181–187.

Farooq M, Basra SMA, Wahid A. 2006b. Priming of field-sown rice seed enhances germination, seedling establishment, allometry and yield. *Plant Growth Regul* 49(2–3): 285–294.

Farooq M, Basra SMA, Tabassum R, Afzal I. 2006c. Enhancing the performance of direct seeded fine rice by seed priming. *Plant Prod Sci* 9(4): 446–456.

Farooq M, Aziz T, Basra SMA, Cheema MA, Rehman H. 2008. Chilling tolerance in hybrid maize induced by seed priming with salicylic acid. *J Agro Crop Sci* 194(2): 161–168.

Farooq M, Wahid A, Kobayashi N, Fujita D, Basra SMA. 2009. Plant drought stress, effects, mechanisms and management. *Agron Sustain Dev* 29(1): 185–212.

Farooq M, Basra SMA, Wahid A, Ahmad N. 2010. Changes in nutrient-homeostasis and reserves metabolism during rice seed priming: Consequences for germination and seedling growth. *Agri Sci China* 9(2): 101–108.

Farooq M, Irfan M, Aziz T, Ahmad I, Cheema SA. 2013. Seed priming with ascorbic acid improves drought resistance of wheat. *J Agro Crop Sci* 199(1): 12–22.

Farooq M, Hussain M, Nawaz A, Lee D-J, Alghamdi SS, Siddique KHM. 2017. Seed priming improves chilling tolerance in chickpea by modulating germination metabolism, trehalose accumulation and carbon assimilation. *Plant Physiol Biochem* 111: 274–283.

Farooq M, Ullah A, Rehman A, Nawaz A, Nadeem A, Wakeel A, Nadeem F, Siddique KHM. 2018. Application of zinc improves the productivity and biofortification of fine grain aromatic rice grown in dry seeded and puddled transplanted production systems. *Field Crops Res* 216: 53–62.

Farooq M, Usman M, Nadeem F, Rehman H, Wahid A, Basra SMA, Siddique KHM. 2019. Seed priming in field crops: Potential benefits, adoption and challenges. *Crop Pasture Sci* 70(9): 731–771.

Fercha A, Capriotti AL, Caruso G, Cavaliere C, Samperi R, Stampachiacchiere S, Laganà A. 2014. Comparative analysis of metabolic proteome variation in ascorbate-primed and unprimed wheat seeds during germination under salt stress. *J Proteome* 108: 238–257.

Fu J, Huang Z, Wang Z, Yang J, Zhang J. 2011. Pre-anthesis non-structural carbohydrate reserve in the stem enhances the sink strength of inferior spikelets during grain filling of rice. *Field Crops Res* 123(2): 170–182.

Ghassemi-Golezani K, Farshbaf-Jafari S, Shafagh-Kolvanagh J. 2011. Seed priming and field performance of soybean (*Glycine max* L.) in response to water limitation. *Not Bot Horti Agrobot Cluj-Napoca* 39(2): 186–189.

Ghezal N, Rinez I, Sbai H, Saad I, Farooq M, Rinez M, Zribi I, Haouala R. 2016. Improvement of Pisum sativum salt stress tolerance by biopriming their seeds using *Typha angustifolia* leaves aqueous extract. *S Afr J Bot* 105: 240–250.

Gonzalez VH, Lee EA, Lukens LL, Swanton CJ. 2019. The relationship between floret number and plant dry matter accumulation varies with early season stress in maize (*Zea mays* L.). *Field Crops Res* 238: 129–138.

Gosling SN, Arnell NW. 2016. A global assessment of the impact of climate change on water scarcity. *Clim Change* 134(3): 371–385.

Gupta B, Huang B. 2014. Mechanism of salinity tolerance in plants: Physiological, biochemical and molecular characterization. *Int J Genomics* 2014: 701596.

Gupta BK, Sahoo KK, Anwar K, Nongpiur RC, Deshmukh R, Pareek A, Singla-Pareek SL. 2021. Silicon nutrition stimulates Salt-Overly Sensitive (SOS) pathway to enhance salinity stress tolerance and yield in rice. *Plant Physiol Biochem* 166: 593–604.

Hameed M, Sheikh MA, Hameed A, Farooq T, Basra SMA, Jamil A. 2014. Chitosan seed priming improves seed germination and seedling growth in wheat (*Triticum aestivum* L.) under osmotic stress induced by polyethylene glycol. *Philipp Agric Sci* 97(3): 294–299.

Hasanuzzaman M, Nahar K, Alam MM, Fujita M. 2012. Exogenous nitric oxide alleviates high temperature induced oxidative stress in wheat (*Triticum aestivum*) seedlings by modulating the antioxidant defense and glyoxalase system. *Aust J Crop Sci* 6: 1314–1323.

Hatfield JL, Prueger JH. 2015. Temperature extremes: Effect on plant growth and development. *Weather Clim Extrem* 10: 4–10.

Himanshu SK, Ale S, Bordovsky JP, Kim J, Samanta S., Omani N, Barnes EM. 2021. Assessing the impacts of irrigation termination periods on cotton productivity under strategic deficit irrigation regimes *Sci Rep* 11(1): 1–16.

Himanshu SK, Ale S, Bell J, Fan Y, Samanta S, Bordovsky JP, Gitz III DC, Lascano RJ, Brauer DK. 2023. Evaluation of growth-stage-based variable deficit irrigation strategies for cotton production in the Texas High Plains. *Agric Water Manag* 280: 108222.

Horie T, Shiraiwa T, Homma K, Katsura K, Maeda S, Yoshida H. 2005. Can yields of lowland rice resume the increases that they showed in the 1980s? *Plant Prod Sci* 8(3): 259–274.

Huang S, Leng G, Huang Q, Xie Y, Liu S, Meng E, Li P. 2017. The asymmetric impact of global warming on US drought types and distributions in a large ensemble of 97 hydroclimatic simulations. *Sci Rep* 7(1): 5891.

Hussain M, Farooq M, Lee DJ. 2017. Evaluating the role of seed priming in improving drought tolerance of pigmented and non-pigmented rice. *Agro Crop Sci* 203(4): 269–276.

Hussain S, Khan F, Cao W, Wu L, Geng M. 2016a. Seed priming alters the production and detoxification of reactive oxygen intermediates in rice seedlings grown under sub-optimal temperature and nutrient supply. *Front Plant Sci* 7: 439.

Hussain S, Khan F, Hussain HA, Nie L. 2016b. Physiological and biochemical mechanisms of seed priming-induced chilling tolerance in rice cultivars. *Front Plant Sci* 7: 116.

Ibrahim EA. 2016. Seed priming to alleviate salinity stress in germinating seeds. *J Plant Physiol* 192: 38–46.

Imran M, Maria K, Romheld V, Neumann G. 2015. Impact of nutrient seed priming on germination, seedling development, nutritional status and grain yield of maize. *J Plant Nutr* 38(12): 1803–1821.

IPCC (Inter-governmental Panel on Climate Change). 2013. Climate Change. 2013: The physical science basis. Contribution of working group I to the fifth assessment report of the intergovernmental panel on climate change. Cambridge and New York.

Iqbal H, Yaning C, Waqas M, Rehman H, Shareef M, Iqbal S. 2018. Hydrogen peroxide application improves quinoa performance by affecting physiological and biochemical mechanisms under water-deficit conditions. *Agro Crop Sci* https://doi.org/10.1111/ jac.12284.

Islam ATMT, Ullah H, Himanshu SK, Tisarum R, Cha-um S, Datta A. 2022. Effect of salicylic acid seed priming on morpho-physiological responses and yield of baby corn under salt stress. *Sci Hort* 304: 111304.

Jamil M, Malook I, Parveen S, Naz T, Ali A, Ullah JS. 2013. Smoke priming, a potent protective agent against salinity: Effect on proline accumulation, elemental uptake, pigmental attributes and protein banding patterns of rice (*Oryza sativa*). *J Stress Physiol Biochem* 9: 169–183.

Janas KM, Posmyk MM. 2013. Melatonin, an under estimated natural substance with great potential for agricultural application. *Acta Physiol Plant* 35(12): 3285–3292.

Jiang X, Li H, Song X. 2016. Seed priming with melatonin effects on seed germination and seedling growth in maize under salinity stress. *Pak J Bot* 48: 1345–1352.

Jisha KC, Puthur JT. 2016. Seed Priming with beta-amino butyric acid improves abiotic stress tolerance in rice seedlings. *Rice Sci* 23(5): 242–254.

Jisha KC, Vijayakumari K, Puthur JT. 2013. Seed priming for abiotic stress tolerance: An overview. *Acta Physiol Plant* 35(5): 1381–1396.

Kalai T, Khasmassi K, Teixeria de Silva JA, Gouia H, Ben-kaab LB. 2014. Cadmium and copper stress affect seedling growth and enzymatic activities in germinating barley seeds. *Arch Agron Soil Sci* 6:60.

Kataria S, Tripathi DK, Jain M, Singh VP. 2020. Role of nitric oxide during germination in regulation of magneto-priming induced alleviation of salt stress in soybean (*Glycine max*). *Physiol Plant* 168: 422–436.

Kavi Kishor PB, Hima Kumari P, Sunita MSL, Sreenivasulu N. 2015. Role of proline in cell wall synthesis and plant development and its implications in plant ontogeny. *Front Plant Sci* 6: 544.

Khan MN, Zhang J, Luo T, Liu J, Rizwan M, Fahad S, Xu Z, Hu L. 2019. Seed priming with melatonin coping drought stress in rapeseed by regulating reactive oxygen species detoxification: Antioxidant defense system, osmotic adjustment, stomatal traits and chloroplast ultrastructure perseveration. *Ind Crops Prod* 140: 111597.

Khan MS, Ahmad D, Khan MA. 2015. Utilization of genes encoding osmoprotectants in transgenic plants for enhanced abiotic stress tolerance. *Electron J Biotechnol* 18(4): 257–266.

Kohli SK, Handa N, Sharma A, Gautam V, Arora S, Bhardwaj R, Wijaya L, Alyemeni MN, Ahmad P. 2018. Interaction of 24-epibrassi-nolide and salicylic acid regulates pigment contents, antioxidative defense responses, and gene expression in *Brassica juncea* L. seedlings under Pb stress. *Environ Sci Pollut Res Int* 25(15): 15159–15173.

Kousar R, Queshi R, Uddin J, Munir M, Shabbir G. 2018. Salicylic acid mediated heat stress tolerance in selected bread wheat genotype of Pakistan. *Pak J Bot* 50(6): 2141-3146.

Kumari P, Rastogi A, Yadav S. 2020. Effects of heat stress and molecular mitigation approaches in orphan legume, Chickpea. *Mol Biol Rep* 47(6): 4659–4670.

Lesk C, Rowhani P, Ramankutty N. 2016. Influence of extreme weather disasters on global crop production. *Nature* 529(7584): 84–87.

Liu J, Wang W, Wang L, Sun Y. 2015a. Exogenous melatonin improves seedling health index and drought tolerance in tomato. *Plant Growth Regul* 77(3): 317–326.

Liu P, Yin L, Wang S, Zhang M, Deng X, Zhang S, Tanaka K. 2015b. Enhanced root hydraulic conductance by aquaporin regulation accounts for silicon alleviated salt induced osmotic stress in Sorghum bicolor L. *Environ Exp Bot* 111: 42–51.

Lohani N, Singh MB, Bhalla PL. 2020. High temperature susceptibility of sexual reproduction in crop plants. *J Exp Bot* 71(2): 555–568.

Lu L, Huang M, Huang Y, Corvini PFX, Ji R, Zhao L. 2020. Mn_3O_4 nanozymes boost endogenous antioxidant metabolites in cucumber (*Cucumis sativus*) plant and enhance resistance to salinity stress. *Environ Sci Nano* 7(6): 1692–1703.

Lutts S, Benincasa P, Wojtyla L, Kubala S, Pace R, Lechowska K, Quinet M, Garnczarska M. 2016. Seed priming: New comprehensive approaches for an old empirical technique, new challenges in seed biology. In: Susana Araújo S, Balestrazzi A (eds) *Basic and Translational Research Driving Seed Technology*. InTech, Open, Rijeka.

Maclean J, Hardy B, Hettel G. 2013. *Rice Almanac: Source Book for One of the Most Important Economic Activities on Earth.* IRRI.

Mahajan, M., Gupta, P. K., Singh, A., Vaish, B., Singh, P., Kothari, R., & Singh, R. P. (2022). A comprehensive study on aquatic chemistry, health risk and remediation techniques of cadmium in groundwater. *Science of The Total Environment, 818*, 151784. https://doi.org/10.1016/j.scitotenv.2021.151784.

Mahakham W, Sarmah AK, Maensiri S, Theerakulpisut P. 2017. Nano-priming technology for enhancing germination and starch metabolism of aged rice seeds using photosynthesized silver nanoparticles. *Sci Rep* 7(1): 8263.

Mahboob W, Khan MA, Shirazi MU, Faisal S, Asma. 2018. Seed priming induced high temperature tolerance in wheat by regulating germination metabolism and physio-biochemical properties. *Int J Agric Biol* 20: 2140–2148.

Marques MC, do Nascimento CWA. 2013. Analysis of chlorophyll fluorescence spectra for the monitoring of Cd toxicity in a bio-energy crop (*Jatropha curcas*). *J Photochem Photobiol B* 127: 88–93.

Marthandan V, Geetha R, Kumutha K, Renganathan VG, Karthikeyan A, Ramalingam J. 2020. Seed priming–A feasible strategy to enhance drought tolerance in crop plants. *Int J Mol Sci* 21(21): 8258.

Molins H, Michelet L, Lanquar V, Agorio A, Giraudat J, Roach T, Kriegerliszkay A, Thomine S. 2013. Mutants impaired in vacuolar metal mobilization identify chloroplasts as a target for cadmium hypersensitivity in Arabidopsis thaliana. *Plant Cell Environ* 36(4): 804–817.

Mostofa MGD, Saegusa M, Fujita L, Tran SP. 2015. Hydrogen sulfide regulates salt tolerance in rice by maintaining Na^+/K^+ balance, mineral homeostasis and oxidative metabolism under excessive salt stress. *Front Plant Sci* 6: 1055.

Mustafa A, Ahmad R, Farooq M, Wahid A. 2018. Effect of seed size and seed priming on stand establishment, wheat productivity and profitability under different tillage systems. *Int J Agric Biol* 20: 1710–1716.

Najafabadi MY, Ehsanzadeh P. 2017. Salicylic acid effects on osmoregulation and seed yield in drought-stressed sesame. *Agron J* 109(4): 1414–1422.

Neeru J, Shaliesh C, Vaishali T, Purav S, Manoherlal R. 2019. Role of orthosilicic acid (OSA) based formulation in improving plant growth and development. *Silicon* 11(5): 2407–2411.

Neupane S, Shrestha S, Ghimire U, Mohanasundaram S, Ninsawat S. 2021. Evaluation of the CORDEX regional climate models (RCMs) for simulating climate extremes in the Asian cities. *Sci Tot Environ* 797: 149137.

Pame AR, Kreye C, Johnson D, Heuer S, Becker M. 2015. Effects of genotype, seed P concentration and seed priming on seedling vigor of rice. *Exp Agric* 51(3): 370–381 https://doi.org/10.1017/S001447 9714000362.

Park JH, Lamb D, Paneerselvam P, Choppala G, Bolan N, Chung J-W. 2011. Role of organic amendments on enhanced bioremediation of heavy metal(loid) contaminated soils. *J Hazard Mater* 185(2–3): 549–574.

Paul S, Roychoudhury A. 2016. Seed priming with spermine ameliorates salinity stress in the germinated seedlings of two rice cultivars differing in their level of salt tolerance. *Trop Plant Res* 3(3): 616–633.

Qi ZY, Kai-Xin W, Meng-Yu Y, Mukesh KK, Li DY, Leonard W, Alyemeni MN, Ahmad P, Zhou J. 2018. Melatonin alleviates high temperature-induced pollen abortion in *Solanum lycopersicum*. *Molecules* 23(2): 386.

Raj AB, Raj SK. 2019. Seed priming: An approach toward agricultural sustainability. *J Appl Nat Sci* 11(1): 227–234.

Rajjou L, Duval M, Gallardo K, Catusse J, Bally J, Job C, Job D. 2012. Seed germination and vigor. *Annu Rev Plant Biol* 63: 507–533.

Rang ZW, Jagadish SVK, Zhou QM, Craufurd PQ, Heuer S. 2011. Effect of high temperature and water stress on pollen germination and spikelet fertility in rice. *Environ Exp Bot* 70(1): 58–65.

Rascio N, Navari-Izzo F. 2011. Heavy metal hyperaccumulating plants: How and why do they do it? And what makes them interesting? *Plant Sci (Shannon Ireland)* 180: 169–181.

Rasool T, Ahmad R, Farooq M. 2019. Seed priming with micronutrients for improving the quality and yield of hybrid maize. *Plant Sci* 71(1): 37–44.

Rehman A, Farooq M, Naveed M, Ozturk L, Nawaz A. 2018. Pseudomonas-aided zinc application improves the productivity and biofortification of bread wheat. *Crop Pasture Sci* 69(7): 659–672.

Rehman A, Farooq M, Nawaz A, Ahmad R. 2014. Influence of boron nutrition on the rice productivity, kernel quality and biofortification in different production systems. *Field Crops Res* 169: 123–131.

Roy SD, Das D, Kabir ME. 2017. Seed germination and seedling growth of sunflower (Helianthus Annuus L) under salt stressed conditions. *Khulna Univ Stud* 14: 39–47.

Roychoudhury A, Banerjee A. 2016. Endogenous glycine betaine accumulation mediates abiotic stress tolerance in plants. *Trop Plant Res* 3: 105–111.

Schwember AR, Bradford KJ. 2010. A genetic locus and gene expression pattern associated with the priming effect on lettuce seed germination at elevated temperature. *Plant Mol Biol* 73(1–2): 105–118.

Shahid M, Pourrut B, Dumat NM, Aslam M, Pinelli E. 2014. Heavy-metal-induced reactive oxygen species: Phytotoxicity and physicochemical changes in plants. *Environ Contam Toxicol* 232: 1–44.

Shrivastava P, Kumar R. 2015. Soil salinity: A serious environmental issue and plant growth promoting bacteria as one of the tools for its alleviation. *Saudi J Biol Sci* 22(2): 123–131.

Singh, R. P., Mahajan, M., Gandhi, K., Gupta, P. K., Singh, A., Singh, P., ... & Kidwai, M. K. (2023). A holistic review on trend, occurrence, factors affecting pesticide concentration, and ecological risk assessment. *Environmental Monitoring and Assessment*, 195(4), 451. https://doi.org/10.1007/s10661-023-11005-2.

Sirisuntornlak N, Ghafoori S, Datta A, Arirob W. 2019. Seed priming and soil incorporation with silicon influence growth and yield of maize under water-deficit stress. *Arch Agron Soil Sci* 65(2): 197–207.

Sytar O, Kumari P, Yadav S, Brestic M, Rastogi A. 2019. Phytohormone priming: Regulator for heavy metal stress in plants. *J Plant Growth Regul* 38(2): 739–752.

Tabassum T, Ahmad R, Farooq M, Basra SMA. 2018a. Improving the drought tolerance in barley by osmopriming and biopriming. *Int J Agric Biol* 20: 1597–1606.

Tabassum T, Farooq M, Ahmad R, Zohaib A, Wahid A, Shahid M. 2018b. Terminal drought and seed priming improves drought tolerance in wheat. *Physiol Mol Biol Plants* 24(5): 845–856.

Tabassum T, Farooq M, Ahmad R, Zohaib A, Wahid A. 2017. Seed priming and transgenerational drought memory improves tolerance against salt stress in bread wheat. *Plant Physiol Biochem* 118: 362–369.

Taiz L, Zeiger E. 2010. *Plant Physiology*, 5th edition. Sinauer Associates Inc., Publishers, Sunderland, MA.

Tamindžić G, Ignjatov M, Miljaković D, Červenski J, Milošević D, Nikolić Z, Vasiljević S. 2023. Seed priming treatments to improve heat stress tolerance of garden pea (*Pisum sativum* L.). *Agriculture* 13(2): 439.

Tang L, Ying R-R, Jiang D, Zeng X, Morel J-L, Tang Y-T, Qiu R-L. 2013. Impaired leaf CO_2 diffusion mediates Cd-induced inhibition of photosynthesis in the Zn/Cd hyperaccumulator Picris divaricata. *Plant Physiol Biochem* 73: 70–76.

Thorne SJ, Hartley SE, Maathuis FJM. 2020. Is silicon a panacea for alleviating drought and salt in crop? *Front Plant Sci* 11: 1221.

Tripathi DK, Singh S, Singh VP, Prasad SM, Chauhan DK, Dubey NK. 2016. Silicon nanoparticles more efficiently alleviate arsenate toxicity than silicon in maize cultivar and hybrid differing in arsenate tolerance. *Front Environ Sci* 4: 46.

Ullah H, Datta A. 2018a. Root system response of selected lowland Thai rice varieties as affected by cultivation method and potassium rate under alternate wetting and drying irrigation. *Arch Agron Soil Sci* 64(14): 2045–2059.

Ullah H, Datta A. 2018b. Effect of water saving technologies on growth, yield and water productivity of lowland rice variety. *Int J Technol* 7(7): 1375–1383.

Ullah H, Santiago-Arenas R, Ferdous Z, Attia A, Datta A. 2019. Improving water use efficiency, nitrogen use efficiency, and radiation use efficiency in field crops under drought stress: A review. *Adv Agro* 156: 109–157.

Waraich EA, Muhammad A, Walid S, Muhammad TM, Zahoor A, Muhammad HUR, Ayman ELS. 2021. Seed priming with sulfhydral thiourea enhances the performance of *Camelina sativa* L. under heat stress conditions. *Agronomy* 11(9): 1875.

Wojtyla L, Lechowska K, Kubala S, Garnczarska M. 2016. Molecular processes induced in primed seeds—Increasing the potential to stabilize crop yields under drought conditions. *J Plant Physiol* 203: 116–126.

Wu XH, Wang W, Yin CM, Hou HJ, Xie KJ, Xie XL. 2017. Water consumption, grain yield, and water productivity in response to field water management in double rice systems in China. *PLoS One* 12(12): e0189280.

Yadav S, Atri N. 2020. Impact of salinity stress in crop plants and mitigation strategies. In: Rakshit A, Singh H, Singh A, Singh U, Fraceto L (eds) *New Frontiers in Stress Management for Durable Agriculture*. Springer, Singapore. https://doi.org/10.1007/978-981-15-1322-0_4.

Yasmeen A, Basra SMA, Farooq M, Rehman H, Hussain N, Athar HR. 2013. Exogenous application of Moringa leaf extract modulates the antioxidant enzyme system to improve wheat performance under saline conditions. *Plant Growth Regul* 69(3): 225–233.

Yatoo, A.M., Ali, M.N., Zaheen, Z., Baba, Z.A., Ali, S., Rasool, S., Sheikh, T.A., Sillanpää, M., Gupta, P.K., Hamid, B. and Hamid, B., (2022). Assessment of pesticide toxicity on earthworms using multiple biomarkers: a review. *Environmental Chemistry Letters*, 20(4), 2573–2596. https://doi.org/10.1007/s10311-022-01386-0.

Yucel NC, Heybet EH. 2016. Salicylic acid and calcium treatments improves wheat vigor, lipids and phenolics under high salinity. *Acta Chim Slov* 63(4): 738–746.

Zhang Y, Zhou X, Dong Y, Zhang F, He Q, Chen J, Zhu S, Zhao T. 2021. Seed priming with melatonin improves salt tolerance in cotton through regulating photosynthesis, scavenging reactive oxygen species and coordinating with phytohormone signal pathways. *Ind Crops Prod* 169: 113671.

Zhao J, Han T, Wang C, Jia H, Worqlul AW, Norelli N, Zeng Z, Chu Q. 2020. Optimizing irrigation strategies to synchronously improve the yield and water productivity of winter wheat under interannual precipitation variability in the North China Plain. *Agric Water Manag* 240: 106298.

Zhou H, Wu H, Zhang F, Su Y, Guan W, Xie Y, Giraldo JP, Shen W. 2021. Molecular basis of cerium oxide nanoparticle enhancement of rice salt tolerance and yield. *Environ Sci Nano* 8(11): 3294–3311.

19 Protected Cultivation
Microclimate-Based Agriculture Under Greenhouse

Vikas Kumar Singh, K. N. Tiwari, Shivam
Gupta, and Vijay Kumar Singh

INTRODUCTION

During the last two decades, the productivity and efficiency of horticultural crop production has received special attention from the Government of India for enhancing fruits and vegetables to achieve nutritional security. Farmers are encouraged to use the farm inputs of land, water, and nutrients more efficiently to attain maximum yield per unit area, as well as for better quality products (Ale et al., 2022, 2023). Protected cultivation makes it possible to obtain increased crop productivity by maintaining a favorable environment for the plants. Therefore, production in a greenhouse has become more popular than in the open field. Due to the protective cover and shape of greenhouses the climatic conditions in the greenhouse get modified in comparison to those in the open field. In the greenhouse radiation and air velocity are reduced, temperature and water vapor pressure of the air increased and fluctuation in CO_2 concentration are much higher. Each of these parameters has its own impact on the growth, production, and quality of the greenhouse crop, therefore, it becomes necessary to study these parameters and keep them in the optimum range (Bakker, 1995). Greenhouse crop production is based on control of the environment in such a way as to provide the conditions that are most favorable for maximum yield. A plant's ability to grow and develop is dependent on the photosynthetic process. In the presence of light, the plant combines carbon dioxide and water to form sugars which are then utilized for growth and fruit production. Optimization of the greenhouse environment is directed at optimizing the photosynthetic process in the plants, the plant's ability to utilize light at maximum efficiency.

Protected cultivation or greenhouse cultivation is a kind of farming system that is used to maintain a controlled or partially controlled environment suitable for maximum crop production. This includes creating an environment suitable for working efficiency as well as for better crop growth (Aldrich and Bartok, 1989). In India, the cultivation of greenhouses for commercial purposes started in 1988 and it has

DOI: 10.1201/9781003441175-19

increased because of liberalization in the economy and emphasis on export promotion (Mahajan and Singh, 2006). Naturally ventilated greenhouse uses solar energy and ventilates warm air to grow crops normally. These naturally ventilated greenhouses play an important role in India's horticultural production and supplication during the winter (Gao et al., 2010; Liang et al., 2014). The main advantages of greenhouse cultivation are that crops can be cultivated successfully throughout the year, getting high productivity with excellent quality. A greenhouse protects crops against extreme climatic conditions and the incidence of pests and diseases (Von Zabeltitz, 1999).

Microclimate control in greenhouses is one of the first priority problems since even despite good genetic properties of crops, the quality of fertilizers and soil, incorrect maintenance of temperature, humidity mode of the greenhouse and poor dosing of the carbon dioxide may result in a considerable drop in productivity up to the loss of crops. The process of variation of the microclimate of the greenhouse is complex, multi-parametric, and depends on a set of external and internal factors. The external factors are: ambient temperature, humidity, intensity of solar radiation, the direction and velocity of wind, etc. The geometric dimensions of the greenhouse, the location of elements of heating and ventilation systems, types of soil, genetic properties, and kinds of crops, etc., are the internal factors.

The microclimate of the greenhouse can be scientifically controlled to an optimum level throughout the cultivation period to increase productivity by several folds. Greenhouse also permits to cultivation of four to five crops in a year with controlled microclimate, efficient use of various inputs like water, fertilizer, and seeds and plant protection chemicals. In addition, automation of irrigation, precise application of other inputs and environmental controls by using computers, and artificial intelligence are possible for the acclimatization of tissue culture plants and high-value crops in greenhouses.

Plants require specific climatic factors to enhance growth resulting from photosynthesis. Some of the important microclimate parameters such as solar radiation, temperature, relative humidity, light, and carbon dioxide are detailed in the sections below.

EFFECTS OF MICROCLIMATE ON PLANT GROWTH

Solar Radiation

The production of plant dry matter decreases almost linearly with the decrease in solar radiation values. The growth comes to a halt at the compensation point, which is at 14–30 W/m^2 light power for plants (0.1 kWh/m^{2}day). Efficient production of dry matter cannot be expected in higher latitudes during winter without artificial lighting. The minimum amount of irradiation necessary to ensure sufficient growth and flowering corresponds to a daily global radiation of 2.0–2.3 kWh/m^2 day. Most important for photosynthesis, i.e., the growth of plants is solar radiation power (W/m^2).

LIGHT INTENSITY

The growth of plants is controlled by three light (photo) processes, namely photosynthesis, photomorphogenesis, and photoperiodism. Every variation in light has a direct effect on these processes. Light is part of the photosynthesis process, by converting carbon dioxide into organic material and then releasing oxygen in the presence of light. Photo morphogenesis is the way plants develop under the influence of different types of light and photoperiodism is how the plant reacts to different day lengths and whether it will seed or flower. The most important process is photosynthesis and light is the primary energy source to enable this process. Light is affected by the latitude of a place, the orientation of the greenhouse, covering material, shadows from the structure and plants, dust and the environment. It is a fact that if light interception by plants decreases by a certain percentage, the yield of the crop will also decrease at the same percentage. A common problem of low-light conditions is flower abortion. Light levels can be improved by using artificial light, especially during the winter. Excessive light intensity can have a negative effect on fruit/flower/leaf quality due to sunburn; this problem can be avoided by installing shade nets. The effect of different microclimatic parameters on plant growth is presented in Table 19.1.

Light limits the photosynthetic productivity of all crops and is the most important variable affecting productivity in the greenhouse. Supplementary lighting does offer the opportunity to increase yield during low-light periods. Photosynthetically Active Radiation (PAR) is defined as radiation in the 400 to 700 nm waveband. Plants use the light in the 400 to 700 nm range for photosynthesis. Visible light (390–700 nanometers) provides essential energy for plant development and growth. Intensity, duration, and spectral distribution of light affect plant response. Ultraviolet light (290–390 nanometers) is generally detrimental to plants. Photosynthesis proceeds only with visible light, of which the red and blue wavelengths are used most efficiently. The change from vegetative to reproductive development in many plants is controlled by red (660 nanometers) and far red (730 nanometers) light.

TABLE 19.1
Influence of Climatic Factors on Plant Growth

Climatic factors	Influence
Radiation/Light	Photosynthesis, Photomorphogenesis, Photoperiodism
Temperature	Cell division and elongation, Respiration, Photosynthesis Water uptake, Transpiration
Relative Humidity	Quality of plant
CO_2 level	Photosynthesis, Respiration
Air circulation	Temperature, Humidity, and CO_2 level in greenhouse
Rainfall	Relative Humidity

Temperature

Temperature has a direct impact on the physiological development phases (flowering, germination, development) of the plant, and regulates the transpiration rate and plant water status through stomatal control during the photosynthesis. Temperature is the single greatest environmental input regulating plant growth, which is then closely followed by light intensity. High or low temperatures have negative effects on plant growth and development. Pest pressure increases as temperature increases. The management of the greenhouse environment is strongly reliant on temperature manipulation. There are optimum temperatures for each crop and for each stage of development (Table 19.2). At the high end of the temperature range, above the optimum, losses in quality can be experienced such as tender stem, thinner stem, fewer flowers, bleaching of flowers, and slower flower bud development. At excessively higher temperatures entire plant damage will occur. Maintaining the optimum temperature for each stage of growth is ideal in greenhouse environmental control. Optimum photosynthesis occurs between 24 to 28°C, this temperature serves as the target for managing temperatures during the day when photosynthesis occurs. The optimum temperature for vegetative growth for greenhouse crops is between 24 to 30°C, with the optimum temperature for yield about 28°C (Bakker, 1995). Fruit set, however, is determined by the 24-hour mean temperature and the difference in day-night temperatures, with the optimum night temperature for flowering and fruit setting at 16 to 18°C.

Relative Humidity

Relative humidity (RH) is the ratio of the actual vapor pressure of water vapor in the air to the vapor pressure that would be present if the air was saturated with moisture at the same temperature. Water vapor moves from one location to another because of vapor pressure differences, so relative humidity affects transpiration from plants by affecting the vapor pressure difference between a plant leaf and surrounding air. Humidity in the greenhouse is controlled for various reasons. The two main reasons are: avoiding too high humidity to avoid fungal infection and regulating transpirations. This environmental parameter has major effects on plant diseases. It is

TABLE 19.2
Climatic Requirement for Various Crops

Name of the crop	Day (°C)	Night (°C)	Humidity (%)	Light intensity (LUX)
Gerbera	20–24	18–21	60–65	40,000–50,000
Rose	24–28	18.5–20	65–70	60,000–70,000
Tomato	22–27	15.5–19	60–65	50,000–60,000
Cucumber	24–27	18–19	60–65	50,000–60,000
Capsicum	21–24	18–20	60–65	50,000–60,000

TABLE 19.3
Measuring Equipment for Different Climatic Parameters

Sr. No.	Parameter	Measuring Equipment
1.	Light/Radiation	Radiation meter, LUX meter
2.	Temperature	Glass thermometer, Bimetal s thermometer, Infra-red thermometer
3.	Relative Humidity	Psychrometer, Hygrometer, Electric moisture meter
4.	CO_2 Concentration	CO_2 meter, CO_2 Analyzer
5.	Wind Speed	Anemometer
6.	Wind Direction	Wind Vane
7.	Rain / Precipitation	Rain gauge, Rain Indicator

desirable to have daytime relative humidity between 60–80%. Normal plant growth will generally occur at relative humidity between 25–80%. A secondary effect of relative humidity is the response of pathogenic organisms. For example, most pathogenic spores will not germinate at a relative humidity below 85%. Botrytis can develop when the relative humidity exceeds 80%, especially in open wounds. During winter, daytime relative humidity is high since ventilation does not occur due to low outdoor temperatures. Combined with short days and low light, these are the ideal conditions for fungal diseases, especially powdery mildew. A psychrometer or hygrometer is used for measuring relative humidity (Table 19.3).

The high value of RH can be reduced by ventilation and/or heating. Minimizing the areas of wet surfaces, including plants, soils and floors, is another strategy to minimize the elevated RH. The methods used for dehumidification are: natural ventilation, condensation on a cold surface, forced ventilation in conjunction with heat exchanger and hygroscopic dehumidification. Control techniques for different microclimatic parameters are presented in Table 19.4.

Vapor Pressure Deficit (VPD)

It is the difference in water vapor pressure at saturation and the actual water vapor pressure, or amount of moisture, at the same temperature. Increasing the VPD will increase plant transpiration, thus plants take up more water and nutrients from the growing medium to supply the demand for water for transpiration. In the greenhouse, it is recommended to have values between 0.3–1.3 kPa for optimum plant growth. By maintaining these values, problems such as the Blossom End Rot of tomatoes and peppers (calcium deficiency) can be avoided.

Carbon Dioxide Supplementation

The carbon dioxide (CO_2) concentration inside a greenhouse can drop significantly below the outside level when a dense crop is growing, even if the greenhouse is well

TABLE 19.4
Control Techniques for Different Micro Climatic Parameters

Parameters	Technical options to control	Equipment required
Light /Radiation	To increase light intensity and day length – Artificial lighting	Incandescent lamp, Fluorescent lamps, High intensity discharge lamps
	To decrease light intensity and day length – Shading/Screening, Whitewashing	Shade nets, Blackout screens, Aluminets, Whitewash, etc.
Temperature	To increase temperature – Heating	Heaters, Hot water heating, Steam heating etc.
	To decrease temperature – Cooling Shading	Misters, Foggers, Roof Sprinklers Shade nets/Screens, Whitewash etc.
	Ventilation (Air circulation)	Vents, Exhaust fans etc.
Relative humidity	To increase humidity – Humidification	Humidifiers, Foggers, Misters
	To decrease humidity – Dehumidification	Heaters, Vents, Dehumidifiers
CO_2 concentration	CO_2 enrichment	CO_2 burner, Pure CO_2, Fuel gases etc.

ventilated. The concentration can drop to less than 200 µmol mol⁻¹ during winter in mild climate regions. As the CO_2 concentration limits the photosynthesis of most vegetable species, productivity decreases. Carbon dioxide (CO_2) is one of the inputs of photosynthesis and as such CO_2 plays an important role in increasing crop productivity. Carbon dioxide is the raw material which, along with water, is required for photosynthesis; it is usually the limiting factor in the greenhouse environment. In a tight greenhouse, carbon dioxide concentration may be 400 parts per million (ppm) before daylight and drop to 150 ppm shortly after light is available. Outside air is about 330 ppm. Optimal CO_2 concentrations for the greenhouse atmosphere fall within the range between 700 to 900 ppm. Crop productivity depends not only on the efficiency of interception of light but also on the efficiency with which light is converted to chemical energy in photosynthesis. Carbon dioxide enrichment to 1200 ppm increases the maximum conversion efficiency by a substantial amount (between 28 to 59%). The combination of high carbon dioxide levels (1500 ppm), elevated day temperatures, and optimum light levels will reduce the time between germination and harvest by as much as 50% for some crops. Increases in carbon dioxide levels result in improved plant quality, yield, and development.

VENTILATION COOLING AND SHADING

Minimizing the heat load is a major concern for greenhouse climate management in hot climate conditions. This can be achieved by reducing incoming solar radiation, removing the extra heat through air exchange and increasing the fraction of energy partitioned into latent heat. Shade nets and whitewash (lime or paint application on cladding material) are the major existing methods used to reduce the income of solar radiation. Greenhouse ventilation is an effective way to remove the extra heat through air exchange between inside and outside, when the outside air temperature is lower. Evaporative cooling is the common technique to reduce sensible heat load by increasing the latent heat fraction of dissipated energy. Other cooling technical solutions are available (heat pumps, heat exchangers), but are not yet widely used, especially in India, because investment costs are very high.

Natural Ventilation

Ventilation affects the temperature inside the greenhouse during times of high solar radiation; it is necessary to circulate air from the outside to the inside of the greenhouse in a homogeneous manner in order to remove excess heat. Poor ventilation has a negative impact on indoor air composition, mainly by reducing the CO_2 concentration. Inadequate ventilation generates overheating and excessive transpiration, leading to problems such as plant water stress and physiological disorders, including fruit cracking and abortion of flowers and fruits. On the other hand, natural ventilation helps to evacuate excess moisture and prevent its accumulation in the air layer near the leaves which can cause condensation, leading to the onset of diseases.

Natural ventilation is the result of pressure differences created by wind and temperature gradients between the inside and outside of a greenhouse. It occurs through openings in the greenhouse structure. It controls humidity and temperature build-up within the greenhouse and can ensure sufficient air exchange. Natural ventilation is a process that directly influences the climate inside the greenhouse and is a decisive factor when it comes to designing. A properly designed ventilation system can improve climate control and consequently, optimize energy use. In addition, ventilation is related to other factors such as temperature, humidity and CO_2 concentration, which directly influence the growth and development of crops. The efficiency of natural ventilation depends on factors such as wind speed and direction, temperature differences between the outside and the inside of the greenhouse, greenhouse design and the presence or absence of crops.

Good ventilation in the greenhouse can be achieved with a combination of a roof vent, front doors, and fans. One of the simplest and most effective ways to reduce the difference between inside and outside air temperature is to improve ventilation. If the greenhouse is equipped with ventilation openings, both near the ground and at the roof, then this type of ventilation replaces the internal hot air with an external cooler one during hot sunny days with weak wind. The external cool air enters the greenhouse through the lower side openings while the hot internal air exits through the roof openings due to density difference between air masses of different temperatures causing the lowering of temperature in the greenhouse. Sufficient ventilation is very important for optimal plant growth, especially in case of high outside temperatures and solar radiation.

Forced Ventilation

The principle of forced ventilation is to create an airflow through the house. Fans suck air out on the one side and openings on the other side let air in. Forced ventilation by fans is the most effective way to ventilate a greenhouse, but consumes electricity. Fans are installed to maintain uniform temperature and humidity inside the greenhouse. The fresh air enters at one side and replaces the hot stale air that moves out at the opposite side of the greenhouse. Ventilation fans should be located on the wind side of the greenhouse and the distance between two fans should not exceed 8–10 m. Furthermore, an inlet opening on the opposite side of a fan should be at least 1.25 times the fan area. The velocity of the incoming air must not be too high. The airspeed should not exceed 0.5 m s-1in the greenhouse with crop. The openings must be closed automatically when the fans are not in operation.

EVAPORATIVE COOLING

One of the most efficient solutions for alleviating the climatic conditions is to use evaporative cooling systems, based on the conversion of sensible heat into latent heat by means of evaporation of water supplied directly into the greenhouse atmosphere (mist or fog system, sprinklers) or through evaporative pads (wet pads). The fogging system is based on spraying the water as small droplets with high pressure into the air above the plants in order to increase the water surface in contact with the air. The freefall velocity of these droplets is slow and the air streams inside the greenhouse easily carry the drops. This can result in high efficiency of water evaporation combined with keeping the foliage dry. Fogging is also used to create high relative humidity along with cooling inside the greenhouse. A wide range of fogging systems are available their cooling efficiency is reported in the literature.

The fan pad cooling system is most commonly used in horticulture. The fan pad system consists of a fan on one gable end and a wet pad on the opposite end. A small stream of water runs over the pad continuously and air is drawn through the pad by the fans, absorbing heat and water vapor in the greenhouse. It is essential that the pad be free of leaks through which air could pass without making contact with the pad. Different pad materials are available, such as wood, wool, swelling clay minerals, and specially impregnated cellulose paper. These installations have shown a reduction in air temperature of up to 12°C, even under very high ambient temperatures. It also increases the humidity of the internal air. Air from outside is blown through pads with as large a surface as possible. Uniform shading of the greenhouse with nets results in desired cooling during summer. The water flow rate, water distribution system, pump capacity, recirculation rate and output rate of the fan pad cooling system must be carefully calculated and designed to provide sufficient wetting of the pad to avoid deposition of dissolved material on it. The advantage of fogging systems over wet pad systems is the uniformity of conditions throughout the greenhouse, therefore eliminating the need for forced ventilation and airtight enclosure. Whereas the disadvantage is that it is an expensive installation with high operation costs, namely, freshwater supply, electricity, and maintenance costs.

Roof evaporative cooling includes spraying water onto the external surface of a roof and this creates a thin water layer on the surface. This decreases the solar radiation transmissivity to the greenhouse and increases the evaporation rate which consequently decreases the water temperature and closely surrounding air. This system works most effectively in hot and dry climate regions.

GREENHOUSE HEATING

Greenhouse heating is required in cold climate regions such as the Himalayan region of the country. Heating costs have a critical influence on the profitability of greenhouse production. Apart from the costs, energy consumption and associated environmental problems through the emission of noxious gases are the constraints in greenhouse operations. The heating system should provide heat to the greenhouse at the same rate at which it is lost. There are several popular heating systems for greenhouses. The most common and least expensive is the unit heater system.

UNIT HEATERS

In this system, the warm air is blown from unit heaters that have self-contained fireboxes. Heaters are located throughout the greenhouse, each heating system has a floor area of 180 to 500 m^2.

CENTRAL HEATING

Heat is half dissipated through radiation and half through convective transfers. Unlike unit heater systems a portion of the heat from central boiler systems is delivered to the root and crown zone of the crop. This can lead to improved growth of the crop and to a higher level of disease control.

SALIENT RESEARCH FINDINGS

STUDY CONDUCTED AT IIT KHARAGPUR

Sawtooth and Quonset shape greenhouses were designed and constructed at the Field Water Management Laboratory of Agricultural and Food Engineering Department, Indian Institute of Technology Kharagpur, India, and were used for the experimental purpose. Both the greenhouses were naturally ventilated and cladded with 200µ UV-stabilized poly film (Transmissivity of PAR is 90% with a diffusivity of 42%). The ground area of the Sawtooth-shaped greenhouse is 84 m^2 (6 m x 14 m), the central height is 4.5 m and the gutter height is 3 m (Figure 19.1a). The Quonset-shaped greenhouse has a 44 m^2 (11m x 4m) floor area and a central height is 2.5 m (Figure 19.1b). Both the greenhouses were oriented toward East–West to get maximum sunshine for photosynthesis during winter. The Quonset greenhouse was provided with a side ventilation and the Sawtooth greenhouse was provided with ridge and side ventilations with a provision to vary the extent of ventilation area through rollup sides. Ventilation openings of 60% of the floor area were provided to remove hot air and to reduce high temperatures during peak summer. Ventilation openings were covered with an insect-proof net of 20 mm-mesh size to prevent the entry of insects. The Sawtooth greenhouse was equipped with a fogging system (foggers of 16 L hr^{-1} discharge) and shade net (75% shading intensity) beneath the roof of the greenhouse cladded with UV-stabilized plastic film, whereas only the shade net (75% shading intensity) was provided in the Quonset greenhouse during summer months only. The Sawtooth greenhouse was equipped with two exhaust fans with

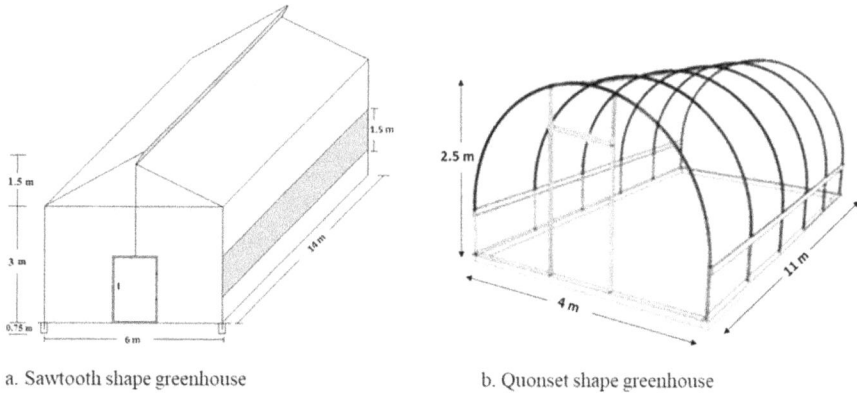

a. Sawtooth shape greenhouse   b. Quonset shape greenhouse

FIGURE 19.1 Isometric view of experimental greenhouses: (a) Sawtooth shape greenhouse (b) Quonset shape greenhouse

a capacity of 1100 m³ min⁻¹ and 2.2 kW power to exchange hot air from the greenhouse with ambient air. The fans were used only when the greenhouse air temperature was extremely high during peak summer for a few hours.

An automatic weather station of M/S Campbell Scientific, Canada comprising a data logger (CR1000) and sensors for measuring air temperature, relative humidity, global radiation, and PAR, was installed inside the greenhouse to monitor microclimate parameters. The temperature and humidity inside the greenhouse were measured by a combined sensor of model HMP 45 C installed at 1.25 m above the ground level. The measuring ranges of temperature and humidity of the sensor are −40 to 60°C and 0 to 100%, respectively. Incoming radiation in the greenhouse (global) and photosynthetically active radiation (PAR) were measured by the two pyranometers SPLITE and PARLITE of Kipp and Zonen with quantum sensor sensitivity of 5.42 μ v/μ m. The spectral range of the pyranometers was 305 to 2800 nm. All the sensors were connected to the CR1000 datalogger of M/S Campbell Scientific, Canada. Although these sensors sense the data every second, however, logging was done at a 30-minute interval. Outside air and soil temperatures, relative humidity and solar radiation were measured manually at 8:30 AM, 12:30 PM and 4:00 PM in a day.

RESULTS

TEMPERATURE VARIATION

The average temperature in the Quonset shape greenhouse was observed to vary from 22 to 36°C, 29 to 45°C, and 23 to 41°C during winter, summer, and rainy seasons, respectively, against ambient temperature (20 to 36°C, 26 to 42°C and 21 to 38°C). The temperature difference in the Quonset shape varied between 1 to 2°C, 2 to 3°C, and 1 to 3°C in the respective seasons against ambient conditions. The temperature variations in the Sawtooth shape were found to vary from 21 to 34°C, 27 to 42°C, and 22 to 39°C in winter, summer, and rainy seasons, respectively (Table 19.5).

TABLE 19.5

Variation in Microclimate Under Sawtooth and Quonset Shape Greenhouses During Different Seasons

Microclimate Parameter	Quonset Greenhouse			Sawtooth Greenhouse		
	Winter	Summer	Rainy	Winter	Summer	Rainy
Temperature (°C)	22–36	29–45	23–41	21 to 34	27 to 42	22 to 39
Relative humidity (%)	20–80	15–85	30–81	19–87	26–93	34–90
Solar radiation (MJ/m²/day)	15.1–20.2	21.3–25.4	20.5–22.9	14.5–18.3	19.8–22.6	19.2–22.3

It was observed that the Sawtooth shape maintained less temperature variation than the Quonset shape and the differences between the shapes during winter, summer and rainy seasons were 2°C, 3°C, and 2°C, respectively.

The findings of the study of temperature revealed that daily variation of air temperature in the two shapes of the experimental greenhouses is different in all the seasons thus highlighting the effect of greenhouse shape and percent of ventilation. Furthermore, it can be seen that in the Sawtooth greenhouse the temperature was sometimes found less than the outside temperature during the early hours of the day. The reason for the lower temperature may be due to the effect of crop transpiration that contributed to less latent heat transfer.

Relative Humidity Variation

Relative humidity in the Quonset shape was observed to vary in range from 20 to 80%, 15 to 85%, and 30 to 81% during winter, summer and rainy seasons, against ambient conditions of 18 to 73%, 13 to 79%, and 21 to 72%, respectively. On the other hand, relative humidity variations in the Sawtooth shape were found to range from 19 to 87%, 26 to 93%, and 34 to 90% during the three seasons with differences of 1 to 7, 8 to 11, and 4 to 8% between the shapes during winter, summer and rainy seasons, respectively. The humidity difference in the Quonset and Sawtooth shape varied between 2 and 7%, 2 and 6%, 8, and 9% and 1 to 14, 13 to 14, and 13 to 18%, respectively, in the respective seasons against ambient conditions.

The results of relative humidity variations in the two experimental greenhouses revealed that on average, the variations are found significant during all three seasons. The relative humidity was always found to be higher in the Sawtooth shape greenhouse than in the Quonset shape greenhouse and ambient. It may be due to the lower air temperatures in the Sawtooth-shaped greenhouse.

Variation in Solar Radiation

The solar radiation in the Quonset shape varied from 15.1 to 20.2, 21.3 to 25.4, and 20.5 to 22.9 MJ/m²/day against ambient conditions (19.1 to 21.5, 24.0 to 28.5, and

22.4 to 23.9 MJ/m^2/day) with average reductions of 29%, 35%, and 26% during winter, summer, and rainy seasons, respectively. The variations in the Sawtooth shape were observed as 14.5 to 18.3, 19.8 to 22.6, and 19.2 to 22.3 MJ/m^2/day with average reductions of 31%, 38%, and 32% during respective seasons.

From the results obtained, it is evident that solar radiation varies significantly between the two shapes of the greenhouse. The Quonset shape always maintained a larger amount of solar energy than the Sawtooth shape during a specified course of time in all three seasons. This could be due to the effect of the shade net provided in the Sawtooth-shaped greenhouse. Solar radiations in both the greenhouses are lesser than ambient because of the absorption properties of UV-stabilized poly film.

CONCLUSIONS

The study of microclimate variation in the two experimental greenhouses (Quonset and Sawtooth) during different seasons of the year indicated that the shape, height, and types of ventilation of the greenhouse have a significant effect on the microclimatic parameters in all the seasons. The Sawtooth greenhouse maintained not only less temperature variation against ambient conditions but also maintained 3°C less temperature than the Quonset greenhouse during summer months. The relative humidity has also varied significantly during all the seasons, especially in summer due to fogging operations. The Quonset shape of the greenhouse maintained less relative humidity variations than the Sawtooth shape. The Quonset greenhouse always received greater solar radiation than that of the Sawtooth greenhouse in all seasons. These microclimate parameters data in the greenhouse are important for the selection of crop(s), greenhouse shape, and cultivation in a particular season.

REFERENCES

Aldrich, R.A. and Bartok, J.W. 1989. *Greenhouse Engineering*. Northeast Regional Agricultural Engineering Service, Cooperative Extension, Ithaca, NY.

Ale, S., Su, Q., Singh, J., Himanshu, S., Fan, Y., Stoker, B., ... & Wall, J. (2022). A Mobile App for Cotton Irrigation Management. *Resource Magazine*, 29(4), 6–8.

Ale, S., Su, Q., Singh, J., Himanshu, S., Fan, Y., Stoker, B., ... & Wall, J. (2023). Development and Evaluation of a Decision Support Mobile Application for Cotton Irrigation Management. *Smart Agricultural Technology*, 5, 100270.

Bakker, J.C. 1995. Greenhouse climate control: Constraints and limitations. *Acta Horticulture* 399: 15–37.

Gao, L.H., Qu, M., Ren, H.Z., Sui, X.L. and Chen, Q.Y. 2010. Structure, function, application and ecological benefit of single-slope, energy-efficient solar greenhouse in China. *Hort Technology* 20(3): 626–631.

Liang, X., Gao, Y., Zhang, X., Tian, Y. and Zhang, Z. 2014. Effect of optimal daily fertigation on migration of water and salt in soil, root growth and fruit yield of cucumber (*Cucumissativus* L.) in Solar-Greenhouse. *PLoS One* 9(1) e86975.

Mahajan, G. and Singh, K.G. 2006. Response of Greenhouse tomato to irrigation and fertigation. *Agricultural Water Management* 84(1–2): 202–206.

Von Zabeltitz. 1999. *Greenhouse Structures, Ecosystems of the World's 20 Greenhouses*. Elsevier, Amsterdam.

20 Estimation of Groundwater Fluctuation and NDVI Using Geospatial Techniques for Chandigarh City, India

Sunny Kumar, Sushindra Kumar Gupta, and Kanwarpreet Singh

INTRODUCTION

Groundwater is a crucial natural resource that is used for drinking, farming, and other sectors. Groundwater supplies have been overused as a result of population growth, high-yield farming methods, industrial development, and numerous other household and recreational water uses. It is a significant source for irrigation and the ecosystem in addition to serving as drinking water. Rainfall, the quality of surface water, as well as recharged water, are a few of the variables that affect water quality (Gupta et al., 2016). Hydrogeochemical processes play a major part in determining the quality of groundwater. As a result of precipitation, infiltration, evapotranspiration, water withdrawal from bore wells, and groundwater discharge into streams and lakes, the water table fluctuates (Sultana & Satyanarayana, 2020). Droughts increase the use of groundwater because they lead to a greater reliance on groundwater reserves, a faster rate of evapotranspiration, and increased irrigation needs, all of which lead to a sharp reduction in the groundwater table (Kumar & Pandey, 2013; Singh et al., 2021).

Groundwater potential is largely determined by natural variables including geology, hydrogeology, and geomorphology; however, because of changes in land use and cover, human activities are now having an increasingly large impact on groundwater variability and quality (Gupta, 2020a,b; Mahajan et al., 2022; Gupta et al., 2020; Gupta et al., 2022; Yatoo et al., 2022; Singh et al., 2023). Urban communities typically lack sufficient surface water resources to meet all of their water needs. Therefore, reliance on groundwater has grown over time. More groundwater has

DOI: 10.1201/9781003441175-20

been withdrawn for agricultural and industrial purposes in the majority of Indian States than can be replenished (Krishan et al., 2021).

The assessment of groundwater level variation proved to be a significant benefit of GIS-based maps. In the study area, five key hydrogeological factors, soil, elevation, slope, drainage, and geology, have been taken into account because they have a big impact on groundwater fluctuation. In general, the groundwater is only accessible in worn and fractured areas in hard rock terrain. The availability of secondary porosity and the thickness of the weathered zone are both necessary for aquifer recharge, and the result of recharge and discharge is a variation in groundwater level (Chandra et al., 2015). Geographical information systems (GIS) are effective tools for handling spatial data and helping decision-makers in a variety of domains, including the geological and environmental sciences, make swift decisions about policy (Stafford, 1991; Ghosh et al., 2015). The concept of utilizing cutting-edge methods in groundwater management research, such as remote sensing and GIS, is relatively new (Chowdhury et al., 2009; Kaur et al., 2018).

The significance of monitoring groundwater levels has been recognized, and numerous research have been conducted in this area. Conventional measurement tapes, electronic water-level indicators, airline pressure methods, acoustic methods, and automatic recording techniques can all be used to measure groundwater levels from the chosen wells in the monitoring network. A dual conductor wire, a probe, and an indicator are components of the commonly used electronic water-level indicators. Light or sound will signify when the probe makes contact with the water table, and the level is then measured. In this work, a sensor-based electronic indication has been used to measure the water level.

As per CGWB 2020, there are 289 deep tube wells in the city for drinking and domestic use, of which 224 are actually in use and provide 21 MGD (3483.20 ham/year) of water, 32 are for commercial or industrial use and are allowed to withdraw groundwater at a rate of 5227.80 m^3/day (190.8 ham/year), and 30 are for irrigation use, with a combined draft of 5.73 MGD (950.80 ham/year) from these wells. Although there are a few small ponds in the countryside, Chandigarh lacks significant natural surface water features. The Sukhna Choe has been dammed in the northeastern part of the city, creating an artificial lake that is approximately 1.99 km^2 in size. The lake, known as Sukhna, can store 5 million cubic meters of water (MCM). The Ghaggar Basin is where Chandigarh's UT is located. The Siwalik Hills ranges give rise to two significant streams, Sukhna Choe and Patiali ki Rao, which serve as the city's natural drainage system. The Sukhna Choe drains the eastern region from north to south before joining the Ghaggar River.

STUDY AREA

India's Chandigarh serves as the joint capital of the neighboring states of Punjab and Haryana. It is a district, a union territory, and a city. Chandigarh is bordered to the north, west, and south by the state of Punjab (Figure 20.1). East of Chandigarh

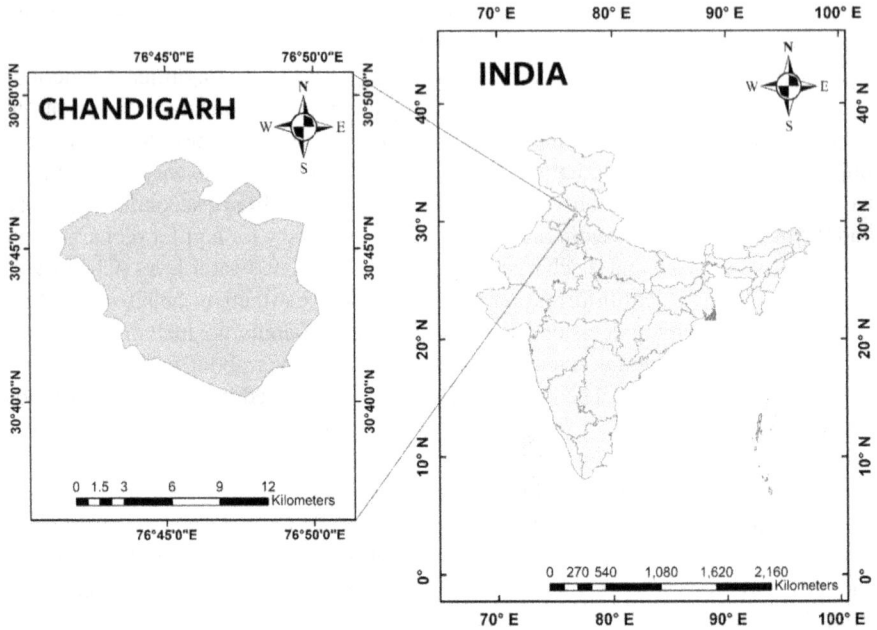

FIGURE 20.1   Location map of the study area

is where the state of Haryana borders. The Greater Chandigarh region, which also encompasses the neighboring satellite cities of Panchkula and Mohali, is largely made up of it. It is located 260 km (162 miles) north of New Delhi and 229 km (143 miles) southeast of Amritsar and 110 km (68 miles) southwest of Shimla. The climate in Chandigarh is humid subtropical, with a yearly rhythm that includes scorching summers, mild winters, unpredictable rainfall, and wide temperature ranges (from 1 to 45°C or 30.2 to 113°F). The average annual rainfall is 43.73 inches, or 1110.7 mm.

CLIMATE

A seasonal rhythm may be observed in Chandigarh's humid subtropical climate, which features extreme heat in the summer, moderate winters, erratic rainfall, and wide temperature swings (from 1 to 45°C or 30.2 to 113°F), with 1110.7 mm, or 43.73 inches, of rainfall per year on average. A western disturbance that originates over the Mediterranean Sea occasionally brings winter showers to the city as well. When the temperature turns colder (between March and April), the western disturbances bring rain that is primarily present from mid-December through the end of April. This is usually terrible for local crops because the rain can occasionally be heavier and accompanied by severe wind and hail. The Himalayas, which are to the north and get winter snowfall, are typically the source of cold winds.

GEOLOGY

Chandigarh, a union territory, is situated on the northern foothills of the Shivalik hill ranges, a part of the delicate Himalayan ecology. Sirowal (Tarai) and alluvial plains make up the majority of the remaining area, with Kandi (Bhabhar) in the northeast. In the subsurface formation, there are beds of boulders, pebbles, gravel, sand, silt, clays, and some kankar. Two seasonal rivers, Patiala-Ki-Rao Choe in the west and Sukhna Choe in the east, drain the area. There are two little streams and a surface water division in the center. N-Choe and Choe Nala are two streams that originate in Sector 29 and flow through the center of the area.

SOIL AND CROP

In most sectors, the soil strata are composed of clayey silt soils in the upper layers, followed by sandy silt to silty sands at depths. While the soil in the southern regions is loamy to silt loam, it is sandy to sandy loam in the northern regions. A 2009 study by G. C. Kandpal et al. found that pebble, pebbly sand, clay, and sand strata are widely distributed in the region. The lithologs of the Chandigarh boreholes show that interlayered sequences of clay and sand are present in the majority of the boreholes.

GEOMORPHOLOGY

In this chapter, we'd like to go over the morphological characteristics that came about as a result of the severe erosion that affected sedimentary strata in the Siwalik region, specifically close to the Masol archaeological site. The Patiali Rao and its tributaries' river erosion significantly contributed to the Masol anticline's entrenchment (excavation). From northeast to southwest, perpendicular to the HFT, streams drain the frontal flank of the heavily eroded Siwalik bedrock. The terrain of the Chandigarh hills is asymmetrical and the hills are sharply divided. Compared to the northeastern flank, the southwest flank of the Chandigarh anticline has a greater amount of erosion. In the Patiali Rao valley, relief is typically 80 m. About 180 m above the Indo-Gangetic plain, near the drainage divide, are hills with peaks. There are many tiny tributaries and a concave longitudinal profile on the Patiali Rao thalweg. The older geological formations and fossil records of the Late Pliocene period in Masol have been exposed as a result of this region's deep erosion of the Chandigarh anticline.

METHODOLOGY

Maps made with ArcGIS need categories grouped as layers. Each layer is spatially recorded, allowing the application to align each layer correctly when it is overlaid on top of another to produce a complicated data map. Depending on the required display, the base layer is nearly always a geographical map that is sourced from a variety of sources (satellite, road map, etc.) Figure 20.2

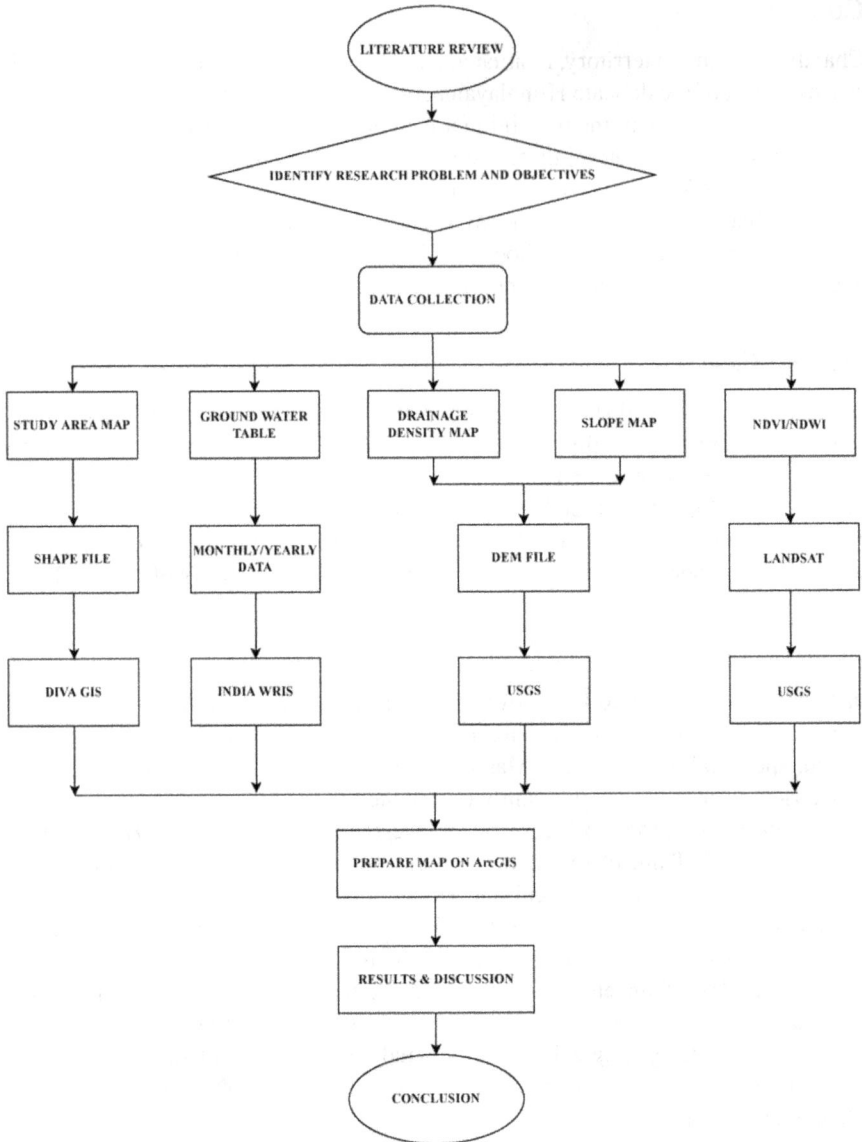

FIGURE 20.2   Methodology flow chart

Ground Water Map

Firstly, Create the data of study region using the WRIS open data web source and save it as an Excel sheet in the "Excel 97-2003 Workbook (.xls)" format. Transform the Excel sheet's X and Y coordinate systems to WGS 1984 in the Geographic coordinate system before importing it into the Arc map window. Use the IDW method to interpolate. Select the Excel sheet as the input point feature in the IDW window. Now carefully choose the year for which you wish to produce a groundwater map in the Z field choice.

Finally, choose the environment variable and set the environment window's processing level to the same as the shape file for the research region. Then choose Chandigarh's shape file under Mask in Raster Analysis. Then click OK. Now select the Output Raster option in the IDW window, enter the path of your save file there, and then click the OK button to start the interpolation process. The groundwater map of your study region is eventually prepared once the map is classified under the properties option. Figure 20.3

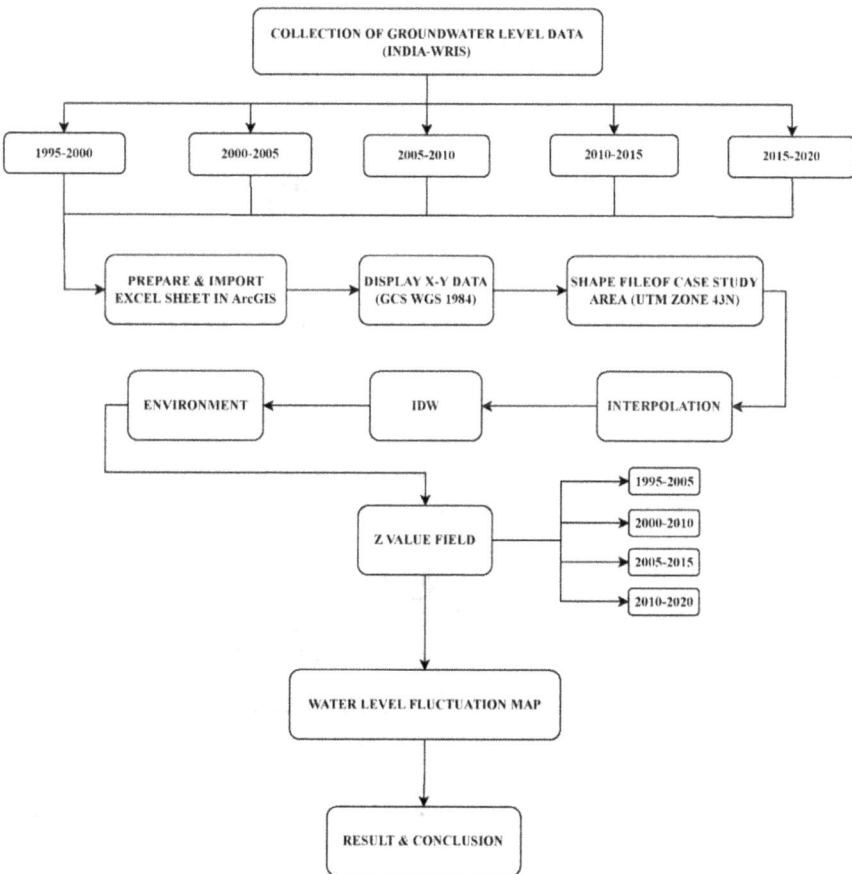

FIGURE 20.3 Methodology flow chart of the groundwater map

NORMALIZED DIFFERENCE VEGETATION INDEX MAP

The Normalized Difference Vegetation Index (NDVI) measures the difference between red and near-infrared light, which vegetation significantly reflects (which vegetation absorbs). Figure 20.4

NDVI is calculated using the formula:
NDVI = (NIR – Red) / (NIR + Red)
NDVI for different data sets:
- Landsat 8 = (Band5 – Band4)/(Band5 + Band4)
- Landsat 5 and 7 = (Band4 – Band3)/(Band4 + Band3)

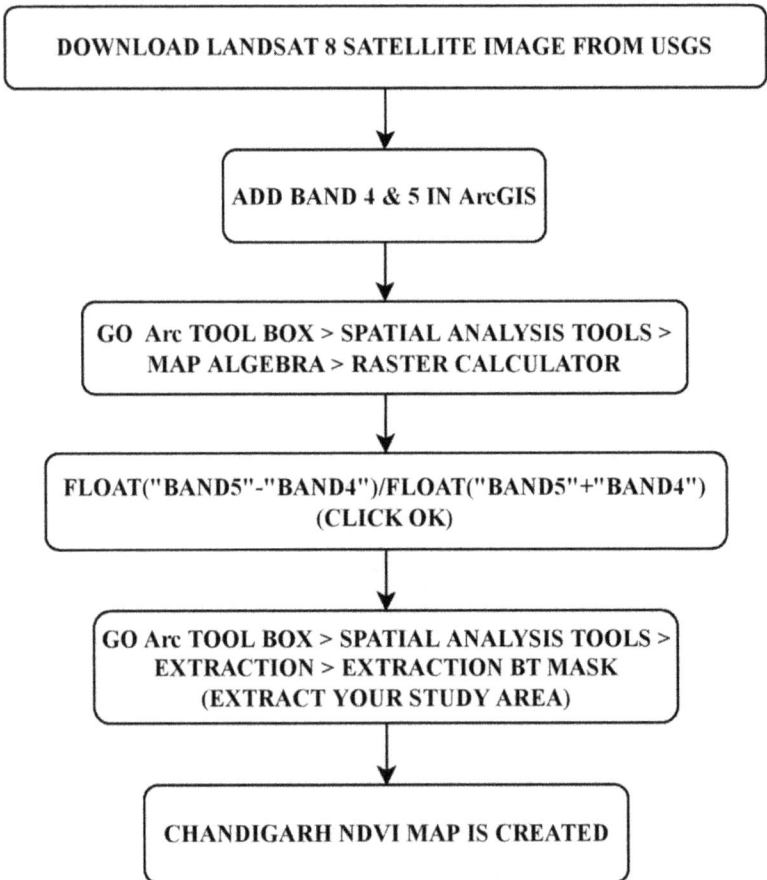

DOWNLOAD LANDSAT 8 SATELLITE IMAGE FROM USGS

ADD BAND 4 & 5 IN ArcGIS

**GO Arc TOOL BOX > SPATIAL ANALYSIS TOOLS >
MAP ALGEBRA > RASTER CALCULATOR**

**FLOAT("BAND5"-"BAND4")/FLOAT("BAND5"+"BAND4")
(CLICK OK)**

**GO Arc TOOL BOX > SPATIAL ANALYSIS TOOLS >
EXTRACTION > EXTRACTION BT MASK
(EXTRACT YOUR STUDY AREA)**

CHANDIGARH NDVI MAP IS CREATED

FIGURE 20.4   Methodology flow chart of NDVI

NORMALIZED DIFFERENCE WATER INDEX MAP

The Normalized Difference Water Index (NDWI) is used for the examination of water bodies. The index makes use of remote sensing data in the green and near-infrared regions. While lower values suggest water stress, higher NDWI levels indicate adequate moisture. Figure 20.5

NDWI is calculated using the formula:
NDWI = (Green – NIR) / (Green + NIR)
NDWI for Landsat 8 Data = (Band3 – Band5)/(Band3 + Band5)

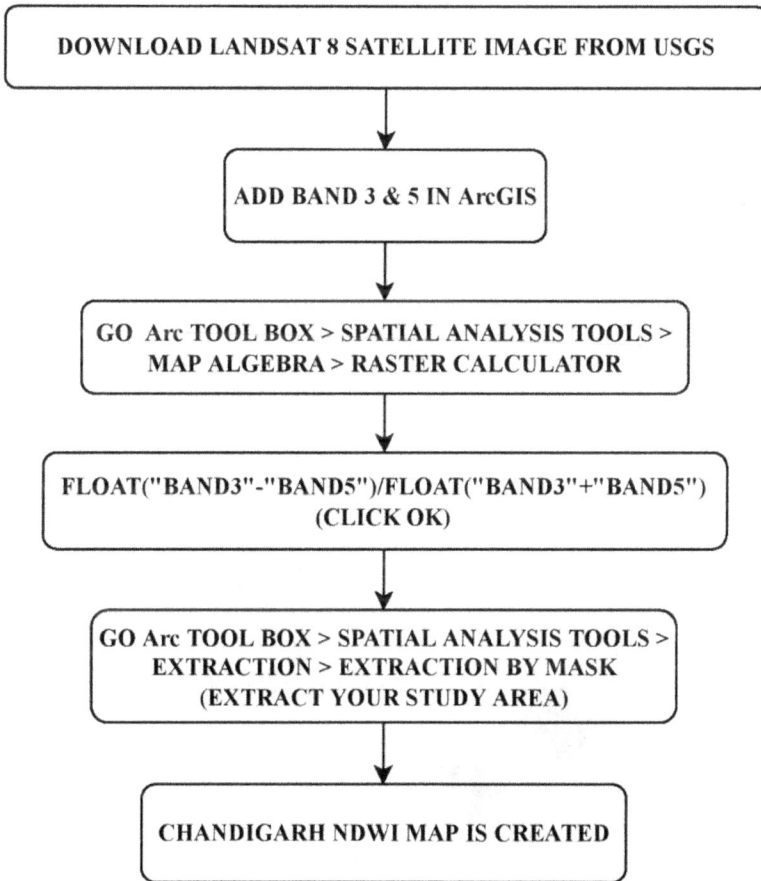

```
┌─────────────────────────────────────────────────────────┐
│      DOWNLOAD LANDSAT 8 SATELLITE IMAGE FROM USGS          │
└─────────────────────────────────────────────────────────┘
                            │
                            ▼
              ┌──────────────────────────────┐
              │   ADD BAND 3 & 5 IN ArcGIS     │
              └──────────────────────────────┘
                            │
                            ▼
        ┌──────────────────────────────────────────┐
        │  GO  Arc TOOL BOX > SPATIAL ANALYSIS TOOLS >│
        │      MAP ALGEBRA > RASTER CALCULATOR        │
        └──────────────────────────────────────────┘
                            │
                            ▼
    ┌──────────────────────────────────────────────────┐
    │ FLOAT("BAND3"-"BAND5")/FLOAT("BAND3"+"BAND5")       │
    │                  (CLICK OK)                         │
    └──────────────────────────────────────────────────┘
                            │
                            ▼
        ┌──────────────────────────────────────────┐
        │  GO Arc TOOL BOX > SPATIAL ANALYSIS TOOLS > │
        │      EXTRACTION > EXTRACTION BY MASK        │
        │        (EXTRACT YOUR STUDY AREA)            │
        └──────────────────────────────────────────┘
                            │
                            ▼
           ┌──────────────────────────────────┐
           │  CHANDIGARH NDWI MAP IS CREATED    │
           └──────────────────────────────────┘
```

FIGURE 20.5   Methodology flow chart of NDWI

RESULTS AND DISCUSSION

The groundwater level information is shown in Table 20.1. The complete information about the five years from 1995 to 2020 and seasonal water levels is shown in Figures 20.6 and 20.7.

The most vital resource on the planet is groundwater. The groundwater level in the Chandigarh area is declining at an alarming rate, according to the current study, which was conducted using the ArcGIS 10 Software. The groundwater table dropped from 2.89 m to 37.20 (mbgl) between 1995 and 2020, and it was discovered that the groundwater level has fallen by more than a meter annually. Water table depletion was an issue that existed prior to 1995, but it started to get worse after 2000. However, because of the farmer's easy access and necessity, the switch from canal irrigation to tube wells resulted in ground depletion. Figures 20.6, 20.7, 20.8, 20.9

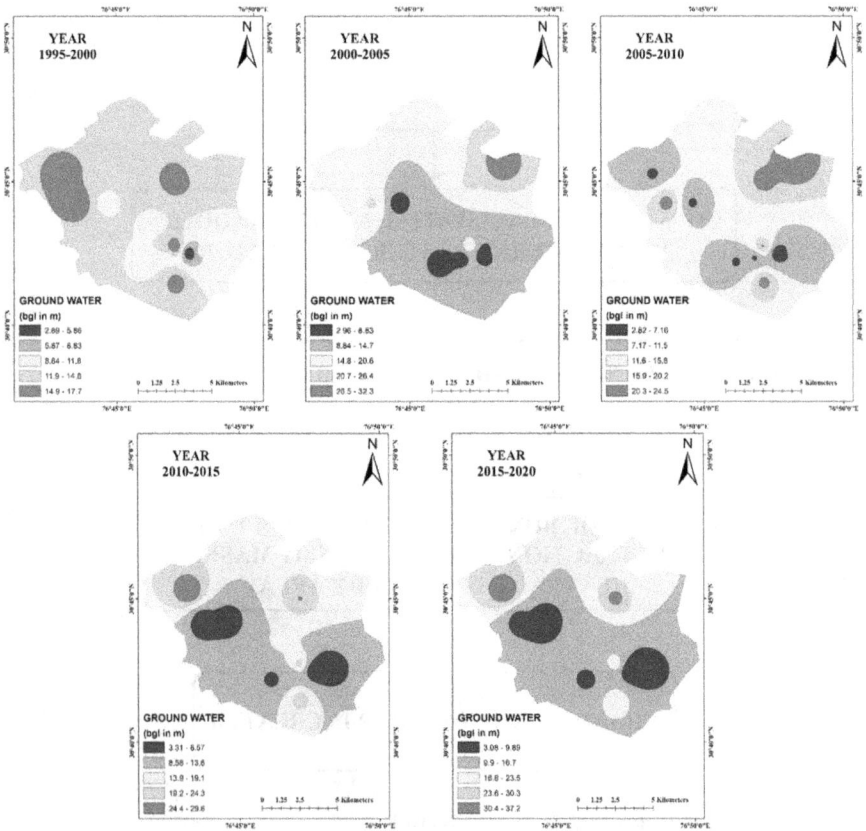

FIGURE 20.6   Average groundwater level from 1995 to 2020

TABLE 20.1
Groundwater Level Data from Chandigarh (in m bgl)

Sr. No.	STATION	Latitude	Longtitude	1995–2000	2000–2005	2005–2010	2010–2015	2015–2020
1	SECT 21D (D)	30.725	76.77361	10.696	12.464	13.762	14.35	10.18
2	CSIO-S	30.7125	76.78528	10.696	12.464	13.762	21.3175	21.656
3	Sector-46 (shallow)	30.70333	76.76944	10.696	6.14	6.606	6.96	7.674
4	SECT 37D (S)	30.7375	76.74306	10.696	6.14	6.606	5.55	4.842
5	SECT 10C (S)	30.75	76.78611	10.696	6.14	6.606	15.1725	22.906
6	SECT 31D (S)	30.70556	76.78056	10.696	6.14	6.606	10.7925	12.042
7	Csio-combined	30.7125	76.78528	17.75	18.385	20.377	20.664	21.4175
8	SECT 39D (D)	30.7375	76.72778	15.45	18.498	6.656	10.246	11.8475
9	Maloya Pz-Deep	30.75417	76.71944	15.45	18.498	6.656	29.597	37.164
10	Sector-46(vs)	30.70333	76.76944	15.45	18.498	6.656	29.597	6.99
11	SECT 44D (D)	30.70972	76.80139	9.376	12.63	7.368	8.7	3.802
12	SECT 44D (S)	30.70972	76.80139	9.376	12.63	7.368	3.42	4.134
13	SECT 21D (S)	30.725	76.77361	9.376	12.63	7.368	9.765	4.134
14	SECT 37D (D)	30.7375	76.74306	12.893	15.908	18.596	18.018	16.852
15	Sec-27, AR Well	30.75972	76.80417	12.893	32.3	24.519	18.018	16.852
16	BURAIL	30.70833	76.79444	2.86	2.93	2.794	3.29	3.052
17	SECT 12 (D)	30.75	76.78611	16.65	21.294	21.703	24.48	32.434
18	SECT 39D (S)	30.7375	76.72778	16.65	21.294	21.703	4.8175	6.935
19	SECT 31D (D)	30.70556	76.78056	11.988	14.29	18.65	18.37	18.62
20	NEW INDUST AREA	30.69167	76.78611	16.515	14.29	21.8	22.086	22.376

(*Source*: India WRIS portal)

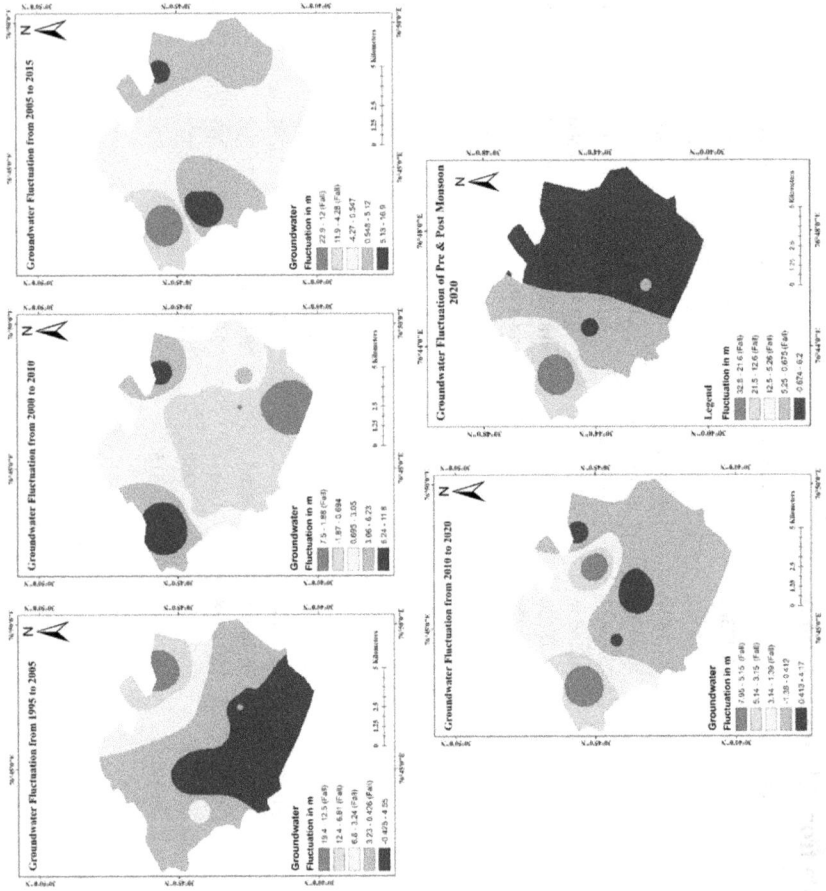

FIGURE 20.7 Average groundwater fluctuation from 1995 to 2020

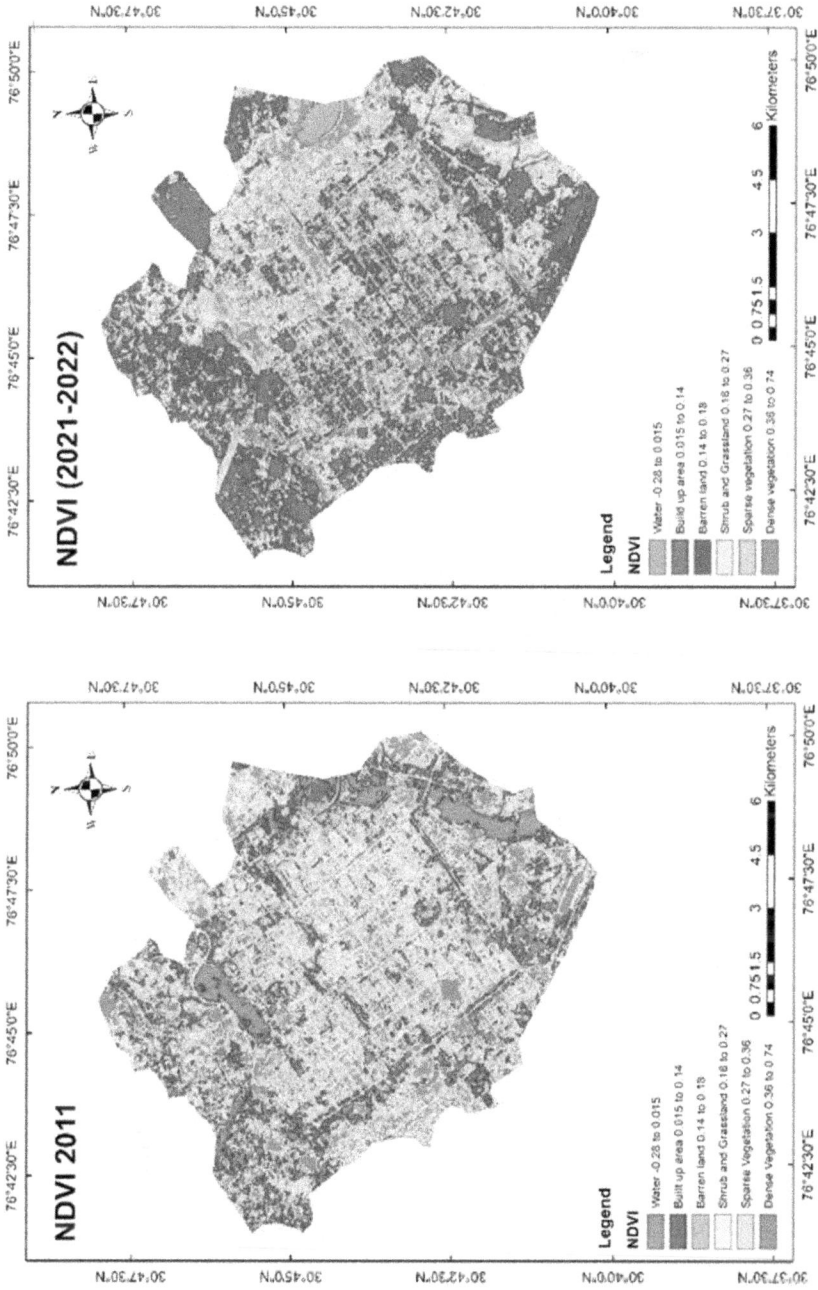

FIGURE 20.8   NDVI of Chandigarh in 2011

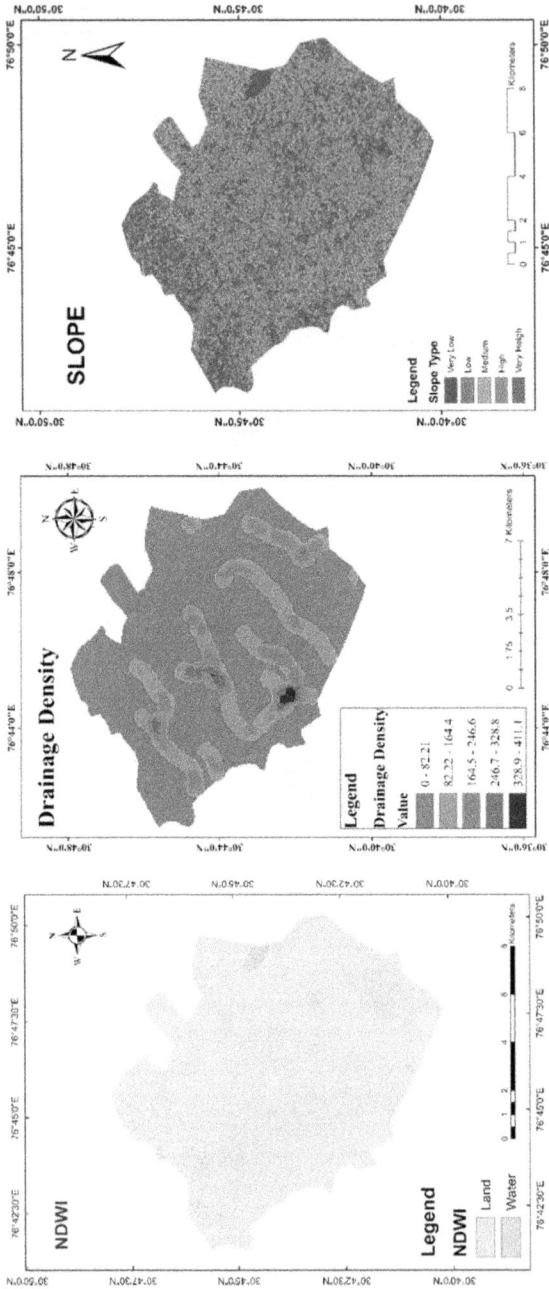

FIGURE 20.9 NDWI, drainage density, and slope map of Chandigarh city

CONCLUSIONS

The following conclusions that can be drawn from the present study are as given below

1. The most crucial resource on the planet is groundwater. According to the current study, which was conducted using the ArcGIS 10 program, the groundwater level in Chandigarh (union territory) is rapidly declining, which is a worrying problem.
2. The groundwater table dropped from 2.89 to 37.2 meters (BGL) between 1995 and 2020, and it was discovered that the groundwater level has fallen by more than a meter annually.
3. Water table depletion was a problem that existed prior to 1995, but it started to get worse after 2000. However, due to the farmer's easy access and necessity, the switch from canal irrigation to tube wells resulted in ground depletion.
4. Although the government has already implemented certain irrigation and underground pipeline systems, groundwater is still being depleted to a greater extent.
5. According to the current groundwater circumstances, the irrigation policy must be updated. Without public involvement, groundwater resource management cannot be successful. Therefore, raising public awareness is absolutely essential.

REFERENCES

Chandra, S., Singh, P. K., Tiwari, A. K., Panigrahy, B. P., & Kumar, A. (2015). Evaluation of hydrogeological factors and their relationship with seasonal water table fluctuation in Dhanbad district, Jharkhand, India. *ISH Journal of Hydraulic Engineering*, 21(2), 193–206.

Chowdhury, A., Jha, M. K., Chowdary, V. M., & Mal, B. C. (2009). Integrated remote sensing and GIS-based approach for assessing groundwater potential in West Medinipur district, West Bengal, India. *International Journal of Remote Sensing*, 30(1), 231–250.

Ghosh, A., Tiwari, A. K., & Das, S. (2015). A GIS based DRASTIC model for assessing groundwater vulnerability of Katri Watershed, Dhanbad, India. *Modeling Earth Systems and Environment*, 1(3), 1–14.

Gupta, P. K. (2020a). Pollution load on Indian soil-water systems and associated health hazards: a review. *Journal of Environmental Engineering*, *146*(5), 03120004. https://doi.org/10.1061/(ASCE)EE.1943-7870.0001693.

Gupta, P. K. (2020b). Fate, transport, and bioremediation of biodiesel and blended biodiesel in subsurface environment: a review. *Journal of Environmental Engineering*, *146*(1), 03119001. https://doi.org/10.1061/(ASCE)EE.1943-7870.0001619.

Gupta, N. K., Jethoo, A. S., & Gupta, S. K. (2016). Rainfall and surface water resources of Rajasthan State, India. *Water Policy*, 18(2), 276–287.

Gupta, P. K., Mustapha, H. I., Singh, B., & Sharma, Y. C. (2022). Bioremediation of petroleum contaminated soil-water resources using neat biodiesel: A review. *Sustainable Energy Technologies and Assessments*, *53*, 102703. https://doi.org/10.1016/j.seta.2022.102703.

Gupta, P. K., Gharedaghloo, B., Lynch, M., Cheng, J., Strack, M., Charles, T. C., & Price, J. S. (2020). Dynamics of microbial populations and diversity in NAPL contaminated peat soil under varying water table conditions. *Environmental Research*, *191*, 110167. https://doi.org/10.1016/j.envres.2020.110167.

Kaur, L., & Rishi, M. S. (2018). Integrated geospatial, geostatistical, and remote-sensing approach to estimate groundwater level in North-western India. *Environmental Earth Sciences*, 77, 1–13.

Krishan, G., Sudarsan, N., Sidhu, B. S., & Vashisth, R. (2021). Impact of lockdown due to COVID-19 pandemic on groundwater salinity in Punjab, India: Some hydrogeoethics issues. *Sustainable Water Resources Management*, 7(3), 1–11.

Kumar, A., & Pandey, A. C. (2013). Spatio-temporal assessment of urban environmental conditions in Ranchi Township, India using remote sensing and geographical information system techniques. *International Journal of Urban Sciences*, 17(1), 117–141.

Mahajan, M., Gupta, P. K., Singh, A., Vaish, B., Singh, P., Kothari, R., & Singh, R. P. (2022). A comprehensive study on aquatic chemistry, health risk and remediation techniques of cadmium in groundwater. *Science of The Total Environment*, *818*, 151784. https://doi.org/10.1016/j.scitotenv.2021.151784.

Singh, B., Kaur, S., Litoria, P. K., & Das, S. (2021). Development of web enabled water resource information system using open source software for Patiala and SAS Nagar districts of Punjab, India. *Water Practice and Technology*, 16(3), 980–990.

Singh, R. P., Mahajan, M., Gandhi, K., Gupta, P. K., Singh, A., Singh, P. … & Kidwai, M. K. (2023). A holistic review on trend, occurrence, factors affecting pesticide concentration, and ecological risk assessment. *Environmental Monitoring and Assessment*, *195*(4), 451. https://doi.org/10.1007/s10661-023-11005-2.

Stafford, D. B. (1991). *Civil Engineering Applications of Remote Sensing and Geographic Information Systems*. New York City: ASCE.

Sultana, S., & Satyanarayana, A. N. V. (2020). Assessment of urbanisation and urban heat island intensities using landsat imageries during 2000–2018 over a sub-tropical Indian City. *Sustainable Cities and Society*, 52, 101846.

Yatoo, A.M. Ali, M.N., Zaheen, Z., Baba, Z.A., Ali, S., Rasool, S., Sheikh, T.A., Sillanpää, M., Gupta, P.K., Hamid, B. and Hamid, B., (2022). Assessment of pesticide toxicity on earthworms using multiple biomarkers: a review. *Environmental Chemistry Letters*, *20*(4), 2573–2596. https://doi.org/10.1007/s10311-022-01386-0.

21 Holistic Approach for Prediction of Total Nitrogen Based on Machine-Learning Techniques

Bibhuti Bhusan Sahoo, Sushindra Kumar Gupta, Mani Bhushan, and Bhabani Shankar Dash

INTRODUCTION

In recent years, the prediction of total nitrogen has become a crucial topic in the field of agriculture and environmental science. Total nitrogen is an essential nutrient for plant growth, and its accurate estimation is crucial for optimizing fertilizer application and improving crop yield (Amiri and Nakane, 2009; Hur and Cho, 2012). Additionally, nitrogen runoff from agricultural fields is a significant source of water pollution, which can lead to environmental degradation (Gupta, 2020a,b; Mahajan et al., 2022; Gupta et al., 2020; Gupta et al., 2022; Yatoo et al., 2022; Singh et al., 2023). Machine-learning (ML) techniques have proven to be effective in predicting total nitrogen accurately (Yan et al., 2021; Ye et al., 2019). The purpose of this study is to compare the performance of three ML models, Support Vector Regression (SVR), Multi-Layer Perceptron (MLP), and Gaussian Process Regression (GPR), for nitrogen prediction in predicting total nitrogen and develop a holistic approach for total nitrogen prediction using ML techniques. The study aims to provide insights into the advantages and limitations of these models and identify the most accurate model for predicting total nitrogen. This research could contribute to the development of more precise nitrogen management practices and aid in the prevention of environmental pollution.

Several studies have used machine-learning models to predict various parameters in soil or water (Chen et al., 2020; Ding et al., 2019). Were et al. (2015) compared the performance of three ML models (SVR, MLP, and Random Forest) in predicting soil organic carbon. They found that the SVR model outperformed the other two models. Similarly, in a study by Xu et al. (2021), they compared the performance of four ML models (SVR, MLP, Random Forest, and K-Nearest Neighbor) in predicting total nitrogen in surface water. They found that the MLP model had the highest accuracy. Wang et al. (2020) used a deep learning model based on convolutional neural networks (CNN) to predict the total nitrogen content in the soil. Geng et al. (2022) explored the use of a hybrid model multiphase attention-based recurrent neural network (MPA-RNN) to predict the nitrogen content of wastewater treatment plants. They found that the MPA-RNN model satisfactorily predicted the total nitrogen content. Rajaee and Shahabi (2016) used wavelet-based machine-learning models such as wavelet-gene expression programming (WGEP) and wavelet-artificial neural network (WANN) hybrid model was assessed in the prediction of total nitrogen concentration in Charlotte harbor marine waters. He et al. (2011) used a feed-forward ANN model to investigate the relationship among land-use, fertilizer, and hydro-meteorological conditions in 59 river basins in Japan and then applied it to estimate the monthly river total nitrogen concentration. From the results obtained, the ANN model gave satisfactory predictions of stream TNC and appears to be a useful tool for further prediction of TNC in Japanese streams. These studies demonstrate the diversity of approaches and models used for predicting total nitrogen, and the continued efforts to improve the accuracy of predictions.

PROPOSED METHODOLOGY

SUPPORT VECTOR REGRESSION

Support Vector Regression is a type of regression analysis that uses Support Vector Machine (SVM) algorithms to perform regression tasks. The main goal of SVR is to find the hyperplane that best fits the data points in the feature space. SVR works by mapping the input variables to a high-dimensional feature space using a nonlinear function. In the feature space, SVR tries to find the hyperplane that best fits the data points while also minimizing the error between the predicted output and the actual output. The hyperplane that best fits the data points is defined by a set of support vectors, which are the data points closest to the hyperplane. These support vectors are used to construct the decision boundary, which separates the data points into different classes. To find the optimal hyperplane, SVR uses a regularization parameter, which controls the trade-off between the complexity of the model and its ability to fit the data (Figure 21.1). This regularization parameter helps to prevent overfitting, which is when the model is too complex and fits the training data too closely, leading to a poor generalization of new data. SVR can use different kernel functions, such as linear, polynomial, radial basis function (RBF), and sigmoid, to map the input variables to the feature space. The choice of kernel function depends on the specific problem and the characteristics of the data. One advantage of SVR is that it can handle both linear and nonlinear input–output relationships. SVR is also robust to outliers and noise in the data, as it only uses a subset of the data points as support vectors. However, SVR can be computationally expensive, especially when dealing with large datasets or complex kernel functions.

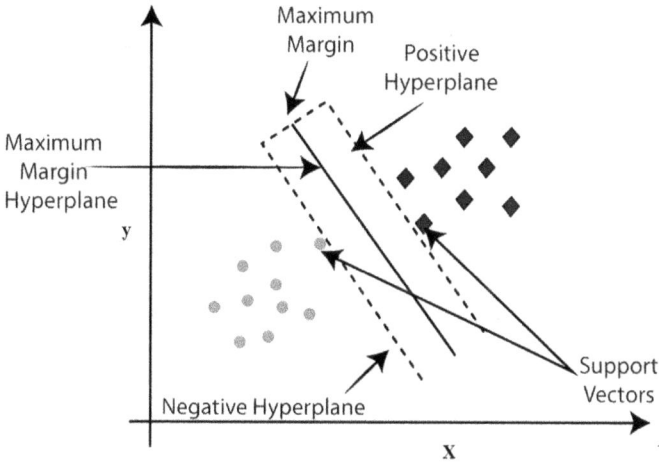

FIGURE 21.1   The principle of Support Vector Regression (SVR)

In this present study, SVM is used for regression analysis therefore its SVR. SVR for regression is defined by the following equations:

Considering a set of training data $\{(x_1, y_1), \ldots, (x_n, y_n)\}$, where $x_i \in R^n$, $y_i \in R$, the decision function is represented by $f(x) = w^T \emptyset(x) + b$ with respect to $w \in R^n$, $b \in R$, where f denotes a nonlinear transformation from R^n to a high-dimensional space 0. The primal optimization problem is given by:

$$R_{reg}(f) = \frac{1}{2}w^2 + c\left(\sum_{i=1}^{n}|y - f(x)|_\epsilon\right) \qquad \text{Minimize, (21.1)}$$

MULTI-LAYER PERCEPTRON

Multi-Layer Perceptron is a type of artificial neural network (ANN) that consists of multiple layers of interconnected nodes. It is a powerful machine-learning model that can be used for both regression and classification tasks. The basic building block of an MLP is the perceptron, which takes a weighted sum of its inputs, applies an activation function to the result, and passes the output to the next layer. The activation function is a nonlinear function that introduces nonlinearity into the model, allowing it to capture complex input–output relationships. An MLP consists of an input layer, one or more hidden layers, and an output layer (Figure 21.2). The input layer takes the input variables and passes them to the first hidden layer, which performs a nonlinear transformation on the input variables. The output of the first hidden layer is then passed to the next hidden layer, and so on until the output layer is reached.

During training, the weights and biases of the MLP are adjusted to minimize the error between the predicted output and the actual output. This is done using an optimization algorithm, such as stochastic gradient descent (SGD), which iteratively updates the weights and biases to minimize the error. One advantage of MLP is that it can capture complex input–output relationships that may be difficult or

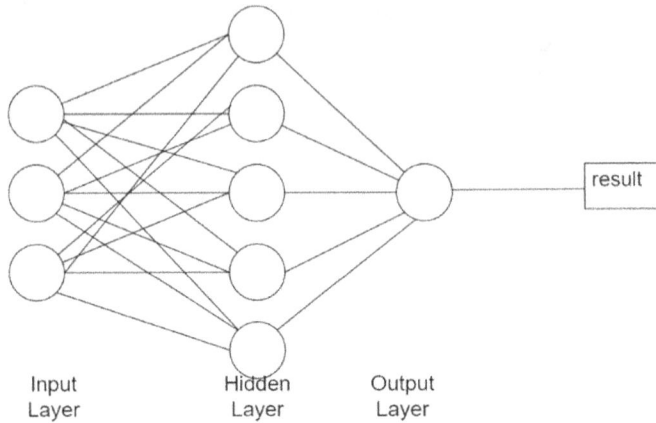

FIGURE 21.2   Structure of an MLP

impossible to model using other machine-learning algorithms. MLP can also handle large amounts of data and is relatively robust to noise in the data. However, MLP can suffer from overfitting, which is when the model is too complex and fits the training data too closely, leading to poor generalization of new data. To address this issue, regularization techniques, such as weight decay and dropout, can be used to reduce the complexity of the model and improve its generalization performance. Overall, MLP is a powerful and versatile machine-learning model that has been successfully applied to a wide range of applications, including image recognition, natural language processing, and financial forecasting.

Gaussian Process Regression

Gaussian Process Regression is a non-parametric machine-learning technique that is used for regression analysis. It is based on the Gaussian process, which is a collection of random variables, any finite subset of which has a joint Gaussian distribution. In GPR, the output variable is modeled as a Gaussian process, which is characterized by a mean function and a covariance function. The mean function represents the expected value of the output variable, while the covariance function determines the degree of similarity between the output values at different input locations. During training, GPR estimates the mean and covariance functions from the observed data points. The mean and covariance functions are then used to make predictions for new input values by computing the joint distribution of the observed and predicted output values. One advantage of GPR is that it can model complex input–output relationships without assuming any particular functional form. GPR can also provide a measure of uncertainty in its predictions, which can be useful in decision-making. However, GPR can be computationally expensive, especially when dealing with large datasets or complex covariance functions. GPR also requires careful selection of the covariance function, which can have a significant impact on the performance of the model. One popular covariance function used in GPR is the radial basis function kernel, which is a type of kernel that measures the similarity between input values using the Euclidean distance between them. Other types of covariance functions, such as the maternal kernel and the periodic kernel, can also be used in GPR. Overall, GPR is a powerful and flexible machine-learning technique that can be used for a wide range of applications, including forecasting, time series analysis, and spatial modeling.

DATA SOURCE

The Goodwater Creek experimental watershed monitored by the United States Department of Agriculture-Agriculture Research Service (USDA-ARS) is extensively used for the measurement of soil and water quality within the larger Salt River Basin (SRB) of north-eastern Missouri (Learch and Blanchad, 2003; Lerch et al., 2011b). The SRB encompasses about 30% of the claypan soils within the Central Claypan Areas of Major Land Resources Area (MLRA) 113. In the present study, the Missouri catchment has been selected and further sub-watershed of the Missouri catchment of Hunter Cave and Devils Icebox used for the prediction of total nitrogen (TN). Both of the watersheds' daily TN data have been used for the prediction of TN. The details of the data availability are as shown in https://data.nal.usda.gov/dataset /data-long-term-agroecosystem-research-central-mississippi-river-basin-goodwater -creek-experimental-watershed-and-regional-herbicide-water-quality-data.

MODEL DEVELOPMENT AND EVALUATION
METRICS FOR MODEL SELECTION

We used several inputs such as (Discharge, NH_4-N, NO_3-N, and Org-N) to predict total nitrogen. We divided the data set into 70% training and 30% testing randomly and five-fold cross-validation was followed while developing the models. Evaluation metrics for model selection are important aspects of machine learning that are used to assess the performance of different models and select the best one for a particular task. The choice of evaluation metrics criteria depends on the specific application and the goals of the analysis. In this study, we used, root mean squared error (RMSE), which is the square root of the MSE and represents the average magnitude of the error (Gupta et al., 2019a, 2019b, 2020).

$$RMSE = \sqrt{\frac{1}{n}\sum_{i=1}^{n}\left(TN_{i(\text{Observed})} - TN_{i(\text{Predicted})}\right)^2} \qquad (21.2)$$

Mean absolute error (MAE) measures the average absolute difference between the predicted and actual output values (Sihag and Gupta, 2022).

$$MAE = \frac{1}{n}\sum_{i=1}^{n}\left|TN_{i(\text{Observed})} - TN_{i(\text{Predicted})}\right| \qquad (21.3)$$

R-squared (R^2): This measures the proportion of the variance in the output variable that is explained by the model.

$$R^2 = 1 - \frac{\sum_{i=1}^{n} \left(TN_{i(\text{Observed})} - TN_{i(\text{Predicted})} \right)^2}{\sum_{i=1}^{n} \left(TN_{i(\text{Observed})} - \overline{TN}_{\text{Observed}} \right)^2} \qquad (21.4)$$

RESULTS AND DISCUSSION

The results of the study showed that all three models, SVR, MLP, and GPR, performed reasonably well in predicting total nitrogen. However, there were some differences in their performance, as indicated by the evaluation metrics (Table 21.1). The table provides a summary of the performance of three models – SVR, GPR, and MLP – in predicting the total nitrogen content in river water at two different monitoring stations – Hunters Cave and Devils Icebox. The table presents the training and testing performance metrics for each model, including the coefficient of determination (R^2), root mean square error, and mean absolute error, with the values provided in units of g/d (grams per day).

Looking at the performance of the models at the Hunters Cave station, we can see that SVR had the highest R^2 values, achieving an R^2 of 0.993 for training and 0.987 for testing. The RMSE and MAE values for SVR were also the lowest, with training values of 12,431 g/d for RMSE and 1219 g/d for MAE, and testing values of 12,673 g/d for RMSE and 1227 g/d for MAE. MLP also performed well at this station, with an R^2 of 0.961 for training and 0.960 for testing, and RMSE and MAE values of 15,250 and 1479 g/d for training, and 15,680 and 1152 g/d for testing, respectively. However, GPR had lower performance metrics than the other two models, with an R^2 of 0.939 for training and 0.928 for testing, and higher RMSE and MAE values of 16,150 and 1589 g/d for training, and 16,168 and 1586 g/d for testing, respectively. At the Devils Icebox station, SVR and MLP had higher R^2 values, achieving R^2 values of 0.994 and 0.984 for training, and 0.995 and 0.995 for testing, respectively. The RMSE and MAE values for both models were also relatively low, with training

TABLE 21.1

Training and Testing Phase of the Proposed Models

		Training			Testing		
		R^2	RMSE g/d	MAE(g/d)	R^2	RMSE(g\d)	MAE(g/d)
	SVR	0.993	12,431	1219	0.987	12,673	1227
Hunters Cave	GPR	0.939	16,150	1589	0.928	16,168	1586
	MLP	0.961	15,250	1479	0.960	15,680	1152
Devils Icebox	SVR	0.994	12,239	1124	0.995	12,159	1259
	GPR	0.561	18,417	1463	0.769	21,985	1658
	MLP	0.984	16,239	1576	0.950	16,168	1589

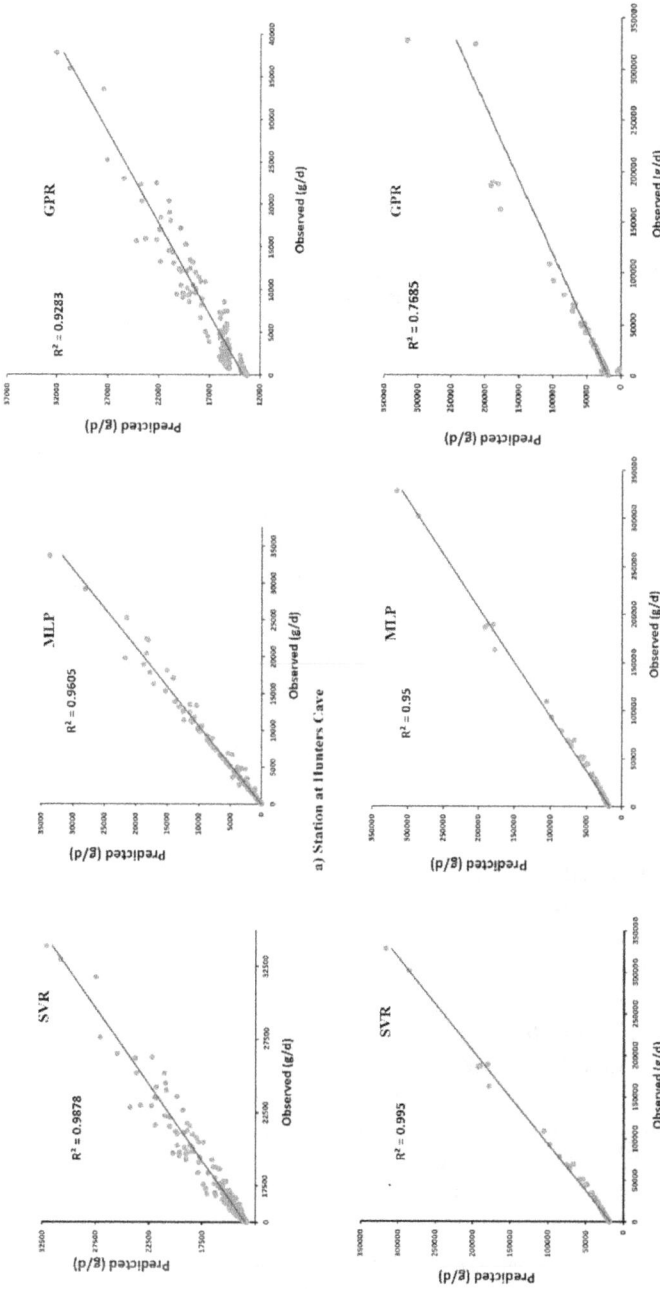

FIGURE 21.3 Plot between observed TN vs predicted TN during the testing period for the station a) Hunters Cave b) Devils Icebox

values of 12,239 and 15,239 g/d for RMSE, and 1124 and 1576 g/d for MAE, respectively, and testing values of 12,159 and 16,168 g/d for RMSE, and 1259 and 1589 g/d for MAE, respectively. However, GPR had significantly lower R^2 values of 0.561 for training and 0.769 for testing, with higher RMSE and MAE values of 18,417 and 1463 g/d for training, and 21,985 and 1658 g/d for testing, respectively. Overall, the results suggest that SVR and MLP are effective models for predicting total nitrogen content in river water, while GPR may not be the best option for this particular application. Figure 21.3

It is important to note that the performance of the models can vary depending on the specific monitoring station, highlighting the need for site-specific calibration and validation of models. The differences in performance between the three models may be due to several factors. For example, the choice of hyperparameters, such as the kernel function or the number of hidden layers, may have a significant impact on the performance of the models. In addition, the quality and quantity of the training data may also affect the performance of the models. Therefore, it is important to carefully select the hyperparameters and train the models on high-quality data to obtain the best possible performance. Overall, the results of the study suggest that machine-learning techniques such as SVR, MLP, and GPR can be effective in predicting total nitrogen in study stations. However, the choice of the most accurate model may depend on the specific application and the goals of the analysis. Further research is needed to explore the potential of other machine-learning techniques and to investigate the factors that affect their performance in predicting TN.

CONCLUSIONS

The following conclusions can be drawn from the present study:

1. The study aimed to compare the performance of three machine-learning models, SVR, MLP, and GPR, in predicting total nitrogen in rivers.
2. The results showed that all three models performed reasonably well in predicting TN, but the SVR model achieved the highest level of accuracy.
3. The study has implications for future research on the use of machine-learning techniques for predicting river water quality parameters. Based on the study findings, it is recommended that future research focus on improving the quality and quantity of the training data, as well as the selection of hyperparameters for the machine-learning models.
4. It is also recommended that researchers consider using a combination of machine-learning techniques to improve the accuracy of TN prediction.
5. Overall, the study highlights the potential of machine-learning techniques for predicting TN and provides insights into the factors that affect their performance.
6. With further research and development, these techniques could have significant implications for river management and agriculture, ultimately leading to more sustainable and efficient use of water resources.

REFERENCES

Amiri, B. and Nakane, K. 2009. Comparative prediction of stream water total nitrogen from land cover using artificial neural network and multiple linear regression. *Polish Journal of Environmental Studies* 18(2), 151–160.

Chen, K., Chen, H., Zhou, C., Huang, Y., Qi, X., Shen, R., Liu, F., Zuo, M., Zou, X. and Wang, J. 2020. Comparative analysis of surface water quality prediction performance and identification of key water parameters using different machine learning models based on big data. *Water Research* 171, 115454.

Ding, L., fan Mao, R., Guo, X., Yang, X., Zhang, Q. and Yang, C. 2019. Microplastics in surface waters and sediments of the Wei River, in the northwest of China. *Science of the Total Environment* 667, 427–434.

Geng, J., Yang, C., Li, Y., Lan, L. and Luo, Q. 2022. MPA-RNN: A novel attention-based recurrent neural networks for total nitrogen prediction. *IEEE Transactions on Industrial Informatics* 18(10), 6516–6525.

Gupta, S. K., Singh, P. K., Tyagi, J., Sharma, G. and Jethoo, A. S. 2020. Rainstorm-generated sediment yield model based on soil moisture proxies (SMP). *Hydrological Processes* 34(16), 3448–3463.

Gupta, S. K., Tyagi, J., Sharma, G., Jethoo, A. S. and Singh, P. K. 2019. An event-based sediment yield and runoff modeling using Soil Moisture Balance/Budgeting (SMB) method. *Water Resources Management* 33(11), 3721–3741.

Gupta, S. K., Tyagi, J., Singh, P. K., Sharma, G. and Jethoo, A. S. 2019. Soil Moisture Accounting (SMA) based sediment graph models for small watersheds. *Journal of Hydrology* 574, 1129–1151.

Gupta, P. K., Gharedaghloo, B., Lynch, M., Cheng, J., Strack, M., Charles, T. C., & Price, J. S. (2020). Dynamics of microbial populations and diversity in NAPL contaminated peat soil under varying water table conditions. *Environmental Research*, *191*, 110167. https://doi.org/10.1016/j.envres.2020.110167.

Gupta, P. K. (2020a). Pollution load on Indian soil-water systems and associated health hazards: a review. *Journal of Environmental Engineering*, *146*(5), 03120004. https://doi.org/10.1061/(ASCE)EE.1943-7870.0001693.

Gupta, P. K. (2020b). Fate, transport, and bioremediation of biodiesel and blended biodiesel in subsurface environment: a review. *Journal of Environmental Engineering*, *146*(1), 03119001. https://doi.org/10.1061/(ASCE)EE.1943-7870.0001619.

Gupta, P. K., Mustapha, H. I., Singh, B., & Sharma, Y. C. (2022). Bioremediation of petroleum contaminated soil-water resources using neat biodiesel: A review. *Sustainable Energy Technologies and Assessments*, *53*, 102703. https://doi.org/10.1016/j.seta.2022.102703.

He, B., Oki, T., Sun, F., Komori, D., Kanae, S., Wang, Y., Kim, H. and Yamazaki, D. 2011. Estimating monthly total nitrogen concentration in streams by using artificial neural network. *Journal of Environmental Management* 92(1), 172–177.

Hur, J. and Cho, J. 2012. Prediction of BOD, COD, and total nitrogen concentrations in a typical urban river using a fluorescence excitation-emission matrix with PARAFAC and UV absorption indices. *Sensors* 12(1), 972–986.

Lerch, R. N. 2011. Contaminant transport in two central Missouri karst recharge areas. *Journal of Cave and Karst Studies* 73(2), 99–113. doi:10.4311/jcks2010es0163.

Lerch, R. N. and Blanchard, P. E. 2003. Watershed vulnerability to herbicide transport in northern Missouri and southern Iowa streams. *Environmental Science and Technology* 37(24), 5518–5527. doi:10.1021/es030431s.

Mahajan, M., Gupta, P. K., Singh, A., Vaish, B., Singh, P., Kothari, R., & Singh, R. P. (2022). A comprehensive study on aquatic chemistry, health risk and remediation techniques of cadmium in groundwater. *Science of The Total Environment*, *818*, 151784. https://doi .org/10.1016/j.scitotenv.2021.151784.

Rajaee, T. and Shahabi, A. 2016. Evaluation of wavelet-GEP and wavelet-ANN hybrid models for prediction of total nitrogen concentration in coastal marine waters. *Arabian Journal of Geosciences* 9, 1–15.

Sihag, P. and Gupta, S. K. 2022. Discussion on prediction of maximum scour depth near spur dikes in uniform bed sediment using stacked generalization ensemble tree- based frameworks. *Journal of Irrigation & Drainage Engineering (ASCE)*. Ms. No. IRENG-9880, Accepted. doi:10.1061/(ASCE)IR.1943-4774.0001740.

Singh, R. P., Mahajan, M., Gandhi, K., Gupta, P. K., Singh, A., Singh, P., ... & Kidwai, M. K. (2023). A holistic review on trend, occurrence, factors affecting pesticide concentration, and ecological risk assessment. *Environmental Monitoring and Assessment*, *195*(4), 451. https://doi.org/10.1007/s10661-023-11005-2.

Wang, Y., Li, M., Ji, R., Wang, M. and Zheng, L. 2020. Comparison of soil total nitrogen content prediction models based on Vis-NIR spectroscopy. *Sensors* 20(24), 7078.

Were, K., Bui, D. T., Dick, Ø. B. and Singh, B. R. 2015. A comparative assessment of support vector regression, artificial neural networks, and random forests for predicting and mapping soil organic carbon stocks across an Afromontane landscape. *Ecological Indicators* 52, 394–403.

Xu, J., Xu, Z., Kuang, J., Lin, C., Xiao, L., Huang, X. and Zhang, Y. 2021. An alternative to laboratory testing: Random forest-based water quality prediction framework for inland and nearshore water bodies. *Water* 13(22), 3262.

Yan, J., Liu, J., Yu, Y. and Xu, H. 2021. Water quality prediction in the Luan River based on 1-drcnn and bigru hybrid neural network model. *Water* 13(9), 1273.

Yatoo, A.M., Ali, M.N., Zaheen, Z., Baba, Z.A., Ali, S., Rasool, S., Sheikh, T.A., Sillanpää, M., Gupta, P.K., Hamid, B. and Hamid, B., (2022). Assessment of pesticide toxicity on earthworms using multiple biomarkers: a review. *Environmental Chemistry Letters*, *20*(4), 2573–2596. https://doi.org/10.1007/s10311-022-01386-0.

Ye, Q., Yang, X., Chen, C. and Wang, J. 2019. *River Water Quality Parameters Prediction Method Based on LSTM-RNN Model*, pp. 3024–3028. IEEE.

22 Resource Conservation Technologies for Sustainable Management of Soil, Water and Energy in Modern Agriculture

Anshu Gangwar, Arvind Kumar Singh,
Tarun Kumar, Bhaskar Pratap Singh,
Ashish Rai, and Jitendra Rajput

INTRODUCTION

Agriculture is of primary importance in India, as it is the largest source of livelihoods and a vital contributor to the country's economy. In 2017, it accounted for 6.1% of India's GDP and still provides an income for 70% of rural households. According to the Land Use Statistics 2014–2015, India's total geographical area is 328.7 million hectares, of which 43% is net sown area. With 8% of the world's population reliant on this essential grain supply, India is the world's second-largest producer of rice and wheat. The area of the rice–wheat cropping system under different states is presented in Figure 22.1. The Indo-Gangetic Plains cover 85% of the land area, making it critical for world food security. One of the greatest concerns confronting the globe today is guaranteeing food and livelihood security for rising populations (Seckler et al., 1996). Agricultural issues include declining factor productivity, depleting human resources and their rising prices, and shifting socioeconomic conditions are becoming more prevalent today (Erenstein, 2011; Gathala et al., 2011a). It is critical to utilize energy and other resources effectively in the agriculture industry due to the rise in fuel, fertilizer, and other input costs. Crop growers must be able to produce more food with fewer resources while maintaining environmental quality if the supply of food is to keep up with the rising demand caused by population and economic development, as well as changing consumer preferences (Foresight, 2011).

More than 80% of the nation's exploitable water resources are used by agriculture, making it the main user of this resource. This availability is, however, decreasing as irrigation water is diverted to domestic, industrial, and energy-related uses. Water

DOI: 10.1201/9781003441175-22

Area under Rice wheat cropping system

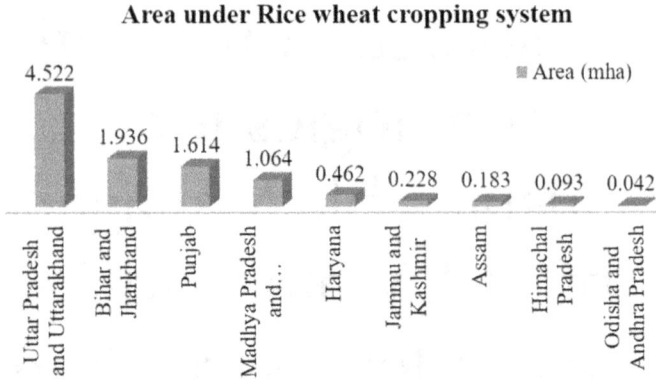

FIGURE 22.1   Area under rice–wheat cropping system in different state in India

stress levels are predicted to reach a crisis level by 2050 as a result of the combination of climate change, global warming, and population growth (UNESCO, 2006).

Agriculture's water production and efficiency must urgently improve in light of the current water crisis. To fulfill the demands of a growing population, intensive rice–wheat cultivation is required; however, this must be done through rapid technological progress that is both beneficial and non-degrading. Therefore, in order to boost crop yields and boost water efficiency, improved resource conservation technologies must be adopted. According to Sharma and Kumar (2000), these technologies have the potential to greatly raise income and productivity.

The Indo-Gangetic Plains' adoption of resource conservation methods has significantly increased in recent years due to the rice–wheat system there. This comprises a sizable area under rotavator, laser land leveler, and zero tillage techniques. Resource conserving technology (RCT) is a broad phrase that refers to all procedures or techniques that boost the efficiency of resources including land, labor, money, and inputs. These techniques include mulching and residue management, no-till/minimal tillage, surface planting, skip furrow irrigation inter-cropping, water harvesting and supplemental irrigation, living fences and vegetative barriers, and mulching. It is crucial to keep in mind that not all RCTs are appropriate for conservation agriculture (CA), such as land leveling with leftovers.

Through the management of soil, water, and biological resources, resource conservation technologies (RCTs) have the potential to improve, preserve, and use natural resources more effectively. The addition of RCTs has increased the land's ability to produce, as well as a notable production response linked to reduced water and energy use (fossil fuels and electricity), decreased greenhouse gas emissions, reduced soil erosion, improved farm incomes, lower cultivation costs, and reduced labor and fertilizer requirements (Pandey et al., 2012). In order to achieve sustainable and profitable production of rice–wheat cropping systems and improve farmer livelihoods, natural resource conservation is crucial.

CONSTRAINTS OF THE RICE–WHEAT CROPPING SYSTEM

- Over mining of nutrients from soil
- Distributed soil aggregates due to puddling in rice
- Decreasing response to nutrients
- Declining groundwater table
- Build-up disease/pests
- Build-up active *Phalaris minor*
- Low input use efficiency in northwestern plains
- Low use of fertilizer in eastern and central India
- Lack of appropriate varietal combination

THE NEED FOR RCTs

In the current state of agricultural science, randomized controlled trials (RCTs) are essential. The difficulty of providing food security without harming the environment is a major topic due to a continually expanding global population. An increase in food production must be accompanied by an increase in pay for those who work in agriculture if agricultural output is to remain sustainable. There is broad consensus on a number of important topics, including the necessity of meeting food demand in order to maintain ecological balance, food security, and nutritional security, as well as to make sure that our resources are available to both present and future generations.

MAJOR ADVANTAGES OF RCTs

1. To conserve the significant quantities of water (30–50%)
2. Reduced seed rates
3. Conserved rainwater facilitated mechanical weed control
4. Minimized lodging in wheat crop
5. Reducing thereby potential nutrient loss
6. Cost reduction and conservation of resources
7. Create a proper balance of soil health
8. Reduces the weed growth

FUTURE THRUST

The challenges listed below must be planned for in order to have a successful RCT in the rice–wheat cropping system.

i. A requirement for reaching agricultural production goals is the availability of mechanized equipment (appropriate zero tillage drills, a standard metering system for seeding/planting, and so on).
ii. Making tools for planting and cultivation that use the least amount of energy possible.

iii. To reduce subsurface compaction caused by the unplanned movement of heavy machinery in fields, research on the use of regulated traffic should be conducted.

iv. Localized solutions for crop diversification, through substitution or intensification, and to match production technologies, should be applied. Identification and standardization of new crops and farming systems incorporating livestock and fisheries.

v. Developing a complete package of practice for RCTs for prominent cropping systems in each agroecological region.

vi. Any successful agricultural strategy should include appropriate water and soil management measures, as well as understanding which crops and kinds require the least amount of water.

vii. Farmers' involvement in participatory research and development trials can accelerate the adoption of RCTs.

TECHNIQUES OF SOIL, WATER AND ENERGY CONSERVATION

Bed planting: Bed planting in a rice–wheat cropping system may be a way to enhance output and improve resource use efficiency. The field is prepared normally with this approach, and raised beds and furrows are made manually or with a raised bed planting machine. Crops are planted in rows on the raised bed's surface, and irrigation water is applied in the furrows between the beds.

Benefits
- To conserve significant quantities
- Reduced seed rates
- Conserved rainwater
- Facilitated mechanical weed control
- Minimized lodging in Wheat crop
- Reducing thereby potential nutrient loss
- Cost reduction and conservation of resources

Direct Seeded Rice (DSR): Rice is sown directly in dry soils (dry seeding) or wet soils (wet seeding), with irrigation provided to maintain the soil moist enough for good growth but never flooded. In many places in India, there are substantial incentives to promote puddled rice farming, such as optimizing system productivity, lowering irrigation requirements, reducing labor requirements, and reducing weed growth. In research studies in NW India, dry seeded rice placed into non-puddled soil with the soil kept near saturation or field capacity saved 35–57% more water than continually flooded (5 cm) PTR (Singh et al., 2002; Sharma et al., 2002). Figure 22.2a shows the sowing of paddy through a seed drill machine under DSR conditions and Figure 22.2b presents the growth of paddy in DSR.

   Leaf Color Chart (LCC): The LCC, a nondestructive way of determining the amount of green color in rice leaves, was invented in Japan and is being standardized with the chlorophyll meter. The LCC typically consists of four or more panels that

FIGURE 22.2 (a) Direct sowing of paddy through seed drill machine (b) Growth of direct seeded rice

range in color from yellowish green to dark green and is fashioned like a ruler and made of plastic. Nitrogen control by LCC outperformed locally advised Nitrogen application in three splits in both hybrid and inbred rice. When nitrogen is applied according to LCC values, 20–30 kg of nitrogenous fertilizer per hectare can be reduced without affecting rice output.

Brown manuring: Sesbania is first established in a standing rice crop and then removed for the purpose of fertilization using the brown manuring technique. This procedure leaves behind plant matter that turns brown, giving brown manuring its name. Sesbania crop at 20 kg/ha is broadcast three days after rice sowing and left to grow for 30 days before being dried by spraying 2–4 D ethyl easter which delivers up to 35 kg/ha nitrogen dry matter management of wide leaf weeds. This is a new brown manuring practice. Due to the addition of organic matter to soils with low fertility, yields increased by 4–5 q/ha.

Zero tillage wheat: Germination and emergence of seedlings are improved in zero tillage systems when soil moisture content is 3–4% higher at sowing time than in traditional tillage. The term "zero-till" refers to crop planting with the least amount of soil disturbance. Seeds are inserted straight into thin slits 2–3 cm wide and 4–7 cm deep produced with a drill fitted with a chisel, "inverted T" coulter, or twin disk openers. This coulter creates a small groove/slit in the soil for seed germination placement; wheat should be planted at slightly more than field capacity soil moisture content. At least 30% of the soil surface is still covered with crop residue. In some ways, zero tillage is a full farm management system that should incorporate a variety of agricultural practices such as planting, plant residue management, weed and pest control, harvesting, and crop rotations (Ekboir, 2003). Most common cropping systems, such as rice–wheat–mungbean, cotton–wheat, rice–maize, peral millet–wheat, sorghum–wheat, rice–pulses, and sugarcane–wheat, allow for zero tillage in one or both crops, etc. Sowing of wheat in a zero tillage environment with the help of Happy Seeder at KVK, Piprakothi Farm (Figure 22.3a) and Figure 22.3b depicts the growth of wheat crop under zero tillage (ZT). Net Savings and yield gain due to the adoption of ZT is illustrated in Table 22.1.

FIGURE 22.3   (a) Sowing of wheat in zero tillage environment, (b) Wheat crop in zero tillage

TABLE 22.1
Net Savings and Yield Gain Due to Adoption of ZT

S.No.	Net Savings	Bihar
1.	Diesel in land preparation (Rs/ha)	635.5*
2.	Saving in Tractor used for tillage	1328*
3.	Cost of Seeds	28.6
4.	Sowing Charges (including drill)	–71.2
5.	Yield gain (%)	8.7*

(Source: Laxmi and Mishra, 2007, Ind. J. of Agri. Econ., 62(1):126–127)

Notes: 1. Statistically significant at a 5% level; adopted from Vijay Laxmi et al. (2003). Users of the system reported a reduction in diesel and tractor use.

Benefits

- Water saving (30%)
- Improve soil quality
- Reduce weed growth
- Enhance crop yield (17%)
- Cost saving and Profitability of ZT
- Enhanced farmer's and livelihood

Furrow Irrigated Raised Bed (FIRB): A FIRB planting system in which crops are sown on ridges or beds. Depending on the crops, the beds are kept between 12 and 15 cm in height and between 37 and 107 cm in width. In a raised bed system with furrow irrigation, water flows horizontally from the furrow into the beds and is drawn upward and downward by capillary, evaporation, and transpiration, as well as upward and downward by gravity, toward the soil surface. Two rows of rice, wheat, maize, or chickpeas are typically planted on raised beds.

Benefits

- Reduced wheat seed usage by 30–50% in comparison to flatbed planting
- Improved crop production even in wet monsoon climates due to better drainage
- Enhanced fertilizer efficiency
- More tillering and increased panicle
- Ability to apply nitrogen and irrigation water at the grain filling stage to increase protein content without lodging issues
- Mechanical control of weeds between beds
- Reduced reliance on herbicides
- Increased yield potential due to improved nutrient lodging

Crop residue cover: Crop residues are the plant or crop parts, such as the protein, that are left in the field after harvesting or that are not used domestically, sold commercially, or thrown away during processing. There is a significant opportunity to effectively minimize crop waste, particularly in the rice–wheat belt of Punjab, Haryana, and Western Uttar Pradesh where it is burned on-site. Each year, more than 340 Mt of crop leftovers from various crops are produced, with rice and wheat contributing the majority of this total. This equates to roughly 6 Mt of important nutrients, at least a third of which can be trapped for recycling. When compared to residue retrieval, crop residue recycling in wheat was found to increase yields in both rice and wheat by 13–18%, decrease cost-effectiveness by 3–5%, and increase energy efficiency by 6–13%. However, yield advantage was only 3–9% when compared to residue burning.

Laser land leveler: In order to obtain accuracy in land leveling, drag buckets fitted with lasers are used to flatten the land surface from its average elevation (Figure 22.4a). With precision land leveling, the fields are modified to maintain a constant slope of 0 to 2%. Land leveling (Figure 22.4b) makes it easier to apply irrigation water uniformly and maximizes the interactions between nutrients and water for greater yield. When compared to conventional methods, laser land leveling in India reduced irrigation water application by about 20–30% while increasing crop yields by 10–20% (Jat et al., 2006). It is a predecessor technology that will improve the advantages of bed planting and zero tillage, among other crop-establishing methods.

FIGURE 22.4 (a) Laser land leveler, (b) Levelling of land through laser land leveler

Rotary Tiller: The Rotary Tiller is a potent tillage tool used to break up pastures, remove weeds, mix in fertilizer and manure, and prepare the soil for planting seeds. It has been discovered to consume 30–35% less time and cost 20–25% less than conventional tillage with a cultivator. It also delivers a speedier seedbed preparation and lower draft. Additionally, compared to a cultivator, it produces work of a higher quality (up to 25–30%). With the Rotavator, you can avoid wheel slide and drastically lower gearbox power loss while transferring engine power straight to the soil.

Micro-irrigation: In comparison to traditional irrigation techniques, micro-irrigation offers a steady supply of water in the crop zone and has been shown to boost crop output and water use efficiency. The micro-irrigation system comprises both drip (Figure 22.5a and 22.5b) and sprinkler irrigation systems. Micro-irrigation systems in rice–wheat cropping systems use 16 to 33% less irrigation water than surface irrigation, water use efficiency has significantly increased. Micro-irrigation is becoming more widely employed as a solution to combat water scarcity and poverty as a result of its high water use efficiency. Its benefits in terms of energy savings (30%), greater yields (up to 100%), fertilizer savings (28%), decreased salinization rate, removed wood and diseases, and decreased labor are well acknowledged.

Alternate Wetting and Drying (AWD): The International Rice Research Institute (IRRI) created an alternate wetting and drying irrigation practice, a low-cost method for growing rice that can conserve a large quantity of irrigation water. AWD is commonly used as "mid-season drainage" on heavy textured soil in China (Zhang et al., 2010a; Ye et al., 2013), as well as in other Asian nations like India, Bangladesh, Vietnam, and the Philippines (Kukal et al., 2005; Bouman et al., 2007). There are three primary parts that make up the AWD system: In order to help seedlings recover from any shock and stop weeds from spreading, seedlings should receive three types of irrigation: 1) shallow flooding right after transplanting or seeding; 2) a thin layer of standing water (2–3 cm) from panicle initiation (PI) until the end of flowering; and 3) an AWD cycle for the remainder of the crop growth stages (Bouman et al., 2006; Yang et al., 2017). Instead of being continuously flooded throughout the season, fields are treated to alternating flooding and drying under the AWD approach (Zhang et al., 2009). The use of AWD has been observed in both transplanted and DSR systems (Sandhu et al., 2017; Kar et al., 2018; Carrijo et al., 2017; Ishfaq et al.,

FIGURE 22.5   (a) Gravity-based drip irrigation system, (b) Filter and fertigation components of micro-irrigation system

TABLE 22.2

Categorization of AWD Condition Based on Field water level (in Water Pipes Below the Soil Surface) and SWP (Soil Water Potential) of the Root Zone

S.No.	Field Water Level (cm)	Soil Water Potential (kPa)	AWD Condition
1	≤ 15	> -10	Safe
2	15–20	≥ -20	Moderate
3	> 20	< -20	Severe

(*Source:* Ishfaq et al., 2020b)

2020a). AWD can help a farm become more productive while releasing fewer green-house emissions into the atmosphere if it is used properly. Whether in the conventional flooded and puddled transplanted rice system or DSR, techniques like AWD or mid-drainage can be used to reduce water input and its possibly negative impacts on the growth, physiology, and production of rice. Both Mishra (2012) and Jabran et al. (2017) hypothesized that the development of rice plants was unaffected by the switch from traditional flooded (CF) conditions to unsaturated ones (AWD or aerobic rice). In conclusion, AWD as a whole uses 26% less water than CF. Irrigation application based on AWD condition is illustrated in Table 22.2.

Organic manure/fertilizer: One method to increase the sustainability of the production system without having a negative impact on the environment and natural resources is organic farming (Stockdale et al., 2001; Ram et al., 2011a). The secret to successful organic farming is the use of copious volumes of organic manure (Swift and Woomer, 1993). The effects of a variety of organic nutrient sources (FYM, crop residue, biofertilizers, and vermicompost) and their combinations on crop yields, NPK uptake, grain quality, and gross and net returns of the cropping system were reported by Davari et al. (2012) and Ram et al. (2011b). Each ton of organic carbon in the plow layer is equivalent to 4.75 kg fertilizer N ha^{-1} (Benbi and Chand, 2007). The organic amendments increased the water use efficiency over the inorganic fertilizer and it also enhanced the water-holding capacity of soil by increasing the porosity.

CONCLUSION

It is crucial to use an interdisciplinary approach for improving land and water productivity, and this can only be done by choosing a resource conservation technique that is suited to the area. By stopping drainage losses, which are undesirable in regions where water resources are quickly decreasing, the majority of resource conservation technologies (RCTs) can result in a significant decrease in the amount of water utilized for irrigation. The fertility status, organic carbon content, and water retention capacity of the soil were improved as a result of the application of these resource conservation technologies on farmers' fields, which improved the overall health of the soil and resulted in the preservation of priceless resources like water, time, energy, and money.

REFERENCE

Benbi, D. K., & Chand, M. (2007). Quantifying the effect of soil organic matter on indigenous soil N supply and wheat productivity in semiarid sub-tropical India. *Nutrient Cycling in Agroecosystems*, 79(2), 103–112.

Bouman, B. A. M., Lampayan, R. M., & Tuong, T. P. (2007). *Water management in irrigated rice: Coping with water scarcity*. International Rice Research Institute, Los Baños (Philippines), 54.

Bouman, B. A., Humphreys, E., Tuong, T. P., & Barker, R. (2007). Rice and water. *Advances in Agronomy*, 92, 187–237.

Carrijo, D. R., Lundy, M. E., & Linquist, B. A. (2017). Rice yields and water use under alternate wetting and drying irrigation: A meta-analysis. *Field Crops Research*, 203, 173–180.

Davari, M., Sharma, S. N., & Mirzakhani, M. (2012). The effect of combinations of organic materials and biofertilisers on productivity, grain quality, nutrient uptake and economics in organic farming of wheat. *Journal of Organic Systems*, 7(2), 26–35.

Ekboir, J. M. (2003). Research and technology policies in innovation systems: Zero tillage in Brazil. *Research Policy*, 32(4), 573–586.

Erenstein, O. (2011). Livelihood assets as a multidimensional inverse proxy for poverty: A district-level analysis of the Indian Indo-Gangetic Plains. *Journal of Human Development and Capabilities*, 12(2), 283–302.

Foresight UK. (2011). The future of food and farming. Final Project Report. The Government Office for Science, London.

Gathala, M. K., Ladha, J. K., Kumar, V., Saharawat, Y. S., Kumar, V., Sharma, P. K., Sharma, S., & Pathak, H. (2011). Tillage and crop establishment affects sustainability of South Asian rice–wheat system. *Agronomy Journal*, 103(4), 961–971.

Ishfaq, M., Akbar, N., Anjum, S. A., & ANWAR-IJL-HAQ, M. (2020a). Growth, yield and water productivity of dry direct seeded rice and transplanted aromatic rice under different irrigation management regimes. *Journal of Integrative Agriculture*, 19(11), 2656–2673.

Ishfaq, M., Farooq, M., Zulfiqar, U., Hussain, S., Akbar, N., Nawaz, A., & Anjum, S. A. (2020b). Alternate wetting and drying: A water-saving and ecofriendly rice production system. *Agricultural Water Management*, 241, 106363.

Jabran, K., Riaz, M., Hussain, M., Nasim, W., Zaman, U., Fahad, S., & Chauhan, B. S. (2017). Water-saving technologies affect the grain characteristics and recovery of fine-grain rice cultivars in semi-arid environment. *Environmental Science and Pollution Research International*, 24(14), 12971–12981.

Jat, M. L., Chandna, P., Gupta, R., Sharma, S. K., & Gill, M. A. (2006). Laser land leveling: A precursor technology for resource conservation. *Rice-Wheat Consortium Technical Bulletin Series*, 7, 48.

Kar, I., Mishra, A., Behera, B., Khanda, C., Kumar, V., & Kumar, A. (2018). Productivity trade-off with different water regimes and genotypes of rice under non-puddled conditions in Eastern India. *Field Crops Research*, 222, 218–229.

Kukal, S. S., Humphreys, E., Yadvinder-Singh, T. J., & Thaman, S. (2005). Performance of raised beds in rice–wheat systems of northwestern India. Evaluation and performance of permanent raised bed cropping systems in Asia, Australia and Mexico. *ACIAR Proceedings*, 121.

Laxmi, V., & Mishra, V. (2007). Factors affecting the adoption of resource conservation technology: Case of zero tillage in rice-wheat farming systems. *Indian Journal of Agricultural Economics*, 62, 902-2016-67372.

Laxmi, V., Gupta, R. K., Swarnalatha, A., & Perwaz, S. (2003, October). Environmental impact of improved technology-farm level survey and farmers' perception on zero tillage (Case Study). *In Roles of Agriculture Workshop*, 20–22.

Mishra, A. (2012). Intermittent irrigation enhances morphological and physiological efficiency of rice plants. *Agriculture*, 58(4), 121.

Pandey, D., Agrawal, M., & Bohra, J. S. (2012). Greenhouse gas emissions from rice crop with different tillage permutations in rice–wheat system. *Agriculture, Ecosystems and Environment*, 159, 133–144.

Ram, M., Davari, M. R., & Sharma, S. N. (2011). Effect of organic manures and biofertilizers on basmati rice (Oryza sativa L.) under organic farming of rice-wheat cropping system. *International Journal of Agriculture and Crop Sciences*, 3(3), 76–84.

Ram, M., Davari, M., & Sharma, S. N. (2011). Organic farming of rice (Oryza sativa L.)-wheat (Triticum aestivum L.) cropping system: A review. *International Journal of Agronomy and Plant Production*, 2(3), 114–134.

Sandhu, N., Subedi, S. R., Yadaw, R. B., Chaudhary, B., Prasai, H., Iftekharuddaula, K., Thanak, T., Thun, V., Battan, K. R., Ram, M., & Venkateshwarlu, C. (2017). Root traits enhancing rice grain yield under alternate wetting and drying condition. *Frontiers in Plant Science*, 8, 1879.

Seckler, D. W. (1996). *The new era of water resources management: From dry "to" "wet" water savings* (Vol. 1). International Water Management Institute, Colombo.

Sharma, R. K., Chhokar, R. S., Chauhan, D. S., Gathala, M. K., Rani, V., & Kumar, A. (2002). *Paradigm tillage shift in rice–Wheat system for greater profitability. Herbicide resistance management and zero tillage in rice–wheat cropping system*. CCSHAU, Hisar, 131–135.

Sharma, V. P., & Kumar, A. (2000). Factors influencing adoption of agro-forestry programme: A case study from North-West India. *Indian Journal of Agricultural Economics*, 55(3), 500–510.

Singh, A. K., Choudhury, B. U., & Bouman, B. A. M. (2002). *Effects of rice establishment methods on crop performance, water use, and mineral nitrogen. Water-wise rice production*. International Rice Research Institute, Los Baños (Philippines), 237–246.

Stockdale, E. A., Lampkin, N. H., Hovi, M., Keatinge, R., Lennartsson, E. K. M., Macdonald, D. W., Padel, S., Tattersall, F. H., Wolfe, M. S., & Watson, C. A. (2001). *Agronomic and environmental implications of organic farming systems*.

Swift, M. J., & Woomer, P. (1993). *Organic matter and the sustainability of agricultural systems: Definition and measurement*.

UNESCO. (2006). Water, a shared responsibility: The United Nations world water development report 2. UNESCO Publishing, Paris/Berghahn Books, Oxford.

Yang, J., Zhou, Q., & Zhang, J. (2017). Moderate wetting and drying increases rice yield and reduces water use, grain arsenic level, and methane emission. *The Crop Journal*, 5(2), 151–158.

Ye, Y., Liang, X., Chen, Y., Liu, J., Gu, J., Guo, R., & Li, L. (2013). Alternate wetting and drying irrigation and controlled-release nitrogen fertilizer in late-season rice. Effects on dry matter accumulation, yield, water and nitrogen use. *Field Crops Research*, 144, 212–224.

Zhang, H., Chen, T., Wang, Z., Yang, J., & Zhang, J. (2010). Involvement of cytokinins in the grain filling of rice under alternate wetting and drying irrigation. *Journal of Experimental Botany*, 61(13), 3719–3733.

Zhang, H., Xue, Y., Wang, Z., Yang, J., & Zhang, J. (2009). An alternate wetting and moderate soil drying regime improves root and shoot growth in rice. *Crop Science*, 49(6), 2246–2260.

Appendix

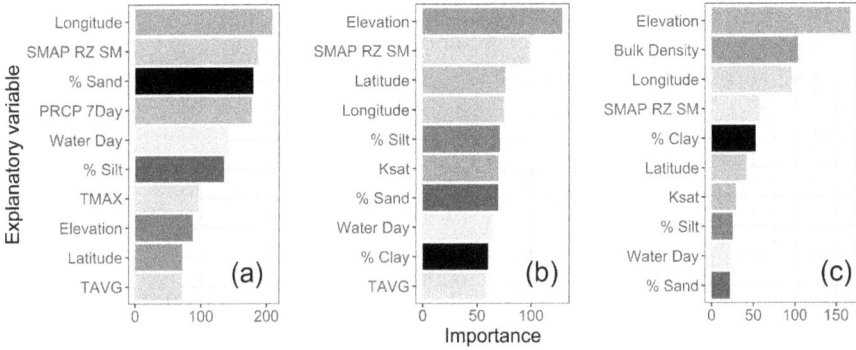

FIGURE A1 Permutation Feature Importance (PFI) of top-ten explanatory variables from A) Random Forest (RF), B) Support Vector Regression (SVR), C) Neural Network (NN) models at 10 cm depth based on the percent increase in Root Mean Squared Error (RMSE) as shown on the x-axis of each bar plot

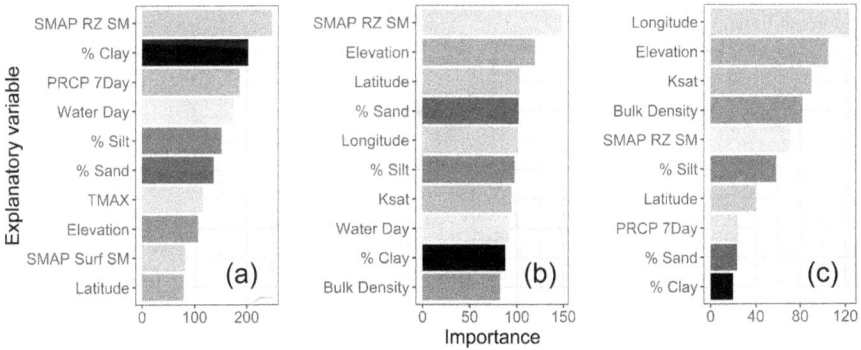

FIGURE A2 Permutation Feature Importance (PFI) of top-ten explanatory variables from A) Random Forest (RF), B) Support Vector Regression (SVR), C) Neural Network (NN) models at 20 cm depth based on the percent increase in Root Mean Squared Error (RMSE) as shown on the x-axis of each bar plot

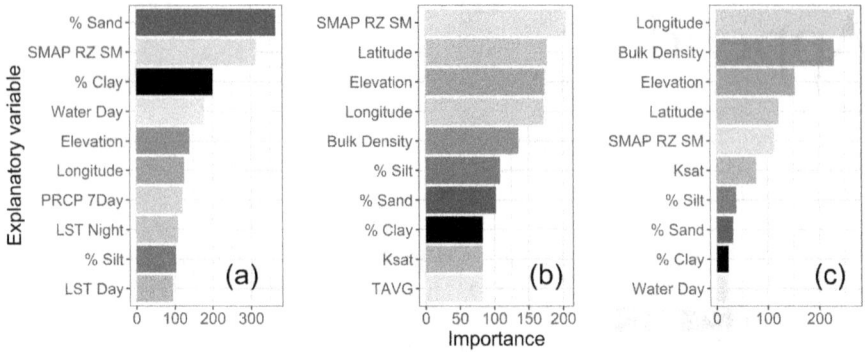

FIGURE A3 Permutation Feature Importance (PFI) of top-ten explanatory variables from (a) Random Forest (RF), (b) Support Vector Regression (SVR), (c) Neural Network (NN) models at 51 cm depth based on the percent increase in Root Mean Squared Error (RMSE) as shown on the x-axis of each bar plot

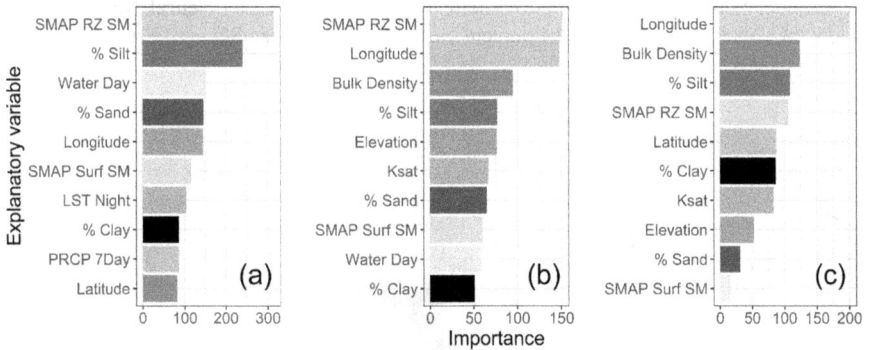

FIGURE A4 Permutation Feature Importance (PFI) of top-ten explanatory variables from (a) Random Forest (RF), (b) Support Vector Regression (SVR), (c) Neural Network (NN) models at 102 cm depth based on the percent increase in Root Mean Squared Error (RMSE) as shown on the x-axis of each bar plot

TABLE A1
Root Mean Square Error (ubRMSE) for Soil Moisture Estimates Using the Support Vector Regression (SVR) Model During the Test Period for Each Soil and Climate Analysis Network (SCAN) Station in the Study

Site	ubRMSE (m³/m³)				
	5 cm	10 cm	20 cm	51 cm	102 cm
Beaumont	0.079	0.082	0.044	0.029	0.119
Crossroads	0.027	0.051	0.054	0.029	0.071
Kingsville	0.028	0.048	0.035	0.081	0.034
Knox City	0.039	0.098	0.037	0.052	0.047
Lehman	0.032	0.035	0.037	0.037	0.021
Levelland	0.025	0.066	0.043	0.061	0.025
Prairie View #1	0.078	0.076	0.074	0.084	0.082
Reese Center	0.031	0.108	0.028	0.050	0.058
Riesel	0.102	0.117	0.105	0.116	0.118
San Angelo	0.051	0.071	0.094	0.083	0.106
Stephenville	0.038	0.053	0.055	0.135	0.093
Uvalde	0.060	0.047	0.059	0.094	0.055
Weslaco	0.036	0.036	0.029	0.040	0.071

TABLE A2
Root Mean Square Error (ubRMSE) for Soil Moisture Estimates Using the Neural Networks (NN) Model During the Test Period for Each Soil and Climate Analysis Network (SCAN) Station in the Study

Site	ubRMSE (m³/m³)				
	5 cm	10 cm	20 cm	51 cm	102 cm
Beaumont	0.079	0.057	0.043	0.023	0.096
Crossroads	0.035	0.037	0.061	0.015	0.028
Kingsville	0.028	0.042	0.038	0.068	0.027
Knox City	0.041	0.073	0.039	0.042	0.040
Lehman	0.037	0.031	0.035	0.020	0.012
Levelland	0.028	0.056	0.044	0.068	0.014
Prairie View #1	0.084	0.066	0.063	0.055	0.078
Reese Center	0.034	0.113	0.026	0.035	0.070
Riesel	0.107	0.107	0.126	0.112	0.096
San Angelo	0.051	0.067	0.087	0.081	0.108
Stephenville	0.039	0.047	0.053	0.119	0.078
Uvalde	0.058	0.047	0.057	0.092	0.060
Weslaco	0.041	0.034	0.035	0.036	0.063

Index

Taylor & Francis Group
an **informa** business

Taylor & Francis eBooks

www.taylorfrancis.com

A single destination for eBooks from Taylor & Francis
with increased functionality and an improved user
experience to meet the needs of our customers.

90,000+ eBooks of award-winning academic content in
Humanities, Social Science, Science, Technology, Engineering,
and Medical written by a global network of editors and authors.

TAYLOR & FRANCIS EBOOKS OFFERS:

A streamlined
experience for
our library
customers

A single point
of discovery
for all of our
eBook content

Improved
search and
discovery of
content at both
book and
chapter level

REQUEST A FREE TRIAL

support@taylorfrancis.com

Routledge
Taylor & Francis Group

CRC Press
Taylor & Francis Group

For Product Safety Concerns and Information please contact our EU
representative GPSR@taylorandfrancis.com
Taylor & Francis Verlag GmbH, Kaufingerstraße 24, 80331 München, Germany

www.ingramcontent.com/pod-product-compliance
Lightning Source LLC
Chambersburg PA
CBHW060751220326
41598CB00022B/2400

9 7 8 1 0 3 2 5 7 8 2 2 4